POLYMER MANUFACTURING

POLYMER MANUFACTURING

Technology and Health Effects

by

Radian Corporation

McLean, Virginia

NOYES DATA CORPORATION

Park Ridge, New Jersey, U.S.A.

1986

Copyright © 1986 by Noyes Data Corporation
Library of Congress Catalog Card Number 86-17977
ISBN: 0-8155-1090-X
Printed in the United States

Published in the United States of America by
Noyes Data Corporation
Mill Road, Park Ridge, New Jersey 07656

10 9 8 7 6 5 4 3 2 1

Library of Congress Cataloging-in-Publication Data

Polymer manufacturing.

   Bibliography: p.
   Includes index.
   1. Polymers and polymerization. 2. Polymers and
polymerization--Toxicology. I. Radian Corporation.
TP1092.P65   1986   668.9   86-17977
ISBN 0-8155-1090-X

Foreword

This book is a detailed analysis of the polymer manufacturing industry. Elements of the analysis include an industry definition, raw materials, products and manufacturers, environmental impacts, and occupational health concerns.

Since their discovery and early use nearly 60 years ago, plastic products have evolved from a few materials that were typically considered to be cheap substitutes to a large group of products that can be processed to produce goods for a wide variety of applications. In some cases, plastic substitutes have virtually replaced the original products.

Raw materials for plastics and resins are industrial organic chemicals used as monomers or plasticizers and specialty chemicals used as additives to modify resin properties. The types of companies involved in plastics production vary, but the principal producers include major chemical, petroleum, paint, tire and rubber, steel, and electrical manufacturing companies.

Plastic and resin production processes generate air emissions, wastewater, and solid waste. Volatile emissions are generally highest in processing steps upstream of the reactor and in mass and solution polymerization monomer and solvent recovery steps. Suspension and emulsion polymerization processes typically generate more particulate emissions. The most significant source of wastewater in plastics production is the water used for emulsion and suspension polymerization. Wastewater may contain monomer, comonomer, additives, and fillers. Solid waste is generated from plastics production in one of two ways: polymer lost from the process (i.e., spills, routine cleaning, particulate collection) and by-product formation. Spent catalyst or additives also may constitute solid waste from some processes.

Some of the chemicals used as raw materials in plastics production are highly toxic and may produce serious adverse health effects in overexposed employees. However, effective engineering controls and personal protective equipment and clothing exist that greatly reduce worker exposure potential. Successful application of these controls depends on plant-specific factors such as plant design, materials handled, process configuration, and management and employee dedication to maintaining a good occupational health program.

This book is organized by types of polymers, and will be a valuable overview of the field.

The information in the book is from *Industrial Process Profiles for Environmental Use: The Plastics and Resins Production Industry,* prepared by Radian Corporation for the U.S. Environmental Protection Agency, July 1985.

The table of contents is organized in such a way as to serve as a subject index and provides easy access to the information contained in the book.

Advanced composition and production methods developed by Noyes Data Corporation are employed to bring this durably bound book to you in a minimum of time. Special techniques are used to close the gap between "manuscript" and "completed book." In order to keep the price of the book to a reasonable level, it has been partially reproduced by photo-offset directly from the original report and the cost saving passed on to the reader. Due to this method of publishing, certain portions of the book may be less legible than desired.

NOTICE

Contents and Subject Index

1. Industry Analysis

INTRODUCTION

Since their discovery and early use nearly 60 years ago, plastic products have evolved from a few materials that were typically considered to be
cheap substitutes to a group of about 100 products that can be processed to
produce goods superior to those made from wood, metal, and leather. In some
cases, plastic substitutes have virtually replaced the original products.
The economic importance of plastics can be illustrated by considering industry statistics. Nearly 250 companies are listed as plastics and resins
manufacturers in the <u>1982 Directory of Chemical Producers</u>. The U.S. producers of plastics and resins employ approximately 54,000 people; in 1981,
shipments of plastic and resin products were valued by the Bureau of Census
at \$20.7 billion for a total of 18×10^6 metric tons (40×10^9 pounds).
The processing of resins to generate plastic end products resulted in an
additional value of \$18.0 billion in 1981, so that the value of all end
products was about \$39 billion. Resins are processed in more that 10,000
establishments with 468,000 employees. The relative importance of plastics
compared with other chemical products, is illustrated by considering
statistics for other industries in Table 1.

The production of plastic products, in addition to being the largest, is the most complex industrial activity involving chemicals. This
industry uses a wide variety of monomer and additive combinations and is
dynamic in nature. Changes in processes, products offered, and production
technology are constant.

INDUSTRY DEFINITION

As suggested by the data above, the production of plastic consumer
products involves processing in two distinctly different industries. The
first comprises companies that employ conventional chemical processing to
convert industrial organic chemicals, using polymerization, into plastic and
resin products. The plastic products from this industry are fabricated into
what is essentially their final form--for example, polyester films. These
may be further processed, as polyester films are processed to manufacture
photographic film, but the processing does not alter the film itself. The
resins are in the form of solids (pellets, powders, or granules) or liquids
that are further processed to produce the wide variety of plastic products
that are so pervasive in our present society.

TABLE 1. VALUE OF PRODUCT SHIPPED BY MAJOR CHEMICAL INDUSTRIES
(1981)

Industry	Value of Product ($\$10^9$)	Employees (10^3)
Plastics and Plastic Products	38.7	52
Paints and Allied Products	8.7	57
Medicinal Chemicals	5.8	35
Pharmaceuticals	14.9	145
Cosmetics	10.0	55
Synthetic Rubbers	3.9	9
Rubber Products (tires, inner tubes, belting, etc.)	11.6	126
Pesticides	4.8	17

Source: Bureau of the Census.

For purposes of incorporation in the EPA's Industrial Process Profiles, the two industries involved have been treated separately. The plastics and resins manufacturing industry is discussed here. Plastics processing is discussed in a companion volume. Though the end products of the polymer manufacturing industry are the solids or liquids that are shipped to processors, the usage discussion in this book has been expanded to include end products where it seemed appropriate.

The plastics industry in the United States is expected to continue to grow, with a compound annual rate of 6 percent predicted into the late 1980's. Although this growth is forecast at a constant rate, there will inevitably be years during which it will exceed or drop below this level. As seen in 1979, production tends to peak, then drop according to the over-all state of the U.S. economy. The 1986 production is forecast to exceed 24 x 10^6 metric tons (54 x 10^9 pounds) in spite of competition from foreign companies. Inexpensive feedstocks in the Middle East have led to new plastics facilities in that region. However, the proximity of U.S. producers to established markets and effective marketing systems will take advantage of the growth in major applications such as pipe, plumbing systems, and films, resulting in an increased advantage over Middle Eastern products.

As the plastics industry grows, interest on the part of Federal regulatory agencies increases in this industry. A number of raw materials and products are either known or suspected carcinogens or are potentially hazardous if handled improperly. The U.S. Environmental Protection Agency, the Occupational Safety and Health Administration, and the Consumer Product Safety Commission are all considering some form of regulation that could potentially affect the industry.

The balance of this section deals with (1) raw materials that are consumed, (2) the nature of the products and markets to which they are supplied, (3) companies that make up the industry, (4) environmental impacts associated with air pollution, water pollution, and solid waste, and (5) health effects information for the industry. After the industry analysis, details pertaining to the technology are presented in 27 sections that deal with the categories of polymers shown in Table 2.

Raw Materials

The principal raw materials for plastics and resins are industrial organic chemicals that are used as monomers or plasticizers and speciality chemicals that are used as additives to modify the properties of the resins or plastic products.

More that 200 industrial organic chemicals used as monomers are listed in Appendix B. The 12 with the greatest percent of use in the plastics

TABLE 2.   INDUSTRY SECTOR DESCRIPTIONS

Acrylic Resins - Acrylic resins are water-clear polymers that are lightweight, durable, and exhibit excellent stability. Methyl methacrylate is
the starting block for nearly all acrylics. Polymethyl methacrylate is the
most common polymer, but many other comonomers may be employed to produce
polymers that range from tacky adhesives to the more common hard plastics
that are used in commercial aircraft windows.

Acrylonitrile-Butadiene-Styrene (ABS) - Variations in the manufacturing
process for ABS polymers can be made to produce a wide range of physical
properties. Typically, ABS is a tough, rigid, heat and chemical resistant
polymer with high impact strength and low temperature property retention
that makes it useful in pipe, business machine housings, automotive parts,
toys, and packaging.

Alkyd Molding Resins - Alkyd resins are typically made from a polyhydric
alcohol (such as ethylene glycol) and a polybasic acid (such as phthalic
anhydride) reacted in the presence of a drying oil such as tung or linseed
oil. Such resins are widely used in surface coatings such as printing inks
and lacquers. Alkyd molding resins are manufactured as solid molding
compounds used in automobile parts, furniture, electrical equipment and
cables, and construction components for buildings.

Amino Resins - Two major types of amino resins are produced: melamine-
formaldehyde and urea-formaldehyde resins. Although they may be produced in
a variety of forms, amino resins exhibit excellent resistance to heat,
solvents, and chemicals which makes them useful in adhesives, lamination,
dinnerware, and molding compounds. Adhesives represent the largest single
market for amino resins.

Modified Polyphenylene Oxide and Polyphenylene Sulfide - Since this category
typically includes fluorocarbons, nylons, polycarbonates, and other polymers
which are presented as separate segments, only modified polyphenylene oxide
and polyphenylene sulfide are included in this segment. These two polymers
exhibit high temperature performance and easy processability. Most engineering thermoplastic uses center around appliances, business machines, and
electrical or electronic uses.

Epoxy Resins - Epoxy resins are tough, hard, thermosetting solids which cure
quickly over a wide temperature range. Epoxy resins may be filled and come
in liquid and solid form for use as coatings, paints, mortar, adhesives,
laminates, and high-performance cast parts. The liquid resins are produced
by a two step reaction between bisphenol-A and epichlorohydrin while the
solid resins are manufactured from these raw materials by the Taffy or
advancement processes.

(continued)

TABLE 2 (continued)

<u>Fluoropolymers</u> - Fluoropolymers, named for the inclusion of fluorine in the polymer chain, are highly corrosive resistant. In addition, they exhibit outstanding electrical properties and maintain self-lubricating surface qualities. These polymers are used in chemical processing equipment, insulation and jacketing for wire and cable, nonstick cookware, and industrial equipment parts.

<u>Phenolic Resins</u> - Heat resistance, chemical resistance, stability, and low cost characterize phenolic resins. They are easily molded, exhibit good electrical properties, and retain surface hardness. Phenolic resins are used in the electrical, automotive, appliance, and consumer industries. One of the special uses for phenolic resins is in the sands used to make foundry shell molds and cores.

<u>Polyacetal</u> - Polyacetals, or acetal resins, are produced from the polymerization of formaldehyde. These plastics are the first with strength properties approaching those of nonferrous metals. Polyacetals are rigid, tough, and resilient and retain these properties over an extended time. They are used primarily as metal replacements in the automotive, plumbing, machinery, and consumer products markets.

<u>Polyamide Resins</u> - Two distinct types of polyamide resins exist: linear nylon polymers and non-linear non-nylon polymers. Nylons contribute approximately 90 percent to the total polyamide market, which is comprised of electrical parts, gears and fasteners, tubing, pipe, fibers, and film. The flame resistance, chemical resistance, toughness, low coefficient of friction, stiffness, outstanding wear resistance, good electrical properties, and high temperature resistance of polyamides enables their use in these varied applications.

<u>Polybutylene</u> - Commercial polybutylene resins are polymers of 1-butene. The major uses of these resins, pipe and film, utilize the high temperature stability, chemical resistance, tear strength, puncture resistance, and moisture impermeability of the polymer. Polybutylene is still in the developmental stage of commercialization.

<u>Polycarbonate</u> - Polycarbonates are a special class of linear polyesters. Although polycarbonates result from many different carbonic acid derivatives and diols, the polycarbonate of bisphenol-A is the only product of major commercial importance. The properties exhibited by polycarbonates, including impact strength, complete transparency, good weathering properties, and self-extinguishing properties, make these polymers suitable replacements for metals, glass, wood, and other thermoplastics. Some uses include electrical applications, appliances, instrumentation, marine, aerospace, photography, and food packaging.

(continued)

TABLE 2 (continued)

<u>Poly(Ester-Imide) and Poly(Ether-Imide) Resins</u> - Poly(ester-imide) and
poly(ether-imide) resins are condensation polymers which have a repeating
imide group as an integral part of the chain structure. High temperature
resistance, good electrical properties, good wear and friction properties,
chemical inertness, radiation and cryogenic temperature stability, and
inherent nonflammability are properties of polyimides which allow their use
in a variety of applications. Poly(ester-imides) are used primarily for
wire insulation. Poly(ether-imides) are used for injection molding,
extrusion, blow molding, and structural foam molding.

<u>Polyester Resins (Saturated)</u> - The saturated polyester resin market is
dominated by two products: polyethylene terephthalate (PET) and
polybutylene terephthalate (PBT). PET is a stiff resin which is a good
CO_2 barrier, making one of the major uses of this polymer bottles for
carbonated beverages. PBT is chemically resistant and is not affected by
soldering, acids, freon, or cleaning solutions. This polymer is used in
electrical, automotive, and industrial applications.

<u>Polyester Resins (Unsaturated)</u> - Unsaturated polyester resins are produced
from the condensation reaction of glycols and dibasic acids, one component
of which is unsaturated. These resins exhibit varying degrees of hardness,
flexibility, and flexural strength depending on the input materials used.
Due to this wide range of properties, unsaturated polyester resins are used
in marine products, automobile parts, bathroom fixtures, electrical
components, furniture, auto coatings, and appliances.

<u>Polyethylene--High Density (HDPE)</u> - High density polyethylene is a highly
ordered structure which is essentially linear. It exhibits increased
stiffness, tensile strength, hardness, and barrier properties when compared
to low density polyethylene. Polyethylene is a lightweight, flexible
polymer which exhibits outstanding electrical insulation properties. HDPE
has the third largest volume of polymers produced in the United States.

<u>Polyethylene--Linear Low Density (LLDPE)</u> - LLDPE is a newcomer to the
polyethylenes. It has the linear structure of HDPE combined with lower
densities, giving the polymer better elongation and environmental stress
crack resistance when compared to LDPE. LLDPE also possesses the high
tensile strength and chemical resistance of other polyethylenes. Its uses
include housewares, toys, pipe, packaging, and bags for a variety of uses.

<u>Polyethylene--Low Density (LDPE)</u> - LDPE has been produced the longest of the
three polyethylenes: HDPE, LLDPE, and LDPE. It is a tough, chemical
resistant, low cost polymer. The partially crystalline nature of LDPE
distinguishes it from LLDPE and HDPE. LDPE is the largest volume plastic in
the United States. Its uses include films, coating, pipes, and housewares.

(continued)

TABLE 2 (continued)

<u>Polypropylene</u> - Commercial polypropylene is a highly crystalline polymer which is resistant to chemicals and moisture as well as exhibiting high tensile strength. Ethylene is commonly added as a comonomer to improve the low temperature performance of the polymer. The largest markets for polypropylene are consumer oriented, including packaging, squeeze bottles, automobile parts, carpet backing, and piping.

<u>Polystyrene (General Purpose and Impact)</u> - Polystyrene is a clear, odor-free polymer with low specific gravity and good thermal stability. Impact, or rubber modified, polystyrene has rubber incorporated into the polymer chain to increase impact strength. Both the modified and unmodified polymer are easily fabricated and exhibit excellent thermal and electrical properties. Uses include doors, air conditioner housings, flower pots, caps, closures, toys, disposable cutlery and drinking cups, and food packaging.

<u>Polyurethane Foam</u> - Polyurethane foams are cellular plastics which are produced by the reaction of a polyol and a polyisocyanate in the presence of a blowing agent. Both rigid and flexible foams are used in the furniture, bedding, seating, refrigeration, construction, and automotive parts industries. Flexible foam must have good recovery from compression while rigid foam is used for insulation or sound deadening.

<u>Polyvinyl Acetate</u> - Polyvinyl acetate is a tasteless, transparent polymer. It resists oxidation and is inert to the effects of both ultraviolet and visible light. These properties make polyvinyl acetate useful in latex paints, adhesives, surface coatings, and textile finishings. Its most familiar use is the white glue used in many homes and schools. Polyvinyl acetate also serves as an intermediate for polyvinyl alcohol production.

<u>Polyvinyl Alcohol</u> - Polyvinyl alcohol is a water sensitive polymer which is derived from polyvinyl acetate since the theoretical monomer (vinyl alcohol) rearranges to acetaldehyde. Many of the properties of polyvinyl alcohol depend on the parent polyvinyl acetate. Major uses for polyvinyl alcohol include textile warp sizing, adhesives, suspension aids, paints, and membranes. Polyvinyl alcohol is also an intermediate in the production of polyvinyl butyral.

<u>Polyvinyl Chloride (PVC)</u> - PVC is the second largest volume plastic produced in the United States. It is a strong, moderately rigid, lightweight replacement for metals. PVC is used mainly in building and construction. Other uses include transporation, upholstery, packaging, electrical tapes, and consumer goods. PVC exhibits good impact resistance when modified.

(continued)

TABLE 2 (continued)

<u>Polyvinylidene Chloride</u> - Polyvinylidene chloride is a thermally stable, chemical resistant, highly impermeable polymer. Only copolymers are commercially produced since the homopolymer is difficult to process due to high crystallinity. Saran was the original trademark for this polymer, but has now become a generic term in the United States. Polyvinylidene chloride is primarly used as a film or coating on flexible surfaces for food packaging.

<u>Styrene-Acrylonitrile (SAN)</u> - Styrene-acrylonitrile combines the high clarity and gloss of styrene with the added chemical resistance and toughness of acrylonitrile. This transparent resin is easily processed and retains good dimensional stability. Polymer properties are partially determined by the molecular weight and acrylonitrile content. SAN is used for vegetable compartments, appliances, dash components, medical products, packaging, and specialty items.

industry are shown in Table 3. The data show order-of-magnitude differences
in the volume used for these top 12 monomers. Most of the other monomers
identified show even greater variation in the amounts used, since these
monomers are consumed in low volumes.

Plasticizers are used principally with vinyl chloride polymers and
copolymers. About 75 percent of the U.S. plasticizer consumption is used to
make PVC, which is otherwise brittle but workable. The largest volume plasticizers are phthalic acid esters. The 1982 consumption of phthalates was
400,000 metric tons, which amounted to 64 percent of the U.S. plasticizer
consumption. Most of these chemicals are used to impart flexibility to
resins; however, other applications include uses as insect repellents,
solvents, and lubricants. Many additives other than plasticizers are used
to modify the properties of polymers and plastic products. Table 4 presents major additive categories that include more than 1,000 specific chemicals.

Products

Table 5 displays the polymer categories and the 1980 production for
each. Polyethylene (combining LDPE, LLDPE, and HDPE - 33.7 percent) and
polyvinyl chloride (15.4 percent) account for approximately 50 percent by
weight of the total plastic resin production. Together with polypropylene
and polystyrene, these four major products account for approximately 70 percent by weight of the plastic resins produced in 1980.

A separate category called specialty plastics is listed in Table 5.
These products are polymers which have highly selective uses and are
generally produced in very low volumes. Table 6 gives a listing of the
products considered to be specialty plastics. The combined production
volume of all the polymers listed is approximately 23,000 metric tons.[143]
Since these 40 plastics have such small production volumes and widely varying technologies associated with their production, they have been excluded
from this analysis.

Table 7 shows the general markets for plastics and gives an indication
of their role in the economy. Although the demand for plastics products is
expected to grow into the late 1980's, it is tied to the vitality of the
economy and is sensitive to demands in the basic industries (automobiles,
housing, construction) and to consumer buying patterns. Production of most
plastics has increased since 1976, peaking in 1979. Since 1979, production
has generally declined with the tightening economy. Available capacities
should be adequate to accomodate any foreseeable growth.

High volume plastics such as low density polyethylene, high density
polyethylene, polyvinyl chloride, polypropylene, and polystyrene are
expected to continue downward trends into 1983. Plastics such as PVC, which
are closely associated with the automobile and construction industries, have
suffered in particular and should lag any economic upturn.

Barring any major moves by the economy, increases in production of
specific plastics will be related to two functions:

TABLE 3.   PRINCIPAL RAW MATERIALS USED IN THE PLASTICS INDUSTRY

Raw Material	1980 Production Thousand Metric Tons	Percent Used In Plastics Industry	Polymers
Ethylene	12,618	61	LDPE, HDPE, LLDPE
Propylene	5,573	44	Polypropylene
Styrene	3,136	72	ABS, SAN, IPS, GPPS
Vinyl Chloride	2,941	100	PVC
Formaldehyde	2,586	76	Amino Resins, Phenolic Resins
Dimethyl Terephthalate	1,529	100	PBT, PET
Butadiene	1,334	15	ABS
Phenol	1,118	73	Phenolic Resins
Vinyl Acetate	873	100	PVA, PVAc
Acrylonitrile	832	12	ABS, SAN
Phthalic Anhydride	389	57	Alkyd Resins, Unsaturated Polyester Resins
Bisphenol-A	239	82	Epoxy Resins

TABLE 4. MAJOR ADDITIVE CLASSES USED IN THE PLASTICS INDUSTRY

<u>Activators</u> - See Blowing Agents and Initiators.

<u>Accelerators</u> - See Catalysts and Curing Agents.

<u>Antiblocking Agents</u> - Also known as adherents, prevent adhesion of films to themselves. This term is specifically used for PVC, while the term slip agent is generally used for polyolefins. Silicate minerals and high melting waxes which migrate to the surface of the plastic are commonly used.

<u>Antioxidants</u> - Retard oxidative degradation of the plastic at processing temperatures and for extended life uses. The most commonly used antioxidants include alkylated phenols, amines, phosphates, and thio compounds for polyolefins, polystyrene, and ABS.

<u>Antistatic Agents</u> - Reduce the accumulation of electronic charge on the surface of polymers. These additives function either as lubricants to reduce friction and therefore reduce charge buildup or by providing a conductive path for dissipation of accumulated charge.

<u>Blowing (Foaming) Agents</u> - Produce large quantities of gases upon heating which form cellular plastics. These additives can be gases, solvents which volatize upon heating, or chemical agents which decompose to form gases. Activators may be used to promote gas formation. They are used for a wide variety of polymers. Cell stabilizers and nucleating agents control cell size and distribution in foamed plastics.

<u>Buffers</u> - Compounds containing both weak acids and weak bases. The pH of solutions containing buffers changes only slightly upon addition of acid or alkali.

<u>Catalysts</u> - Affect the rate of chemical reactions without themselves being consumed or undergoing chemical change. The catalyst can be either an inorganic or organic compound or an element. The designation of catalyst is used rather loosely in plastics processing, since some compounds designated as catalysts are consumed by the reaction, for example peroxides. Olefin polymerizations, for example, use organotransition metal compounds and Ziegler-Natta catalysts. Because the catalyst is not consumed in the reaction, catalyst neutralizers are sometimes added to deactivate or limit the catalyst's activity during product use.

<u>Catalyst Neutralizers</u> - See Catalysts.

<u>Cell Stabilizers</u> - See Blowing Agents.

<u>Chain Transfer Agents</u> - React with a growing polymer molecule to remove its free radical site and terminate polymer growth. The chain transfer agent is then a free radical capable of initiating the growth of another polymer chain.

(continued)

TABLE 4 (continued)

<u>Coalescing Agents</u> - Are used primarily in suspension polymerization to promote the formation of large droplets (beads) which will readily settle out of the aqueous phase.

<u>Colorants</u> - Impart hue (shade), value (brightness), and chroma (strength of color) to plastics. They increase the asthetic appeal of a product, and may enhance other properties including strength and UV stability. This broad classification includes dyes, organic and inorganic pigments, delusturants, pearlescents, and metallics. Titanium dioxide and carbon black constitute almost 90 percent of the consumption of colorants in plastics.

<u>Coupling Agents</u> - Enhance polymer-mineral surface bonds and the ability of a composite to retain its original properties after prolonged exposure to moisture. Silane and titanate compounds are high volume chemicals in this class.

<u>Craze Resistant Agents</u> - Reduce the tendency of a polymer to crack. These are commonly used in amino resins.

<u>Crosslinking Agents</u> - Form a bridge between individual polymer chains, changing a thermoplastic to a thermosetting resin.

<u>Cure Retardants</u> - Reduce the cure rate for thermosetting resins. They are most commonly amines used in amino resins.

<u>Curing Agents</u> - Improve the curing of thermosetting resins upon exposure to heat. The largest class is the amine compounds for curing of epoxy, amino, and phenolic resins. Organotin and amine compounds are commonly used for polyurethane foams.

<u>Defoamers</u> - Reduce the tendency of solutions to emulsify.

<u>Diluents</u> - See Solvents.

<u>Dispersing Aids</u> - See Emulsifiers.

<u>Elastomers and Flexibilizers</u> - Increase the pliability of polymers, and reduce their tendency to break.

<u>Emulsifiers</u> - Reduce the surface tension of water or the interfacial tension between two liquids or a liquid and a solid. They are commonly used in emulsion polymerization, and include various soaps and detergents.

<u>Fillers and Extenders</u> - Increase the bulkiness and decrease the total cost of plastic formulations. These compounds generally tend to reduce the desirable properties of the plastic, although some fillers are added to also reinforce, harden, insulate, and improve the appearance of the plastics they fill.

(continued)

TABLE 4 (continued)

Flame Retardants - Act chemically or physically as insulators; by creating endothermic cooling reactions, by coating the plastic to exclude oxygen, or by influencing combustion through reaction with materials that have different physical properties. Antimony, halogen, boron, nitrogen, and phosphorus based compounds are most common. Smoke suppressants and self-extinguishing agents are also used to reduce the fire hazard of polymers.

Flow Control Agents - Either promote or inhibit the flow properties of thermosetting resins.

Fungicides - See Preservatives.

Hardeners - Can be either internal or external. These compounds improve the firmness and set of resins. Also see Curing Agents.

Heat Stabilizers - Are primarily used in PVC and other vinyl formulations. These additives prevent the degradation of the plastic during process heating and for extended life uses. A variety of organic and organometallic compounds are used to react with chlorine radicals formed during heating and thus limit further degradation of the polymer.

Impact Modifiers - Decrease the plastic's tendency to break or crack upon impact. These additives include polymers such as acrylics and chlorinated polyethylene, and are primarily used in PVC formulations.

Initiators - Form free radicals upon heating or activation to begin polymerization. They are used in the production of thermoplastic resins, unsaturated polyester thermosetting resins and in crosslinking. Peroxy compounds such as benzoyl peroxide and methyl ethyl ketone peroxides are the highest volume initiators. Other peroxides, azo compounds and some inorganics are also used. Activators (also called promotors) may be used to lower the temperature at which the free radical forms.

Inhibitors - Prevent the polymerization of resins during storage and blending.

Lubricants - Enhance resin processibility and appearance of the end product. Desirable properties for lubricants include compatibility with the resin, ease of mixing, and FDA approval (where applicable). The lubricant should not adversely affect the end product. Common lubricants include fatty acid esters, hydrocarbon oils and paraffin wax, and amides.

Mold Release Agents - Coat the mold to prevent sticking of newly formed parts and to reduce static buildup. These additives can be sprays, powders, or paint-on liquids. Silicone, fluorocarbons, mineral oil, waxes, fatty acids, mica, and talc are used as mold release agents.

Neutralizers - Are either acids or bases used to bring aqueous solutions to a neutral pH.

(continued)

TABLE 4 (continued)

<u>Peroxide Decomposers</u> - See Initiators.

<u>Plasticizers</u> - Impart flexibility, resiliency, and melt flow to polymers.
These additives act by reducing the intramolecular forces between polymer
chains. Seventy to 80 percent of plasticizers are used in PVC formulation.
The major chemical classes include phthalates, adipates, trimellitates,
glycolates, epoxies, polyesters, fatty acid esters and organic phosphates.

<u>Preservatives and Biocides</u> - Inhibit biological degradation of polymers.
This class of additives includes fungicides and bacteriostats.

<u>Promoters</u> - See Initiators.

<u>Protective Colloids</u> - Are special surface active substances which prevent
the dispersed phase from coalescing. These are commonly used in emulsion
polymerization, and include polyvinyl alcohol and various cellulose
derivatives.

<u>Reinforcing Agents</u> - Impart tensile, flexural, and compressive strength to
plastics. These reinforcers can be synthetic fibers, carbon fibers, glass
fibers, asbestos or silica or other combinations of materials. The most
common reinforcers are glass fibers which are available in chemical and
electrical grades to provide increased chemical resistance and electrical
insulating properties, respectively, to the plastic.

<u>Reodorants</u> - Improve the olefactory appeal of polymers by masking their
odor.

<u>Sizing Agents</u> - Coatings which protect the polymer surface. Waxes are
commonly used.

<u>Slip Promoters</u> - Provide surface lubrication during processing and use.
They have minimal compatibility with the polymer, are added internally, and
flow to the surface immediately following processing. Since compatiblity is
minimal, slip promoters may exude from the product long before the end of
its useful life. This term is commonly applied to abherents for
polyolefins.

<u>Solution Modifiers</u> - Include a wide variety of chemicals used for
plastisols, coatings, and latexes. These additives have a wide variety of
uses, but are mainly incorporated for ease of handling and for applcation to
the selected processing step.

<u>Solvents</u> - Are liquids capable of dissolving various substances. They are
frequently used in polymerization and include water, hydrocarbons, and
alcohols.

(continued)

TABLE 4 (continued)

<u>Stabilizers</u> - Include a wide variety of compounds which improve the stability of plastics during processing and use. See also, Antioxidants, Heat Stabilizers, and UV Stabilizers.

<u>Surfactants</u> - See Emulsifiers.

<u>Suspension Aids</u> - See Emulsifiers and Protective Colloids.

<u>Thixotropic Agents</u> - Used for thermosetting resins to promote gelling upon standing, and liquefaction upon shaking.

<u>UV Stabilizers</u> - Function either to absorb ultraviolet radiation and reradiate it at a harmless wavelength or as quenchers which consume the free radicals generated by UV light. Benzotriazoles and benzophenones constitute the major compounds used as UV absorbers. Nickel compounds are commonly used as quenchers.

<u>Viscosity Aids</u> - Reduce or increase the viscosity of plastisols (PVC-plasticizer-solvent mixtures) either during normal processing or for plastisols which have been subjected to prolonged storage.

Source: <u>Plastics Processing</u>, by Radian Corporation, Noyes Publications, Park Ridge, NJ, 1986.

TABLE 5.   PLASTICS AND RESINS PRODUCTION - 1980

Category	U.S. Production Thousand Metric Tons	Percent of Total
Acrylic Resins	247	1.5
Acrylonitrile-Butadiene-Styrene	416	2.6
Alkyd Resins	210	1.3
Amino Resins	589	3.7
Engineering Thermoplastics	56	0.4
Epoxy Resins	145	0.9
Fluoropolymers	6	<0.1
Phenolic Resins	686	4.3
Polyacetal	40	0.3
Polyamide Resins	158[a]	1.0
Polybutylene	7	<0.1
Polycarbonate	106	0.7
Poly(ester-imide) and Poly(ether-imide) Resins	5	<0.1
Polyester Resins (Saturated)	480	3.0
Polyester Resins (Unsaturated)	438	2.7
Polyethylene--High Density	1,997	12.5
Polyethylene--Linear Low Density	336	2.1
Polyethylene--Low Density	3,058	19.1
Polypropylene	1,689	10.6
Polystyrene (General Purpose and Impact)	1,599	10.0
Polyurethane Foam	762	4.8
Polyvinyl Acetate	304	1.9
Polyvinyl Alcohol	64	0.4
Polyvinyl Chloride	2,458	15.4
Polyvinylidene Chloride	78	0.5
Styrene-Acrylonitrile	47	0.3
Specialty Plastics	N/A	N/A
	15,981	100.0

[a]The most recent polyamide production figures available are for 1979.

N/A - Not available.

TABLE 6. SPECIALTY PLASTICS

Acrylamide-Acrylic Acid Copolymer
Acrylamide Resins
Alkylphenol-Acetylene Resins
Cellophane
Cellulose Acetate
Cellulose Acetate Butyrate
Cellulose Acetate Phthalate
Cellulose Acetate Propionate
Coumarone-Indene and Hydrocarbon Resins
Cresol-Formaldehyde Resins
Cyclohexanone Formaldehyde Resins
Dicyandiamide Resins
Dimethylhydantoin-Formaldehyde Resins
Furan Resins
Glyoxal-Formaldehyde Resins
Hydrocarbon Resins
Ionomer Resins
Ketone-Aldehyde Resins
Maleic Resins
Methyl Vinyl Ether-Maleic Anhydride Copolymer
  Resins
Methyl Vinyl Ether-Monobutyl Maleate Copolymer
  Resins
Methyl Vinyl Ether-Monoethyl Maleate Copolymer
  Resins
Nitrocellulose
Nonyl Phenol-Formaldehyde Resin
Phenoxy Resins
Polysulfone Resins
Polyterpene Resins
Miscellaneous Polyurethane Resins
Polyvinyl Butyral Resins
Polyvinyl Formal Resins
Resorcinol Formaldehyde Resins
Rosin and Rosin Ester Resins
Silicone Resins
Styrene-Allyl Alcohol Resins
Styrene-Divinylbenzene Copolymer Resins
Styrene-Maleic Anhydride Copolymer Resins
Triazone Resins
Vinyl Polybutadiene Resins
1-Vinyl-2-Pyrrolidinone-Styrene Copolymer
  Resins
Vinyl Toluene Copolymer Resins

TABLE 7. U.S MARKETS FOR PLASTICS - 1980

Market	Thousand Metric Tons	% of Plastics Market	Plastics Used	%
Transporation	729.5	4.6	ABS/SAN	11.5
			Nylon	5.0
			Phenolic	8.4
			Polyester	16.3
			Polyethylene--High Density	5.3
			Polypropylene	14.3
			Polyvinyl Chloride	12.2
			Other Thermoplastics	15.2
			Other Thermosets	11.8
Packaging	4,546.8	28.4	Polyester (Thermoplastic)	4.1
			Polyethylene--High Density	20.5
			Polyethylene--Low Density	45.8
			Polypropylene	7.2
			Polystyrene	13.5
			Polyvinyl Chloride	4.1
			Other Themoplastics	4.4
			Thermosets	0.4
Building and Construction	2,920.0	18.3	ABS/SAN	3.6
			Amino	13.0
			Phenolic	13.4
			Polyester	6.0
			Polyethylene--Low Density	2.9
			Polyethylene--High Density	6.7
			Polystyrene	4.3
			Polyvinyl Chloride	41.9
			Other Thermoplastics	7.6
			Other Thermosets	0.6
Electrical and Electronic	1,115.0	7.0	ABS/SAN	10.8
			Epoxy	2.2
			Phenolic	7.4
			Polypropylene	8.0
			Polyethylene--High Density	6.8

(continued)

TABLE 7 (continued)

Market	Thousand Metric Tons	% of Plastics Market	Plastics Used	%
Electrical and Electronic (Continued)			Polyethylene--Low Density	17.0
			Polystyrene	12.3
			Polyvinyl Chloride	22.0
			Other Thermoplastics	12.0
			Other Thermosets	1.5
Furniture and Furnishings	748.2	4.7	Amino	10.2
			Polypropylene	44.2
			Polystyrene	4.5
			Other Styrene Polymers	18.8
			Polyvinyl Chloride	13.1
			Other Thermoplastics	6.9
			Other Thermosets	2.3
Consumer and Institutional	1,615.0	10.1	ABS/SAN	2.6
			Polyester	2.8
			Polyethylene--High Density	13.9
			Polyethylene--Low Density	16.0
			Polypropylene	14.2
			Polystyrene	21.9
			Polyvinyl Chloride	12.0
			Other Thermoplastics	14.2
			Other Thermosets	2.4
Industrial and Machinery	177.7	1.0	ABS/SAN	3.3
			Nylon	5.1
			Phenolic	5.4
			Polypropylene	13.3
			Polyvinyl Chloride	4.1
			Other Thermoplastics	53.0
			Other Thermosets	15.8

(continued)

TABLE 7 (continued)

Market	Thousand Metric Tons	% of Plastics Market	Plastics Used	%
Adhesives, Inks, and Coatings	1,085.0	6.8	Amino	4.1
			Epoxy	6.3
			Phenolic	7.8
			Polyvinyl Chloride	3.3
			Other Vinyl Polymers	27.2
			Other Thermoplastics	34.7
			Other Thermosets	16.6
Other	1,388.2	8.7		
Export	1,668.2	10.4		
TOTAL	15,993.6	100.0		

Source: _Facts and Figures of the Plastics Industry_, 1981.

- Improved technologies which make products of a specific plastic or specific manufacturing route more desirable or cost effective than alternative production or process procedures; or

- An increased demand resulting from a basic switch in consumer buying patterns, a new product, or the displacement of a non-plastic material used in the manufacture of an item. (In either case a new market is created.)

The anticipated growth of linear low density polyethylene is an example of the former function. The increase in production of PET and PBT through the 1980 to 1981 time frame and the anticipated continued growth of engineering thermoplastics are examples of the latter phenomena.

<u>Companies</u>

The companies involved in the production of plastics and resins are varied, being comprised of major oil and chemical, paint, tire and rubber, steel, electrical manufacturing, and miscellaneous manufacturing companies. This diversity is a result of both horizontal and vertical integration within the economy and the general economic attractiveness that has marked plastic and resin production during the 1970's. The horizontal and vertical integration of plastics production involves a large number of extremely different products, thus requiring the participation of a range of companies from large to small. For example, paint manufacturers produce the emulsions used for their products.

Companies that produce six or more types of plastic and resin are shown in Table 8. Another 211 companies produce five or fewer products and 120 of these have a single product. A complete listing of producers and their products are shown in Appendix C. From Table 8 it can be seen that the diversified producers are major chemical companies and oil companies with chemical production operations. The firms that produce only one or two products are generally small companies competing for a very limited share of the lower volume plastics market such as amino, acrylic, alkyd, phenolic, and polyvinyl acetate. They are not involved with the high volume plastics, such as polyvinyl chloride and low density polyethylene, which have experienced not only a change in technology but also a shakedown in prices due to a strongly competitive market. Smaller volume plastics, such as amino resins and phenolic resins, are not experiencing the flux associated with an emerging technology, and are also not experiencing the competition from other plastics which can provide the same, or similar, properties. PVC and polyethylenes have started to experience competition from other thermoplastics, such as polybutylene and polypropylene, which has forced these resins into a very competitive cross-product market.

<u>Environmental and Occupational Health Impacts</u>

Plastic production processes generate air emissions, wastewater, and solid waste. Although every process does not generate all three waste streams, most processes have air emissions, solid wastes, and wastewaters other than routine cleaning water. Table 9 presents a summary of the

TABLE 8.   DIVERSIFIED PLASTIC AND RESINS PRODUCERS

	Number of Products	Number of Plants
E. I. du Pont de Nemours	10	16
Dow Chemical	9	9
Monsanto Co.	9	9
Celanese	8	4
Gulf Oil	8	11
Reichhold Chemicals	8	20
Borden Inc.	7	14
General Electric	7	6
Mobil Corporation	7	9
Union Carbide	7	12
American Hoescht	6	7
Occidental Petroleum	6	6

TABLE 9. PLASTIC PRODUCTION EMISSIONS SUMMARY

Plastic	Air Emissions		Waste-water	Solid Waste
	Volatiles	Particulates		
Acrylic Resin	X	X	X	X
Acrylontirile-Butadiene-Styrene	X	X	X	X
Alkyd Resins	X		X	X
Amine Resins	X	X	X	X
Engineering Thermoplastics	X	X	X	X
Epoxy Resins	X		X	X
Fluoropolymers	X	X	X	X
Phenolic Resins	X	X	X	X
Polyacetal	X	X	X	X
Polyamide Resins	X	X	X	X
Polybutylene	X	X	X	X
Polycarbonate	X	X	X	X
Poly(Ester-Imide) and Poly(Ether-Imide) Resins	X		X	X
Polyester (Saturated)	X	X	X	X
Polyester (Unsaturated)	X	X	X	X
Polyethylene--High Density	X	X	*	X
Polyethylene--Linear Low Density	X	X	*	X
Polyethylene--Low Density	X	X	*	X
Polypropylene	X	X	X	X
Polystyrene	X	X	X	X
Polyurethane Foam	X	X	*	X
Polyvinyl Acetate	X	X	X	X
Polyvinyl Alcohol	X	X[a]	X	X
Polyvinyl Chloride	X	X	X	X
Polyvinylidene Chloride	X	X	X	X
Styrene-Acrylonitrile	X	X	X	X

[a]Only if product is dried for shipment.
*Routine cleaning water only.

related emissions and waste streams generated by the production of the plastics listed in Table 2. A detailed description of the emissions from each polymerization process can be found in the section describing the plastic of interest.

Occupational health concerns in plastics and resins manufacturing vary widely with the type of plastic being produced. Production processes for polyethylene and polypropylene, for example, use input materials of substantially lower toxicity than polyurethane, polyvinyl chloride or epoxy resins. Correspondingly, the most rigid control measures are typically applied in the latter processes. Chemical substances and physical agents of particular concern in each process are discussed in the section describing that process.

Air Emissions

The air emissions from plastics production are of two types: volatile and particulates. Volatile emissions for most processes are greatest in the processing steps upstream of the reactor. While this is generally true, mass (bulk) polymerization and solution polymerization processes experience significant volatile emissions during the monomer recovery and solvent recovery steps which are downstream of the reactor. Lower conversion for mass polymerization make monomer recovery necessary. Solvents added during solution polymerization are typically volatile organics, which increase the volatile emissions for this process. Typical volatile emissions contain mostly monomers, comonomers, and solvents with small quantities of plasticizers, initiators, emulsifiers, and suspension stabilizers (protective colloids).

Though they generally exhibit fewer volatile emissions than mass or solution processes, suspension and emulsion polymerization processes typically generate more particulates. Since these processes use water as a heat transfer medium, the polymer produced must be separated and dried. The filtration and drying steps generate particulates which may be controlled by using cyclones or filters. Some particulates are generated during pelletizing the product of mass polymerization and drying the solution polymerization product; however, the bulk of the emissions from these processes will be volatile. The particulates from plastic production are mainly polymer particles, but they may include any non-reactive fillers or additives incorporated during the polymerization process or reactive additives prior to pelletization for mass polymerization.

Venting emissions outside the polymerization area, providing adequate ventilation, operating the polymerization area at a slightly negative pressure, and collecting particulates are examples of good engineering practices which aid in the control of air emissions. Instituting a routine system for inspection and detection of leaks also helps reduce the air emissions from these processes.

Variability in the nature of materials handled, the involvement of many types of large and small companies, and the large number of establishments involved all combine to make it difficult to generalize regarding the

overall adequacy of air pollution control in the plastics and resins
industry. Where known potential problems exist (e.g., exposure to vinyl
chloride) highly effective measures are taken to control emissions. Other
considerations such as process economics contribute to the use of control
equipment in some instances. However, in the absence of more information on
toxicity and other potential hazards of many of the materials that may be
discharged and background on processes and practices in the industry, potential for discharge of significant amounts of air pollution must be considered a possibility even though specific problems have not been identified.

Wastewater

The most significant source of wastewater generated from plastic
production is the water used for emulsion and suspension polymerization.
The water used in these processes as a heat transfer medium allows the monomer to be polymerized without the difficulties associated with the increasing viscosity of the polymerizing mass (mass polymerization) or the added
emissions and hazards of an organic solvent (solution polymerization).
Another wastewater source arises from the polymerization reaction itself
when the polymer is produced by a condensation reaction. Water used to wash
the polymer, collected condensate from steam stripping for solvent recovery,
and routine cleaning water are other sources of wastewater associated with
plastics production.

These wastewaters may contain monomer, comonomer, filler, and additives. Emulsifiers, suspension stabilizers (protective colloids), and
initiators are present if water soluble, but found in very small amounts.
Polymer particles are present in polymer wash waters and routine cleaning
water. Some of the plastics have been selected by the Environmental Protection Agency for Effluent Limitations Guidelines. Specific regulations may
be found in the section concerning the plastic of interest.

The treatment of wastewaters generated by these processes is highly
specific. The different variables which effect the treatment include
plastics produced, additives used, and process operating parameters. Therefore, no general form of wastewater control is applicable to this industry.

Solid Waste

Solid waste occurs from plastics production in one of two ways:
polymer is lost or by-products are formed. The largest portion of solid
waste for plastics production may be attributed to lost polymer. Spillage,
routine cleaning, and collected particulates are classified as polymer lost
during the production process. Substandard polymer which cannot be blended
is also defined as lost polymer.

By-products formed during the production process are another source of
solid waste. They are found in fewer processes and constitute a smaller
portion of the solid waste than polymer losses. Finally, spent catalyst,
spent filter material, and low molecular weight polymers are included as
solid wastes.   However, their small volume and applicability to only

select processes make them less important than the lost polymer portion. As
with wastewater, solid waste disposal and generation is highly process
specific.

Health Effects

As indicated in the preceding discussion, the manufacture of plastics
and resins involves the use of a large number of organic and inorganic
chemicals for process raw materials, processing aids, property enhancers,
and other specialty uses. Worker exposure to these chemicals, some of which
are potentially dangerous, can occur through accidental ingestion, inhalation of volatile or particulate emissions, and skin contact.

To delineate better the potential for adverse health effects or harm to
the environment from the use of chemicals in the plastics and resins industry, a chemical hazardous screening task was undertaken to identify chemicals that had a history of health effects investigations. This involved
conducting a comprehensive literature search and screening of information
for 476 chemical raw materials in three on-line data banks operated by the
National Library of Medicine—the Registry of Toxic Effects of Chemical
Substances (RTECS), Toxicology Data Bank (TDB), and Toxline.* The chemicals were screened for toxic, carcinogenic, mutagenic, and teratogenic
effects and classified according to hazard potential.

The chemicals screened included specialty chemicals used as additives
as well as major industrial organic chemicals input to the industry. Almost
200 of the 476 chemicals screened had little or no history of investigation.
As a result, the hazard potential of these chemicals could not be determined.

Slightly more than 200 chemicals were considered to have either very
low or slight to moderate hazard potential. The balance of 71 chemicals was
found to have some record of study to suggest they could be considered
chemicals of concern with regard to hazard potential.

It should be noted that the results of this screening cannot be considered comprehensive because they are limited in terms of industry cov-
erage. Chemicals which were not screened include: (1) chemicals which were
not identified as input materials to the plastics industry due to the proprietary nature of some processes; (2) several hundred input chemicals (most
believed to be of minor importance) which have not yet been assigned a CAS
identification number, e.g., dihydroxy diphenyl sulfone, a crosslinking
agent in the manufacture of polyvinyl alcohol; and (3) an additional several
hundred input materials which were identified only as classes of compounds,
e.g., aldehydes or alkyl acrylates. Despite these limitations, the results
of the screening do represent a useful step in putting into perspective the
potential hazards of exposure to chemicals used in the industry.

*The searches were keyed to the Chemical Abstracts Service identification
number (CAS number).

Table 10 provides details on the reported health effects associated with the 71 chemicals of concern as well as pertinent information on OSHA, NIOSH, and ACGIH regulations. Recommendations have been made for 12 of these chemicals for which no OSHA standard has been set as yet. They are: acrylic acid, antimony oxide, cadmium sulfide, ferrous sulfate, mercuric bromide, mercuric chloride, methyl ethyl ketone peroxide, pyrocatechol, resorcinol, thiophenol, vinylcyclohexene dioxide, and vinylidene chloride. Neither a standard nor a recommendation has been made for 18 chemicals of interest, most of which pose a possible hazard due to carcinogenic potential rather than acute toxicity. In many cases, sufficient health effects and exposure data do not exist on which to base a standard or recommendation.

There is considerable variation among the 71 chemicals of concern in factors such as number and types of application. About one third are industrial organic chemicals used in relatively large volumes. The balance are specialty chemicals used in smaller amounts as additivies. More than one-half of these chemicals are used by one industry segment only. Others are used by several industry segments, sometimes for varied purposes. Styrene, for example, is a monomer in the production of ABS, SAN, and poly-styrene; a comonomer in the production of acrylic resins and polyvinylidene chloride; a crosslinking agent in the production of unsaturated polyester resins; a modifier in the production of alkyd resins; a diluent in the pro-duction of epoxy resins; and a polymer for blending in the production of engineering thermoplastics. The acrylic resins segment of the industry uses the largest number of chemicals of concern (21), followed by epoxy resins (15) and polyurethane foam (12). However, more than half of the plastics and resins industry segments use four or fewer of these chemicals.

Employee Exposure Controls

Good industrial hygiene practice and OSHA policy dictate that personnel exposures to workplace hazards should first be controlled by engineering and administrative means, and then by personal protective equipment where neces-sary. The full use of both control types in plastics manufacturing is best demonstrated in polyvinyl chloride production. For this reason, that indus-try will be used as a model in the discussion which follows.

Polymerization processes are such that a control system for one process may be applicable to an entirely different process. Thus, with minor modi-fication (e.g. for low volatility monomers such as acrylonitrile), exposure controls for polyvinyl chloride production may be applied to most of the plastics and resins industry. Similarly, administrative measures such as real-time and personnel monitoring programs, fugitive emissions reduction measures, and personal protective equipment and clothing programs developed for one industry segment very likely can be successfully used throughout the industry.

Controls, monitoring programs and protective measures must be tailored to meet the needs of individual plants. With sufficient time, resources and technical expertise, however, worker exposure to hazardous materials can be successfully controlled.

TABLE 10.   REGULATIONS AND RECOMMENDATIONS FOR AIRBORNE EMISSIONS OF CHEMICALS OF CONCERN

Chemical	Polymers	Chemical Function	Major Concern	OSHA Air Regulation	NIOSH (*) or ACGIH (+) Recommendations
Acrylamide	Amino Resins	Heat resistance additive	Neurotoxin; potential mutagen and teratogen	TWA 300 g/m^3	TLV-TWA 300 g/m^{3+} (skin) STEL 600 g/m^{3+} TWA 0.3 mg/m^{3*}
	Unsaturated Polyester Resins	Crosslinking agent			
	PVA	Comonomer			
Acrylic Acid	PVAc; Acrylic Resins; Polyvinylidene Chloride	Comonomer	Severe skin, eye and respiratory irritant; ingestion may produce circulatory collapse, and in severe cases, death due to shock; potential tumorigen and teratogen	No Standard	TLV-TWA 10 ppm^+
	ABS: SAN	Suspension stabilizer			
Acrylonitrile	ABS;SAN	Monomer	Suspected human carcinogen; potential mutagen and teratogen; toxic; causes cyanosis, nausea, and convulsions; eye and respiratory irritant	TWA 2 ppm CL 10 ppm/15M	TLV-TWA 2 ppm^+ (skin) CL 4 ppm^*
	PVC; PVA; Acrylic Resins; Polyvinylidene Chloride	Comonomer			
	Alkyl Resins	Modifier			
	Epoxy Resins	Diluent			
Allyl Glycidyl Ether	Epoxy Resins	Diluent	Potential tumorigen and mutagen; causes severe local reactions; currently being tested for carcinogenesis	CL 10 ppm	TLV-TWA 5 ppm^+ (skin) STEL 10 ppm^+ CL 45 $mg/m^3/15M^*$
Antimony Oxide	PS; Unsaturated Polyester Resins; Polyurethane Foam	Flame Retardant	Suspected carcinogen; potential mutagen	TWA 0.5 mg/m^3 (as Sb)	TWA 0.5 mg/m^{3*}
	PET; PBT	Catalyst			
	Epoxy Resins	Colorant			
Antimony Pentachloride	Polyurethane Foam	Catalyst	Extremely toxic by ingestion and inhalation; can cause systemic damage upon acute exposure; potential mutagen	TWA 0.5 mg/m^3 (as Sb)	TWA 0.5 mg/m^{3*}
Antimony Trichloride	Polyurethane Foam	Catalyst	Extremely toxic; produces upper respiratory tract irritation, abdominal pain, loss of appetite	TWA 0.5 mg/m^3 (as Sb)	TWA 0.5 mg/m^{3*}

+IARC Determination.

(continued)

TABLE 10 (continued)

Chemical	Polymers	Chemical Function	Major Concern	OSHA Air Regulation	NIOSH (*) or ACGIH (+) Recommendations
Asbestos	Amino Resins; Phenolic Resins; Epoxy Resins; Polyamide Resins; Engineering Thermoplastics; PTFE; PCTFE	Filler	Known human carcinogen†	TWA 0.5 fb/cc	TLV-TWA 0.2 fb/cc[+] TWA 100,000 fb/m^3* CL 500,000 fb/m^3*
Benzene	PVC; LDPE; Acrylic Resins	Solvent	Suspected human carcinogen; potential mutagen and teratogen; toxic to the central nervous system by ingestion, inhalation, and skin absorption	TWA 10 ppm CL 25 ppm PK 50 ppm/10M/8H	TLV-TWA 10 ppm[+] STEL 25 ppm[+] CL 10 ppm/60M*
p-Benzoquinone	PS	Retardant	Potential tumorigen and mutagen; extremely toxic by ingestion; apparently has a direct effect on medulla and oxygen-carrying capacity of blood	TWA 0.1 ppm	TLV-TWA 0.1 ppm[+] STEL 0.3 ppm[+]
	Unsaturated Polyester Resins	Polymerization inhibitor			
t-Butyl Hydroperoxide	PS; LLDPE	Radical initiator	Extremely toxic; produces symptoms of severe depression, cyanosis, and respiratory arrest; severe local irritant	No Standard	No Recommendation
Cadmium Sulfide	PMMA; Polyamide Resins	Colorant	Known animal carcinogen; potential mutagen; highly toxic by ingestion and inhalation; affects respiratory tract and kidneys	TWA 0.2 mg/m^3 CL 0.6 mg/m^3 (as Cd)	TWA 0.05 mg/m^3 (as Cd) TWA 40 g/m^3* CL 200 g/m^3/15M*
Carbon Tetrachloride	PVC; PVAc; PS; Acrylic Resins	Chain transfer agent	Suspected human carcinogen; extremely toxic by inhalation or ingestion; destroys tissue membranes	TWA 10 ppm CL 25 ppm PK 200 ppm/5M/4H	TLV-TWA 5 ppm[+] (skin) STEL 20 ppm[+] CL 2 ppm/60M*
Chlorine	Fluoropolymers	Suspension stabilizer	Highly toxic to the respiratory tract by inhalation; strong local irritant; potential mutagen	CL 1 ppm	TLV-TWA 1 ppm[+] STEL 3 ppm[+] CL 0.5 ppm/15M*
Chloroform	PMMA	Chain transfer agent	Suspected human carcinogen; potential mutagen and teratogen; moderately toxic; causes respiratory depression, dizziness, nausea, intracrania pressure, and in high concentrations, cardiac arrest	CL 50 ppm	TLV-TWA 10 ppm[+] STEL 50 ppm[+] CL 2 ppm/60M*
	Fluoropolymers	Promoter			
Chloroprene	Polyvinylidene Chloride	Comonomer	Suspected human carcinogen; potential mutagen and teratogen	TWA 25 ppm	TLV-TWA 10 ppm[+] (skin) CL 1 ppm/15M*

†IARC Determination.

(continued)

TABLE 10 (continued)

Chemical	Polymers	Chemical Function	Major Concern	OSHA Air Regulation	NIOSH (*) or ACGIH (+) Recommendations
Chromium Oxide	HDPE; LDPE; LLDPE	Catalyst	Known animal carcinogen†; potential mutagen and teratogen; toxic upon chronic exposure; affects liver and central nervous system	TWA 0.5 mg/m^3 (Cr^{+6})	TWA 0.05 mg/m^3 (Cr^{+6}) TWA 25 g/m^3 (Cr^{+6}) CL 50 g/m^3/15M (Cr^{+6})
Cumene Hydroperoxide	Unsaturated Polyester Resins	Catalyst; cure agent	Acutely toxic by ingestion, inhalation, and skin absorption; produce headache and throat irritation; potential mutagen and teratogen	No Standard	No Recommendation
	LDPE; PMMA	Radical initiator system			
	ABS	Redox initiation system			
Diazomethane	PET; PBT	Carboxy and group control	Known animal carcinogen†; highly toxic by inhalation upon acute or chronic exposure; can cause chest pain, cough, cyanosis, shock, and death	TWA 0.2 ppm	TLV-TWA 0.2 ppm$^+$
1,2-Dichloroethane	PVC	Solvent	Suspected human carcinogen†; potential mutagen; acutely toxic by ingestion and inhalation; can affect gastrointestinal tract, central nervous system, and liver, kidney, and adrenal systems	TWA 50 ppm CL 100 ppm PK 200 ppm/5M/3H	TLV-TWA 10 ppm$^+$ STEL 15 ppm$^+$ TWA 1 ppm* CL 2 ppm/15M*
Diglycidyl Ether	Epoxy Resins	Diluent	Highly toxic by ingestion and inhalation; moderately toxic by skin absorption; affects central nervous system and liver	CL 0.5 ppm	TLV-TWA 0.1 ppm$^+$ CL 1 mg/m^3/15M*
N,N-Dimethylaniline	PMMA	Cure Agent	Acutely toxic by ingestion, inhalation, and skin absorption; can produce methemoglabinemia; potential tumorigen	TWA 5 ppm	TLV-TWA 5 ppm$^+$ (skin) STEL 10 ppm$^+$
	Unsaturated Polyester Resins	Accelerator			
1,1-Dimethylhydrazine	PB	Catalyst	Highly toxic by ingestion, inhalation, and skin absorption; can cause damage to respiratory system, liver, and blood; known animal carcinogen†; potential mutagen	TWA 0.5 ppm	TLV-TWA 0.5 ppm$^+$ (skin) STEL 1 ppm$^+$ CL 0.15 mg/m^3/2H*
Divinyl Sulfone	PVA	Crosslinking agent	Severe skin and eye irritant; toxic; can cause enzyme inhibition	No Standard	No Recommendation

†IARC Determination.

(continued)

TABLE 10 (continued)

Chemical	Polymers	Chemical Function	Major Concern	OSHA Air Regulation	NIOSH (*) or ACGIH (†) Recommendations
Epichlorohydrin	Epoxy Resins; Polyamide Resins	Monomer	Highly toxic by ingestion, inhalation, and skin absorption; acute exposure may cause respiratory paralysis and chronic exposure can cause kidney injury; known animal carcinogen†; potential mutagen and teratogen	TWA 5 ppm	TLV-TWA 2 ppm† (skin) STEL 5 ppm† TWA 2 mg/m³† CL 19 mg/m³/15M*
3,4-Epoxy-6-Methyl cyclohexylmethyl-3,4-Epoxy-6-Methyl cyclohexane Carboxylate	Epoxy Resins	Diluent	Known animal carcinogen	No Standard	No Recommendation
Ethyl Acrylate	PVAc; Acrylic Resins; Polyvinylidene Chloride	Comonomer	Highly toxic by ingestion, inhalation, and skin absorption upon acute exposure; produces hypersomnia and convulsions	TWA 25 ppm	TLV-TWA 5 ppm† (skin) STEL 25 ppm†
Ethyleneimine	PVA	Comonomer	Known animal carcinogen*; severe skin, eye, and respiratory tract irritant; extremely toxic by ingestion	TWA 0.5 ppm	TLV-TWA 0.5 ppm† (skin)
Ethylene Oxide	PVA; Polyacetal	Comonomer	Toxic upon acute exposure; affects gastrointestinal tract and pulmonary system; potential tumorigen, mutagen, and teratogen	TWA 50 ppm	TLV-TWA 1 ppm†
Ferrous Sulfate	ABS	Redox initiation system	Highly toxic upon acute exposure; affects central nervous system; tumorigen and mutagen	No Standard	TLV-TWA 1 mg/m³† STEL 2 mg/m³† (as Fe)
Formaldehyde	Amino Resins; Phenolic Resins; Polyacetal PVA	Monomer Comonomer	Highly toxic by ingestion, inhalation, and skin absorption upon acute exposure; suspected carcinogen; potential mutagen and teratogen	TWA 3 ppm CL 5 ppm PK 10 ppm/30M/8H	TLV-TWA 1 ppm† STEL 2 ppm† CL 1.2 mg/m³/30N*
Furfuryl Alcohol	Amino Resins	Craze resistance agent	Acutely toxic; affects respiration and body temperature; potential tumorigen and mutagen	TWA 50 ppm	TLV-TWA 10 ppm† (skin) STEL 15 ppm† TWA 200 mg/m³*
Hydrocyanic Acid	Polyurethane Foam	Blocking agent	Highly toxic by ingestion, inhalation, and skin absorption upon acute exposure; produces cyanosis, unconsciousness, and convulsions	TWA 10 ppm	CL 10 ppm† (skin) CL 5 mg(CN)/m³/10M*

†IARC Determination.

(continued)

TABLE 10 (continued)

Chemical	Polymers	Chemical Function	Major Concern	OSHA Air Regulation	NIOSH (*) or ACGIH (+) Recommendations
Hydroquinone	Unsaturated Polyester Resins	Polymerization inhibitor	Highly toxic by ingestion upon acute exposure; potential tumorigen, mutagen, and teratogen	TWA 2 mg/m^3	TLV-TWA 2 mg/m^{3+} STEL 4 mg/m^{3+} CL 2 mg/m^3/15M*
	Polyurethane Foam	Suspension stabilizer			
Lead Acetate	Polyamide Resins	Nucleating agent	Suspected human carcinogent; potential mutagen and teratogen	TWA 0.05 mg/m^3 (as Pb)	TLV-TWA 0.15 mg/m^3 (as Pb)$^+$ TWA 0.10 mg(Pb)/m^{3*}
Lead Oxide	PET; PBT	Catalyst	Highly toxic by ingestion or inhalation of lead oxide dust; potential tumorigen and mutagen	TWA 200 g(Pb)/m^3	TLV-TWA A 0.15 mg/m^3 (as Pb)$^+$ TWA 0.10 mg(Pb)/m^{3*}
Maleic Anhydride	Unsaturated Polyester Resins	Dibasic anhydride	Highly toxic by ingestion and inhalation; causes pulmonary edema, abdominal pain and vomiting; severe skin and eye irritant; potential tumorigen	TWA 0.25 ppm	TLV-TWA 0.25 ppm$^+$
	Epoxy Resins	Cure agent			
	Alkyl Resins	Polybasic acid			
2-Mercaptoethanol	Acrylic Resins	Chain transfer agent	Highly toxic by ingestion, inhalation, and skin absorption; causes myocardial diseases, cardiovascular diseases, and arterslosclerosis in test animals	No Standard	No Recommendation
Mercuric Bromide	Polyamide Resins	Nucleating agent	Highly toxic by ingestion, inhalation, and skin absorption upon acute or chronic exposure; critical organs are the kidney and intestional tract	No Standard	TLV-TWA 0.1 mg/m^3 (as Hg)$^+$ TWA 0.05 mg(Hg)/m^{3*}
Mercuric Chloride	Polyamide Resins	Nucleating agent	Highly toxic by ingestion, inhalation, and skin absorption upon acute or chronic exposure; critical organs are the kidney and intestinal tract	No Standard	TLV-TWA 0.1 mg/m^3 (as Hg)$^+$ TWA 0.05 mg(Hg)m^{3*}
Methyl Acrylate	Polyvinyl Chloride	Comonomer	Highly toxic by ingestion, inhalation, and skin absorption upon acute exposure; may cause lethargy and convulsions; severe skin, eye, and respiratory irritant	TWA 10 ppm	TLV-TWA 10 ppm$^+$ (skin)
Methyl Ethyl Ketone Peroxide	Unsaturated Polyester Resins	Catalyst; cure agent	Highly toxic by ingestion and inhalation upon acute exposure; affects gastrontestinal tract; potential tumorigen	No Standard	CL 0.2 ppm$^+$

†IARC Determination.

(continued)

TABLE 10 (continued)

Chemical	Polymers	Chemical Function	Major Concern	OSHA Air Regulation	NIOSH (*) or ACGIH (†) Recommendations
Methylene Bisphenyl Isocyanate (MDI)	Polyurethane Foam	Isocyanate	Toxic by ingestion, inhalation, and skin absorption; may cause systemic damage; severe local irritant	TWA 0.02 ppm	TLV-TWA 0.02 ppm†
Ozone	Fluoropolymers	Suspension stabilizer	Acutely toxic to lungs by inhalation; severe irritant to skin, eyes, and mucous membranes; potential tumorigen, mutagen, and teratogen	TWA 200 g/m^3	TLV-TWA 100 ppb† STEL 300 ppb†
Phenol	Phenolic Resins	Monomer	Acutely and chronicly toxic by ingestion, inhalation, and skin absorption; may cause renal and heptatic damage; potential tumorigen, mutagen, and teratogen	TWA 5 ppm	TLV-TWA 5 ppm† STEL 10 ppm† TWA 20 mg/m^3* CL 60 mg/m^3/15M*
	Polyurethane Foam	Blocking agent			
	Epoxy Resins	Cure accelerator			
	Polycarbonate	Molecular weight regulator			
Phosgene	Polycarbonate	Monomer	Strong local irritant; extremely toxic by inhalation—short exposures at low concentrations can cause death	TWA 0.1 ppm	TLV-TWA 0.1 ppm† TWA 0.1 ppm* CL 0.2 ppm/15M*
Picric Acid	Unsaturated Polyester Resins	Polymerization inhibitor	Extremely toxic by ingestion, inhalation, and skin contact of dust of picric acid or its salts; may cause nausea, gastroenteritis, nephritis, and acute hepatitis	TWA 0.1 mg/m^3	TLV-TWA 0.1 mg/m^3† STEL 300 g/m^3†
Piperidine	Epoxy Resins	Cure agent	Highly toxic by ingestion and skin absorption; moderately toxic by inhalation; may affect central nervous systems, pulponary system, liver, and kidney; potential mutagen and teratogen	No Standard	No Recommendations
Polypropylene	PP	Monomer	Known animal carcinogen	No Standard	No Recommendation
	Polybutylene	Transformation aid			
	Amino Resins	Self extinguishing agent			
	Epoxy Resins	Filler			

†IARC Determination.

(continued)

TABLE 10 (continued)

Chemical	Polymers	Chemical Function	Major Concern	OSHA Air Regulation	NIOSH (*) or ACGIH (+) Recommendations
Polytetrafluoro-ethylene	Polyamide Resins	Self extinguishing agent	Known animal carcinogen†	No Standard	No Recommendation
Polyurethane	Polyamide Resins	Comonomer	Known animal carcinogen†	No Standard	No Recommendation
Polyvinyl Alcohol	PVC; PS; ABS; Acrylic Resins; Polyvinyli-dene Chloride	Suspension stabilizer	Known animal carcinogen†	No Standard	No Recommendation
	PVAc	Protective colloid			
Polyvinyl Chloride	Acrylic Resins	Self extinguishing agent	Suspected human carcinogen†	No Standard	No Recommendation
	Epoxy Resins	Modifier			
Pyrocatechol	Polyurethane Foam	Blocking agent	Highly toxic depending on route of administration; affects similar to affects of phenol	No Standard	TLV-TWA 5 ppm†
Pyrogallol	Unsaturated Polyester Resins	Polymerization inhibitor	Highly toxic by ingestion, inhalation, and skin absorp-tion; causes methemoglobinemia, hemolysis, and renal injury; potential tumorigen and mutagen	No Standard	No Recommendation
Resorcinol	Phenolic Resins	Comonomer	Highly toxic by ingestions; oral exposure of 29 mg/kg can cause death; potential tumori-gen and mutagen	No Standard	TLV-TWA 10 ppm† STEL 20 ppm†
	Polyurethane Foam	Antioxidant			
Rhodamine B	Acrylic Resins	Colorant	Known animal carcinogen; potential mutagen	No Standard	No Recommendation
Rhodamine 6GDN	Acrylic Resins	Colorant	Known animal carcinogen; potential mutagen	No Standard	No Recommendation
Styrene	ABS; SAN; PS	Monomer	Highly toxic; affects central nervous system; potential tumorigen, mutagen, and tera-togen	TWA 100 ppm CL 200 ppm PK 600 ppm/5M/3H	TLV-TWA 50 ppm† STEL 100 ppm†
	Acrylic Resins; Poly-vinylidene Chloride	Comonomer			
	Unsaturated Polyester Resins	Crosslinking agent			

†IARC Determination.

(continued)

TABLE 10 (continued)

Chemical	Polymers	Chemical Function	Major Concern	OSHA Air Regulation	NIOSH (*) or ACGIH (†) Recommendations
Styrene (Continued)	Alkyd Resins	Modifier			
	Epoxy Resins	Diluent			
	Engineering Thermo-plastics	Polymer for blending			
Tetrahydrofuran	PVC	Solvent	Highly toxic by ingestion and inhalation; affects central nervous system, liver, and kidney	TWA 200 ppm	TLV-TWA 200 ppm† STEL 250 ppm†
Thiophenol	Acrylic Resins	Chain transfer agent	Highly toxic by ingestion, inhalation, and skin absorption; may cause behavioral symptoms, pulmonary system effects, coma and death	No Standard	TLV-TWA 0.5 ppm† CL 0.5 mg/m^3/15M*
Thiourea	PVA	Comonomer	Known animal carcinogen; potential mutagen; toxic; death has occurred when the human exposure level reached 147 mg/kg	No Standard	No Recommendation
Toluene	Acrylic Resins; Epoxy Resins; Engineering Thermoplastics; Poly-vinylidene Chloride	Solvent	Potential tumorigen, mutagen, and teratogen; toxic; causes psychotropic and central nervous system effects	TVA 200 ppm CL 300 ppm PK 500 ppm/10M	TLV-TWA 100 ppm† STEL 150 ppm† TWA 100 ppm* CL 200 ppm/10M*
Toluene-2,4-diisocyanate	Polyurethane Foam	Isocyanate	Toxic by ingestion, inhalation and skin absorption; severe skin and respiratory tract irritant; severe respiratory sensitizer.	CL 0.02 ppm	TLV-TWA 0.005 ppm† STEL 0.02 ppm† TWA 0.005 ppm* Cl. 0.02 ppm*
Trichloroethylene	Polyvinylidene Chloride	Comonomer	Known animal carcinogen; potential mutagen and terato-gen; toxic; may cause death from respiratory failure, liver and kidney lesions, reversible nerve damage; long biological half-life	TWA 100 ppm CL 200 ppm PK 300 ppm/5M/2H	TLV-TWA 50 ppm† STEL 150 ppm† TWA 100 ppm* Cl. 150 ppm/10M*
Tri(2-Chloroethyl) Phosphate	Polyurethane Foam	Flame retardant	Suspected carcinogen	No Standard	No Recommendation

†IARC Determination.

(continued)

TABLE 10 (continued)

Chemical	Polymers	Chemical Function	Major Concern	OSHA Air Regulation	NIOSH (*) or ACGIH (+) Recommendations
Triethylamine	Polyurethane Foam	Catalyst	Highly toxic by ingestion and inhalation; moderately toxic by skin absorption; strong irritant to tissue; potential mutagen	TWA 25 ppm	TLV-TWA 10 ppm[+] STEL 15 ppm[+]
Trimethyl Phosphate	Amino Resins	Cure agent	Known animal carcinogen; potential mutagen and teratogen; toxic by ingestion and inhalation; strong irritant to skin and eyes	No Standard	No Recommendation
Tris(2,3-Dibromo-propyl) Phosphate	Polyurethane Foam	Flame retardant	Suspected human carcinogen†; potential mutagen and teratogen; irritant to skin and eyes; moderately toxic by ingestion	No Standard	No Recommendation
Vinyl Chloride	PVC	Monomer	Known human carcinogen†; highly toxic; may cause death or permanent injury after inhalation even if exposure is short	TWA 1 ppm CL 5 ppm/15M	TLV-TWA 5 ppm[+] TWA 1 ppm[*] CL 5 ppm/15M[*]
Vinylcyclohexene Dioxide	Epoxy Resins	Diluent	Suspected carcinogen†; potential mutagen; strong irritant to skin and tissue; moderately toxic by ingestion and skin absorption	No Standard	TLV-TWA 10 ppm[+]
Vinylidene Chloride	Polyvinylidene Chloride	Monomer	Known animal carcinogen†; potential mutagen and teratogen; highly toxic; causes systemic effects of low concentrations	No Standard	TLV-TWA 5 ppm[+] STEL 20 ppm[+] TWA 1 ppm[*] CL 5 ppm/15M[*]
	Acrylic Resins; PVC	Comonomer			

†IARC Determination.

"Skin" notation indicates material is highly absorbable through the skin.

Source: Registry of Toxic Effects of Chemical Substances, On-Line File, U.S. Department of Health and Human Services, Public Health Service, Center for Disease Control, National Institute for Occupational Safety and Health, Cincinnati, Ohio, updated monthly. Data retrieved September and October 1982.

Engineering Controls

A recent NIOSH study documented engineering controls in use at several polyvinyl chloride manufacturing operations employing each of the five polymerization processes discussed in Section 26.[280] Most polymer production operations are carried out in closed vessels and reactors. The closed process, in itself, is an effective engineering control. Additional effective engineering controls include:

Computerized process control--by reducing the potential for operator error during manual operations, the possibility of significant monomer escape is reduced. Computer control reduces the number of operators needed to run the plant and the amount of time operators actually spend in potential exposure areas.

Specialized equipment--use of certain types of valves, pumps, and agitator, compressor and pump seals may reduce the occurrence of leaks and thus reduce employee exposure potential. Reportedly successful equipment types include butterfly valves with a double blind and bleed system, pressurized, grease-packed agitator seals, nitrogen-pressurized compressor seals, double mechanical seals for pumps and compressors and dual rupture disks.

Enclosed control room--locating process controls in an enclosed, positive pressure control room, supplied with clean, filtered air is a most effective control measure. Employees minimize time spent in the plant and maximize time spent in a clean environment.

Local exhaust ventilation--a local exhaust system flexible enough to control multiple, periodic leaks may be an effective control for highly toxic substances such as vinyl chloride. Permanent exhaust hoods or flexible duct openings may be located near compressors, pumps, valves or other equipment with a high leak potential. Alternatively, portable exhaust systems may be used.

Administrative Controls

Rapid leak detection is an important administrative means of reducing employee exposure to hazardous fugitive emissions. A portable hydrocarbon detector may be used to regularly inspect potential leak sources (e.g. valve stems, pump and compressor seals, blind flanges) to facilitate early leak detector and repair. Continuous, real-time leak detection may be accomplished via a process gas chromatograph monitoring system. Air samples are taken periodically at several locations throughout a plant; individual readings greater than a predetermined level activate an alarm in the control room.

Another effective administrative control is the development and use of standard operating procedures. In performing any operation in the reaction area, employees must follow a checklist step-by-step. Each step must be signed off, with critical steps verified by at least two employees.

Employee training is very important. Workers should be knowledgeable about the nature of the safety and health hazards in the plant and the provisions that are available to them for protection against the hazards. They must be educated in the proper use and reason for each protective device or procedure required. Employee knowledge has particular significance for hazards involving skin contact with carcinogenic compounds, where good personal hygiene is essential.

Actual contaminant levels in the plant can be determined only through implementation of a comprehensive employee monitoring program. Employees typically wear personal sampling pumps or monitoring badges throughout an average work shift.

Personal sampling pumps can be used to collect samples for many different airborne contaminants. The sampling pump is worn by the employee and is connected to an appropriate sample collection medium which is positioned in the employee's breathing zone. NIOSH has developed sampling techniques with recommendations for collection medium, pump flow rate, and sample volume. Additional sampling techniques can be found in the Hygienic Guide Series for individual substances, published by the American Industrial Hygiene Association, and in NIOSH criteria documents for individual substances.

Monitoring badges (passive dosimeters) are available for measuring employee exposure to organic vapors and some inorganic gases. These badges are easy to use and allow samples to be collected by non-technical personnel. Subsequent analysis is done by gas chromatography (organic vapors) or colorimetric methods (inorganic gases).

Personal Protective Equipment and Clothing

Use of personal protective equipment and clothing minimizes employee exposure to hazardous materials in situations where engineering or administrative controls cannot reduce exposure to reported safe levels. The degree of protection necessary depends on the types of hazardous materials in the workplace environment and the conditions under which an employee may be exposed.

Dermal protection minimizes or prevents contact of the skin with potentially hazardous materials which may be absorbed into the body, cause contact burns (chemical or physical), skin lesions, or carcinomas. Use of a removable protective layer also provides a means for removing contaminants from employees before they leave work, thus minimizing exposure to persons outside the plant. Coveralls, gloves, aprons and other dermal protection devices must be constructed of materials which will not allow hazardous material penetration. Eye protection is necessary for employees involved in operations such as handling input chemicals, taking process samples, performing maintenance and machining or grinding. ANSI Z87.1 gives specifications for protective eyewear used in a variety of situations.

Respiratory protection is important because the potential exists in some plastics manufacturing plants for conditions resulting in immediate respiratory failure or long term damage to internal organs. Use of respiratory protection is not recommended as a substitute for other control methods, but should be used in emergency conditions or in nonroutine operations where protection can be afforded in no other way.

Selection of the proper type of respiratory protection may be a complex task. This is due to multiple potential hazards encountered in particular activities, as well as the large number and variety of respiratory protective devices available.

Respirators are of either the air-purifying type or atmospheric supplying type. Air-purifying respirators are limited in they can only be used in atmospheres with sufficient oxygen. Additionally, they are specific for particular substances or classes of substances. In contrast, atmosphere supplying respirators provide the breathing atmosphere and are not dependent on either of the two above factors. However, atmosphere supplying respirators are sometimes cumbersome because they must have an air supply attached.

Common air-purifying respirators draw air through cannisters or cartridges to clean the contaminant from the air before it is taken into the facepiece for breathing. Typical cartridges or cannisters are specific for organic vapors, acid gases, or dusts, mists, and fumes. Air purifying respirators with various facepiece styles (full facepiece, half facepiece) have upper limits of contaminant concentrations, established by NIOSH and used in areas or operations where the exposure level is known to be below the ceiling allowed for the respirator, and where no chance exists of encountering an oxygen deficient atmosphere. Additionally, this type of respirator should be used only in areas where no significant concentrations of toxic gases are present for which the respirator may be ineffective.

In oxygen-deficient atmospheres, employees should wear atmosphere supplying respirators with full facepieces. Either a supplied air respirator or self contained breathing apparatus (SCBA) is an appropriate choice. Supplied air (air line) respirators may be preferrable for this type of work if an air supply connection is available and if the work is limited to a single area without need for movement over great distances. A standard supplied air respirator (preferrably positive pressure or pressure demand), certified by NIOSH or MSHA for work in this type of atmosphere is recommended for use in areas where exposure levels remain below those "immediately dangerous to life or health".

A typical SCBA unit is bulky, heavy, and usually can supply air for a maximum of 30-60 minutes. This type of respirator should be used for emergency responses to unknown conditions, or entry and work where movement over large areas is required.

Summary

The plastics and resins producers comprise a relatively large segment of the United States chemical manufacturing capability. Although the economic downturn of the late seventies and early eighties has reduced the production of most polymers, the market is still strong for most plastic materials. Their ease of handling, energy efficient processability, and superior qualities have established a place for them in the market.

The plastics and resins industry generates air emissions, wastewaters, and solid wastes. Air emissions are either particulates or volatiles. Solution polymerization processes typically generate more volatile emissions than mass (bulk), suspension, or emulsion. Wastewaters are primarily from suspension and emulsion polymerization processes, both of which use large quantities of water. Solid wastes are typically polymer lost, such as particulates, or by-products formed during polymerization. Due to the variability in the chemicals used in each of these processes and the wide range of operating parameters, no general form of environmental control for any of these processes is applicable.

A wide range of chemicals is used in the manufacture of plastics and resins. The potential hazards to workers depends upon process type, engineering controls, practices, and, of course, the chemicals used. The methods outlined in the previous section list the types of control and practices used to minimize any worker health problems.

2. Acrylic Resins

Acrylic resins are water-clear polymers which exhibit excellent stability upon aging. Acrylic monomers are very versatile, polymerizing and copolymerizing with a variety of other monomers with relative ease. Polymethyl methacrylate is the major polymer produced from the acrylic family. Methyl methacrylate is the starting block for almost all acrylic polymers. [3, 140, 286] Other acrylics include: methyl acrylate, ethyl acrylate, butyl acrylate, 2-ethylhexyl acrylate, 2-ethoxyethyl acrylate, acrylic acid, ethyl methacrylate, butyl methacrylate, 2-hydroxypropyl methacrylate, ethyl 2-hydroxymethyl acrylate, t-butylaminoethyl methacrylate, dimethylaminoethyl methacrylate, glycidyl methacrylate, and methacrylic acid.

Polymethyl methacrylate (PMMA) is a crystal clear, lightweight, durable polymer. PMMA properties can be varied to form extremely tacky adhesives, rubbers, tough plastics, or hard powders. When produced in thin sections, PMMA is flexible. The polymer may be cut, turned on a lathe, sawed, drilled, polished, or formed.[247] PMMA is the most resistant of all transparent plastics to ultraviolet radiation, and moisture. PMMA polymers with varying degrees of flexibility are achieved by incorporating various copolymers into the resin.

The outstanding transparency, light piping qualities, colorability, and dimensional stability of PMMA coupled with the retention of these properties in outdoor applications over long periods of time make this resin useful in many industries. PMMA is used for aviation parts, including pilots' canopies and windows on commercial aircraft, because it is free from optical distortion, resistant to light and weathering, resistant to shattering, and able to withstand sharp pressure and temperature differentials.[140] Other uses include safety glass interlaying, glazing, dentures, contact lenses, and various coatings and finishes.

PMMA properties are in part a function of the processing step used to produce the finished plastic product. PMMA in bulk form starts an auto-acceleration reaction at about 50 percent conversion. Therefore, complete conversion is not attained until the polymer is processed, for example cast or molded. Properties of cast and molded PMMA are listed in Table A-1 in Appendix A.

Modifiers are used in PMMA production to improve impact strength, give higher service temperatures, increase toughness, and add extinguishing properties. Impact modified resins exhibit lessened stiffness, transparency, weather stability, and abrasion resistance when compared to unmodified resins. However, in addition to improved impact strength, the modified resins also have harder surfaces, lower water absorption, greater resistance to staining, and better color stability. Table A-2 of Appendix A lists properties of impact modified compounds.

PMMA is not affected by alkalis, oils, dilute acids, hydrochloric acid, sulfuric acid, or dilute alcohols. This resin is soluble in polar solvents, such as ketones, esters, ethers, and aromatic hydrocarbon/alcohol mixtures.

Thermosetting PMMA are used for bake coatings. The PMMA resin is postreacted with formaldehyde and butanol. A cure agent, such as styrene, is added to provide crosslinking during curing. These resins vary from very flexible coatings, which may be applied by rollcoating to tinplate or aluminum stock used for cans or siding, to sprayable, chemically resistant, very hard enamels for household appliances. Uncured, these resins are hard, brittle, weak, and lack chemical and solvent resistance. Upon curing, they become tough, flexible, impact resistant, very hard, very insoluble, and resistant to solvents, alkalis, acids, and detergents.

PMMA is produced by mass, solution, suspension, or emulsion polymerization. Bulk polymerization is carried to less than 50 percent conversion due to the autoacceleration that occurs between 20 and 50 percent. The resulting resin is usually cast. Solution resins are used in the preparation of coatings, adhesives, impregnates, and laminates. Suspension resins are used as molding powders or ion exchange resins, while emulsion resins are used in the paint, paper, textile, floor polish, and leather industries for coatings and binders.

PMMA solid resins are molded, extruded, or cast sheet. Finishing processes used include hot stamping, spray painting, and vacuum metalizing. Higher molecular weight molding resins have a lower melt flow rate, greater hot strength during processing, and improved resistance to cracking during ejection. Lower molecular weight molding resins have a higher melt flow rate and are designed for complex parts with hard to fill molds.[3] When heated, PMMA is a tough, pliable polymer that is easily bent or formed into complex shapes that can be molded or extruded.

INDUSTRY DESCRIPTION

Fifty-four producers with 104 sites in 27 states comprise the acrylic resin industry; PMMA is the major acrylic product. A preponderance of sites (33) are located in the Great Lakes States (EPA Region V). Over 10 percent are located in each of the following regions: New England (EPA Region I), New York and New Jersey (EPA Region II), Mid-Eastern States (EPA Region III), Southeastern States (EPA Region IV), and the Southwestern States (EPA Region IX). Acrylic resin producers and locations are listed in Table 11. Plant capacities are considered to be proprietary.

TABLE 11. U.S. ACRYLIC RESIN PRODUCERS

Producer	Location
ADCO Chemical Co., Inc.	Newark, NJ
American Cyanamid Co. Industrial Chemicals Division	 Wallingford, CT
AZS Corp. AZS Chemical Co. Division	 Atlanta, GA

Beatrice Foods Co.
 Beatrice Chemical Division
  Farbroil Co. Division
  Polyvinyl Chemical Industry Division
  Stahl Finish Division
	
	Baltimore, MD
	Wilmington, MA
	Peabody, MA

Borden Inc. Borden Chemical Division Thermoplastics Products	 Compton, CA Illiopolis, IL Leominster, MA
Celanese Corp. Celanese Plastics and Specialties Co. Division Celanese Specialty Resins Division	 Los Angeles, CA Louisville, KY
Chemical Products Corp.	Elmwood Park, NJ
Cook Industrial Coatings Co.	Detroit, MI
Cook Paint and Varnish Co.	Milpitas, CA North Kansas City, MO
Crown-Metro, Inc.	Greenville, SC
The Derby Co., Inc.	Ashland, MA
DeSoto, Inc.	Chicago Heights, IL Garland, TX
The Dexter Corp. Midland Division	 Cleveland, OH Hayward, CA Rocky Hill, CT Waukegan, IL

(continued)

TABLE 11 (continued)

Producer	Location
Dock Resins Corp.	Linden, NJ
E. I. DuPont de Nemours & Co., Inc. Fabrics and Finishes Department	Chicago, IL Flint, MI Fort Madison, IA Front Royal, VA Parlin, NJ Philadelphia, PA S. San Francisco, CA Toledo, OH
Polymer Products Department (Lucite®)	Parkersburg, WV
Dutch Boy, Inc. Coatings Group	Baltimore, MD
H. B. Fuller Co.	Atlanta, GA Blue Ash, OH
General Latex and Chemical Corp.	Ashland, OH Cambridge, MA Charlotte, NC Dalton, GA
The BF Goodrich Co. BF Goodrich Chemical Group	Avon Lake, OH

W. R. Grace & Co.
  Industrial Chemicals Group
  Organic Chemicals Division
	Owensboro, KY
	South Acton, MA

Guardsman Chemicals, Inc.	Grand Rapids, MI

Gulf Oil Corp.
  Millmaster Onyx Group, subsidiary
    Lyndal Chemical Division
	Lyndhurst, NJ

Hanna Chemical Coatings Corp.	Columbus, OH
Hart Products Corp.	Jersey City, NJ
Henkel of America, Inc. Henkel Corp., subsidiary	Kankakee, IL

(continued)

TABLE 11 (continued)

Producer	Location
Hercules Inc.	Clairton, PA
Hugh J.-Resins Co.	Long Beach, CA
S. C. Johnson & Son, Inc.	Racine, WI
Komac Paint, Inc.	Denver, CO
Minnesota Mining and Manufacturing Co. Chemical Resources Division	St. Paul, MN
Mobay Chemical Corp. Dyes and Pigments Division	Bayonne, NJ

Mobil Corp.
  Marcor Inc.
    Montgomery Ward & Co., Inc., subsidiary
      Standard T Chemical Co., Inc., subsidiary
	Chicago Heights, IL
	Staten Island, NY

Mobil Oil Corp.
  Mobil Chemical Co. Division
    Chemical Coatings Division
	Rochester, PA

Morton-Norwich Products, Inc. Morton Chemical Division	Ringwood, IL

National Starch and Chemical Corp.
  Resins Division
    Charles S. Tanner Division
	Meredosia, IL
	
	Enoree, SC

Norris Paint Co., Inc.	Salem, OR

The O'Brien Corp.
  The O'Brien Corp.-Central Region
  The O'Brien Corp.-Western Region
	South Bend, IN
	S. San Francisco, CA

Phillip Morris, Inc.
  Polymer Industries, Inc., subsidiary
    Adhesives and Liquid Coatings Division
    Textile Chemicals Division
	Stamford, CT
	Greenville, SC

(continued)

TABLE 11 (continued)

<u>Producer</u>	<u>Location</u>
PPG Industries, Inc.	
Coatings and Resins Division	Circleville, OH
	Oak Creek, WI
	Torrance, CA
Purex Corp.	Carson, CA
K. J. Quinn & Co., Inc.	
Polymer Division	Malden, MA
	Seabrook, NH
Raffi and Swanson, Inc.	
Polymeric Resins Division	Wilmington, MA
Reichhold Chemicals, Inc.	Detroit, MI
	Elizabeth, NJ
	S. San Francisco, CA
Emulsion Polymers Division	Cheswold, DE
Reichhold Chemicals Del Caribe, Inc., subsidiary	Rio Piedras, PR
H. H. Robertson Co.	
Freeman Chemical Corp., subsidiary	Chatham, VA
	Saukville, WI
Rohm & Haas Co.	Bristol, PA
	Croydon, PA
Rohm & Haas California Inc., subsidiary	Hayward, CA
Rohm & Haas Kentucky Inc., subsidiary	Louisville, KY
Rohm & Haas Tennessee Inc., subsidiary	Knoxville, TN
SCM Corp.	
Glidden Coatings & Resins Division	Huron, OH
	Reading, PA
	San Francisco, CA
The Sherwin-Williams Co.	
Coatings Group	Chicago, IL
Sun Chemical Corp.	
Chemicals Group	
Chemicals Division	Chester, SC
Sybron Corp.	
Chemical Division	
Ionac Chemical Co. Division	Birmingham, NJ
Jersey State Chemical Co. Division	Haledon, NJ

(continued)

TABLE 11 (continued)

Producer	Location
Textron Inc. Spencer Kellogg Division	Newark, NJ
Union Oil Co. of California Union Chemicals Division	Bridgeview, IL Charlotte, NC Kearny, NJ Lemont, IL
United Merchants & Manufacturers, Inc. Valchem-Chemical Division	Langley, SC
United Technologies Corp. Inmont Corp., subsidiary	Anaheim, CA Cincinnati, OH Detroit, MI Greenville, OH
Valspar Corp. McWhorter, Inc., subsidiary	Baltimore, MD Carpentersville, IL
Yenkin-Majestic Paint Corp. Ohio Polychemicals Co. Division	Columbus, OH

Source: _Directory of Chemical Producers_, 1982.

Acrylic resins are used in many consumer products. From 1975 to 1979, acrylic resin production increased from 352,000 metric tons to 519,000 metric tons.[65] The production fell from this peak to 247,000 metric tons in 1980, with 1981 production increasing to 256,000 metric tons.[143] The excess capacity exhibited by 1979 production figures of at least 263,000 metric tons is sufficient to accommodate any industry growth in the near future.

Low level additives, including surfactants used in the manufacture of PMMA latexes, may be toxic. The following inputs to the PMMA process are listed as hazardous: acetone, acrylic acid, acrylonitrile, benzene, carbon tetrachloride, chlorobenzene, chloroform, cumene, cyclohexane, ethyl acetate, ethyl acrylate, isobutyl alcohol, methyl methacrylate, toluene, trichloroethylene, vinyl chloride, and vinylidene chloride.

The comonomer vinyl chloride is a known human carcinogen. Suspected human carcinogens used in this process include the comonomer acrylonitrile, the self-extinguishing agent polyvinyl chloride, the solvent benzene, and the chain transfer agents chloroform and carbon tetrachloride. Potential human carcinogens include the suspending agent polyvinyl alcohol, the chain transfer agent trichloroethylene, the comonomer vinylidene chloride, and the colorants cadmium sulfide, Rhodamine B, and Rhodamine 6GDN. Other inputs which pose a significant health risk are styrene, ethyl acrylate, acrylic acid, N,N-dimethylaniline, toluene, triethylamine, 2-mercaptoethanol, and thiophenol.

Solution polymers, molding powders, and plastic sheet are considered to be flammable materials; dispersion polymers are not when water is present. Closed kettle reactors reduce both the exposure and flammability hazard associated with the polymerization process. Dust control also reduces the explosive potential in the handling of powders or the manufacture of plastic sheet. Acrylic products require the same fire precautions as wood; they must comply with building codes, meet applicable Underwriters Laboratory (UL) standards, and be installed while observing established principles of fire safety.[3] Lighting fixture lenses and diffusers are freely mounted so that they fall prior to ignition in case of fire, preventing fire spread to the ceiling.[3]

PRODUCTION AND END USE DATA

In 1981, acrylic resin production totaled 256,000 metric tons.[143] These resins have many uses, including:[3, 120, 121, 140, 247]

- Airplane windows, cockpit closures, aviation canopies, radar plot boards, instrument panels, and landing light covers;

- Sighting devices, photographic equipment, wide-angle TV screens, storm doors and windows, solar collector panels, safety glass interlaying, complex reflex lenses, bank tellers' windows, panels around hockey rinks, bath and shower enclosures, show cases, cast plastic eyeglass lenses, Fresnel lenses and Fresnel lens film, windshields for small boats, safety guards for tools and machinery, push buttons and panels for business and vending machines;

- Automotive parts, including tail lights, dials, instrument panels, and lenses;

- Lighting refractors for fluorescent and mercury-vapor lamp sources, and indoor and outdoor lighting and signs;

- Thermosetting resins for upgrading polyester systems, impregnating varnishes, electrical and electronic encapsulating and plotting compounds, automotive glazes, and glazes for windows;

- Domes over stadiums, pools, and tennis courts, and glazed archways between buildings;

- Cosmetic and food packaging, dentures and denture bases, tooth filling materials, orthodontic sealants, dental and surgical bindings and filling agents which adhere to bones and teeth, and hard and soft contact lenses;

- Plastic sanitary fixtures, thermoformed bathtubs, toys, transparent drawers for storage and display of goods, racks and holders, countertop displays, buttons and bars on typewriters, business machine parts, pump parts, and piano keys;

- Leather finishes for furniture, auto upholstery, garments, and shoes;

- Coatings for stucco, cinder block, concrete, asbestos shingle, and exterior wood shingle, heat resistant enamels, aluminum paints, fumeproof enamels, luminescent and phosphorescent finishes, furniture lacquers, vinyl printing inks, vinyl film topcoats, and military aircraft paint;

- Paper uses, including pigment binders, saturants, fiber binders, coatings for paperboard and letter-press, offset, and rotogravure painting papers (such as frozen-food containers, bread wrap, folding boxboard, greaseproof paper, wallpaper, business machine paper, and reproductive paper), artificial leather, pressure sensitive tape, window shades, and gaskets;

- Floor polishes, flooring products, ceramics, mortar cements, protective and decorative overlay on wood, masonite, plywood, and other surfaces, and wire coatings; and

- Cast furniture, skylights, and outdoor table tops.

Table 12 lists 1981 consumption of acrylic resins.

TABLE 12.   1981 CONSUMPTION OF ACRYLIC RESINS

Market	Thousand Metric Tons
Cast Sheet	82
Molding and Extrusion Compounds	71
Other Grades	32
Coatings	46
Other (Emulsion Polymers, Transesterification Resins, etc.)	25
TOTAL	256

Source: _Modern Plastics_, January 1982.

PROCESS DESCRIPTIONS

PMMA is produced by mass, solution, suspension, or emulsion polymerization. The process used depends upon the desired end product. Mass polymerization resins are used mainly for cast products; solution polymerization is used to manufacture coatings and binders. Suspension polymerization resins are tiny beads which are used as molding compounds while emulsion polymerization latexes produce ingredients for paints.

Acrylic monomers have the general formula $CH_2=CHCOOR$. Methyl methacrylate resins are a specific type of acrylic resin which have a methyl group substituted for the α-hydrogen, $CH_2=CCH_3COOR$. This α-methyl group gives methacrylate polymers increased stability, hardness, and stiffness due to the stereochemistry of the molecule.[121]

PMMA is produced by polymerization of the methyl methacrylate monomer with a radical initiator ($R\bullet$), commonly benzoyl peroxide. The initiator combines with a methyl methacrylate monomer to produce a methyl methacrylate radical. This new radical then combines with other monomer molecules to form the PMMA chain. The reaction is terminated either when two radical groups combine or by disproportionation. This series of reactions is shown below.

<u>Initiation</u>

$$CH_2=\underset{\underset{COOCH_3}{|}}{\overset{\overset{CH_3}{|}}{C}} \quad + \quad \overset{\bullet}{R} \longrightarrow R-CH_2-\underset{\underset{COOCH_3}{|}}{\overset{\overset{CH_3}{|}}{C}}\bullet$$

<u>Propagation</u>

$$R-CH_2-\underset{\underset{COOCH_3}{|}}{\overset{\overset{CH_3}{|}}{C}}\bullet \quad + \quad CH_2=\underset{\underset{CH_3}{|}}{\overset{\overset{COOCH_3}{|}}{C}} \longrightarrow R-CH_2-\underset{\underset{COOCH_3}{|}}{\overset{\overset{CH_3}{|}}{C}}-CH_2-\underset{\underset{CH_3}{|}}{\overset{\overset{COOCH_3}{|}}{C}}\bullet$$

<u>Termination</u>

Radical Combination

$$R-CH_2-\underset{\underset{COOCH_3}{|}}{\overset{\overset{CH_3}{|}}{C}}-CH_2-\underset{\underset{CH_3}{|}}{\overset{\overset{COOCH_3}{|}}{C}}\cdot \quad + \quad R-CH_2-\underset{\underset{COOCH_3}{|}}{\overset{\overset{CH_3}{|}}{C}}\cdot \longrightarrow$$

$$R-CH_2-\underset{\underset{COOCH_3}{|}}{\overset{\overset{CH_3}{|}}{C}}-CH_2-\underset{\underset{CH_3}{|}}{\overset{\overset{COOCH_3}{|}}{C}}-CH_2-\underset{\underset{COOCH_3}{|}}{\overset{\overset{CH_3}{|}}{C}}-R$$

<u>Disproportionation</u>

$$R-CH_2-\underset{\underset{COOCH_3}{|}}{\overset{\overset{CH_3}{|}}{C}}-CH_2-\underset{\underset{CH_3}{|}}{\overset{\overset{COOCH_3}{|}}{C}}\cdot \quad + \quad R-CH_2-\underset{\underset{COOCH_3}{|}}{\overset{\overset{CH_3}{|}}{C}}-CH_2-\underset{\underset{CH_3}{|}}{\overset{\overset{COOCH}{|}}{C}}\cdot \longrightarrow$$

$$R-CH_2-\underset{\underset{COOCH_3}{|}}{\overset{\overset{CH}{|}}{C}}-CH_2-\underset{\underset{CH_3}{|}}{\overset{\overset{COOCH}{|}}{C}}-R \quad + \quad CH_3-\underset{\underset{COOCH_3}{|}}{\overset{\overset{CH}{|}}{C}}-CH=\underset{\underset{CH_3}{}}{\overset{\overset{COOCH}{}}{C}}$$

This series of reactions is inhibited by oxygen. Oxygen becomes a comonomer
in the polymerization process, decreasing the rate of reaction and the
molecular weight of the polymer. This copolymerization reaction is shown
below.

$$R-CH_2-\underset{\underset{COOCH_3}{|}}{\overset{\overset{CH_3}{|}}{C}}\cdot \quad + \quad O_2 \longrightarrow R-CH_2-\underset{\underset{COOCH_3}{|}}{\overset{\overset{CH_3}{|}}{C}}OO\cdot$$

PMMA production is performed in an inert gas atmosphere to minimize this
copolymerization.

Tables 13 and 14 list typical input materials and operating parameters for the four PMMA production processes. Other input materials used in PMMA production such as buffers, lubricants, pigments, and comonomers are listed in Table 15.

Mass Polymerization

Mass polymerization is the major process used for the production of PMMA casting resin. Mass polymerized PMMA may be discharged directly from the polymerization reactor to the casting process. In contrast to solution, suspension, or emulsion polymerization, no solvent removal or dewatering is necessary prior to casting the resin.

Mass polymerization using a batch process is depicted in Figure 1. Methyl methacrylate, any comonomer desired, UV absorbers, pigments or dyes, additives, and an initiator are added to the reactor. Higher acrylates are often added to soften and improve the flexibility of the resin without decreasing durability. An inert gas is sometimes introduced to prevent oxygen from slowing the reaction while heat is added to increase the reaction rate. When the monomer/polymer mixture has reached the desired conversion, the mass is discharged from the reactor and processed to form the desired product, such as cast sheets, rods, or tubes.

Methyl methacrylate starts an autoacceleration reaction when the PMMA conversion reaches 20 to 50 percent. There is a sudden increase in molecular weight, increase in viscosity, and decrease in termination. Therefore, conversions for this process are kept low. The monomer/polymer mixture is then processed with the processing temperature completing the polymerization.

A typical recipe for mass polymerization of PMMA is given below [73]:

Material	Parts by Weight
Methyl Methacrylate	2,703
Benzoyl Peroxide (initiator)	13.5
Dibutyl Phthalate (plasticizer)	81.1
UV Absorber	27
Pigment or Dyes	
Lubricant	
Ethylene Glycol Dimethacrylate (comonomer)	13.8

Solution Polymerization

Solution polymerization is used in the preparation of PMMA for coatings, adhesives, impregnates, and laminates. The solvent, such as benzene or toluene, is chosen on the basis of cost, toxicity, flammability, volatility, and chain transfer. Thiols and chlorohydrocarbons exhibit high chain transfer; therefore, they are rarely used as solvents but are used in concentrations of 0.1 to 5 percent as chain transfer agents.

TABLE 13.   TYPICAL INPUT MATERIALS TO PMMA PRODUCTION PROCESSES
IN ADDITION TO MONOMER AND INITIATOR

Process	Chain Transfer Agent	Suspending Agent	Water	Solvent	Emulsifier
Mass	X				
Solution	X			X	
Suspension	X	X	X		
Emulsion	X		X		X

TABLE 14.   TYPICAL OPERATING PARAMETERS FOR PMMA PRODUCTION PROCESSES

Process	Temperature	Reaction Time	Conversion
Mass	30 – 90°C	8 – 24 hours	20 – 50%
Solution	136 – 143°C	6 – 8 hours	>95%
Suspension	30 – 75°C	3 – 6 hours	>95%
Emulsion	25 – 90°C	>2.5 hours	>95%

Sources: Encyclopedia of Chemical Technology, 3rd Edition.

Encyclopedia of Polymer Science and Technology.

TABLE 15. INPUT MATERIALS AND SPECIALTY CHEMICALS USED IN PMMA
MANUFACTURE IN ADDITION TO MONOMER

Function	Chemical
Buffer	Sodium hydrogen phosphates

Chain Transfer Agents	Chlorohydrocarbons
	  Chlorobenzene
	  Butyl chloride
	  Chloroform
	  Carbon tetrachloride
	Carbon tetrabromide
	Lauryl mercaptan
	Mercapto acids
	t-dodecyl mercaptan
	Thiols
	  Ethanethiol
	  2-propanethiol
	  1-butanethiol
	  2-methyl-2-propanethiol
	  2-mercaptoethanol
	  Ethyl mercaptoacetate
	  Thiophenol
	  2-naphthalenethiol
	Trichloroethylene

Colorants White	Titanium dioxide (alone or with barium sulfate, zinc oxide, or pearl essence)
Black	Channel black Furnace black
Red, Orange	Aminoanthraquinone dyes Cadmium sulfoselenides Iron oxides Rhodamine B Rhodamine 6 GDN
Yellow	Cadmium sulfide Hansa yellow R Quinoline yellow D
Green	Phthalocyanide greens
Blue	Alizarine sky blue B Indanthrene blue GCD Phthalocyanide blue (noncrystalline) Ultramarine blue

(continued)

TABLE 15 (continued)

Function	Chemical
Comonomers	Acrylic acid
	Acrylic esters
	Acrylonitrile
	Allyl esters of dicarboxylic acids
	Allyl ethers of dihydric alcohols
	Allyl methacrylate
	α-methyl styrene
	Butadiene
	t-butylaminoethyl methacrylate
	1,4-butylene dimethacrylate
	Dialkyl fumarates
	Dialkyl maleates
	Diallyl phthalate (DAP)
	Dimethylaminoethyl methacrylate
	Divinyl benzene
	Ethyl acrylate
	Glycidyl methacrylate
	Glycol dimethacrylates
	2-hydroxyethyl acrylate
	N-hydroxymethyl acrylamide
	N-hydroxymethyl methacrylamide
	Itaconic acid
	Methacrylate esters
	Methacrylic acid
	Styrene
	Triallyl cyanurate
	Unsaturated polyesters
	Vinyl acetate
	Vinyl chloride
	Vinylidene chloride
	Vinyl toluene
Cure Agent	Benzoyl peroxide and dimethyl aniline
Emulsifiers	
Anionic Surfactants	Alkyl sulfates
	Alkylarene sulfates
	Phosphates
Nonionic Surfactants	Alkyl polyoxyethylenes
	Aryl polyoxyethylenes
Filler	Glass fiber

(continued)

TABLE 15 (continued)

Function	Chemical
Initiators	Ammonium persulfate Azo compounds 2,2'-Azobisisobutyronitrile (AIBN) Benzoyl peroxide Butyl peroxide Cumene hydroperoxide Dicumyl peroxide Di-t-butyl peroxide Hydrogen peroxide
Lubricants	Cetyl alcohol Lauryl alcohol Stearic acid
Redox System Initiators	Peroxydisulfate and sodium bisulfite, sodium thiosulfite, sodium hydrosulfite, or sodium formaldehydesulfoxylate
Self Extinguishing Agent	Polyvinyl chloride (PVC)
Solvents	Acetic acid Acetone Benzene Butanol 2-butanone t-butylbenzene Cumene Cyclohexane Ethyl acetate Ethylbenzene Isobutyric acid Isobutyl alcohol Methyl isobutyrate 3-pentanone 2-propanol s-butyl alcohol t-butyl alcohol Toluene Triethylamine

(continued)

TABLE 15 (continued)

Function	Chemical
Suspending Agents (Protective Colloids)	Cellulose derivatives Disodium phosphate Gelatin Monosodium phosphate Polyacrylate salts Polyvinyl alcohol (PVA) Starch
Suspension or Emulsion Modifier	Glycerol Glycols Inorganic salts Polyglycols
UV Light Stabilizer	o-hydroxybenzophenone

Sources: Encyclopedia of Chemical Technology, 3rd Edition.
          Encyclopedia of Polymer Science and Technology.
          Seymour S. Schwartz and Sidney H. Goodman, Plastics Materials and
          Processes, 1982.

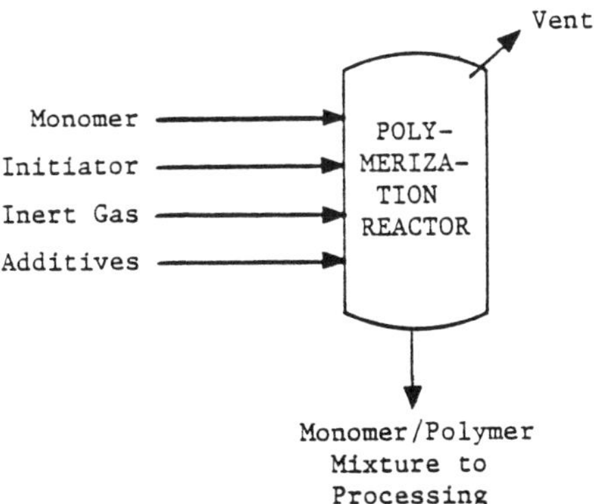

Figure 1. Mass polymerization process.

Source: *Encyclopedia of Chemical Technology*, 3rd Edition.

Solution polymerization is typically a batch process. As shown in
Figure 2, the solvent and an inert gas are charged to the reactor and heated
to the desired temperature. The monomer and initiator (typically 0.01 to 7
percent of the charge) are added via a small continuous stream. When the
monomer addition is complete, the reaction temperature is maintained until
the resin reaches a high conversion (>95 percent). The solvent is refluxed
to the reactor during this process to provide better heat transfer.[247]
When the reaction is complete, the solvent and resin are discharged from the
reactor and diluted, if necessary, to the desired resin concentration. Any
emissions from the resin will result during the production of the end product, such as solvent emissions during solution casting of the resin.

A typical recipe for solution polymerization of an acrylic resin is
given below [73]:

Material	Parts by Weight
Ethyl Acrylate	7,265
Toluene (solvent)	10,897
Benzoyl Peroxide (initiator)	91

Suspension Polymerization

Suspension polymerization resins are produced as tiny beads which are
used as molding powders and ion exchange resins. Acrylate resins are often
added for reduced resin brittleness and improved processibility of molding
compounds. Amino functional monomers, acid functional monomers, divinyl
monomers, or trivinyl monomers are added for crosslinking in ion exchange
resins.

As shown in Figure 3, the polymerization reactor is first charged with
water and an inert gas. The monomer and suspending agent (or protective
colloid) are then added. The monomer is stabilized as tiny droplets in the
water with the suspending agent. When the suspension is stabilized, a
monomer soluble initiator is added to start the reaction. The water acts as
both a dispersion medium and a heat transfer agent. When the beads have
reached a high conversion (greater than 95 percent), the suspension is
cooled, the beads are dewatered, washed, and dried.

A typical recipe for suspension polymerization of PMMA is shown below
[73]:

Material	Parts by Weight
Methyl Methacrylate	2,876
Ethyl Acrylate	508
Water	11,032
Sodium Polyacrylate (protective colloid)	29.25
Anhydrous Disodium Phosphate (suspending agent)	62.25
Stearic Acid (lubricant)	70
Ethyl Crotonate (chain transfer agent)	7
Benzoyl Peroxide (initiator)	35

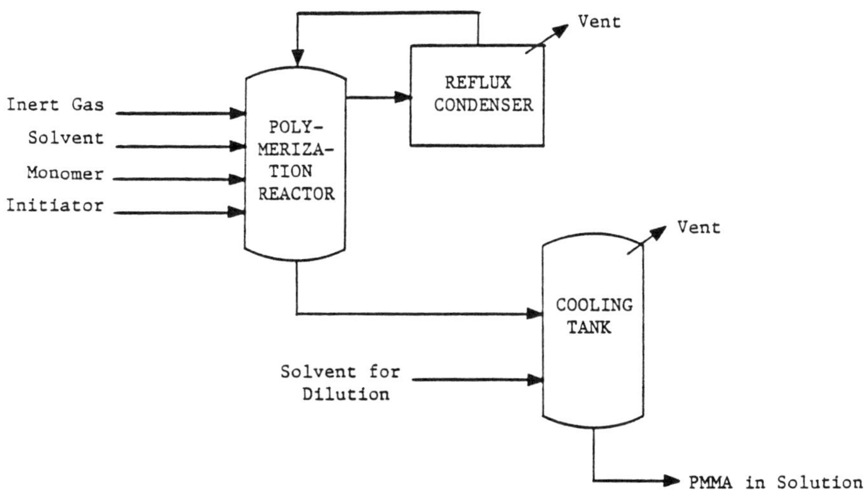

Figure 2.   Solution polymerization process.

Source: *Encyclopedia of Chemical Technology*, 3rd Edition.

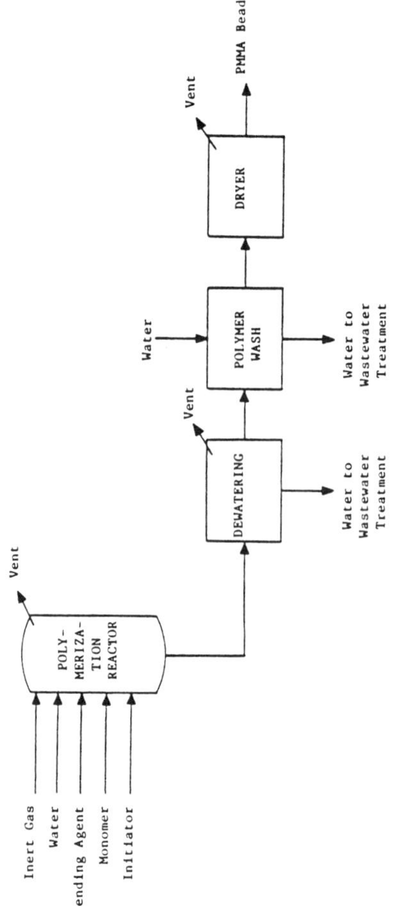

Figure 3. Suspension polymerization process.

Source: Encyclopedia of Chemical Technology, 3rd Edition.

Emulsion Polymerization

Emulsion polymerization using batch processing is the most important industrial method for the preparation of acrylic polymers.[120] This process is used to produce coatings and binders for the paint, paper, textile, floor polish, and leather industries. The safety hazards and the expense associated with the flammable solvents used in solution polymerization are eliminated.

The emulsion polymerization process, illustrated in Figure 4, uses a reactor charged first with water and an inert gas. In a separate mixing tank, water, an emulsifier (typically 1 to 5 percent of the monomer charge), methyl methacrylate, and any comonomer are mixed. The monomer emulsion is then metered into the polymerization reactor where a water soluble initiator has been added. When the monomer addition is complete, the reaction is allowed to proceed to a high conversion (greater than 95 percent) before it is cooled and filtered.

A typical recipe for emulsion polymerization of PMMA is shown below [73]:

Material	Parts by Weight
Ethyl Acrylate	1,460
Methyl Methacrylate	720
Methacrylic Acid (98%) (comonomer)	34
Potassium Persulfate (initiator)	5
Sodium Bisulfite (reducing agent)	5
Water	2,175
Sodium Lauryl Sulfate (30%) (emulsifier)	539

The resulting emulsion typically contains 30 to 60 percent solids and contains particles with size ranging from 0.1 to 1.0 microns. Anionic surfactants give finer particles while nonionic surfactants produce resins with better shelf stability and resistance to salts and freeze-thaw. Blends of surfactants are often used to obtain advantages of both types. Additives such as thickeners, flow improving agents, freeze-thaw stabilizers, pigment dispersants, and preservatives are typically added after polymerization of the resin.[247]

Energy Requirements

No data listing the energy requirements for acrylic polymer production were found in the literature consulted.

ENVIRONMENTAL AND INDUSTRIAL HEALTH CONSIDERATIONS

Acrylic resins are considered to be nontoxic and are used in food packaging, medicine dispensing, and medical applications such as contact lenses and dentures. The resin producers would be adversely impacted by any toxic effect. Acrylic monomers are mild to moderate toxic chemicals, but

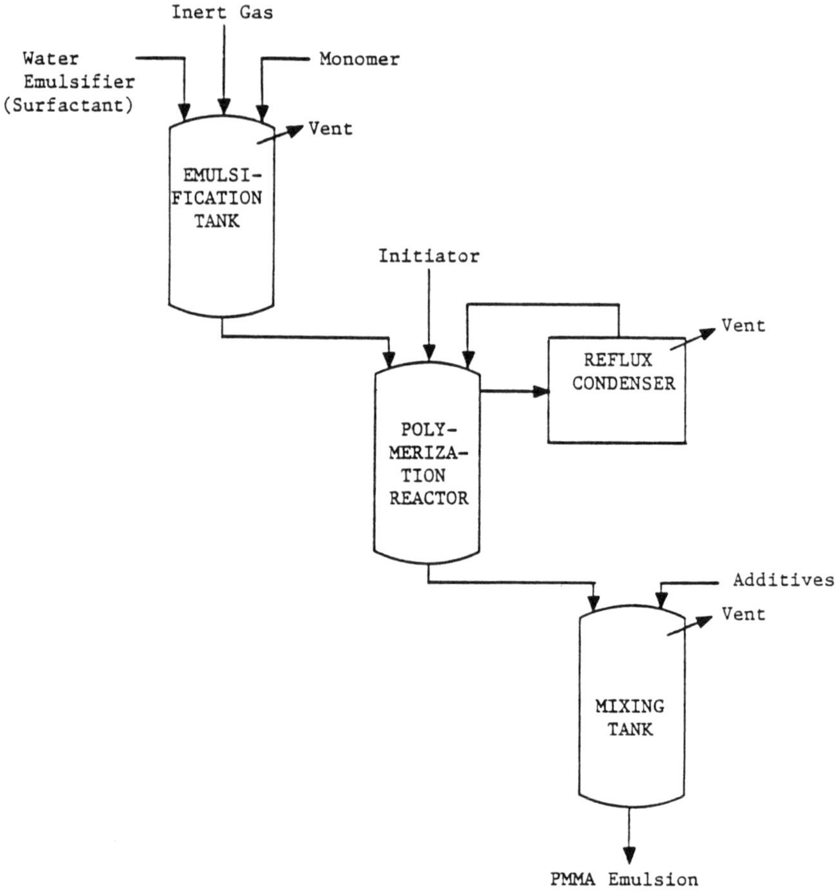

Figure 4. Emulsion polymerization process.

Source: <u>Encyclopedia of Chemical Technology</u>, 3rd Edition.

they may be handled safely by trained personnel.[140] The following compounds are listed as hazardous under RCRA: acetone, acrylic acid, acrylonitrile, benzene, carbon tetrachloride, chlorobenzene, chloroform, cumene, cyclohexane, ethyl acetate, ethyl acrylate, isobutyl alcohol, methyl methacrylate, toluene, trichloroethylene, vinyl chloride, and vinylidene chloride. In addition to the hazardous listing, vinyl chloride is a confirmed human carcinogen, and the suspected human carcinogens acrylonitrile, polyvinyl chloride, benzene, chloroform, and carbon tetrachloride are inputs to this process. Potential human carcinogens used include polyvinyl alcohol, trichloroethylene, vinylidene chloride, and three colorants, Rhodamine B, Rhodamine 6GDN, and cadmium sulfide. Several other chemicals pose a significant health risk: cumene hydroperoxide, styrene, ethyl acrylate, acrylic acid, N,N-dimethylaniline, toluene, triethylamine, 2-mercaptoethanol, and thiophenol.

Solvent, suspension, and emulsion polymerization are carried to over 95 percent conversion. Due to this high conversion, the sources upstream of and including the reactor contribute significantly more to the total VOC emissions than downstream sources. This is not the case, however, with mass polymerization. Since the conversion for this process is between 20 and 50 percent, a significant amount of VOC emissions result from polymer casting, molding, or finishing.

Solution polymerization has added VOC emissions due to the solvent used. Although monomer emissions are reduced because of high conversion, solvent emissions are also present after the reactor.

<u>Worker Distribution and Emissions Release Points</u>

We have estimated worker distribution for acrylic resin production by correlating major equipment manhour requirements with the process flow diagrams in Figures 1 through 4. Estimates for each polymerization process are shown in Table 16.

The closed process conditions associated with acrylic resin production minimize employee exposure potential to hazardous chemical substances and physical agents. Fugitive emissions from the sources listed in Table 17 and from leaks in valves, pump and compressor seals, and drains are the major sources of workplace contamination.

As Table 17 shows, principal contaminants from acrylic resin manufacture are methyl methacrylate, polymer particulates (nuisance dust) and solvent (probably toluene) from solution polymerization. Toxicity information for these and other potential input materials to the process is discussed in the "Health Effects" subsection which follows.

Additionally, employees may be exposed to fugitive emissions of nitrogen gas or other inert gases used to blanket the reactor during polymerization. Inert gases are classified as simple asphyxiants because their presence limits the available oxygen in the atmosphere.

TABLE 16.   WORKER DISTRIBUTION ESTIMATES FOR ACRYLIC RESIN PRODUCTION

Process	Unit	Workers/Unit/8-hour Shift
Mass Polymerization	Batch Reactor	1.0
Solution Polymerization	Batch Reactor	1.0
	Reflux Condenser	0.125
	Cooling Tank	0.125
Suspension Polymerization	Batch Reactor	1.0
	Dewatering	0.25
	Polymer Wash	0.25
	Dryer	0.5
Emulsion Polymerization	Batch Reactor	1.0
	Reflux Condenser	0.125
	Batch Mixer	1.0

TABLE 17. SOURCES OF FUGITIVE EMISSIONS FROM PMMA MANUFACTURE

Source	Constituent	Mass	Solution	Suspension	Emulsion
Polymerization Vent	Methyl Methacrylate	X		X	
Condenser Vent	Methyl Methacrylate Solvent		X X		X
Cooling Tank Vent	Methyl Methacrylate Solvent		X X		
Dewatering Vent	Methyl Methacrylate PMMA Particulates			X X	
Dryer Vent	Methyl Methacrylate PMMA Particulates			X X	
Emulsification Tank Vent	Methyl Methacrylate				X
Mixing Tank Vent	Methyl Methacrylate				X

TABLE 18. SOURCES OF WASTEWATER FROM PMMA MANUFACTURE

Source	Mass	Solution	Suspension	Emulsion
Polymer Dewatering			X	
Polymer Bead Wash			X	
Routine Cleaning Water	X	X	X	X

No data were located in the literature from which worker exposure potential may be estimated.

Health Effects

More than 100 input materials and specialty chemicals are used in the manufacture of acrylic polymers. The health effects of exposure to over one-half of these chemicals cannot be evaluated because sufficient data do not exist in the available literature. Input materials for which data have been compiled include a large number of tumorigenic, mutagenic, and teratogenic agents and highly toxic chemicals.

The comonomer vinyl chloride is a confirmed human carcinogen. Five other chemicals are suspected human carcinogens, including the comonomer acrylonitrile; polyvinyl chloride, a self extinguishing agent; the solvent benzene; and the chain transfer and terminating agents chloroform and carbon tetrachloride. Six additional chemicals have produced positive results in animal tests for carcinogenesis and thus are also potential carcinogens. They are polyvinyl alcohol, a suspending agent; trichloroethylene, a chain transfer agent; the comonomer vinylidene chloride; and three colorants, Rhodamine B, Rhodamine 6GDN, and cadmium sulfide.

Several chemicals in addition to these potential carcinogens pose a significant health risk to workers in acrylic polymer manufacturing plants due to high toxicities or to mutagenic or teratogenic potential. They include: cumene hydroperoxide, an initiator in the process; the comonomers styrene, ethyl acrylate, and acrylic acid; N,N-dimethylaniline, a cure agent; the solvents toluene and triethylamine; and the chain transfer and termination agents 2-mercaptoethanol and thiophenol.

A brief synopsis of the reported health effects resulting from exposure to each of these substances follows.

Acrylic Acid is a severe skin and respiratory irritant [170] and causes serious eye injury.[157] Ingestion may produce epigastric pain, nausea, vomiting, circulatory collapse, and in severe cases, death due to shock.[85] Although the data are not sufficient to make a carcinogenic determination, the chemical has been associated with tumorigenic and teratogenic effects in laboratory animals.[233]

Acrylonitrile, a suspected human carcinogen [107], causes irritation of the eyes and nose, weakness, labored breathing, dizziness, impaired judgment, cyanosis, nausea, and convulsions in humans. Mutagenic, teratogenic, and tumorigenic effects have been reported in the literature.[233] The OSHA air standard is 2 ppm (8 hour TWA) with a 10 ppm ceiling.[4]

Benzene is a suspected human carcinogen [101] which also exhibits mutagenic and teratogenic properties. It poses a moderate toxic hazard for acute exposures and a high hazard for chronic exposures through injestion, inhalation, and skin adsorption.[242] Effects on the central nervous system have been observed in humans upon exposure to concentrations of 100 ppm. [98] Intermittent exposure to 100 ppm over 10 years has been linked with

the development of cancers in humans.[271] The OSHA standard in air is 10
ppm (8 hour TWA) with a ceiling of 25 ppm.[67]

Cadmium Sulfide appears to be a mutagenic agent and has produced
positive results in animal tests for carcinogenicity.[99] It is also highly
toxic by ingestion or inhalation following acute or chronic exposures.[242]
Sulfides of heavy metals are generally insoluble and usually have little
toxic action except through liberation of hydrogen sulfide.[242] However,
inhalation of dust or fumes of cadmium compounds affects the respiratory
tract and may also involve kidney damage.[149] Even brief exposures to high
concentrations may result in pulmonary edema and death.[149]

Carbon Tetrachloride is a suspected human carcinogen [108] which is
also extremely toxic after inhalation or injection of small quantities.[241]
The toxic hazard posed by skin absorption is slight for acute exposures but
high for chronic exposures.[241] The toxicity of carbon tetrachloride
appears to be primarily due to its fat solvent action which destroys the
selective permeability of tissue membranes and allows escape of certain
essential substances such as pyridine nucleotides.[78] The OSHA air standard is 10 ppm (8 hour TWA) with a 25 ppm ceiling.[67]

Chloroform produces tumorigenic, mutagenic, terotogenic, and carcinogenic effects in laboratory animals.[233] It has been designated a suspected human carcinogen by the International Agency for Research on Cancer.
[108] Chloroform is also moderately toxic and in high concentrations may
cause narcosis and death.[157] Rapid death is attributable to cardiac
arrest, while delayed death has been associated with liver and kidney
damage.[108] Symptoms of chloroform exposure at sublethal concentrations
include respiratory depression, dizziness, nausea, and intracranial
pressure; after-effects include fatigue and headache.[5] The OSHA air
standard is 50 ppm (8 hour TWA).[67]

Cumene Hydroperoxide is highly toxic after short exposures involving
ingestion, inhalation, or skin absorption of relatively small quantities.
[242] Animal test data indicate that the substance is also a tumorigenic
and a mutagenic agent. Prolonged inhalation of vapors results in headache
and throat irritation while prolonged skin contact with contaminated
clothing may cause irritation and blistering.[278]

Ethyl Acrylate is currently being tested by the National Toxicology
Program for carcinogenesis by standard bioassay protocol. Previous test
data led to indefinite results.[107] It is very toxic by injestion,
inhalation, or skin absorption [242], producing symptoms of hypersomnia and
convulsions.[149] The OSHA air standard is 25 ppm (8 hour TWA).[67]

2-Mercaptoethanol is highly toxic by injestion, inhalation, and skin
contact according to animal test data.[233] The substance has induced
myocardial effects on isolated frog hearts, hamster lung cells [89, 262] and
chromosome abberations in laboratory animals. However, very little human
toxicity data exist in the available literature.

N,N-Dimethylaniline is highly toxic by ingestion, inhalation, and skin absorption following acute exposure.[242] It can be absorbed through intact skin to produce dangerous methemoglobinemia [85], and has been lethal to humans by ingestion at a dose of 50 mg/kg.[156] N,N-Dimethylamiline has exhibited tumorigenic potential in animal testing and is currently being tested for carcinogenisis by the National Toxicology Program.[233] OSHA has set an air standard of 5 ppm (8 hour TWA).[67]

Polyvinyl Alcohol has produced positive results in animal testing for carcinogenicity.[107] Its single dose toxicity, however, is presumably low.[85] Implantation of PVA sponge as a breast prosthesis has been associated only with foreign body type of reaction [107]; neutral solutions of PVA have also been used in eyedrops on human eyes without difficulty.[87]

Polyvinyl Chloride is a suspected human carcinogen.[107] Although the finished foam resin may cause allergic dermatitis, it is generally free from health hazard unless comminuted or strongly heated.[3] It burns readily and gives off objectionable smoke.[285] Several cases of angiosarcoma of the liver have been detected among long-time workers in plants that make polyvinyl chloride from monomeric vinyl chloride.[85]

Rhodamine B [(9-(o-carboxyphenyl)-6-(Diethylamino)-3H-xanthen-3-yliden) diethylammonium chloride] has induced mutagenic [233] and carcinogenic [106] effects in laboratory animals. To date, no data exist in the available literature on the human health effects of exposure to this colorant.

Rhodamine 6GDN [o-(6-(ethylamino)-3-(ethylimino)-2,7-dimethyl-3H-xanthen-9-YL)-ethylester, monohydrochloride benzoic acid] produced positive results for carcinogenicity in animal testing.[106] It is currently undergoing additional tests by the National Toxicology Program.[233] Some evidence also exists for mutagenic potential.[106] There is no human toxicity data in the available literature with which to evaluate the toxic effects of this colorant on humans.

Styrene has been linked with increased rates of chromosomal aberrations in persons exposed in an occupational setting. [107] Animal test data strongly support epidemiological evidence of its mutagenic potential.[233] Styrene also produces tumors and affects reproductive fertility in laboratory animals.[233] Toxic effects of exposure to styrene usually involve the central nervous system.[233] The OSHA air standard is 100 ppm (8 hour TWA) with a ceiling of 200 ppm.[67]

Thiophenol has produced toxic effects in laboratory animals at very low doses administered orally, by skin contact, and by inhalation.[233] Observed effects include behavioral symptoms, pulmonary system effects, coma, and death. No human toxicity data exist in the available literature, however, with which to confirm the effects on humans.[233]

Toluene exhibits tumorigenic, mutagenic, and teratogenic potential in laboratory tests.[233] Human exposure data indicate that toluene is also very toxic. Exposures at 100 ppm have produced psychotropic effects and

central nervous system effects have been observed at 200 ppm.[233] Symptoms
of exposure include headache, nausea, vomiting, fatigue, vertigo, paresthe-
sias, anorexia, mental confusion, drowsiness, and loss of consciousness.
The OSHA air standard is 200 ppm (8 hour TWA) with a 300 ppm ceiling.[67]

Trichloroethylene has produced positive results in animal testing for
carcinogenicity.[108] It has also produced mutagenic and teratogenic
effects in laboratory animals [233] and is classified as an acute narcotic
which causes death from respiratory failure if exposure is severe and
prolonged.[5] Since trichloroethylene has a rather long biologic half-life,
major consideration must be given to cumulative effects of this compound.
[31] Sublethal exposure may cause liver and kidney lesions, reversible
nerve damage, and psychic disturbances.[85] OSHA has set an air standard of
100 ppm (8 hour TWA) with a 200 ppm ceiling.[67]

Triethylamine is highly toxic by ingestion and inhalation and moderately toxic via dermal routes.[243] It is a strong irritant to tissue. At
least one animal study has shown that triethylamine produced mutagenic
effects in rats at the low dose of 1 mg/m^3.[96] The OSHA air standard is
25 ppm (8 hour TWA).[67]

Vinyl Chloride is a proven human carcinogen [106] which may cause permanent injury or death after inhalation, even if the exposure is relatively
short.[242] Human reproductive effects occur at concentrations as low as 30
mg/m^3 [126]; both tumorigenic and mutagenic effects have been observed in
humans as well as test animals. The OSHA air standard for vinyl chloride is
1 ppm (8 hour TWA) with a 5 ppm ceiling.[68]

Vinylidene Chloride has produced mutagenic, teratogenic, and carcinogenic effects in laboratory animals.[106] In addition, concentrations as
low as 25 ppm have been associated with systematic effects in humans.[37]
If ignited, the highly toxic hydrogen chloride will likely evolve.[157]

Air Emissions

There are no major air emission point sources associated with the PMMA
process. However, fugitive emission sources may pose a significant environmental and/or worker health problem depending on stream constituents,
process operating parameters, engineering and administrative controls, and
maintenance programs.

Sources of VOC process emissions from PMMA manufacture are summarized
in Table 17 by process and constituent type. All of these process sources
are vents of some type. Control technologies used for the VOC emissions
from vents are:[285]

- Incineration of the hydrocarbons in the vented stream by use of a
  flare; and/or

- Routing the vent stream to a blowdown.

Nonprocess sources of VOC emissions include leaks from valves, flanges, pumps, compressors, agitators, relief valves, drains, air cooling towers. Although equipment modification (for example, closing the atmospheric side of open-ended valves) may be used to reduce VOC emissions, a regular inspection and maintenance program to detect and repair leaks is the most effective control.

Sources of particulates from PMMA processing are also listed in Table 17. Particulate emissions may be controlled by:

o Venting the dryer and dewatering vent streams to a baghouse or electrostatic precipitator for collection of particulates.

According to EPA estimates, the population exposed to acrylic resin emissions in a 100 km^2 area around an acrylic resin plant is: 41 persons exposed to hydrocarbons and two persons exposed to particulates.[286]

Wastewater Sources

There are several wastewater sources associated with the various PMMA polymerization processes shown in Table 18. The major wastewater source results from suspension polymerization which uses water in the polymerization of the PMMA resin. Emulsion polymerization, which also uses water to carry the monomer through the process, does not generate wastewater of this magnitude since the polymer is not dewatered.

Ranges of several wastewater parameters for wastewaters from acrylic resin production are shown below. Values for the wastewater from the processes presented were not distinguished by process type by EPA for the purpose of establishing effluent limitations for the acrylic resin industry.[284]

Acrylic Resin Wastewater Characteristic	Unit/Metric Ton of Acrylic Resin
Production	2.50 - 50.87 m^3
BOD_5	10 - 40 kg
COD	10 - 70 kg
TSS	0.1 - 1.7 kg

Solid Wastes

The solid wastes generated by this process are mostly PMMA. Solid waste streams include substandard resin which cannot be blended and resin lost during routine cleaning, product blending, and spillage. PMMA shows a high conversion of polymer to monomer at temperatures greater than 350°C; therefore, most PMMA solid waste can be recovered and recycled to the polymerization process.[121]

Environmental Regulation

Effluent limitations guidelines have been set for the acrylic resin
industry. BPT, BAT, and NSPS call for the pH of the effluent to fall
between 6.0 and 9.0 (41 Federal Register 32587, August 4, 1976).

New Source Performance Standards (NSPS) (proposed by EPA on January 5,
1981) for volatile organic carbon (VOC) fugitive emissions include:

- Safety/release valves must not release more than 200 ppm above
  background, except in emergency pressure releases, which should not
  last more than five days; and

- Leaks (which are defined as VOC emissions greater than 10,000 ppm)
  must be repaired within 15 days.

The following compounds are listed as solid wastes (46 Federal Register
27476, May 20, 1981):

Acetone	– U002
Acrylic acid	– U008
Acrylonitrile	– U009
Benzene	– U019
Carbon Tetrachloride	– U221
Chlorobenzene	– U037
Chloroform	– U044
Cumene	– U055
Cyclohexane	– U056
Ethyl acetate	– U112
Ethyl acrylate	– U113
Isobutyl alcohol	– U140
Methyl methacrylate	– U162
Toluene	– U220
Trichloroethylene	– U228
Vinyl chloride	– U043
Vinylidene chloride	– U078

All disposal of these materials or PMMA which contains any residuals of
these materials must comply with the provisions set forth in the Resource
Conservation and Recovery Act (RCRA).

3. Acrylonitrile-Butadiene-Styrene (ABS)

INTRODUCTION

Acrylonitrile-butadiene-styrene (ABS) polymers can be manufactured to
provide a wide range of physical properties. Typically, ABS is tough,
rigid, heat and chemical resistant with high impact strength, low temperature property retention, and high gloss. In combination with these properties, the ease of fabrication makes ABS well suited for the manufacture of
drain, waste, and vent pipes, business machine housings, refrigerator crisper trays, automotive parts, television cabinets, telephones, electrical and
electronic equipment, and other uses (e.g., toys, luggage, and packaging).

ABS, which is not a random terpolymer, is manufactured by grafting
styrene and acrylonitrile monomers onto particles of polybutadiene. Varying
ABS physical properties are obtained by altering the degree of grafting,
molecular weight of the resin, and the ratio of acrylonitrile to styrene.
Table A-3 in Appendix A presents typical ABS properties according to resin
grade. The rubber content of the resin determines:[153, 270]

- Resistance to abrasion and heat,

- Impact strength,

- Tensile strength,

- Tensile modulus,

- Hardness,

- Deflection temperature,

- Elongation,

- Specific gravity, and

- Coefficient of expansion.

Generally, ABS exhibits good chemical resistance. ABS is very resistant to water, inorganic salts, weak bases, strong bases, and weak acids
while possessing good resistance to strong acids and poor resistance to some
solvents (such as esters, ketones, aldehydes, and some chlorinated hydrocarbons).[130, 270]

ABS is produced by emulsion, suspension, or mass polymerization. Although each process has advantages and disadvantages, emulsion polymerization offers the greatest flexibility and can produce ABS with a superior balance of properties.[270] In general, emulsion polymerization is used to manufacture ABS with higher impact strength while suspension and mass polymerization are utilized to produce resins with lower impact strength. [153] Heat resistant grades are produced by the addition of α-methyl styrene to the polymerization reactor.

INDUSTRY DESCRIPTION

Four chemical manufacturers comprise the ABS industry. These four producers have nine sites located in eight states: California, Connecticut, Illinois, Iowa, Massachusetts, Michigan, Ohio, and West Virginia. Table 19 lists these ABS producers and their locations.

ABS is used in consumer hard-goods markets and construction, which reflect the general trends of the U.S. economy. In 1980, the use of ABS in electrical and electronic applications, including television cabinets and piping, fell 23 percent.[35, 74] U.S. automobile markets have also declined, causing a decline in ABS automotive markets. Production of ABS for the industry has ranged from 65 to 68 percent of the total capacity available during the period from 1976 to 1979. Therefore, the existing capacity, combined with the planned Borg-Warner expansion, will accomodate any industry growth in the near future.

PRODUCTION AND END USE DATA

In 1980, total ABS production was 441,000 metric tons. [143] ABS has many uses, which include:[130, 153, 270]

- Power tool and appliance housing;

- Protective safety equipment;

- Functional and decorative automotive parts;

- Business machine components;

- Telephones;

- Toys;

- Electronic components such as computer terminals, business machine housings, and consumer electronic goods;

- Electronic housings;

- Consumer items such as faucets, shower heads, and refrigerator crisper trays;

TABLE 19. U.S. PRODUCERS OF ABS

Producer	January 1, 1982 Capacity Thousand Metric Tons
Borg Warner Corp.[a]	
Borg Warner Chemicals USA	
Plastics Division	
Ottawa, IL	136.4
Washington, WV	159.1
COMPANY TOTAL	295.5
Dow Chemical USA	
Gales Ferry, CT	29.5
Ironton, OH	29.5
Midland, MI	34.1
Torrance, CA	34.1
COMPANY TOTAL	127.2
Mobay Chemical Corp.[b]	22.7
Monsanto Co.	
Monsanto Plastics and Resins Co.	
Addyston, OH	145.5
Muscatine, IA	63.6
Springfield, MA	18.2
COMPANY TOTAL	227.3
U.S. Steel Corp.	
USS Chemicals Division	
Scotts Bluff, LA	68.2
	740.9[c]

[a]Company plans to build a 68,200 metric ton per year ABS plant at Fort Bienville, MS to be completed by the end of 1982.

[b]Resin sold by Mobay is produced by B. F. Goodrich at Louisville, KY.

[c]Carl Gordon Industries has a 4,500 metric ton per year plant at Worchester, MS on standby.

Sources: Chemical Economics Handbook, updated annually, 1981 data.
Directory of Chemical Producers, 1982.

- Drain, waste, and vent applications;

- Refrigerator inner liners and door liners;

- Boats, canoes, camper tops, luggage, and construction products; and

- Specialty applications such as video display housing, public transportation components, and the aircraft industry.

Table 20 lists ABS consumption by processing step and Table 21 presents the ABS consumption by end use.

PROCESS DESCRIPTIONS

ABS is produced by emulsion, suspension, or mass polymerization. Emulsion and suspension processes have historically dominated the ABS industry; however, the mass polymerization process has recently gained commercial importance.[153] Mass polymerization generates a minimum amount of wastewater and eliminates the need for dewatering and drying. These advantages are offset by the lessened product flexibility, greater mechanical complexity of the process, and lowered conversion of monomer to polymer which requires devolatilization for this process.

ABS is produced by grafting acrylonitrile and styrene onto a polybutadiene chain. Although the exact mechanism is not known, the polybutadiene chain probably reacts with a free radical initiator ($R \cdot$) to start the polymerization reaction. The polymer radical then combines with either acrylonitrile (CH_2=CHCN) or styrene $\left[\; \underset{\bigcirc}{CH{=}CH_2} \; \right]$ to propagate the

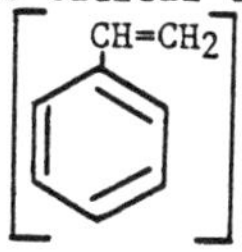

reaction. Polystyrene and acrylonitrile may be polymerizing simultaneously, and the reaction is terminated by either two radical groups combining or the introduction of a chain transfer agent (R'R"). The polymerization reaction is shown below:

<u>Initiation</u>

$$\sim\!\!\sim CH_2{-}CH{=}CH{-}CH_2\!\!\sim\!\! + \; R \cdot \;\longrightarrow\; \sim\!\!\overset{\bullet}{C}H{-}CH{=}CH{-}CH_2\!\!\sim\!\! + \; RH$$
polybutadiene

TABLE 20. ABS CONSUMPTION ACCORDING TO RESIN USE

Resin Use	1981 Thousand Metric Tons
Extrusion	182
Molding	205
Export	20
Other	34
TOTAL	441

Source: _Modern Plastics_, January 1982.

TABLE 21. ABS CONSUMPTION BY END USE MARKET

Market	1981 Thousand Metric Tons
Appliances	85
Business Machines, Telephones	35
Consumer Electronics	25
Furniture	6
Luggage	13
Modifiers	11
Packaging	4
Building and Construction	100
Recreation	25
Transportation	56
Export	20
Other	61
TOTAL	441

Source: _Modern Plastics_, January 1982.

<u>Propagation</u>

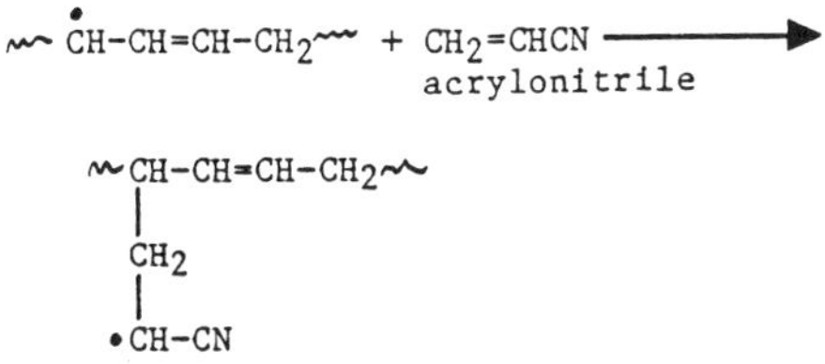

$\sim$ $\overset{\bullet}{C}H-CH=CH-CH_2\sim$ + $CH_2=CHCN$ $\longrightarrow$
acrylonitrile

$\sim CH-CH=CH-CH_2\sim$
|
CH_2
|
$\bullet CH-CN$

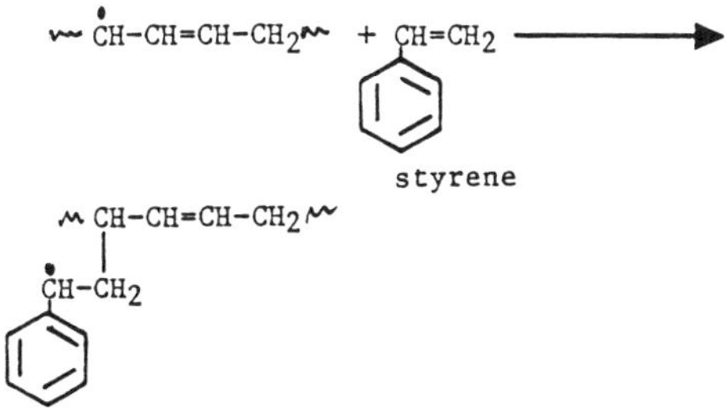

$\sim \overset{\bullet}{C}H-CH=CH-CH_2\sim$ + $\overset{}{C}H=CH_2$ $\longrightarrow$

styrene

$\sim CH-CH=CH-CH_2\sim$
|
$CH-CH_2$

<u>Termination</u>

Radical Combination

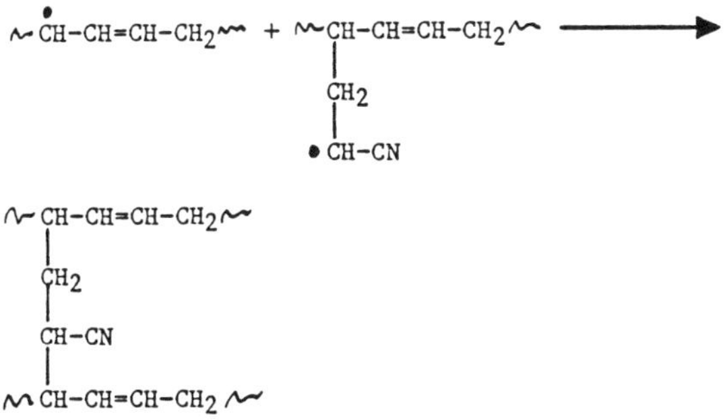

$\sim \overset{\bullet}{C}H-CH=CH-CH_2\sim$ + $\sim CH-CH=CH-CH_2\sim$ $\longrightarrow$
|
CH_2
|
$\bullet CH-CN$

$\sim CH-CH=CH-CH_2\sim$
|
CH_2
|
$CH-CN$
|
$\sim CH-CH=CH-CH_2 \sim$

Chain Transfer

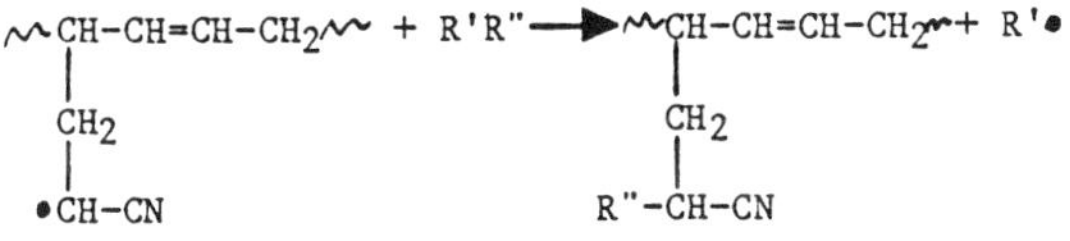

$\sim CH-CH=CH-CH_2\sim$ + $R'R''$ $\longrightarrow$ $\sim CH-CH=CH-CH_2\sim$ + $R'\bullet$
| |
CH_2 CH_2
| |
$\bullet CH-CN$ $R''-CH-CN$

Oxygen reacts with the free radical initiators used, thus making a peroxide
radical shown below:

$$R\bullet + O_2 \rightarrow RO_2\bullet$$

which slows the reaction rate. Polymerization is performed in a nitrogen
atmosphere to minimize this reaction.

Tables 22 and 23 list typical input materials and operating parameters
for the three ABS production processes. Other input materials used in ABS
production such as free radical initiators, redox initiation systems, emulsifiers, coagulants, chain transfer agents, suspension stabilizers, and
specialty chemicals (e.g. antioxidants) are listed in Table 24. Styrene and
acrylonitrile monomers are typically used; however, some ABS processes use
SAN as an input to the polymerization process (see the section on SAN).

<u>Emulsion Polymerization</u>

Emulsion polymerization of ABS resins can be broken into four basic
steps: polybutadiene polymerization, ABS polymerization, polymer coagulation, and product separation. In the first step, shown in Figure 5, polybutadiene is produced in a batch reactor using emulsion polymerization. The
initiator, activators, and emulsifiers are fed to a reactor which has been
purged of oxygen. Water and butadiene are added and the temperature is
increased to initiate the reaction. The polybutadiene produced may be
either a homopolymer or copolymer which includes up to 35 percent styrene
or acrylonitrile. The temperatures for this reaction range from 5 to 70°C.
Free radical initiators or a redox initiation system may be used to promote
the polybutadiene polymerization. A typical recipe for polybutadiene
production using a redox initiation system follows:[153]

Material	Parts by Weight
butadiene	175.00
cumeme hydroperoxide (initiator)	0.30
sodium pyrophosphate (activator)	2.50
dextrose (activator)	1.00
ferrous sulfate (activator)	0.05
sodium oleate (emulsifier)	4.00
water	200.00

When the polybutadiene polymerization has reached the desired conversion, the polymer latex is fed into the ABS polymerization reactor where
acrylonitrile, styrene, emulsifiers, and initiators are added. The ABS
produced in this second step has a polybutadiene content which ranges from
10 to 60 percent. The graft polymerization reaction may be either batch or
semi-batch processes.

TABLE 22. TYPICAL INPUT MATERIALS TO ABS PRODUCTION PROCESSES IN ADDITION
TO STYRENE, ACRYLONITRILE, POLYBUTADIENE, AND INITIATOR

Process	Water	Chain Transfer Agent	Suspension Stabilizer	Emulsifier	Coagulant
Emulsion	X	X		X	X
Suspension	X	X	X		
Mass		X			

TABLE 23. TYPICAL OPERATING PARAMETERS FOR ABS PRODUCTION PROCESSES

Process	Temperature	Pressure	Reaction Time	Conversion
Emulsion	55 - 75°C	atmospheric	1-6 hrs	99%
Suspension	100 - 170°C	350 kPa	8-16 hrs	99%
Mass	120 - 180°C	atmospheric	1-5 hrs	50 - 80%

Source: Encyclopedia of Chemical Technology, 3rd Edition.

TABLE 24.   INPUT MATERIALS AND SPECIALTY CHEMICALS USED IN ABS MANUFACTURE
            (IN ADDITION TO MONOMER AND POLYBUTADIENE)

Free Radical Initiators	Azo compounds
	Di-tert-butyl peroxide
	Other organic peroxides
	Potassium peroxydisulfate (aqueous potassium persulfate)
	Tert-butyl peracetate
Redox Initiation Systems	
Initiator	Cumeme hydroperoxide
Activators	Dextrose
	Ferrous sulfate
	Sodium pyrophosphate
Emulsifiers	Sodium oleate
Antioxidants	Di-tert-butyl-p-cresol
	Tris(nonylphenyl) phosphite
Coagulants	Calcium chloride
	Hydrochloric acid
	Sodium chloride
	Sulfuric acid
Chain Transfer Agents	Mercaptans
	Terpenes
	Terpinolene
Suspension Stabilizers	Acrylic acid-2-ethylhexyl acrylate copolymer
	Carboxymethylcellulose
	Polyvinyl alcohol
	Water soluble acrylic polymers

Sources: _Encyclopedia of Chemical Technology_, 3rd Edition.
           Encyclopedia of Polymer Science and Technology.

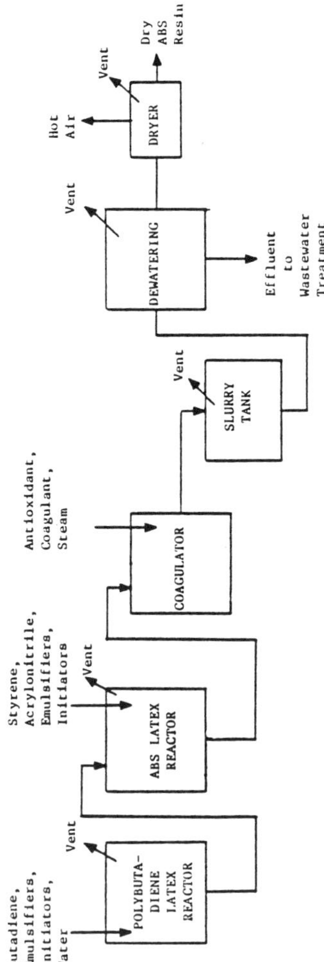

Figure 5. Emulsion ABS process.

Source: Encyclopedia of Chemical Technology, 3rd Edition.

A typical recipe for ABS with a 40 percent polybutadiene content follows:[153]

Material	Parts by Weight
polybutadiene-acrylonitrile (93:7) latex (50% solids)	900.0
water	1055.0
emulsifier	2.0
initiator (2% aqueous solution of potassium persulfate)	240.0
styrene	455.0
acrylonitrile	235.0
terpinolene (chain transfer agent)	4.8

After the reaction reaches the desired conversion, the mixture is cooled and sent to the coagulation system, which is the third step. ABS is recovered from the emulsion by coagulation with dilute salt or acid solutions. Those emulsions using a detergent as an emulsifying agent are coagulated with salt while those using soaps may use either salts or acid solutions.[153] Agglomeration of particles is aided by increasing the temperature to the range from 80 to 100°C.[153]

The resulting slurry is sent to a holding tank before dewatering and drying. In these last two processes which constitute the product separation step the polymer is first separated from any remaining water. The ABS is dried while the effluent is sent to wastewater treatment.

Reactors used to produce the polybutadiene latex are designed to withstand pressures up to 1.0×10^6 Pa, have capacities that range from 13 to 30 m^3, and are constructed of stainless steel or glass-lined carbon steel. ABS polymerization reactors are stainless steel or glass-lined vessels with capacities of up to 20 m^3.[153]

Mass Suspension Polymerization

There are two major differences between emulsion polymerization and suspension polymerization. First, the polybutadiene used for suspension polymerization must be soluble in the organic phase (styrene and acrylonitrile). Polybutadiene must have very little crosslinking in order to attain this desired solubility. Secondly, suspension polymerization does not require a polybutadiene latex as input to the polymerization reactor. Therefore, the polybutadiene need not be manufactured on-site and may be added as a dried resin.

Mass suspension polymerization consists of three basic steps: prepolymerization, polymerization, and product separation. As shown in Figure 6, polybutadiene is dissolved in styrene before entering the prepolymerization reactor. The styrene/polybutadiene mixture in the reactor is combined with acrylonitrile and initiators and heated. Temperatures for the prepolymerization step range from 80° to 120°C and reaction times are typically between two and eight hours.

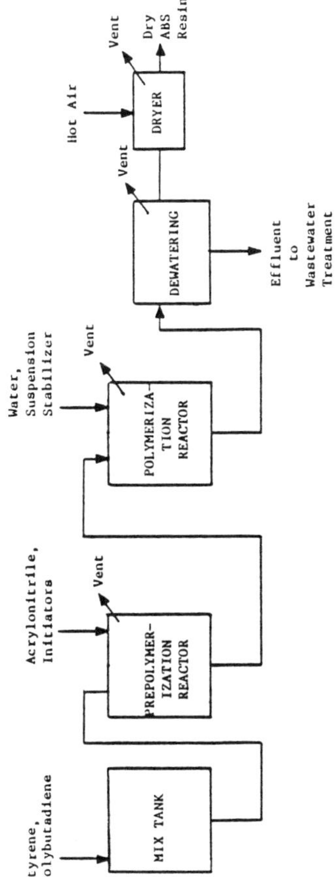

Figure 6. Mass suspension ABS process.

Source: Encyclopedia of Chemical Technology, 3rd Edition.

When the conversion in the prepolymerization reactor reaches 25 to 35
percent, the reaction mass is transferred to a polymerization reactor filled
with water. In this step, a suspending agent is added and the reactor is
heated. The reaction proceeds until the desired conversion is reached.

After polymerization, the polymer beads are separated from the suspension medium. The effluent is sent to wastewater treatment while the beads,
which enter the dryer at 10 percent moisture, are dried to a moisture
content of less than 1 percent.

Suspension polymerization ABS beads range from 0.4 to 1.2 mm in diameter. The reactors used have capacities of up to 41 m^3 and are constructed of stainless steel, stainless-clad carbon steel, or glass-coated
carbon steel. A typical suspension polymerization recipe follows:[153]

Material	Parts by Weight
soluble polybutadiene rubber	14.00
styrene	62.00
acrylonitrile	26.00
tert-butyl peracetate (initiator)	0.07
di-tert-butyl peroxide (initiator)	0.05
terpinolene (chain transfer agent)	0.90
water	120.00
acrylic acid-2-ethylhexyl acrylate copolymer (suspension stabilizer)	0.30

Mass Polymerization

Mass polymerization of ABS differs from the emulsion and suspension
polymerization processes used since water is not used as the polymerization
medium. Therefore, the polybutadiene must be soluble in the styrene and
acrylonitrile monomers. In this aspect, mass and suspension polymerization
use polybutadiene that has similar solubility properties.

Mass polymerization consists of four basic processing steps: prepolymerization, polymerization, devolatilization, and extrusion. As shown
in Figure 7, the polybutadiene is dissolved in styrene before it enters the
prepolymerization reactor. In this first step acrylonitrile and initiators
are combined with the styrene/butadiene mixture in the prepolymerization
reactor and heated. When the polymerization reaches about 30 percent, the
reaction mass is fed to a bulk polymerization reactor.

The reaction continues during the second step in the polymerization
reactor until it reaches the desired conversion, typically 50 to 80 percent.
[153] The resulting monomer/polymer mixture is then fed to a devolatilizer.

In the third step, unreacted styrene and acrylonitrile are removed in a
devolatilizer and recycled to the feed stream. These recycled monomers
constitute between 5 and 30 percent of the input to the polymerization
reactor.

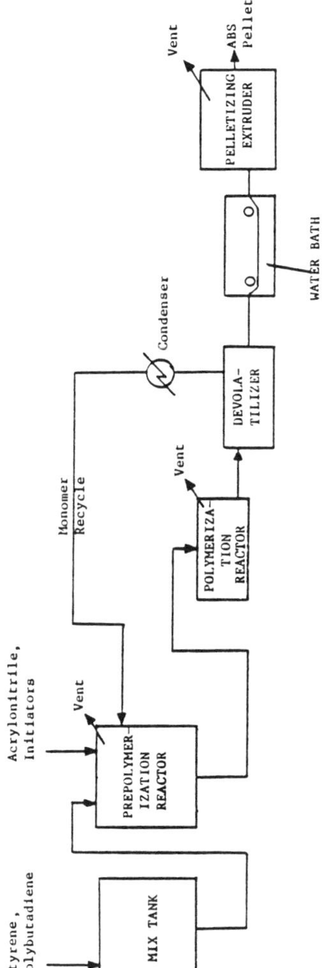

Figure 7.   Mass ABS process.

Source: Encyclopedia of Chemical Technology, 3rd Edition.

The devolatilized polymer is cooled in a water bath before it is pelletized in an extruder. The resulting ABS pellets are then ready for bagging and shipping.

A typical mass polymerization recipe for ABS was not found in the literature.

Energy Requirements

There were no data giving the energy requirements for ABS production in the literature consulted.

ENVIRONMENTAL AND INDUSTRIAL HEALTH CONSIDERATIONS

Although ABS is considered to be a nontoxic compound, used as pipe for potable water, some input materials to ABS processing are considered to pose a significant health risk. Styrene is considered to be a relatively safe organic chemical [17]; however, prolonged or repeated contact with the skin may cause skin irritation and over exposure to vapors may cause eye and nasal irritation. Styrene odors are detectable at 60 ppm and are quite strong at 100 ppm.[17]

Emulsion and suspension polymerization processes for ABS carry the polymerization reaction to 99 percent. Due to this high conversion, the sources upstream of and including the reactor contribute significantly more to the total VOC emissions than the sources downstream of the reactor.

The low conversion in the mass polymerization process (50 to 80 percent) necessitates recycling the unreacted monomers to the reactor. This low conversion also makes the reactor vent suspect as a source of acrylonitrile, styrene, and butadiene emissions. The lower levels of conversion which are achieved by the mass polymerization process require a devolatilization step. Although the mass polymerization process generates significantly less wastewater, the added devolatilization step adds additional fugitive VOC emissions.

Worker Distribution and Emissions Release Points

Worker distribution estimates have been made by correlating major equipment manhour requirements with the process flow diagrams shown in Figures 5 through 7. Estimates for each polymerization process are shown in Table 25.

No major air emission point sources are associated with ABS polymerization processes. Fugitive emission sources, however, may significantly impact environmental residuals and/or worker health. The degree to which these impact the worker's environment are determined by the stream constituents, process operating parameters, engineering and administrative controls, and maintenance programs.

TABLE 25.   WORKER DISTRIBUTION ESTIMATES FOR ABS RESIN PRODUCTION

Process	Unit	Workers/Unit/8-hour Shift
Emulsion Polymerization	Batch Reactor	1.0
	Coagulator	1.0
	Slurry Tank	0.125
	Dewatering	0.25
	Dryer	0.5
Mass Suspension Polymerization	Batch Mixer	1.0
	Batch Reactor	1.0
	Dewatering	0.25
	Dryer	0.5
Mass Polymerization	Batch Mixer	1.0
	Batch Reactor	1.0
	Devolatilizer	0.25
	Water Bath	0.25
	Extruder	1.0

Sources of fugitive emissions are shown in Table 26. As the table shows, principal contaminants to which workers may be exposed include butadiene, styrene, acrylonitrile, and polymer particulates (nuisance dust). Additionally, employees may be exposed to a variety of product additives, stabilizers, polymerization initiators and emulsifiers available. Toxicity information for these contaminants is discussed below in the "Health Effects" subsection of this section.

Health Effects

Animal test data indicate that several input materials and specialty chemicals used in ABS manufacture are tumorigenic agents; however, only one, the suspension stabilizer polyvinyl alcohol, tests positive for carcinogenicity. Other substances that pose a significant health risk to plant employees by virtue of high toxicities include cumene hydroperoxide and ferrous sulfate, which are used in the redox initiation system; and the suspension stabilizer, acrylic acid. The reported health effects of exposure to these substances are summarized below.

Acrylic Acid is a severe skin and respiratory irritant [170] and is one of the most serious eye injury chemicals.[157] Ingestion may produce epigastric pain, nausea, vomiting, circulatory collapse, and in severe cases, death due to shock.[85] Although the data are not sufficient to make a carcinogenic determination, the chemical has been associated with tumorigenic and teratogenic effects in laboratory animals.

Acrylonitrile, a suspected human carcinogen [106], causes irritation of the eyes and nose, weakness, labored breathing, dizziness, impaired judgment, cyanosis, nausea, and convulsions in humans. Mutagenic, teratogenic, and tumorigenic effects have been reported in the literature. The OSHA air standard is 2 ppm (8 hour TWA) with a 10 ppm ceiling.[67]

Cumene Hydroperoxide is highly toxic after short exposures involving ingestion, inhalation, or skin absorption of relatively small quantities. [242] Animal test data indicate that the substance is also a tumorigenic and a mutagenic agent. Prolonged inhalation of vapors results in headache and throat irritation, prolonged skin contact with contaminated clothing may cause irritation and blistering.[278]

Ferrous Sulfate has produced toxic effects involving the central nervous system in children at doeses as low as 20 mg/kg.[111] Death has occurred in humans at exposures of 200 mg/kg.[21] Although iron salts differ considerably in astringency, lethal doses appear to be clearly related to total iron content.[85] Animal test data suggest that ferrous sulfate is also a tumorigen and a mutagen.

Polyvinyl Alcohol has produced positive results in animal testing for carcinogenicity.[107] Its single dose toxicity, however, is presumably low.[85] Implantation of PVA sponge as a breast prosthesis has been associated only with foreign body type of reaction [107]; neutral solutions of PVA have also been used in eyedrops on human eyes without difficulty.[87]

TABLE 26. SOURCES OF FUGITIVE EMISSIONS FROM ABS MANUFACTURE

Source	Constituent	Emulsion	Suspension	Mass
Reactor Vents	Butadiene	X	X	X
	Styrene	X	X	X
	Acrylonitrile	X	X	X
	Initiators	X	X	X
	Emulsifiers	X		
	Suspension Stabilizers		X	
Slurry Tank	Styrene	X		
	Acrylonitrile	X		
	Butadiene	X		
Dewatering	ABS and Monomer Emissions and Particulates	X	X	
Dryer Vent	ABS and Monomer Emissions and Particulates	X	X	
Extruder	ABS and Monomer Emissions and Particulates			X

TABLE 27. SOURCES OF WASTEWATER FROM ABS MANUFACTURE

Source	Emulsion	Suspension	Mass
Polymer Dewatering	X	X	
Spent Water Bath			X
Routine Cleaning Water	X	X	X

Styrene has been linked with increased rates of chromosomal aberrations in persons exposed in an occupational setting.[106] Animal test data strongly support epidemiological evidence of its mutagenic potential.[233] Styrene also produces tumors and affects reproductive fertility in laboratory animals.[233] Toxic effects of exposure to styrene usually involve the central nervous system.[233] The OSHA air standard is 100 ppm (8 hour TWA) with a ceiling of 200 ppm.[67]

Air Emissions

VOC emission sources for ABS processing are summarized in Table 26. Both the constituent and the process type are indicated for each source. Most of these sources are vents, which use the following control technologies:[285]

- Venting the stream to a flare where hydrocarbons are incinerated; and/or

- Venting the stream to a blowdown.

The VOC emissions from dewatering may be controlled by:

- Enclosing any atmospheric dewatering process; and/or

- Venting any centrifuge streams to either a flare or a blowdown.

Similarly, VOC emissions from the extruder may be controlled by:

- Extrusion under reduced pressure followed by routing the offgases to the reactor; and/or

- Venting the stream to a flare or blowdown.

Sources of particulates for the ABS process are also listed in Table 26. Particulate emissions from dewatering may be reduced by:

- Enclosing any atmospheric dewatering sources; and/or

- Venting any centrifugal streams to either an electrostatic precipitator or a baghouse.

Particulate emissions from the extruder and the dryer vent may also be controlled by venting the streams to either an electrostatic precipitator or a baghouse.

Wastewater Sources

There are several wastewater sources associated with the various ABS polymerization processes shown in Table 27. The major wastewater sources result from emulsion and suspension polymerization which use water to polymerize the ABS resin. The water separated from the polymer in emulsion

polymerization is reported to contain: 0.25 kg styrene; 0.5 kg acrylonitrile; 0.25 kg sulfuric acid; 2.5 kg emulsifier; 0.25 kg sodium bisulfite;
and 0.75 kg calcium chloride per ton of ABS produced.[93]

Ranges of several wastewater parameters for wastewaters from ABS
production are shown below. Values for the wastewater from the processes
presented were not distinguished by process type by EPA for the purpose of
establishing effluent limitations for the ABS industry:[284]

ABS Wastewater Characteristics	Unit/Metric Ton of ABS
Production	$1.67 - 24.03$ m^3
BOD$_5$	$2 - 20.7$ kg
COD	$5 - 33.5$ kg
TSS	$0 - 30$ kg

Solid Wastes

The solid wastes generated by this process are mostly ABS. Solid waste
streams originate from reactor cleaning, product blending, particulate
removal, and spillage.

Environmental Regulation

Effluent limitations guidelines have been set for the ABS industry.
BPT, BAT, and NSPS call for the pH of the effluent to fall between 6.0 and
9.0 (41 Federal Register 32587, August 4, 1976).

New source performance standards (proposed by EPA on January 5, 1981)
for volatile organic carbon (VOC) fugitive emissions include:

- Safety/release valves must not release more than 200 ppm above
  background, except in emergency pressure releases, which should not
  last more than five days; and

- Leaks (which are defined as VOC emissions greater than 10,000 ppm)
  must be repaired within 15 days.

Acrylonitrile is listed as a hazardous waste (46 Federal Register
27476, May 20, 1981) and designated as U009. All disposal of acrylonitrile
or ABS which contains any residual acrylonitrile must comply with the
provisions set forth in the Resource Conservation and Recovery Act (RCRA).

4. Alkyd Molding Resins

INTRODUCTION

Alkyd resins comprise a segment of the plastics industry whose major use (90 percent) is in the manufacture of coatings. The other 10 percent produced as molding compounds will be the subject of this section. Alkyd molding resins exhibit high arc resistance, track resistance, good high temperature dielectric properties, and dimensional stability up to 232°C (450°F). Specialty grades offer self-heating arc resistance as well as X-radiation barrier and flame retardant properties. Alkyd molding resins may be pigmented in a wide range of colors which exhibit relatively good heat and light stability. Table A-4 in Appendix A lists typical properties of alkyd resins.

Molding compounds are manufactured as solids, classified as granular, putty, rope, or bulk.[247] Alkyd molding resins cure rapidly at low molding temperatures and exhibit heat resistance, dimensional stability, and electrical properties over a wide temperature range. Fillers are added to alkyd resin molding compounds to control shrinkage as well as internal stresses and strains, reduce cost, control flow properties, impart heat resistance, provide an improved surface appearance, increase electrical resistance, reduce moisture absorption, or provide other, more specific properties. Typical fillers for alkyd molding resins and the properties they provide are listed in Table 28.

Alkyd resins may be classified in three different ways: by the alkyd ratio (polyhydric alcohol to phthalate ratio); by the oil length or percent oil; or by the percent phthalic anhydride.[128] The oil length, or type of oil, describes a general classification of oils which are used for a specific alkyd resin application. The following table gives a list of classifications with the fatty acid (or oil) content and the phthalic anhydride content.[128, 154]

Classification (Type of Oil)	Fatty Acid (Oil) Content, %	Phthalic Anhydride Content, %
short	30 – 45	35 – 46
medium	43 – 55	30 – 37
long	55 – 70	20 – 30
very long	>71	<20

TABLE 28.   TYPICAL FILLERS FOR ALKYD MOLDING RESINS AND THE
PROPERTIES THEY PROVIDE

Filler	Properties
Antimony Oxide	Used to impart flame resistance, usually in conjunction with a chlorinated or brominated resin
Asbestos	Fibrous in nature, forms a strong bond with the resin to impart toughness and stiffness, and is a good flow control agent
Barium Sulfate	High specific gravity and opaque to X-rays
Calcium Carbonate	Near white appearance, high filler loadings possible with good flow, imparts excellent high frequency electrical properties
Clay	Lowers moisture absorption, raises hardness and flexural strength, excellent in high voltage applications
Hydrated Alumina	Raises heat, arc-tracking, and flame resistance
Mica	Good flow control agent, imparts high dielectric strength
Silica and Talc	Inert, inexpensive fillers
Titanium Dioxide	Very white appearance, raises dielectric constant
Wollastonite	Improves hardness

Source: Seymour S. Schwartz and Sidney H. Goodman, _Plastics Materials and Processes_, 1982.

Alkyd resins are currently manufactured by one of two processes: the alcoholysis method or the fatty acid method. These two processes differ in that the fatty acid method uses direct esterification while the alcoholysis method uses an alcoholizing triglyceride oil and a polyol to form partial esters before the esterification is completed with the addition of dibasic acids. Alcoholysis is currently the most widely used process.[128]

INDUSTRY DESCRIPTION

The alkyd resin industry is comprised of 61 producers with 117 sites in 24 different states. Although these plants are distributed across the continent, a preponderance of sites (28 percent) are located in the Great Lakes States, EPA Region V. New York and New Jersey (EPA Region II), the West Coast (EPA Region IX), and the North Central States (EPA Region III) contain 17, 17, and 14 percent of the alkyd resin sites, respectively. Alkyd resin producers and their locations are listed in Table 29. Plant capacities are considered to be proprietary.

Alkyd resins are used primarily as coatings. Over 90 percent of all alkyd resins produced are used to manufacture paint, varnish, or lacquer for industrial and home consumption. Over half of the alkyd resins produced are sold to formulators while the remaining resin is formulated by the alkyd producer. Most alkyd resins used for coatings are pigmented.[154]

Since 1976, alkyd resin production has increased from 306,000 metric tons to 320,000 metric tons, 342,000 metric tons, and 343,000 metric tons in 1977, 1978, and 1979, respectively. Unlike other polymers which suffered from declining markets in 1980, alkyd resin production remained at 343,000 metric tons.[65]

Processing alkyd resins requires adherence to stringent safety practices; volatility and flash points of the solvents and vinyl crosslinking agents used pose potential fire and explosive hazards. Alkyd resin equipment should be separated from storage areas using firewalls and automatic fire doors. Other safety precautions are listed below:[154]

- The use of explosion-proof electric motors, switches, and wiring. The electrical installation should conform to the Class 1, Group D outline in the National Electrical Code.

- The removal of causes of static electricity and grounding of all equipment in accordance with NFPA Pamphlet No. 77, "Static Electricity."

- Adequate ventilation to minimize buildup of an explosive or toxic solvent concentration as a result of pumping or thinning.

- The protection of kettles with rupture discs designed to relieve excess pressure. The rupture disc should be vented to a safe location, so that the hot vapors will not be ignited.

TABLE 29.   U.S. ALKYD RESIN PRODUCERS

Producer	Location
ADCO Chemical Co., Inc.	Newark, NJ

AZS Corporation
  AZ Products, Inc. Division
  AZS Chemical Co. Division
	
	Eaton Park, FL
	Atlanta, GA

Ball Chemical Co. Electrical Insulation Resins	 Glenshaw, PA

Beatrice Foods Co.
  Beatrice Chemical Division
    Farboil Co. Division
	
	Baltimore, MD

Bennett's	Salt Lake City, UT
Bisonite Co., Inc.	Tonawanda, NY
M'. A. Bruder & Sons, Inc.	Philadelphia, PA
California Resin and Chemical Co., Inc.	 Vallejo, CA
Cargill, Inc. Chemical Products Division	 Carpentersville, IL Forest Park, GA Lynwood, CA

Celanese Corporation
  Celanese Plastics & Specialties Co.
  Division
  Celanese Specialty Resins Division
	
	Los Angeles, CA
	Louisville, KY

Chemical Products Corporation	Elmwood Park, NJ
Degen Oil & Chemical Co.	Jersey City, NJ
DeSoto, Inc.	Chicago Heights, IL Garland, TX
The Dexter Corporation Midland Division	 Cleveland, OH Hayward, CA Rocky Hill, CT Waukegan, IL

(continued)

TABLE 29 (continued)

Producer	Location
Dock Resins Corporation	Linden, NJ
Emkay Chemical Co.	Elizabeth, NJ
Foy-Johnston, Inc.	Cincinnati, OH
General Electric Co. Engineered Materials Group Electromaterials Business Department	Chelsea, MA Schenectady, NY
The P. D. George Co.	St. Louis, MO
Guardsman Chemicals, Inc.	Grand Rapids, MI
Handschy Chemical Co. Farac Oil & Chemical Co. Division	Riverdale, IL
Hugh J.-Resins Co.	Long Beach, CA
Insilco Corporation The Enterprise Companies Division	Wheeling, IL
Iovite Chemicals, Inc.	Matteson, IL
Jones-Blair Co. Texas Resins Corporation, subsidiary	Dallas, TX
Lawter International, Inc.	South Kearny, NJ

Mobil Corporation
 Marcor Inc.
  Montgomery Ward & Co. Inc.,
  subsidiary
   Standard T Chemical Co., Inc.
   subsidiary
	Chicago Heights, IL

The O'Brien Corporation The O'Brien Corporation-Central Region	South Bend, IN
The O'Brien Corporation-Eastern Region	Baltimore, MD
The O'Brien Corporation-Southwestern Region	Houston, TX
The O'Brien Corporation-Western Region	South San Francisco, CA

(continued)

TABLE 29 (continued)

Producer	Location
C. J. Osborn Chemicals, Inc.	Pennsauken, NJ
Perry & Derrick Co.	Dayton, KY
Phillips Petroleum Co. Interplastic Corporation, subsidiary Commercial Resins Division	Minneapolis, MN
Polychrome Corporation Cellomer Corporation, subsidiary	Newark, NJ
PPG Industries, Inc. Coatings and Resins Division	Circleville, OH East Point, GA Houston, TX Oak Creek, WI Springdale, PA Torrance, CA
Reichhold Chemicals, Inc.	Azusa, CA Detroit, MI Elizabeth, NJ Houston, TX Jacksonville, FL South San Francisco, CA Tuscaloosa, AL
Sterling Division	Sewickley, PA
H. H. Roberston Co. Freeman Chemical Corporation, subsidiary	Burlington, IA Chatham, VA Saukville, WI
Schenectady Chemicals, Inc.	Schenectady, NY
SCM Corporation Glidden Coatings & Resins Division	Reading, PA San Francisco, CA
Sybron Corporation Chemical Division Jersey State Chemical Co. Division	Haledon, NJ
Synray Corporation	Kenilworth, NJ

(continued)

TABLE 29 (continued)

Producer	Location
Textron Inc. Spencer Kellogg Division	Baltimore, MD Newark, NJ Pensacola, FL San Carlos, CA Valley Park, MO

Tyler Corporation
  Reliance Universal Inc., subsidiary
    Specialty Chemicals and Resins
      Division
	Louisville, KY

U.S. Polymers, Inc.	St. Louis, MO
United States Steel Corporation U.S.S. Chemicals Division	Colton, CA
United Technologies Corporation Inmont Corporation, subsidiary	Anaheim, CA Cincinnati, OH Detroit, MI Greenville, OH
Valspar Corporation McWhorter, Inc., subsidiary	Baltimore, MD Carpentersville, IL Rockford, IL
Westinghouse Electric Corporation Insulating Materials Division	Manor, PA
Yenkin-Majestic Paint Corporation Ohio Polychemicals Co. Division	Columbus, OH

Source: Directory of Chemical Producers, 1982.

- When a charge is removed from the kettle, the introduction of an
  inert gas at a rate sufficient to relieve the displacement, in order
  to avoid the entry of air through the vent on the condensing system
  and thus the formation of a potentially explosive atmosphere.

PRODUCTION AND END USE DATA

In 1980, alkyd resin production totaled 343,000 metric tons.[65] Alkyd
molding resins constitute about 10 percent of the market. Finished products
include automobile parts, furniture, electrical equipment and cables,
packaging, and construction components for buildings. Other uses for alkyd
molding resins include:[128, 154, 247, 263]

- Automotive ignition parts where high track resistance and low
  dielectric losses at high temperatures are desired as well as brush
  holders, coils, distributors, housings, solenoids, and switches;

- Electrical and electronic parts where mechanical strength is needed,
  including bobbins, capacitors, coils, relays, encapsulated components, molded circuitry, resistors, and transformers;

- Appliances and parts which require both color and high temperature
  stability such as connectors, pump impellers, relays, switches,
  thermostats, and timers;

- Circuit breakers, contactors, and overload protection for motors;

- Starters, interlock switches, and switch covers;

- Television parts, including coils, corona caps, sockets,
  transformers, terminal boards, and tuner strips; and

- Phonograph records.

PROCESS DESCRIPTIONS

Alkyd resins are produced by either the alcoholysis method or the fatty
acid method. The alcoholysis method combines an alcoholizing triglyceride
oil, such as soya or linseed, with a polyol, usually pentaerythritol or
glycerol. The partial esters produced are then esterified using a dibasic
acid, such as phthalic anhydride. The fatty acid method uses direct
esterification of the reactants to form the resin.

Alkyd resin production processes are batch due to the difficulties
encountered with continuous processes, including buildup of crosslinked
monomer on the reactor walls. The procedure used for each process depends
upon the properties desired in the alkyd resin. A solvent may be used to
carry the resin through the reaction process. Solvent is added directly to
the reactor for the fatty acid method and to the reactor after the dibasic
acid in the alcoholysis method. Using a solvent provides resins of better
color and uniformity, higher resin yield due to lower phthalic anhydride or

polyol (polyhydric alcohol) losses through the condenser, lower esterification temperatures, faster esterification cycles, and easier kettle cleaning.
If these parameters are not important for the processor, the fusion process,
which uses no solvent, may be used. Currently, this process has a large
investment in existing plants, makes alkyds with isophthalic acids more
easily, represents a lower investment cost for new plants, and requires less
stringent safety precautions.[154]

Alkyd resins produced by partial alcoholysis combine an oil with a
polyol to form a partial ester which in turn is esterified with a dibasic
acid. The alcoholysis reaction is shown below.

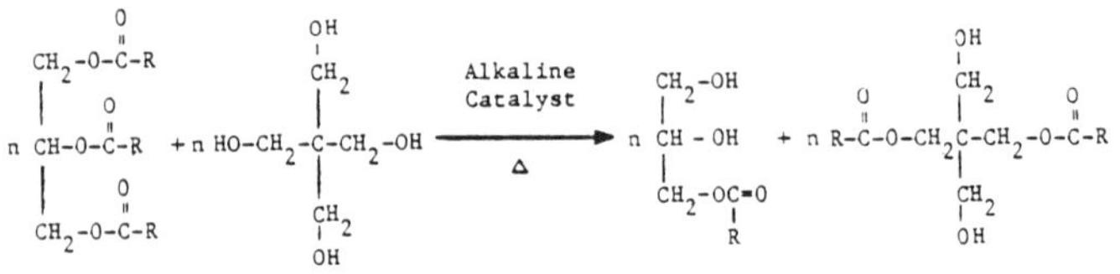

Triglyceride   Pentaerythritol   Partial Esters
                    (Polyhydric Alcohol)

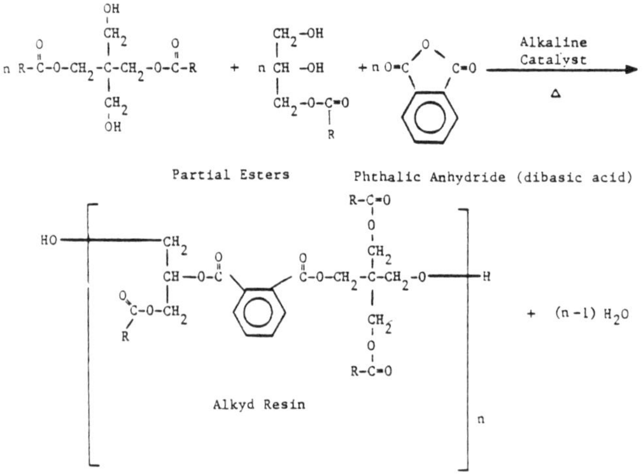

Partial Esters   Phthalic Anhydride (dibasic acid)

Alkyd Resin   + (n-1) H$_2$O

Alcoholysis is used in the preparation of alkyds which have an oil
substituted for the fatty acid. Monomers and peroxide catalysts are added
to crosslink the polymer.[247]

Using the fatty acid process, alkyd resins are produced from the
reaction of a polyol, a polyacid, and an unsaturated fatty acid. This
reaction yields a complex branched structure or network which is an
infusible, insoluble compound at 80 percent esterification. This overall
reaction is shown below.

Unsaturated   Glycerol   Phthalic Anhydride
Fatty Acid

Alkyd Resin

Table 30 lists typical operating parameters for the two alkyd resin
production processes. Alkyd resin properties, however, are determined
primarily by the input materials used instead of the process performed. For
example, unmodified phthalate/glycol resins exhibit poor solubility characteristics. Modification with oils and resins as well as using more than one
alcohol or acid will upgrade the resin properties to the desired level.[263]
Input materials for these processes are listed in Table 31.

Modifiers and blending agents are added to alkyd resins to provide
desired properties, including better hardness, faster drying speed, toughness, corrosion and chemical resistance, heat resistance, and improved color
retention. Modifiers are typically added during alkyd preparation. Blending agents cannot withstand the high temperatures of the preparation step,
usually over 200°C, and are added after the alkyd is produced. Tables 32
and 33 give the effects of various modifiers and blending agents, respectively, on alkyd resin properties.

Alkyd resins which are fabricated into molding compounds are usually
combined with fillers. Asbestos, sisal fibers, cellulose fibers, synthetic
fibers (such as polyester, nylon, or acrylic), or glass fibers are used.
Solvents, including toluene, xylene, and coal-tar naphthas, may also be
combined with the alkyd resin. Other additives used are drying oils, pigments, and thixotropes. Organic acid salts of cobalt, lead, and manganese
may be added to enhance resin drying.

TABLE 30. TYPICAL OPERATING PARAMETERS FOR ALKYD RESIN
PRODUCTION PROCESSES

Process	Temperature	Reaction Time
Alcoholysis Method	210 - 260°C[a]	5 - 16 hours
Fatty Acid Method	210 - 280°C	5 - 16 hours

[a]With catalyst. Without a catalyst, reaction temperatures must exceed
280°C.

Sources: Encyclopedia of Chemical Technology, 3rd Edition.
Encyclopedia of Polymer Science and Technology.
Seymour S. Schwartz and Sidney H. Goodman, Plastics Materials and
Processes, 1982.

TABLE 31. TYPICAL INPUT MATERIALS FOR ALKYD RESIN PRODUCTION PROCESSES

Function	Compounds
Polybasic Acids	Adipic acid
	Azelaic acid
	Camphoric acid
	Chlorendic anhydride
	Citric acid
	Cyclopentadiene
	Diallyl phthalic acids and anhydrides
	Dimerized fatty acid
	Fumaric acid
	Glutaric acid
	Hexahydrophthalic anhydride
	Isophthalic acid
	Maleic acid
	Maleic anhydride
	Phthalic anhydride
	Pimelic acid
	Sebacic acid
	Succinic acid
	Tartaric acid
	Terephthalic acid
	Tetrachlorophthalic anhydride
	Tetrahydrophthalic anhydride
	Trimellitic anhydride
Oils	China wood
	Coconut
	Cottonseed
	Dehydrated castor
	Fish
	Linseed
	Oiticica
	Safflower
	Soya
	Soybean
	Sunflower
	Tung
	Walnut

(continued)

TABLE 31 (continued)

Function	Compounds
Monobasic Acids	Benzoic acid
	Fatty acids and fractionated fatty acids obtained from oils
	eleostearic
	lauric
	licanic
	linoleic
	linoleic (conjugated)
	linolenic
	oleic
	ricinoleic
	palmitic
	stearic
	p-tert-butylbenzoic acid
	Synthetic saturated fatty acids
	2-ethyl hexanoic
	isodecanoic
	isononanoic
	isooctanoic
	pelargonic
	Tall oil fatty acids
Polyols (Polyhydric Alcohols)	Diethylene glycol
	Dipentaerythritol
	Dipropylene glycol
	Ethylene glycol
	Glycerol
	Mannitol
	Neopentylene glycol (2,2-dimethyl-1,3-propanediol)
	Pentaerythritol
	Polyethylene glycol
	Propylene glycol
	Sorbitol
	Trimethylolethane (2-(hydroxymethyl)-2-methyl-1,3-propanediol)
	Trimethylolpropane (2-ethyl-2-(hydroxymethyl)-1,3-propanediol)

(continued)

TABLE 31 (continued)

Function	Compounds
Modifiers	Acid modifiers
	Rosin
	Benzoic Acid
	Acrylonitrile
	Amines
	Amino resins
	Epoxides
	Formaldehydes
	Hydroxyl containing modifiers
	Hydroxylabietyl alcohol
	Isocyanates
	Methyl methacrylate
	Phenolic resins
	Polyamides
	Resins and ester gums
	Silicones
	Vinyl monomers
	Styrene
	Vinyl acetate
	Vinyl toluene
Blending Agents	Amino resins
	Cellulose nitrate
	Chlorinated rubber
Catalysts	Calcium acetate
	Calcium hydroxide
	Calcium naphthenate
	Cerium naphthenate
	Lanthanum naphthenate
	Lime
	Lithium salts
	Lithium carbonate
	Sodium bicarbonate
	Sublimed litharge (PbO)
Free Radical Initiators	Oxygen
Inert Gas	Carbon dioxide
	Mixture of CO_2 and N_2
	Nitrogen

Sources: Encyclopedia of Chemical Technology, 3rd Edition.
         Encyclopedia of Polymer Science and Technology.
         Seymour S. Schwartz and Sidney H. Goodman, Plastics Materials and
            Processes, 1982.

TABLE 32. EFFECT OF MODIFIERS ON ALKYD RESINS

Modifier	Advantages	Disadvantages
Epoxides	Improved adhesion Better alkali resistance Better detergent resistance Better solvent resistance	Poorer color retention Rapid chalking
Hydroabietyl Alcohol (Abitol, trademark Hercules Powder Co.)	Better brushing Reduces alkyd functionality and acts as gelation inhibitor Better solubility (in aliphatic solvent) Better gloss Better flow Greater hardness	Slightly more yellowing Slightly decreased durability when used in excess
Isocyanates	Better water resistance Faster dry Better abrasion resistance	Greater yellowing Toxicity problem (in manufacture)
Phenolic	Greater hardness Better water resistance Better alkali resistance Better solvent resistance	More yellowing Poorer stability
p-tert-butylbenzoic acid, benzoic acid	Reduces alkyd functionality and acts as a gelation inhibitor Greater hardness Higher viscosity Faster dry Improved color and gloss Improved chemical resistance	Poorer solubility Poorer flexibility
Rosin or Rosin Ester	Faster dry Better brushing Greater hardness Better mar resistance Better adhesion	More yellowing Decreased exterior durability when used in excess

(continued)

TABLE 32 (continued)

Modifier	Advantages	Disadvantages
Silicones	Improved heat resistance Greater hardness More resistance to thermal shock	Higher cost Higher curing temperature
Styrene, Vinyl-toluene, Methyl Methacrylate, Acrylonitrile	Faster dry Improved color and gloss Improved color and gloss retention Improved chemical resistance	Poorer solvent resistance

Source: Encyclopedia of Polymer Science and Technology.

TABLE 33.   COATING PROPERTIES OF BLENDED ALKYDS

Blending Agent	Advantages	Disadvantages
Cellulose Nitrate	Faster dry Better chemical resistance Better sovlent resistance Greater hardness	Poorer durability Poorer flexibility Lower solids
Chlorinated Rubber	Greater toughness and hardness Better abrasion resistance Better chemical resistance Faster dry	Poorer solvent resistance
Urea-formaldehyde and Melamine- formaldehyde Resins	Better chemical resistance Better color and color retention Greater hardness Faster cure Higher heat resistance	Poorer flexibility Higher cost Poorer adhesion

Source: Encyclopedia of Polymer Science and Technology.

Inert gases may be used in these processes for several reasons. These sparges provide more rapid reaction, more uniform molecular weight distribution, fewer side reactions, and better colors in alkyd resins in addition to increased safety.

Alcoholysis Method

The alcoholysis method is currently the most commonly used process for alkyd resin production.[128] Alcoholizing triglyceride oils, such as soya or linseed, are combined with either pentaerythritol or glycol. When partial esters are produced, the esterification is completed by the addition of a dibasic acid, such as phthalic anhydride. A typical recipe for production of alkyd resins via alcoholysis includes the following:[128, 154]

Material	Parts by Weight
Glycerol (polyol, monomer)	553
Phthalic Anhydride (polybasic acid, monomer)	296
Soybean Oil	843
Sublimed Litharge (catalyst)	376

The alcoholysis method, shown in Figure 8, may be performed with or without a catalyst. If a catalyst is not used, the reaction temperature must be raised to 280°C or above.[154] Basic catalysts are used in small amounts (typically 0.002 moles catalyst per kg of oil) to promote redistribution of the fatty acid groups on the triglyceride oils. Alkali metal hydroxides may be used, but they give the resin a dark color and render alkyd films more water sensitive. Lead compounds are the most efficient catalysts. Sublimed litharge (PbO) is used commercially since it gives a faster esterfication rate, equivalent resin color, and higher resin viscosity than calcium or lithium salts at a concentration of 0.01 to 0.05 percent.[128, 154] When a catalyst is used, the oil is heated first until the temperature reaches 225 to 250°C. The catalyst and polyol are then added to the hot oil and this mixture is reheated to 230 to 250°C.

After the fatty acid groups have been redistributed, the partial esters produced contain free hydroxyl groups. The dibasic acid is then added and the esterification is completed at temperatures between 210 and 260°C.

An azeotropic liquid (xylene), typically 3 to 10 percent of the batch, may be added to the reactor to facilitate water removal. The refluxed vapors are then sent from the condenser to a decanter where the water is removed and the solvent is recycled to the reactor. If no solvent is used, the water vapor may be vented to the atmosphere.[154]

Conversion is determined by the acid number or the viscosity of the mixture.[247] When the reaction is completed, any solvent desired is added to a mixing tank where the bottoms stream containing the alkyd resin is fed. The resulting resin mixture is then filtered and stored.

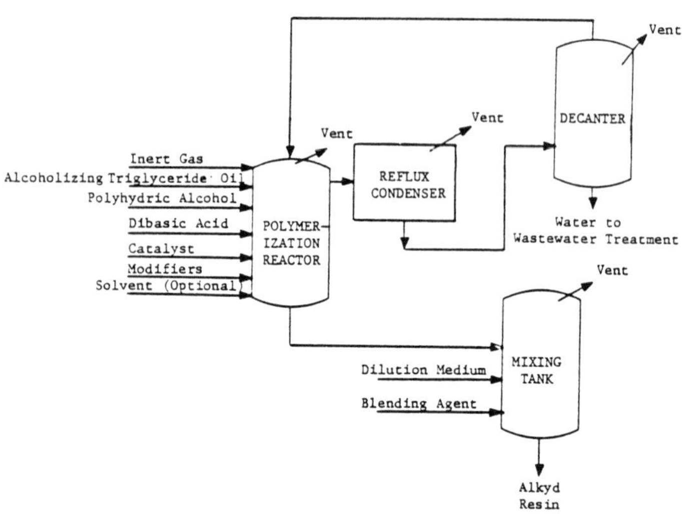

Figure 8.   Alkyd resin production using the alcoholysis method.

Sources: Encyclopedia of Chemical Technology, 3rd Edition.
           Encyclopedia of Polymer Science and Technology.

Fatty Acid Method

The fatty acid method is the second of two processes used to manufacture alkyd resins. This process allows greater freedom of formulation than does the alcoholysis method since a wider range of fatty acids are available for reactions of this type.[128] In this process, presented in Figure 9, the entire charge of raw materials is fed into the reactor where an inert gas is used to provide a safe atmosphere for the reaction. Tall oil, pelargonic acid, or 2-ethylhexanoic acid are most commonly used. If both an oil and a fatty acid are used, the alkyd coating resin produced exhibits more rapid top drying and slower through drying than resins made by the alcoholysis method.[154]

The temperature is typically raised to between 210 and 260°C, although temperatures up to 280°C have been used. The water generated by the condensation reaction is removed overhead from the reflux condenser. An azeotropic liquid, such as xylene, may be used to promote water removal. The refluxed vapors are sent to a decanter where the azeotropic solvent is recycled to the reactor while the water is removed and treated. If no solvent is used, the water vapors may be vented to the atmosphere. Information regarding a typical recipe for alkyd resin production via the fatty acid method was not found in the literature.

After the reaction proceeds to the desired end point as indicated by either the acid number or viscosity, the bottoms stream containing the resin is quickly fed into a jacketed kettle containing a dilution solvent before it is filtered and stored.

Energy Requirements

Energy requirements for this process were not found in the literature consulted.

ENVIRONMENTAL AND WORKER HEALTH CONSIDERATIONS

There are seven compounds used in alkyd resin production processing which have been listed as hazardous under RCRA: acrylonitrile, formaldehyde, maleic anhydride, methyl methacrylate, phthalic anhydride, toluene, and xylene. The solvents used in alkyd resin production are very volatile; therefore, solvent reflux washes are reused and recovered by distillation. Caustic washes used in kettle cleaning are also reused until the caustic is spent.[128]

Alkyd resins are typically carried to high conversions. Removal of the water generated by the condensation reaction forces the resin to complete polymerization, making the reactor vent most suspect for fatty acid, oil, polyol, and polybasic acid emissions. Recycling the solvent and other nonwater vapors to the reactor also serves to reduce emissions from the downstream vents. Overall system losses for all streams is reported to be 3 to 10 percent.[154]

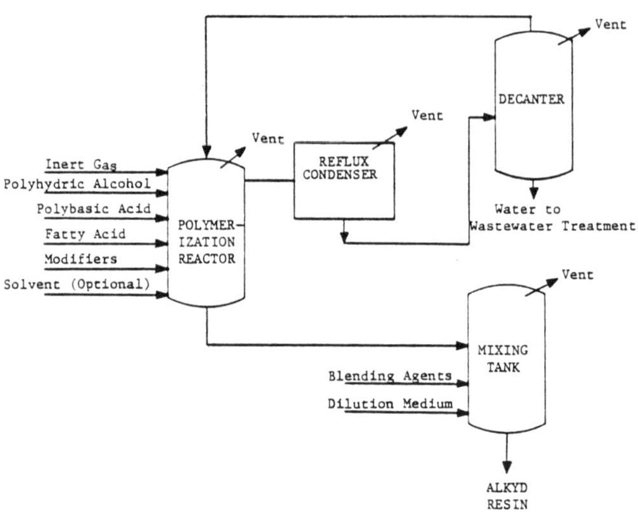

Figure 9. Alkyd resin production using the fatty acid method.

Sources: _Encyclopedia of Chemical Technology_, 3rd Edition.
Encyclopedia of Polymer Science and Technology.

Worker Distribution and Emissions Release Points

Worker distribution estimates have been made by correlating major equipment manhour requirements with the process flow diagrams in Figures 8 and 9. Estimates for both the alcoholysis process and the fatty acid method are shown in Table 34.

The closed process conditions associated with alkyd resin production minimize employee exposure potential to hazardous chemical substances and physical agents. However, fugitive emissions from the sources listed in Table 35 and from leaks in pump and compressor seals, valves, and drains may result in elevated workplace levels of fatty acids, polyols, polybasic acids, solvents and particulate matter (nuisance dust). Available toxicity information for major contaminants is given in the "Health Effects" discussion below and summarized in Table 10 in Section 1 of this document.

The additives used in alkyd molding resin formulation vary with product specifications. Asbestos, a known human carcinogen, is commonly used as a reinforcing agent. Chromium- and lead-based organic acid salts may be used as drying agents.

Employees in alkyd resin production may be exposed to inert gases (i.e. carbon dioxides or nitrogen) which are used to enhance reaction time and color development. Such gases are simple asphyxiants which are potentially hazardous because they reduce the available oxygen in the atmosphere.

Maintenance employees risk overexposure to sodium hydroxide, which is used periodically to clean polymerization reactors. Sodium hydroxide is extremely irritating to the skin and mucous membranes, and even short periods of overexposure can result in severe chemical burns.

Health Effects

Sufficient data do not exist in the available literature with which to evaluate the health effects of exposure to more than two-thirds of the typical input materials in the production of alkyd resin, excluding oils and inert gases. The chemicals for which sufficient data do exist include several highly toxic substances and chemicals noted for mutagenic and/or teratogenic potential. Although a number of input materials are reported tumorigenic agents, only two, the modifier acrylonitrile, and the catalyst lead oxide, are suspected human carcinogens. Styrene, also a modifier in the process, and maleic anhydride, a polybasic acid, also may pose a significant health risk to plant employees due to high toxicities and other properties. The reported health effects of exposure to these three substances are summarized below.

Acrylonitrile, a suspected human carcinogen [107], causes irritation of the eyes and nose, weakness, labored breathing, dizziness, impaired judgment, cyanosis, nausea, and convulsions in humans. Mutagenic, teratogenic, and tumorigenic effects have been reported in the literature. The OSHA air standard is 2 ppm (8 hour TWA) with a 10 ppm ceiling.[69]

TABLE 34.   WORKER DISTRIBUTION ESTIMATES FOR ALKYD RESIN PRODUCTION

Process	Unit	Workers/Unit/8-hour Shift
Alcoholysis	Batch Reactor	1.0
	Reflux Condenser	0.125
	Decanter	0.25
	Batch Mixer	1.0
Fatty Acid Method	Batch Reactor	1.0
	Reflux Condenser	0.125
	Decanter	0.25
	Batch Mixer	1.0

TABLE 35.   SOURCES OF FUGITIVE EMISSIONS FROM ALKYD RESIN MANUFACTURE

		Process	
Source	Constituent	Fatty Acid	Alcoholysis
Reactor Vent	Fatty Acid (Oil)	X	X
	Polyol	X	X
	Polybasic Acid	X	X
	Solvent	X[a]	X[a]
	Modifiers	X[a]	X[a]
Reflux Condenser Vent	Polyol	X	X
	Polybasic Acid	X	X
	Solvent	X[a]	X[a]
	Modifiers	X[a]	X[a]
Mixing Tank Vent	Dilution Medium	X	X
	Blending Agents	X[a]	X[a]
Alcohol Recovery Column Vent	Polyol	X	X
	Solvent	X[a]	X[a]
	Modifiers	X[a]	X[a]

[a]If used.

Lead oxide is a suspected human carcinogen which cause severe acute and chronic toxicological effects. Acute overexposure results in fatigue, colic, anemia and neuritis. Chronic overexposure can produce anemia and damage to the liver, kidneys, and nervous system. Instances of sterility, fetal toxicity, and mutagenicity have been reported.[170] The OSHA air standard is 0.05 mg/m^3 (8 hour TWA) (as lead).[69]

Maleic Anhydride is highly toxic by ingestion and inhalation [242] and is a powerful irritant to the skin and eyes.[149] Inhalation can cause pulmonary edema [149]; other effects of exposure include conjunctivitis, corneal damage, cough, bronchitis, headache, abdominal pain, nausea, and vomiting.[238] Animal data also suggest that maleic anhydride is a tumorigenic agent.[233] The OSHA air standard is 0.25 ppm (8 hour TWA).[67]

Styrene has been linked with increased rates of chromosomal aberrations in persons exposed in an occupational setting.[107] Animal test data strongly support epidemiological evidence of its mutagenic potential.[233] Styrene also produces tumors and affects reproductive fertility in laboratory animals.[233] Toxic effects of exposure to styrene usually involve the central nervous system.[233] The OSHA air standard is 100 ppm (8 hour TWA) with a ceiling of 200 ppm.[67]

Air Emissions

Fugitive VOC emission sources for alkyd resin production are listed in Table 35. These emission sources are all vents, including condenser vents, reactor vents, and mixing tank vents. Control technologies available for these vent streams include:[285]

- Venting the stream to a flare to incinerate noncondensible hydrocarbons; and/or

- Venting the stream to blowdown.

Other sources include fugitive VOC emissions which result from flanges, valves, pumps, compressors, relief valves, and agitators. Althoguh equipment modification is helpful in reducing VOC emissions from these sources in some cases, a regular inspection and maintenance program is the best control available.

There are no sources of particulate emissions from the alkyd resin production process.

Wastewater Sources

There are only two sources of wastewater associated with alkyd resin production: water generated by the condensation reaction used to manufacture the resin and routine cleaning water. These wastewater streams may contain trace amounts of alkyd resin, regardless of which process is used. Wastewater from the alcoholysis method may contain small amounts of catalyst residues. The caustic cleaning water is reused until the caustic is spent

before being sent to wastewater treatment. There are no data listing the
values for wastewater generation or characteristics in the literature
consulted.

<u>Solid Wastes</u>

The solid wastes generated by this process are mostly alkyd resins.
Solid waste streams result from substandard product which cannot be blended
as well as product lost during reactor cleaning, product blending, and
spillage.

<u>Environmental Regulation</u>

Effluent limitations guidelines have been set for the alkyd resin
industry. BPT, BAT, and NSPS call for the pH of the effluent to fall
between 6.0 and 9.0 (41 Federal Register 32587, August 4, 1976).

New source performance standards (proposed by EPA on January 5, 1981)
for volatile organic carbon (VOC) fugitive emissions include:

- Safety/release valves must not release more than 200 ppm above
  background, except in emergency pressure releases, which should not
  last more than five days; and

- Leaks (which are defined as VOC emissions greater than 10,000 ppm)
  must be repaired within 15 days.

The following compounds are listed as hazardous wastes (46 Federal
Register 27476, 20 May 1981):

```
Acrylonitrile         - U009
Formaldehyde          - U122
Maleic anhydride      - U147
Methyl methacrylate   - U162
Phthalic anhydride    - U190
Toluene               - U220
Xylene                - U239
```

All disposal of these compounds and alkyd resins which contain residuals of
these compounds must comply with the provisions set forth in the Resource
Conservation and Recovery Act (RCRA).

5. Amino Resins

INTRODUCTION

Amino resins are produced by reacting formaldehyde and a compound containing an amino group ($-NH_2$); melamine-formaldehyde and urea-formaldehyde are the major amino resins produced. Amino resins may be manufactured in a variety of forms: as liquids, spray dried solids, or filled molding compounds. Although the end uses differ, there is essentially only one process used for producing these compounds.

Amino resins are used in a wide variety of applications. Their resistance to heat, solvents, chemicals, and discoloration is combined with extreme surface hardness, making amino resins an excellent choice for use in adhesives, lamination, paper coating, dinnerware, and molding compounds.

Table A-5 in Appendix A presents typical properties of filled amino resin molding compounds. Fillers impart strength and moldability, improve dimensional stability, and reduce internal stresses. Coating, laminating, and adhesive resins form hard, solvent resistant bonds or finishes. Properties for these resins are formulation specific; that is, specific to end use requirements.

There are four major amino resin products: molding resins, coating resins, laminating resins, and adhesive resins. Adhesives represent the largest single market for amino resins, for which the uses include plywood, chipboard, and sawdust board manufacture.[279] Other amino resin applications include textile coatings, automobile tire adhesives, paper coatings, molded dinnerware, automotive topcoats, and buttons.[279]

INDUSTRY DESCRIPTION

In 1980, amino resin sales accounted for 3.7 percent of the total sales for the plastics industry.[65] The amino resin industry is comprised of 52 producers with 112 sites distributed between 27 states. There is a preponderance of plants in the southeast and great lakes areas (EPA Regions IV and V). Of the 112 sites, 73 sites produce melamine-formaldehyde resins, 105 sites produce urea-formaldehyde resins, and 66 sites produce both. Major U.S. amino resin producers are listed in Table 36 as well as the amino compound used for resin manufacture and resin end uses for each producer. The capacities of the amino resin plants listed are not available in the literature.

TABLE 36. U.S. AMINO RESIN PRODUCERS

Producer	Urea-Formaldehyde Resins	Melamine-Formaldehyde Resins	End Uses
American Cyanamid Co.			
Formica Corp., subsidiary			
Evendale, OH		X	Molding compounds,
Industrial Chemicals Division			fibrous and granu-
Kalamazoo, MI		X	lated wood, ply-
Mobile, AL	X	X	wood, laminates,
Wallingford, CT	X	X	protective coat-
Organic Chemicals Division			ings, paper treat-
Charlotte, NC	X	X	ing, textile
			treating
American Hoechst Corp.			
Industrial Chemicals Division			
Mount Holly, NC	X	X	Laminates, textile
			treating
Apex Chemical Corp., Inc.			
Elizabethport, NJ	X		Paper treating,
			textile treating
Auralux Chemical Associates, Inc.			
Hope Valley, RI	X	X	Textile treating
The Bendix Corp.			
Friction Materials Division			
Green Island, NY	X		Paper treating,
			other
Borden Inc.			
Borden Chemical Division			
Adhesives & Chemicals			
Division East			
Demopolis, AL	X		Fibrous and gran-
Diboll, TX	X	X	ulated wood, ply-
Fayetteville, NC	X		wood, laminates,
Louisville, KY	X		paper treating,
Sheboygan, WI	X	X	other markets

(continued)

TABLE 36 (continued)

Producer	Urea-Formaldehyde Resins	Melamine-Formaldehyde Resins	End Uses
Borden Inc.			
Borden Chemical Division			
Adhesives and Chemicals			
Division West			
Freemont, CA	X		
Kent, WA	X	X	
La Grande, OR	X		
Missoula, MT	X		
Springfield, OR	X	X	
Cargill Inc.			
Chemical Products Division			
Carpentersville, IL	X	X	Protective
Forest Park, GA	X	X	coatings
Lynwood, CA	X	X	
Celanese Corp.			
Celanese Plastics and Specialties			
Co. Division			
Celanese Speciality Resins			
Division			
Louisville, KY	X	X	
Chagrin Valley Co. Ltd.			
Nevamar Corp., subsidiary			
Odenton, MD		X	Laminates
Clark Oil & Refining Corp.			
Clark Chemical Corp., subsidiary			
Blue Island, IL	X	X	
C.N.C. Chemical Corp.			
Providence, RI	X	X	Textile treating
Commercial Products Co., Inc.			
Hawthorne, NJ	X		
Consolidated Papers, Inc.			
Consoweld Corp., subsidiary			
Wisconsin Rapids, WI	X		Laminates

(continued)

TABLE 36 (continued)

Producer	Urea-Formaldehyde Resins	Melamine-Formaldehyde Resins	End Uses
Cook Paint and Varnish Co. Detroit, MI	X	X	Protective coatings
Cook Industrial Coatings Co. North Kansas City, MO	X	X	
Crown-Metro, Inc. Greenville, SC	X	X	

Dan River, Inc.
  Chemical Products Division
   Danville, VA
	X	X	Textile treating

DeSoto Inc. Garland, TX	X		Protective coatings
Dock Resins Corp. Linden, NJ	X	X	Protective coatings, textile treating
Eastern Color and Chemical Co. Providence, RI	X	X	Textile treating

General Electric Co.
  Engineered Materials Group
   Electromaterials Business
   Department
    Coshocton, OH
		X	Laminates

Schenectady, NY		X	

Georgia-Pacific Corp.
  Chemical Division
   Albany, OR
	X	X	Fibrous and granulated wood, plywood, paper treating

Columbus, OH	X	X	
Conway, NC	X	X	
Coos Bay, OR	X	X	
Crossett, AR	X	X	
Eugene, OR	X	X	
Louisville, MS	X	X	

(continued)

TABLE 36 (continued)

Producer	Urea-Formaldehyde Resins	Melamine-Formaldehyde Resins	End Uses
Georgia-Pacific Corp. (Cont.)			
Chemical Division			
Lufkin, TX	X	X	
Newark, OH	X	X	
Peachtree City, GA	X		
Port Wentworth, GA	X	X	
Richmond, CA	X	X	
Russellville, SC	X	X	
Taylorsville, MS	X	X	
Ukiah, CA	X	X	
Vienna, GA	X	X	
Getty Oil Company			
Chembond Corp., subsidiary			
Andalusia, AL	X		Fibrous and granu-
Springfield, OR	X	X	lated wood,
Winnfield, LA	X	X	plywood
Guardsman Chemicals Inc.			
Grand Rapids, MI	X	X	Protective coatings
Gulf Oil Corporation			
Gulf Oil Chemicals Co.			
Industrial Chemicals Division			
High Point, NC	X		
West Memphis, AR	X		
Milmaster-Onyx Group			
Lyndhurst, NJ	X		
Hanna Chemical Coatings Corp.			
Columbus, OH	X	X	Protective coatings
Hercules, Inc.			
Chicopee, MA	X		
Hattiesburg, MS	X		
Milwaukee, WI	X		
Portland, OR	X		
Savannah, GA	X		

(continued)

TABLE 36 (continued)

Producer	Urea-Formaldehyde Resins	Melamine-Formaldehyde Resins	End Uses
H & N Chemical Co.			
Totowa, NJ	X		
Libby-Owens-Ford Co.			
LOF Plastic Products,			
subsidiary			
Auburn, ME	X	X	Laminates
Mobil Oil Corp.			
Mobil Chemical Co. Division			
Chemical Coatings Division			
Kankakee, IL	X	X	Protective coatings
Monsanto Co.			
Monsanto Plastics and Resins Co.			
Addyston, OH	X		Molding compounds,
Chocolate Bayou, TX	X		fibrous and granu-
Eugene, OR	X		lated wood, ply-
Santa Clara, CA	X	X	wood, laminates,
Springfield, MA	X	X	protective coat-
			ings, paper treat-
			ing, textile
			treating
National Casein Co.			
Chicago, IL	X		
Tyler, TX	X		
National Casein of CA			
Affiliate of National Casein Co.			
Santa Anna, CA	X		
National Casein of NJ			
Affiliate of National Casein Co.			
Adhesives Division			
Riverton, NJ	X		
National Starch & Chemical Corp.			
Proctor Chemical Co., Inc.,			
subsidiary			
Salisbury, NC	X	X	Paper treating, textile treating

(continued)

TABLE 36 (continued)

Producer	Urea-Formaldehyde Resins	Melamine-Formaldehyde Resins	End Uses
Perstorp Inc.			
Owned by Perstorp, AB Sweden			
Florence, MA	X	X	Molding compounds
Plaskon Products, Inc.			
Toledo, OH	X		
Plastics Manufacturing Co.			
Dallas, TX	X	X·	Molding compounds, laminates
PPG Industries Inc.			
Coatings and Resin Division			
Circleville, OH		X	Protective
Oak Creek, WI	X	X	coatings
Reichhold Chemicals Inc.			
Andover, MA	X	X	Fibrous and granu-
Azusa, CA	X		lated wood, ply-
Detroit, MI	X	X	wood, protective
Moncure, NC	X		coatings, paper
South San Francisco, CA	X	X	treating, textile
Tacoma, WA	X	X	treating
Tuscaloosa, AL	X	X	
White City, OR	X	X	
Vacuum Division			
Niagara Falls, NY	X		
Scott Paper Co.			
Packaged Products Division			
Chester, PA	X	X	Paper treating
Everett, WA	X		
Fort Edward, NY	X		
Marinette, WI	X		
Mobile, AL	X	X	
Southeastern Adhesives Co.			
Lenoir, NC	X		

(continued)

TABLE 36 (continued)

Producer	Urea-Formaldehyde Resins	Melamine-Formaldehyde Resins	End Uses

The Standard Oil Co. (Ohio)
  Sohio Industrial Products
  Co. Division
  Dorr-Oliver, Inc. Unit
    Niagra Falls, NY
	X		

Sun Chemical Corp.
  Chemicals Group
  Chemicals Division
    Chester, SC
	X	X	Paper treating, textile treating

Sybron Corp.
  Chemical Division
  Jersey State Chemical Co.
  Division
    Haledon, NJ
	X	X	Protective coatings, textile treating

Synthron, Inc. Ashton, RI	X	X	Textile treating
Morganton, NC	X	X	

Tyler Corporation
  Reliance Universal Inc.,
  subsidiary
  Specialty Chemicals and
  Resins Division
    Louisville, KY
	X	X	

United Merchants and Manufacturers, Inc.
  Valchem Chemical Division
    Langley, SC
	X	X	Textile treating

U.S. Oil Company East Providence, RI	X	X	Textile treating

Southern U.S. Chemical Co.,
  Inc., subsidiary
    Rock Hill, SC
	X	X	

(continued)

TABLE 36 (continued)

Producer	Urea-Formaldehyde Resins	Melamine-Formaldehyde Resins	End Uses

Valspar Corporation
  McWhorter, Inc., subsidiary
    Baltimore, MD
	X	X	

Westinghouse Electric Corp.
  Insulating Materials Division
    Manor, PA
		X	Laminates

West Point-Pepperell, Inc.
  Grifftex Chemical Co., subsidiary
    Opelika, AL
	X	X	Textile treating

Weyerhaeuser Co. Marshfield, WI	X		Fibrous and granulated wood, plywood

Sources: Chemical Economics Handbook, updated annually, 1980 data.
         Directory of Chemical Producers, 1982.

The amino resin industry is consumer oriented. Adhesives, coatings, and laminates are used in the construction, transportation, and consumer goods industries. Urea-formaldehyde resin production peaked in 1979 at 621,400 metric tons, then fell to 529,500 metric tons in 1980. This production level, however, represents an 11 percent compound growth rate from 1975 to 1980.[65] Melamine-formaldehyde resins have not fared as well. After a 1978 production peak of 91,800 metric tons, 1980 production fell to 75,900 metric tons which is 9,400 metric tons below the 1976 production level. Compound growth for melamine-formaldehyde resin production from 1975 to 1980 was 7.7 percent.[65] Total amino resin production for 1981 was 645,000 metric tons.[143]

PRODUCTION AND END USE DATA

In 1981, U.S. production of amino resins totaled 645,000 metric tons. Amino resins have many uses, including:[136, 279, 288]

- Adhesives for the manufacture of plywood, particle board, chip board, and sawdust board, and furniture assembly glues;

- Adhesives to bond rubber to tire cord and binders for metal coatings;

- Protective surface coatings, durable baked enamels for applicances, wood furniture coatings, and automotive primer and topcoats;

- Textile treating;

- Coatings for floor finishes, metals, primer coats, and paper coating;

- Decorative laminated plastic sheets for counter and table tops;

- Flexible backing on carpets and draperies;

- Wiring devices, including circuit breakers, wall plates, and receptacles;

- Molded products, including cutlery and telephone handles, ceiling panels, staircases, bottle and jar caps, dinnerware, buttons, cosmetic caps, electric blanket control housings, toothpaste tube inserts, toilet seats, housings, knobs, handles, ashtrays, lavatory bowls, utensil handles, electric shaver housings, mixing bowls, and appliance components;

- Automotive parts, including ignition plugs and connector plug inserts;

- Foam for insulation in commercial buildings, private residences, ships, air frames, and tunnels, and for supporting floral displays;

- Oil absorbers for ocean spills;

- Dressing for wounds;

- Leather tanning;

- Cure of other resins; and

- Plant nutrients and conditioners.

Table 37 lists amino resin consumption by market, and Tables 38 and 39 present consumption of urea-formaldehyde and melamine-formaldehyde molding resins, respectively.

PROCESS DESCRIPTIONS

Amino resins are produced by essentially one process. Variations in the process enable the production of the variety of products which constitute the amino resin industry. For example, coating resins are manufactured by adding methanol or butanol to the reactor; these compounds then contribute to the condensation reaction to form the desired resin. Adhesives are the largest single segment of the amino resins market, and most of the adhesives produced are used for the manufacture of plywood, particle board, and other wood products.

Amino resins are produced in a two step reaction. In the first step, intermediate compounds are formed. Intermediate formation may be catalyzed by either acidic or alkaline conditions. Reactions for the formation of urea-formaldehyde and melamine-formaldehyde intermediates are shown below:

<u>Urea-Formaldehyde</u>

$$H_2N-\overset{\overset{O}{\|}}{C}-NH_2 \;+\; H-\overset{\overset{O}{\|}}{C}-H \longrightarrow H_2N-\overset{\overset{O}{\|}}{C}-NHCH_2OH$$

Urea   Formaldehyde   Monomethylolurea

$$H_2N-\overset{\overset{O}{\|}}{C}-NHCH_2OH \;+\; H-\overset{\overset{O}{\|}}{C}-H \longrightarrow HOCH_2NH-\overset{\overset{O}{\|}}{C}-NHCH_2OH$$

Monomethylolurea Formaldehyde Dimethylolurea

<u>Melamine-Formaldehyde</u>

$$\text{Melamine} \;+\; 3H-\overset{\overset{O}{\|}}{C}-H \longrightarrow \text{Trimethylol Melamine}$$

Melamine   Trimethylol Melamine

TABLE 37. 1981 PATTERN OF CONSUMPTION FOR AMINO RESINS

Market	Thousand Metric Tons
Bonding and Adhesive Resins for:	
Fibrous and Granulated Wood	427
Laminating	16
Plywood	36
Molding Compounds	39
Paper Treatment and Coating Resins	37
Protective Coatings	36
Textile Treatment and Coating Resins	39
Export	10
Other	5
TOTAL	645

Source: _Modern Plastics_, January 1982.

TABLE 38.   1981 MOLDING POWDER MARKETS FOR UREA-FORMALDEHYDE RESINS

Market	Thousand Metric Tons
Closures	6
Electrical	15
Other	1
TOTAL	22

Source: *Modern Plastics*, January 1982.

TABLE 39.   1981 MOLDING POWDER MARKETS FOR MELAMINE-FORMALDEHYDE RESINS

Market	Thousand Metric Tons
Buttons	1
Dinnerware	14
Sanitary Ware	1
Other	1
TOTAL	17

Source: *Modern Plastics*, January 1982.

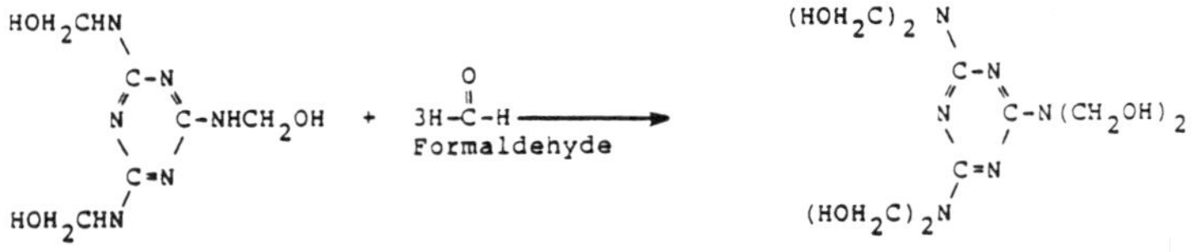

$$\text{Trimethylol Melamine} + 3\,H\!-\!\underset{\text{Formaldehyde}}{\overset{O}{\overset{\|}{C}}}\!-\!H \longrightarrow \text{Hexamethylol Melamine}$$

The second reaction step results in the formation of low molecular weight polymers by several condensation reactions. These reactions form methylene bridges ($-CH_2-$) between a methylol and an amino group and either a methylene bridge ($-CH_2-$) or an ether linkage ($-CH_2-O-CH_2-$) between two methylol groups. Example condensation reactions are shown below:

(1) Methylol and amino group methylene bridge

$$\text{General:}\quad RNHCH_2OH + H_2NR' \longrightarrow RNHCH_2NHR' + H_2O$$

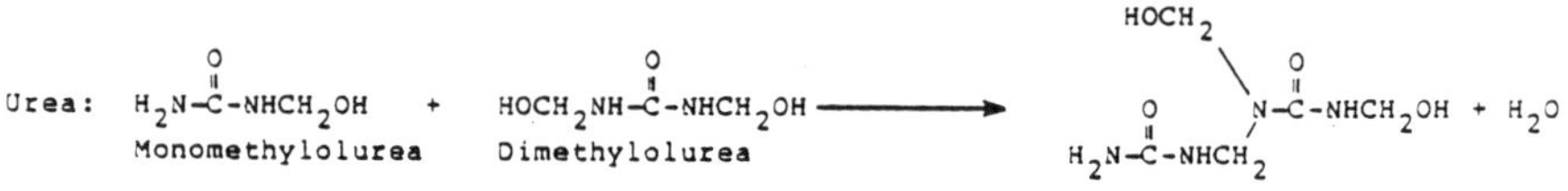

Urea: $H_2N\!-\!\overset{O}{\overset{\|}{C}}\!-\!NHCH_2OH$ (Monomethylolurea) + $HOCH_2NH\!-\!\overset{O}{\overset{\|}{C}}\!-\!NHCH_2OH$ (Dimethylolurea) $\longrightarrow$ product $+ H_2O$

Melamine (Product):

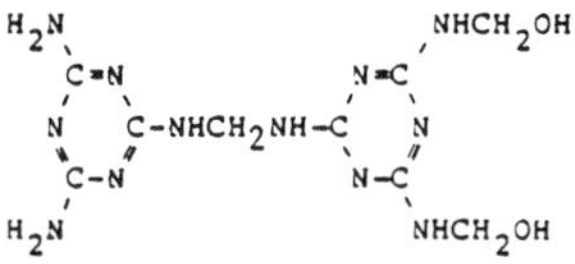

(2) Two methylol groups methylene bridge

General: $2RNHCH_2OH \longrightarrow RNHCH_2NHR + H_2O + HCHO$

Urea:

```
        O                           O
        ||                          ||
H2N-C-NHCH2OH    +   HOCH2NH-C-NHCH2OH  ----->
Monomethylolurea     Dimethylolurea

         O               O
         ||              ||
H2N-C-NHCH2NH-C-NHCH2OH  +  H2O  +  H-C-H
```

Melamine (Product):

```
HOH2CHN                          NHCH2OH
        C=N              N=C
     N      C-NHCH2NH-C      N
        C-N              N-C
HOH2CHN                          NHCH2OH
```

(3) Two methylol groups ether linkage

General: $RNHCH_2OH + HOCH_2NHR' \longrightarrow RNHCH_2OCH_2NHR' + H_2O$

Urea:

```
        O                           O
        ||                          ||
H2N-C-NH-CH2OH   +  HOCH2NH-C-NH-CH2OH  ----->
Monomethylolurea    Dimethylolurea

         O                    O
         ||                   ||
H2N-C-NHCH2-O-CH2NH-C-NHCH2OH  +  H2O
                Ether
```

Melamine (Product):

```
HOH2CHN                              NHCH2OH
        C=N                  N=C
     N      C-NHCH2-O-CH2NH-C      N
        C-N                  N-C
HOH2CHN                              NHCH2OH
              Ether
```

Urea-formaldehyde and melamine-formaldehyde coating resins incorporate an alkyl group to increase the resin's solubility in organic solvents and render it more stable for increased shelf life. The hydrogen of the methylol group is replaced with an alkyl group (alkylation) from either methanol or butanol. The reaction is catalyzed by acids and carried out with excess alcohol. The excess alcohol helps suppress condensation reactions which compete with the alcohol reaction. Following are examples of the alkylation:

Urea-Formaldehyde

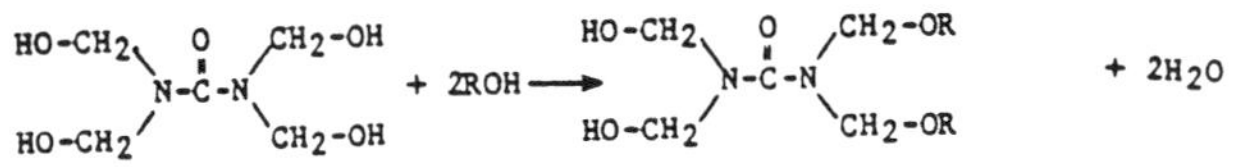

$$HO-CH_2\diagdown \quad \overset{O}{\underset{}{C}} \quad \diagup CH_2-OH \qquad + \ 2ROH \longrightarrow \qquad HO-CH_2\diagdown \quad \overset{O}{\underset{}{C}} \quad \diagup CH_2-OR \qquad + \ 2H_2O$$

$$HO-CH_2\diagup \overset{N-C-N}{} \diagdown CH_2-OH \qquad\qquad HO-CH_2\diagup \overset{N-C-N}{} \diagdown CH_2-OR$$

where R includes CH_3 and $CH_3CH_2CH_2CH_2$.

Melamine-Formaldehyde

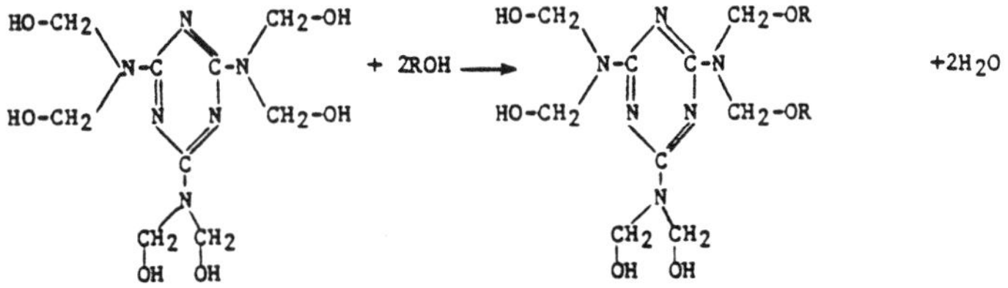

where R includes CH_3 and $CH_3CH_2CH_2CH_2$.

Urea-formaldehyde resins are essentially monomeric with five or six units in a polymer. Melamine-formaldehyde resins have 20 to 30 units per polymer. This difference is due to urea having two sites which react, making it difunctional, while melamine contains three active sites, rendering it trifunctional.

Tables 40 and 41 list typical input materials and operating parameters for the four types of amino resins produced. Other input materials used such as fillers, cure catalysts, stabilizers, and self-extinguishing agents for foam, such as polypropylene, are listed in Table 42.

Adhesives

Adhesive resins are produced using either melamine, urea, or both compounds. Melamine is more costly but produces a more durable glue which

TABLE 40. TYPICAL INPUT MATERIALS TO AMINO RESIN PRODUCTION
IN ADDITION TO UREA, MELAMINE, AND FORMALDEHYDE

Product	Alcohol	Modi-fying Agent	Filler	pH Buffer	Curing Agent	Dyes	Cata-lyst	Mold Release Agent
Adhesives			X	X			X	
Coatings	X[a]	X						
Laminates				X				
Molding Resins	X[b]		X		X	X	X	X

[a]For production.
[b]For dilution.

TABLE 41. TYPICAL OPERATING PARAMETERS FOR AMINO RESIN PRODUCTION

Product	Temperature	pH	Reaction Time	Urea: Formaldehyde Ratio	Melamine: Formaldehyde Ratio
Adhesives	20°C	7.5 -8[a]	8 hours	1:1.5-1:2.0	1:3
Coatings	24-66°C[b]	7-8[b]	4 hours	1:1-1:2	1:2-1:6
Laminates	N/A	7-10	N/A	Not used	1:2
Molding Resins	24-66°C[b] 50-60°C[c]	7-8[b] 6.0-6.5[c]	4 hours	1:1.3-1:1.5	1:2-1:3

N/A - Not available.

[a]Polymerization pH starts at 7.5-8.0, is reduced to 6.0, then
 raised to 8.0 when the reaction is completed.
[b]Polymerization.
[c]Mixing.

TABLE 42. INPUT MATERIALS AND SPECIALTY CHEMICALS USED IN AMINO RESIN
MANUFACTURE (IN ADDITION TO UREA, MELAMINE, AND FORMALDEHYDE)

Function	Compound
Heat Resistance Additive	Acrylamide Hafnium chloride Melamine Zirconium chloride
Impact Resistance Additive	Ammonium polysulfide Glycidyl methacrylate
Odor Reducing Additive	Methylene acetoacetate Methylene phosphate
Crosslinking Agent	Dipropylene glycol
Internal Hardener	2-amino-2-ethyl-1,3-propane- diol hydrochloride
Self-Extinguishing Agent	Polypropylene
Catalyst Neutralizer	Tricalcium phosphate Triethanolamine

External Hardeners	Chlorides of organic amides
	Formic acid
	Phthalic acid
	Pyrridine monochloroacetic
	  acid
	p-toluene sulfonic acid
	Sulfates of organic amides
	Sulfites
	  Sodium disulfite
	  Sodium sulfite
	Sulfur dioxide
	Weak organic acids

Cure Retardant	Ammonia tetramine Hexamethyl tetramine
Buffer	Sodium formate

(continued)

TABLE 42 (continued)

Function	Compound
Fillers	Asbestos
	α-cellulose
	Bleached kraft cellulose
	Bleached wood pulp
	Cellulose filler
	Cereals
	Chopped cotton yarn
	Chopped nylon
	Coconut shell powder
	Cotton linters
	Diced cotton fabric
	Dimethyl formamide
	Furfurylaldehyde
	Glass fiber
	Ground silica
	Maize
	Mica
	Potato, tapioca or wheat starches
	Powdered phenolic molding waste
	Retch (a weed in cereals)
	Root flour
	Rye
	Rye flour
	Sulfite cellulose
	Walnut shell powder
	Wood flour
Sizing Agent	Wax
Modifying Agent	Amide polymer
	Carboxyl polymer
	Hydroxyl polymer
Craze Resistance Agent	Benzyl alcohol
	Furfuryl alcohol
Mold Release Agent	Calcium stearate
	Dicresyl glyceryl ether
	Glyceryl monostearate
	Magnesium stearate
	Oxidized paraffin wax
	Sulfonated castor oil
	Zinc stearate

(continued)

TABLE 42 (continued)

Function	Compound
Flow Control Agent	Melamine-formaldehyde spray dried resin Urea-formaldehyde spray dried resin Water
Stabilizer	Diethanolamine Hexamethylenetetramine (HMTA) Triethanolamine
Curing Catalysts	Ammonium chloride Ammonium sulfate Dichlorohydrin Hexamine thiocyanate Trimethyl phosphate Zinc sulfate Zinc sulfite
Flow Promoters and Plasticizers	Aromatic monoethers of glycerol α-methyl-D-glucoside Benzamide p-toluenesulfonamide p-toluenesulfonamide-formaldehyde
Pigments	Azo pigments Cadmium sulfides Hansa yellows Iron oxides Phthalocyanine blues Titanium dioxide Ultramarines Zinc oxides
Comonomers	Benzoguanamine Dihydroxyethyleneurea Phenol Vinyl acetate

Sources: John F. Blais, _Amino Resins_, 1959.
Encyclopedia of Chemical Technology, 3rd Edition.
Encyclopedia of Polymer Science and Technology.
Beat Meyer, _Urea-Formaldehyde Resins_, 1979.
Modern Plastics Encyclopedia, 1981-1982.
C. P. Vale and W. G. K. Taylor, _Aminoplastics_, 1964.

is waterproof, rendering it useful for exterior and marine applications.
One alternative to lower the cost of using melamine is to combine melamine
and urea in the same formulation.

Adhesives are produced by reacting formaldehyde with urea or melamine
under carefully controlled conditions. Fillers are often added to the glue,
as are acid catalysts to harden the glue after application and furfuryl or
benzyl alcohol which renders the glue resistant to cracking and crazing.
Urea glues are sold as a liquid resin with a separate powdered hardener
which contains a catalyst, a pH buffer, and a cure retardant.

As illustrated in Figure 10, adhesive manufacture begins with formaldehyde and the amino compound being fed into a stirred tank reactor where the
pH is carefully controlled. Reactor inputs for the production of adhesive
amino resins typically include:[281]

Material	Parts by Weight
Urea (amino monomer)	273.0
Formaldehyde (30%) (monomer)	908.7
Boric Acid (catalyst)	45.4
Sodium Hydroxide (pH adjustment)	small amount

The water produced by the condensation reaction is removed in a separator
and the resulting resin mixture is filtered. Additives used in the specific
adhesive formulation desired are blended into the resin in a mixer before
the adhesive is subsequently sold in liquid form.

Amino adhesives have low shrinkage, are craze resistant, and form thick
glue lines. Urea adhesives are sold as a liquid resin with a separate
powdered hardener while melamine resins are sold as a liquid resin with the
hardener suspended in the liquid. The solids content of these resins is 50
to 60 percent.

<u>Coatings</u>

Coating resins are manufactured using an alcohol, such as n-butanol or
methanol, in an alkylation reaction. This alkylation step renders the
liquid product soluble in the common organic solvents used in most coating
processes. Both urea and melamine resins are used. Urea resins have poor
water resistance but cure rapidly while melamine resins are more costly, yet
exhibit better overall performance. If methanol is used in the alkylation,
the resulting coating is water soluble and is used in textile treating.
Amino resins create a hard, solvent resistant coating upon heating with
hydroxyl, carboxyl, and amide polymers.

As illustrated in Figure 11, the amino compound and formaldehyde are
fed to a polymerization reactor to start the production process. The water
is removed in a separator before the polymer mixture is filtered and diluted
with either methanol or butanol. An azeotrope may be formed by adding
xylene to the reactor which aids in removing any remaining water from the
resin. The xylene is not removed from the coating resin. Modifying agents

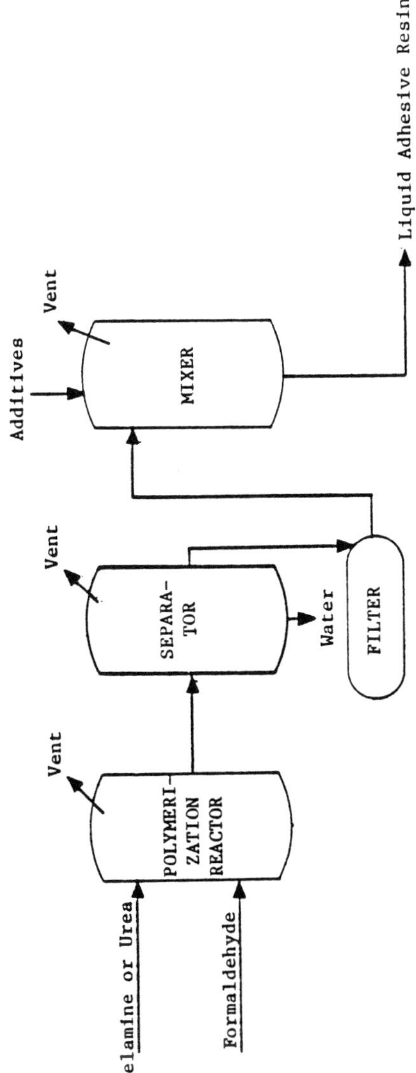

Figure 10. Amino adhesive resin production process.

Sources: Encyclopedia of Chemical Technology, 3rd Edition.
Encyclopedia of Polymer Science and Technology.
Modern Plastics Encyclopedia, 1981-1982.

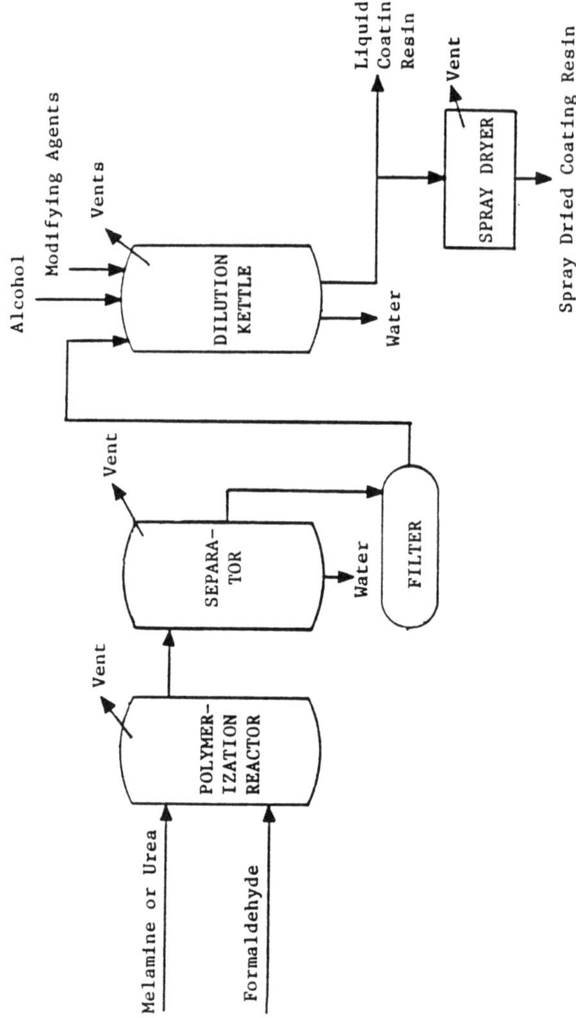

Figure 11.   Amino coating resin production process.

Sources: Encyclopedia of Chemical Technology, 3rd Edition.
Encyclopedia of Polymer Science and Technology.
Modern Plastics Encyclopedia, 1981–1982.

(e.g., hydroxyl, carboxyl, or amide polymers) are added to the dilution
kettle to enable a hard, solvent resistant coating upon application of the
resin. If the resin is desired in liquid form, no further processing is
necessary; the resin content of the liquid ranges from 50 to 60 percent.
Dried resins are produced via spray drying. Typical quantities of input
material for the production of amino resin coatings were not found in the
literature.

Laminates

Laminating resins are based on melamine-formaldehyde products; urea is
not used. Melamine-formaldehyde resin laminates exhibit excellent hardness,
clarity, stain resistance, and freedom from yellowing.

Laminating resin manufacture is presented in Figure 12. A typical production recipe is as follows:

Material	Parts by Weight
Melamine (amino monomer)	100
Formaldehyde (37%) (monomer)	133
Catalyst	2-12
Sodium Hydroxide (pH adjustment)	small amount
Formic Acid (pH adjustment)	small amount

Melamine and formaldehyde are combined in a reactor and polymerized. Water
produced by the condensation reaction is removed in a separator before the
polymer mixture is filtered. Laminating resins may be sold as a viscous
liquid or as a dried resin following spray drying. Spray dried resins
exhibit a longer shelf life than the syrups which have a resin content of 60
to 65 percent. Ethanol or isopropyl alcohol may be added after the syrup is
manufactured for specific laminating uses.

Molding Resins

Both melamine and urea are used to manufacture molding resins. Fillers
are added to impart strength and moldability, improve dimensional stability,
and reduce internal stresses. Asbestos is used as a filler for compounds
requiring high arc, heat, and flame resistance and high dielectric strength.
Glass fiber is used to give high impact strength, heat and arc resistance,
and good electrical properties. A typical production recipe for molding
amino resins is as follows:[236, 257]

Material	Parts by Weight
Melamine (amino monomer)	100
Formaldehyde (37%) (monomer)	133
Alpha Cellulose (filler)	
Dyes, Pigments	

The amounts of filler or dyes and pigments used were not found in the
literature.

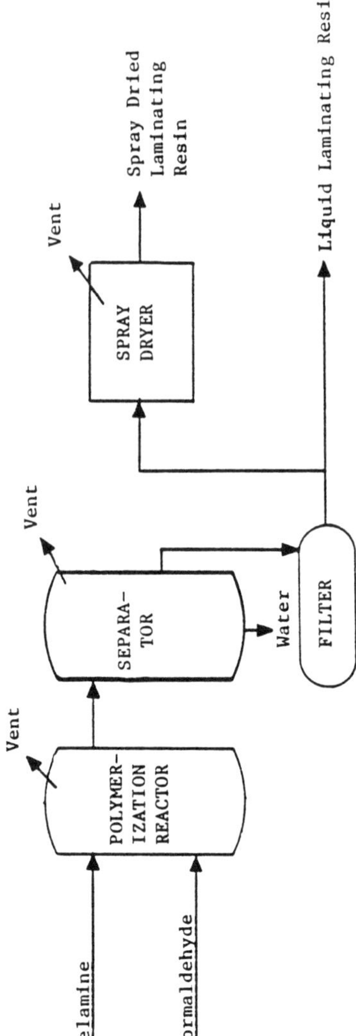

Figure 12.   Amino laminating resin production process.

Sources: Encyclopedia of Chemical Technology, 3rd Edition.
Encyclopedia of Polymer Science and Technology.
Modern Plastics Encyclopedia, 1981–1982.

Molding resin production is depicted in Figure 13. Formaldehyde and the amino compound are added to a stirred reactor to start the production process. Water from the condensation reaction is separated from the polymer mixture before filtration. The resin may be diluted with alcohol prior to mixing where the fillers appropriate for the desired resin end use are added. After mixing, the resin is screened, dried, and compounded in a ball mill. Mold release agents, catalysts, dyes, and curing agents are added before the formulated polymer is extruded.

Energy Requirements

No data were found giving the energy requirements for amino resin production in the literature consulted.

ENVIRONMENTAL AND INDUSTRIAL HEALTH CONSIDERATIONS

Urea-formaldehyde and melamine-formaldehyde resins are considered to be nontoxic in the cured state. Uncured resins, however, contain a small amount of free formaldehyde; therefore, care should be taken in the handling and storage of these resins. Since most recipes for amino resins call for a limiting amount of formaldehyde, the majority of the formaldehyde emissions will occur in the reactor vent and other upstream operations. Four input materials have been listed as hazardous under RCRA: asbestos, formaldehyde, formic acid, and phenol. Input materials which pose significant health effects include: asbestos, a known human carcinogen; polypropylene and tri-methyl phosphate, potential human carcinogens; and acrylamide and furfuryl alcohol.

Worker Distribution and Emissions Release Points

We have estimated worker distribution for amino resin production processes by correlating major equipment manhour requirements with the process flow diagrams in Figures 10 through 13. Estimates for adhesive, coating, laminate and molding resin production are shown in Table 43.

No major air emission point sources are associated with amino resin production. Fugitive emission sources, however, may significantly impact environmental residuals and/or worker health. The degree to which these impact the worker's environment are determined by the stream constituents, process operating parameters, engineering and administrative controls, and maintenance programs.

Sources of fugitive emissions and the probable contaminants emitted are listed in Table 44. In addition to these contaminants, numerous other minor additives (see Table 42) may be emitted, especially around the batch mixer (all processes), dryer (coating and laminate production), and ball mill and extruder vents (molding resin production).

Available toxicity data for principal input materials and additives are discussed below and summarized in Table 10 in Section 1 of this document.

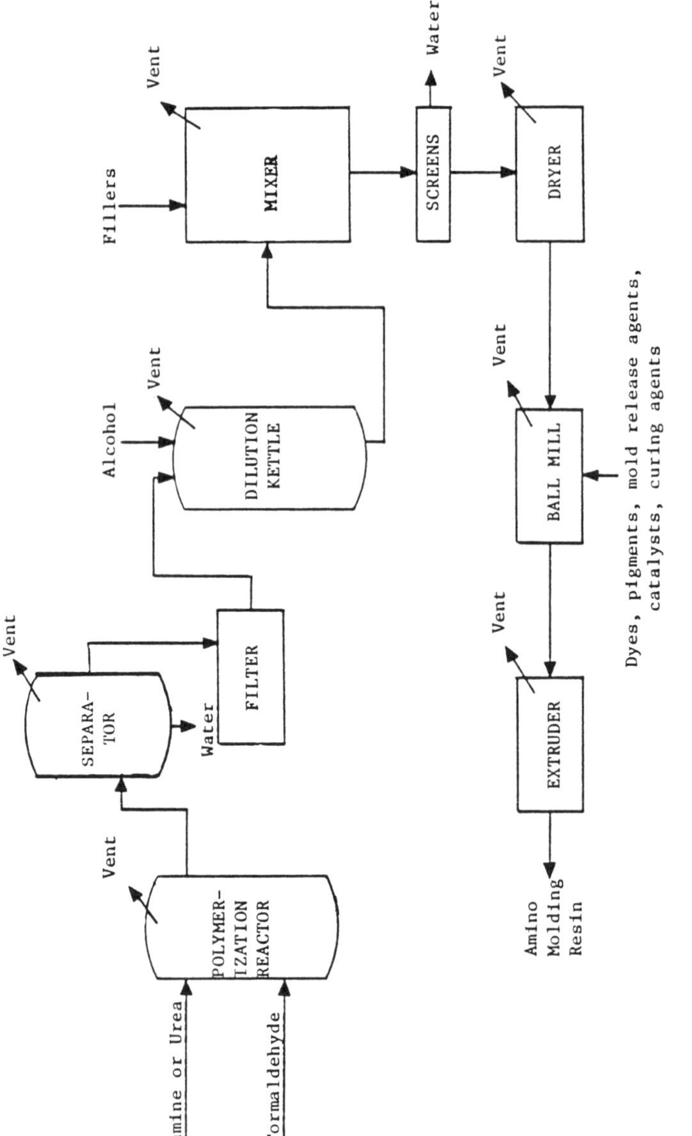

Figure 13. Amino molding resin production process.

Sources: Encyclopedia of Chemical Technology, 3rd Edition.
Encyclopedia of Polymer Science and Technology.
Modern Plastics Encyclopedia, 1981–1982.

TABLE 43.   WORKER DISTRIBUTION ESTIMATES FOR AMINO RESIN PRODUCTION

Process	Unit	Workers/Unit/8-hour Shift
Adhesive Manufacture	Batch Reactor	1.0
	Separator	0.25
	Filter	0.25
	Batch Mixer	1.0
Coating Manufacture	Batch Reactor	1.0
	Separator	0.25
	Filter	0.25
	Dilution Kettle	1.0
	Spray Dryer	1.0
Laminate Manufacture	Batch Reactor	1.0
	Separator	0.25
	Filter	0.25
	Spray Dryer	1.0
Molding Resin Production	Batch Reactor	1.0
	Separator	0.25
	Filter	0.25
	Dilution Kettle	1.0
	Batch Mixer	1.0
	Screens	0.25
	Dryer	0.5
	Ball Mill	0.25
	Extruder	1.0

TABLE 44.   SOURCES OF FUGITIVE EMISSIONS FROM AMINO RESIN MANUFACTURE

Source	Constituent	Product			
		Adhe-sives	Coat-ings	Lami-nates	Molding Resins
Reactor Vent	Formaldehyde	X	X	X	X
	Urea or Melamine	X	X	X	X
Separator Vent	Formaldehyde	*	*	*	*
	Urea or Melamine	X	X	X	X
Mixer Vent	Formaldehyde	*			*
	Urea or Melamine	X			X
Dilution Kettle Vent	Formaldehyde		*		*
	Urea or Melamine		X		X
	Alcohol		X		X
Spray Dryer or Dryer Vent	Formaldehyde, Urea or Melamine, and Resins Emissions and Particulates		X	X	X
	Alcohol		X		X
Ball Mill Vent	Formaldehyde, Urea or Melamine, and Resins Emissions and Particulates				X
	Alcohol				X
Extruder Vent	Formaldehyde, Urea or Melamine, and Resins Emissions and Particulates				X
	Alcohol				X

*Expected in trace amounts if present in this stream due to high conversion and limiting amount of formaldehyde in the reactor.

Little data are available from which employee exposure potential in amino resin production processes may be estimated. However, one source gives the following emission factors [286]:

- Adhesives--25 g hydrocarbon/kg product
- Molding--75 g particulate matter/kg product

Health Effects

Formaldehyde, one of the major raw materials in amino resin manufacture, is highly toxic and exhibits tumorigenic, mutagenic, and teratogenic potential. Three other input materials and specialty chemicals in this industry are known or potential carcinogens. They are: the filler asbestos, which is a known human carcinogen; polypropylene, a self-extinguishing agent, and trimethyl phosphate, a curing agent, both of which are known animal carcinogens. Acrylamide, used as a heat resistance additive, and furfuryl alcohol, a craze resistance agent in the process, also pose a significant risk to the health of plant employees if overexposure occurs. The following paragraphs provide a brief synopsis of the reported health effects of exposure to these substances.

Acrylamide is a neurotoxin, causing symptoms of ataxia, hypersomnia, and vertigo.[170] Although the polymer is nontoxic, absorption of acrylamide through skin or dusts is associated with serious neurological consequences.[88] Animal test data indicate that exposure to acylamide might also cause mutagenic and teratogenic effects. The OSHA air standard is 30 ug/m^3 (8 hour TWA).[67]

Asbestos is classified as a human carcinogen by the International Agency for Research on Cancer [104] and by the American Conference of Governmental Industrial Hygienists.[4] It is currently undergoing additional tests by the National Toxicology Program.[233] Many deaths from exposure to absestos are due to lung cancer.[104] Not only is the lung cancer risk dose-related, but there is also an important enhancement of the risk in those exposed to asbestos who also smoke cigarettes.[104] The OSHA emergency air standard is 0.5 fiber >5 um/cc (8 hour TWA).

Formaldehyde poses a high toxic hazard upon ingestion, inhalation, or skin contact [242] with human effects reported at doses as low as 8 ppm. [127] This suspected carcinogen is a potent mutagen and teratogen. The OSHA air standard is 3 ppm (8 hour TWA) with a 5 ppm ceiling.[67]

Furfuryl Alcohol exhibits tumorigenic and mutagenic potential in laboratory testing.[233] Animal experiments also suggest high toxicity.[242] Human data indicate that small doses stimulate respiration, while large doses depress respiration, reduce body temperature, and produce nausea, salivation, diarrhea, dizziness, and diuresis.[5] The OSHA air standard is 50 ppm (8 hour TWA).[67]

Polypropylene has produced positive results in animal testing for carcinogenicity.[107] It poses a very low toxic hazard, however. For example, polypropylene medical devices have produced little or no irritant response when placed into connective or muscle tissues for short periods of time.[31] The propensity for diffusion of polypropylene from the material is so small that biological response cannot be detected.[31]

Trimethyl phosphate recently produced positive results in a National Cancer Institute carcinogenesis bioassay.[233] Solid evidence also exists of its mutagenic and teratogenic potential.[233] The substance is also toxic by ingestion and inhalation and is a strong irritant to skin and eyes.[42]

Air Emissions

Sources of VOC emissions from amino resin production are listed in Table 44. The sole sources of process VOC emissions are vents. Emissions from these vents may be controlled by:[285]

- Routing the vented stream to a flare to incinerate the hydrocarbons; and/or

- Routing the vented stream to blowdown.

Other VOC emissions result from leaks or joints associated with flanges, valves, relief valves, pumps, compressors, and drains. Enclosing the atmospheric side of open-ended valves and covering open drains are methods which may be used to control the emissions. However, a routine maintenance and inspection program can be a very effective means of control. Air emissions for amino resin production are estimated as 35 kg of formaldehyde and 35 kg of urea per metric ton of urea-formaldehyde resin produced and 20 kg formaldehyde and 20 kg of melamine per metric ton of melamine-formaldehyde resin produced.[93]

Sources of particulates are also listed in Table 43. Particulates from the dryer, ball mill, and extruder may be controlled by:

- Venting the stream to a baghouse or electrostatic precipitator for particulate collection.

According to EPA estimates, the population exposed to amino resin emissions in a 100 km^2 area surrounding an amino resin plant is 40 persons exposed to hydrocarbons and 8 persons exposed to particulates.[286]

Wastewater Sources

There are several wastewater sources associated with the amino resins produced, as shown in Table 45. The production of molding resins contains the most sources of wastewater since these resins are available only in dried form. All amino resin products generate wastewater from the condensation reaction required to produce them.

TABLE 45.   SOURCES OF WASTEWATER FROM AMINO RESIN MANUFACTURE

Source	Product			
	Adhesives	Coatings	Laminates	Molding Resins
Condensation Reaction	X	X	X	X
Dilution Kettle		X		X
Screens				X
Routine Cleaning Water	X	X	X	X

Wastewater production from these processes are shown below. Values
presented were not distinguished by product type by EPA for the purpose of
establishing effluent limitations for the amino resin industry:[93, 284]

	Urea- Formaldehyde Resin	Melamine- Formaldehyde Resin
Production, m^3/metric ton	0.9 - 1.8	0.6 - 1.3

<u>Solid Wastes</u>

The solid wastes generated by amino resin processing include
substandard resins which cannot be blended, collected particulates, and
resin lost due to spillage and reactor cleaning. These wastes are mostly
amino resins with very little contribution from other sources.

<u>Environmental Regulation</u>

Effluent limitations guidelines have been set for the amino resin
industry. BPT, BAT, and NSPS call for the pH of the effluent to fall
between 6.0 and 9.0 (41 Federal Register 32587, August 4, 1976).

New source performance standards (proposed by EPA on January 5, 1981)
for volatile organic carbon (VOC) fugitive emissions include:

- Safety/release valves must not release more than 200 ppm above
  background, except in emergency pressure releases, which should not
  last more than five days; and

- Leaks (which are defined as VOC emissions greater than 10,000 ppm)
  must be repaired within 15 days.

The following compounds are listed as hazardous wastes (46 Federal
Register 27476, May 20, 1981):

```
Asbestos      - U013
Formaldehyde - U122
Formic Acid   - U123
Phenol        - U188
```

All disposal of these compounds or amino resins which contain any residual
amounts of these compounds (i.e., uncured resins) must comply with the
provisions set forth in the Resource Conservation and Recovery Act (RCRA).

6. Modified Polyphenylene Oxide and Polyphenylene Sulfide

INTRODUCTION

Resins considered to be engineering thermoplastics possess high performance characteristics. Polymers generally reported in this category are: fluoropolymers, thermoplastic polyesters, polycarbonates, polyacetals, polyamides, ABS, SAN, modified polyphenylene oxide, polyphenylene sulfide, polyimides, polyamide-imides, and polysulfones. The first seven of these resins are discussed in other sections of this document. Of the remaining resins listed, only modified polyphenylene oxide and polyphenylene sulfide are presented in this section because they represent a major portion of the remainder of the engineering thermoplastic market.

Polyphenylene oxide is the product of oxidative coupling using phenolic monomers, typically 2,6-xylenol. The homopolymer is combined with polystyrene to give an unusually compatible polymer blend. High temperature properties of polyphenylene oxide are complimented by the easy processability of polystyrene, making modified polyphenylene oxide (MPO) an excellent choice for electrical and electronic applications.

Polyphenylene sulfide (PPS) is characterized by a repeating group containing benzene rings para linked with sulfur atoms. This polymer is remarkably stable, exhibiting outstanding high-temperature performance, inherent flame resistance, and chemical resistance second only to PTFE. Typical properties of polyphenylene sulfide and modified polyphenylene oxide are listed in Table A-6 in Appendix A.

These polymers are each produced by only one manufacturer. Therefore, the polymerization processes used for these resins are presented according to the polymer produced.

INDUSTRY DESCRIPTION

General Electric Company is the sole producer of modified polyphenylene oxide. No capacity information was available in the literature consulted for the Selkirk, New York facility. Similarly, polyphenylene sulfide is produced by the Phillips Petroleum facility in Borger, Texas. In 1981, the capacity was expanded to 4,080 metric tons.[94]

Engineering thermoplastics are experiencing an increase in demand. Their high performance characteristics and strength have made these polymers excellent replacements for more expensive metals in automotive and processing applications. The market for these resins is expected to grow steadily into the 1990's.[86]

PRODUCTION AND END USE DATA

MPO and PPS are used for appliances, business machines, and electrical and electronic uses. In 1980, consumption of PPS totaled 1,720 metric tons. The most recent information for MPO consumption shows 58,000 metric tons for 1981. The uses of these polymers include:[14, 92, 94, 151, 247, 258, 295]

Modified Polyphenylene Oxide

- Electrical applications, including use in appliances, electrical construction products, and business machine housings;

- Automobile applications, including dashboards, electrical connectors, filler panels, grilles, wheel covers, and exterior body parts;

- Television market for tuner and deflection yoke components as well as cabinetry;

- Liquid handling uses, such as pumps, underwater components, and shower heads;

- Photographic film processing tanks and housings; and

- Small appliances and personal care products, especially those involving exposure to heat and moisture such as food processors, hair dryers, steam irons, and hot combs.

Polyphenylene Sulfide

- Telecommunications and computer components, such as connector, coil forms, and bobbins;

- Electrical components for high-voltage applications and structural components requiring Underwriter's Laboratory direct support capability;

- Chemical processing equipment, including submersible, centrifugal, vane, and gear-type pumps;

- Under the hood automotive applications;

- Hair dryers, other personal care appliances, small cooking appliances, and range components;

- Release coatings for cookware and metal tire-mold surfaces; and

• Protective coatings for valves, pipes, pipe fittings, heat
  exchangers, and electromotive coils.

Table 46 lists major markets for MPO and PPS resins.

PROCESS DESCRIPTIONS

Modified polyphenylene oxide is formed by blending the polymer product
of the oxidative coupling of 2,6-xylenol with polystyrene and impact modifiers. Although low molecular weight polyphenylene oxides (used for heat
transfer fluids) and brominated polyphenylene oxides (used as flame-
retardants for other polymers) are also produced, the production of MPO for
engineering uses is the only process of commercial importance.

Polyphenylene sulfide was first identifed in the late 1800's but commercial production did not begin until 1973.[94] Dichlorobenzene is reacted
with sodium sulfide in a polar organic solvent, producing PPS and by-product
sodium chloride.

<u>Modified Polyphenylene Oxide Reaction Chemistry</u>

Polyphenylene oxide, blended with polystyrene for MPO, is formed by
oxidative coupling of 2,6-xylenol. The overall reaction is:

$$n \ HO\!-\!\underset{CH_3}{\overset{CH_3}{\bigcirc}} + \frac{n}{2} \ O_2 \ \xrightarrow{\text{Catalyst}} \ \left[-O\!-\!\underset{CH_3}{\overset{CH_3}{\bigcirc}} \right]_n + nH_2O$$

The xylenol forms an aryloxy radical when combined with a copper-amine
complex, such as copper (I) chloride and pyridine. The resulting aryloxy
radicals couple to form cyclohexadienones, which then enolize and redistrib-
ute the double bonds. Dimers, trimers, and other oligomers combine until
phenolic groups are no longer available for oxidation. These reactions are
depicted below.

TABLE 46. U.S. CONSUMPTION OF MODIFIED POLYPHENYLENE OXIDE

Market	1981 Thousand Metric Tons
Appliances	15
Business Machines	15
Electrical/Electronics	11
Plumbing and Hardware	3
Transporation	10
Other	4
TOTAL	58

Source: Modern Plastics, January 1982.

U.S. CONSUMPTION OF POLYPHENYLENE SULFIDE

Market	1980 Thousand Metric Tons
Electrical/Electronics	1.45
Other	0.27
TOTAL	1.72

Source: Chemical Economics Handbook, updated annually, January 1982 Data.

The resulting polymer is blended with polystyrene to form a polymer composite.

Polyphenylene Sulfide Reaction Chemistry

Although the overall reaction which produces PPS is known, the details of the complex chemical mechanism are not fully understood.[94] Dichlorobenzene and sodium sulfide react to form the PPS chain and by-product sodium chloride. This overall reaction is shown below.

$$Cl-\langle C_6H_4 \rangle-Cl + Na_2S \longrightarrow \left[\langle C_6H_4 \rangle - S \right]_n + 2nNaCl$$

Table 47 lists typical input materials for MPO and PPS processing. Operating parameters for these processes are considered to be proprietary. Therefore, any available information will be presented in the appropriate processing section.

TABLE 47. TYPICAL INPUT MATERIALS FOR MPO PRODUCTION

Function	Compound
Monomer	2,6-xylenol (2,6-dimethylphenol)
Comonomers	2,6-diphenylphenol 2-methyl-6-phenylphenol
Polymer for Blending	polystyrene

Catalyst	copper-amine complex
	  copper halide and aliphatic amines
	  copper halide and pyridine

Fillers	glass minerals
Solvent	methanol methyl chloride toluene

TYPICAL INPUT MATERIALS FOR PPS PRODUCTION

Function	Compound
Monomer	p-dichlorobenzene sodium sulfide
Solvent	polar organic solvent*
Filler	asbestos chopped glass iron oxide

*Not specified.

Sources: <u>Encyclopedia of Chemical Technology</u>, 3rd Edition.
<u>Encyclopedia of Polymer Science and Technology</u>.
Seymour S. Schwartz and Sidney H. Goodman, <u>Plastics Materials and
Processes</u>, 1982.

<u>Modified Polyphenylene Oxide Production Process</u>

The oxidative coupling reaction used to produce polyphenylene oxide for MPO is a one-step polymerization process. Xylenol is reacted with oxygen in the presence of a copper-amine catalyst complex. The following is a typical recipe for the preparation of polyphenylene oxide:[92]

<u>Material</u>	<u>Quantity</u>
Nitrobenzene (solvent)	0.200 1
Pyridine (catalyst)	0.070 1
Copper (I) Chloride (catalyst)	0.001 kg
2,6-Xylenol (monomer)	0.015 kg
Chloroform (diluent solvent)	0.600 1
Methanol (solvent)	3.050 1
Concentrated Hydrochloric Acid (solvent)	0.016 1
Oxygen	

Reaction times of seven (7) minutes at 20 to 46°C are reported.[247] Other processing details were not available in the literature consulted. A typical reaction train is pictured in Figure 14. The polyphenylene oxide, once formed, is separated from the water by-product and washed with solvent to purify the polymer prior to blending with polystyrene and impact modifiers to produce MPO.

<u>Polyphenylene Sulfide Production Process</u>

Polyphenylene sulfide production is represented by the process flow diagram in Figure 15. Although it is not shown in the figure, Phillips Petroleum Company produces sodium sulfide for input to this process. No specific processing details or production recipes were available in the literature consulted.

Sodium sulfide, p-dichlorobenzene, and solvent are fed into a polymerization reactor. Once the polymerization is complete, the solvent is stripped from the polymer-solvent mixture and recycled to the reactor. The polymer is then washed with water to remove the sodium chloride by-product and dried. Molding resins differ from virgin resins since they are cured before pelletization and packaging.

<u>Energy Requirements</u>

No data concerning energy requirements for this process were found in the literature consulted.

ENVIRONMENTAL AND INDUSTRIAL HEALTH CONSIDERATIONS

MPO and PPS, which are considered to be nontoxic compounds, are used in food contact as well as potable water applications. However, several inputs to these processes have been identified as hazardous under RCRA: asbestos, a filler; methanol and toluene solvents; and 1,4-dichlorobenzene and 2,6-xylenol (2,4-dimethylphenol) monomers. MPO processing uses asbestos,

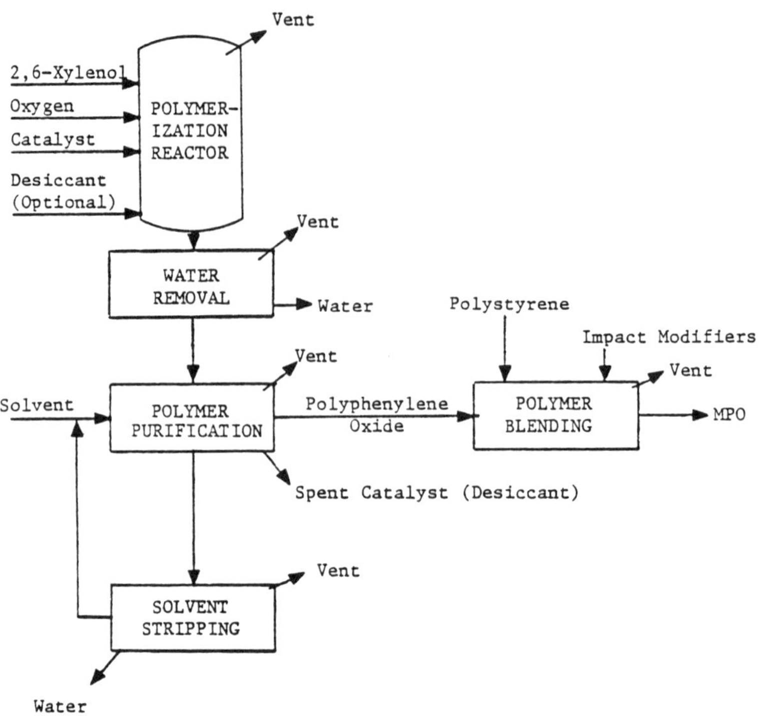

Figure 14.   MPO production process.

Sources: Encyclopedia of Chemical Technology, 3rd Edition.
           Seymour S. Schwartz and Sidney H. Goodman, Plastics Materials and
               Processes, 1982.

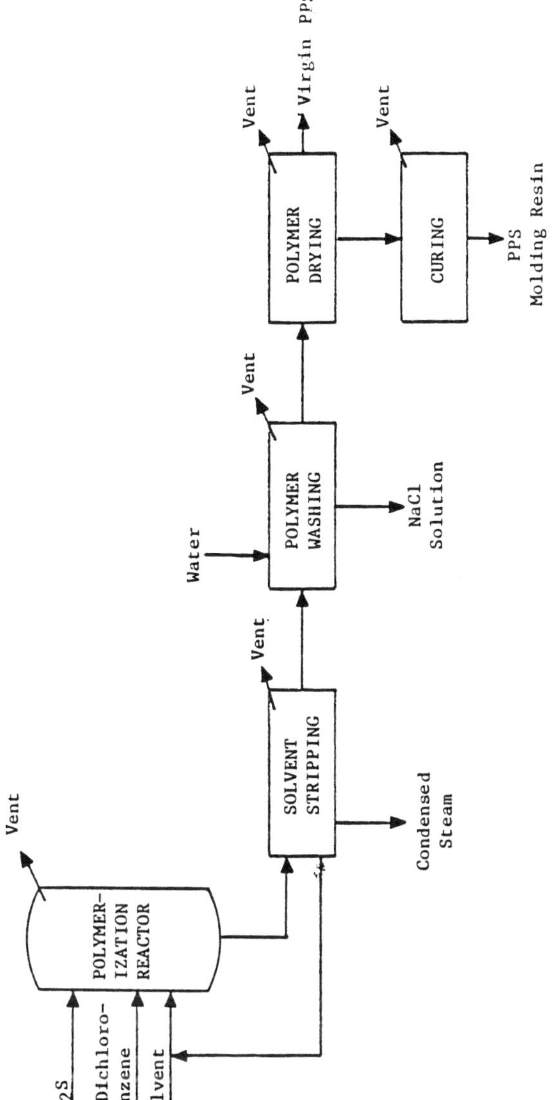

Figure 15. PPS production process.

Source: Encyclopedia of Chemical Technology, 3rd Edition.

methanol, toluene, and 2,6-xylenol while PPS processing involves only
1,4-dichlorobenzene.

Worker Distribution and Emissions Release Points

Estimates of worker distribution for PPS and MPO production are shown
in Table 48. These estimates were derived by correlating major equipment
manhour requirements with the process flow diagrams in Figures 14 and 15.

There are no major point source air emissions associated with either
MPO or PPS processing. The only air emissions are fugitive emissions from
such sources as reactor vents, dryer vents, and condenser vents. The impact
of these emissions depends on the process operating parameters, engineering
and administrative controls, and input materials used.

Sources of fugitive emissions are shown in Table 49. Solvents typically used in both processes include methanol, methylene chloride or toluene. Particulates emitted from blending, washing, drying and curing vents
may result in overexposure to nuisance dust or to particular additives such
as asbestos. Available toxicity information for major process reactants and
additives is summarized under "Health Effects" below.

Health Effects

The production of PPS involves the use of asbestos as a filler. Asbestos is a known human carcinogen whose reported health effects are summarized
below.

Asbestos is classified as a human carcinogen by the International
Agency for Research on Cancer [104] and by the American Conference of
Governmental Industrial Hygienists.[5] It is currently undergoing additional tests by the National Toxicology Program.[233] Most deaths from
exposure to asbestos are due to lung cancer.[107] Not only is the lung
cancer risk dose-related, but there is also an important enhancement of the
risk in those exposed to asbestos who also smoke cigarettes.[107] The OSHA
emergency air standard is 0.5 fiber >5 um/cc (8 hour TWA).

The production of MPO involves the use of an animal carcinogen, polystyrene, as a polymer for blending. Toluene, a highly toxic substance, is
used as a solvent. The reported health effects of exposure to these two
substances are briefly summarized in the following paragraphs.

Polystyrene has produced positive results in animal tests for carcinogenicity.[107] The available animal test data is quite limited, however,
and other potential adverse effects cannot be evaluated.[233]

Toluene exhibits tumorigenic, mutagenic, and teratogenic potential in
laboratory tests.[233] Human exposure data indicate that toluene is also
very toxic. Exposures at 100 ppm have produced psychotropic effects and
central nervous system effects have been observed at 200 ppm.[233] Symptoms
of exposure include headache, nausea, vomiting, fatigue, vertigo, paresthe-
sias, anorexia, mental confusion, drowsiness, and loss of consciousness.
The OSHA air standard is 200 ppm (8 hour TWA) with a 300 ppm ceiling.[67]

TABLE 48. WORKER DISTRIBUTION ESTIMATES FOR ENGINEERING
THERMOPLASTICS PRODUCTION

Process	Unit	Workers/Unit/8-hour Shift
Modified Polyphenylene Oxide	Batch Reactor	1.0
	Centrifuge	0.25
	Solvent Bath	0.25
	Batch Mixer	1.0
	Solvent Stripping	0.25
Polyphenylene Sulfide	Batch Reactor	1.0
	Solvent Stripping	0.25
	Washing	0.25
	Drying	0.5
	Curing	0.25

TABLE 49. SOURCES OF AIR EMISSIONS FROM MPO AND PPS PROCESSING

Source	Constituent	Process	
		MPO	PPS
Reactor Vent	2,6-xylenol	X	
	p-dichlorobenzene		X
	Solvent		X
Water Removal Vent	2,6-xylenol	*	
Polymer Purification Vent	2,6-xylenol	*	
	Solvent	X	
Solvent Stripping Vent	Solvent	*	*
Polymer Blending Vent	2,6-xylenol	*	
	Solvent	*	
	Particulates	X	
Polymer Washing	Solvent		*
	Particulates		X
Polymer Drying	Solvent		*
	Particulates		X
Polymer Curing	Solvent		*
	Particulates		*

*Trace amounts are expected due to removal upstream.

Air Emissions

Sources of fugitive emissions VOC from these processes are summarized in Table 49 by process and constituent type. These vented sources may be controlled by routing the stream to a flare to incinerate any remaining hydrocarbons or routing the stream to blowdown.[285] Other sources of VOC emissions include leaks from process equipment. Although equipment modifications are available, a regular maintenance and inspection program may be the best control for these sources.

Particulate fugitive emissions are also listed in Table 49. These sources may be controlled by venting to a baghouse or an electrostatic precipitator.

Wastewater Sources

The sources of wastewater associated with MPO and PPS processing are listed in Table 50. No data concerning wastewater production or characteristics for these processes were found in the literature consulted.

Solid Waste

The solid wastes from MPO and PPS production include polymer lost due to spillage and cleaning, collected particulates, and substandard polymer which cannot be blended. The solid waste found only in MPO processing is the spent catalyst and desiccant while PPS production generates the by-product sodium chloride as a solid waste stream.

Environmental Regulation

Effluent limitations guidelines have not been set for the MPO or PPS industry.

New source performance standards (proposed by EPA on January 5, 1981) for volatile organic carbon (VOC) fugitive emissions include:

- Safety/release valves must not release more than 200 ppm above background, except in emergency pressure releases, which should not last more than five days; and

- Leaks (which are defined as VOC emissions greater than 10,000 ppm) must be repaired within 15 days.

The following compounds have been listed as hazardous wastes: (46 Federal Register 27476, May 20, 1981).

Asbestos	– U013
1,4-Dichlorobenzene	– U072
2,4-Dimethylphenol (2,6-xylenol)	– U101
Methanol	– U154
Toluene	– U220

Disposal of these compounds and any MPO or PPS resin containing residual amounts of these compounds must comply with the provisions set forth in the Resource Conservation and Recovery Act (RCRA).

TABLE 50.   SOURCES OF WASTEWATER FROM MPO AND PPS PROCESSING

| | Process | |
Source	MPO	PPS
Water Removal	X	
Solvent Stripping	X	X
Polymer Washing		X
Routine Cleaning Water	X	X

7. Epoxy Resins

Epoxy resins are tough, hard, thermosetting solids which quickly cure at temperatures ranging from 5° to 150°C.[247] They are available in a variety of forms which vary from low-viscosity clear liquids to highly filled viscous pastes to free flowing molding powders. Epoxy resins rearrange very little during reaction, in most cases produce no by-products, and maintain surface adhesion and freedom from voids during cure. Their use in metal replacement is enhanced by their moldability at low pressure, unlimited colorability, high gloss, and excellent chemical resistance.

Epoxy resins have a unique combination of properties. Their toughness, adhesive strength, chemical resistance, and superior electrical properties make them suitable for the manufacture of many different products. Resin properties may be modified by blending resin types, by selecting applicable cure agents, and by using modifiers and fillers. The casting resins exhibit excellent strength and low shrinkage during cure. Table A-7 in Appendix A lists some typical properties of epoxy resins.

Over 90 percent of the commercial epoxy resins produced are manufactured from the reaction of epichlorohydrin and bisphenol-A.[131] Other epoxy resins include novolaks (i.e., two-step phenolic resins) epoxidized with epichlorohydrin and unsaturated monomers epoxidized by reaction with hydrogen peroxide or peracetic acid. Unsaturated monomers used include both cyclic aliphatic monomers such as vinylcyclohexane and acyclic aliphatic monomers such as butadiene.

Epoxy resins are manufactured either as low molecular weight liquid resins or higher molecular weight solid resins. Liquid resins are produced using a two step reaction while higher molecular weight resins are produced by either the Taffy process or the advancement process.[247]

INDUSTRY DESCRIPTION

Six chemical producers comprise the epoxy resin industry. These producers have 10 sites located in seven states: California, Kentucky, Louisiana, Massachusetts, Michigan, New Jersey, and Texas. Table 51 lists the epoxy resin producers and their capacities.

TABLE 51.   U.S. EPOXY RESIN PRODUCERS

Producer and Location	1982 Capacity Thousand Metric Tons

Celanese Corporation
 Celanese Plastics & Specialties Co.
  Division
  Celanese Specialty Resins Division
   Louisville, KY
	11

Ciba-Geigy Corporation
 Plastics and Additives Division
  Resins Department
   Toms River, NJ
	27

Dow Chemical U.S.A Freeport, TX	104
Reichhold Chemicals, Inc. Andover, MA Azusa, CA Detroit, MI Houston, TX	15
Shell Chemical Co. Deer Park, TX	122[a]
Union Carbide Corporation Coatings Materials Division Bound Brook, NJ	5[b]
Specialty Chemicals and Plastics Division Taft, LA	9[c]
TOTAL	293

[a]Includes vinyl ester resins.

[b]Phenoxy resins.

[c]Cycloaliphatic epoxy resins.

Sources: Chemical Economics Handbook, updated annually, 1980 data.
Directory of Chemical Producers, 1982.

Epoxy resins are used in coatings for many applications as well as encapsulating electronic equipment. U.S. production of epoxy resins has increased from 111,000 metric tons in 1976 to 143,000 metric tons in 1980. A production peak occurred in 1979 with 164,000 metric tons of resin produced and only 151,000 metric tons sold.[65] Sales for 1981 remained the same as the 1980 level, 145,000 metric tons.[143] The current unused epoxy resin production capacity (293,000 metric tons) is sufficient to accommodate growth in the near future.

PRODUCTION AND END USE DATA

In 1981, epoxy resin sales totaled 145,000 metric tons.[143] Epoxy resins, due to their versatility, have many uses which include:[131, 219, 229, 247, 249, 289]

- Photocurable coatings and printing inks, concrete coatings, wire coatings, marine coatings, coatings for steel pipes, chemically resistant and electrical grade coatings, can and drum linings, military aircraft coatings, floor coatings, masonry coatings, swimming pool paints, formulations for automotive primers, and maintenance paints;

- FDA approved coatings for the interiors of beer and beverage cans;

- Stainless steel and anticorrosive paints, electrical resistor inks, and textile and paper binders and finishes;

- Solution coatings used for maintenance and product finishes, laboratory furniture, lawn furniture, appliances, hardware, and magnetic wire;

- Flowable mortar for traffic areas, chemically resistant mortars and floor topping compounds, monolithic concrete topping applications, decorative applications such as thin-coat terrazzo, and highway maintenance compounds;

- Solid coating compounds for small motor stacks, refrigerator liners, oil filters, hospital equipment, primers, shelving, automobile springs, fire extinguishers, and protection of metal window frames;

- Adhesives for aircraft honeycomb structures and paint brush bristles; adhesion and grouting in buildings, roads, and bridges; adhesives to bond polyester and phenolic laminates to metal, attach metal studs to concrete, bond reflective markers to airport runways, bond solar cells in satellites, bond chemically resistant films to tanks and piping, bond copper sheet to laminates for printed circuit boards, mount gems, restore museum pieces, and bond stained glass windows to replace lead; metal bonding and sealing in the aerospace industry; and highway paving materials, grouts, and adhesives for segmental bridge construction;

- Airframe laminates for wing fairing surfaces, fin tips, tail cones, leading wing panels, double floor constructions, propeller cuff covers, fuel ducting, and radomes;

- Filament wound glass reinforced pipe for use in oil fields, mining, chemical plants, water distribution, electrical conduits, pressure bottles, and rocket motor casings;

- Body solders and caulking compounds for the repair of plastics, metal boats, and automobiles, and caulking and sealant compounds in building, highway construction applications, and in applications where high orders of chemical resistance are required;

- Insulation in motors and generators, distribution switchgear, transformers, cable jointing, high- and low-voltage lines, cross-arms for towers, and railway overhead lines and field coil insulation, form coil tying, blocking, impregnation, and reinforcement;

- Casting compounds for the fabrication of short-run prototype molds, stamping dies, patterns, tooling, brushings, and low-voltage cable splices; encapsulation of high voltage coils for automobiles, power saws, small components in the telephone industry, and random wound motors; post insulators, bus bar supports, switchgear components, instrument transformers, distribution transformers, high- and low-voltage bushings, outdoor insulators, switchgear components, instrument transformers, contact and match molds, stretch blocks, vacuum forming tools, and foundry patterns; and thin sheet castings between television tubes and their contoured safety plates;

- Encapsulation of solid state devices, such as diodes, transistors, and integrated circuits, and passive electrical components, such as coils, capacitors, transformers, and ignition coils, and to manufacture pump housings, impellers, pipe fittings, and valves;

- Manufacture of golf clubs, arrow shafts, vaulting poles, fishing rods, polo sticks, and racetrack railings; and

- Use in printing inks, fabric treating, dental, surgical, and prosthetic applications, breaking petroleum emulsions, lightweight chemically resistant foams, and additives for a variety of plastics, such as vinyl and acrylic resins as well as natural and synthetic fibers.

Table 52 lists epoxy resins markets for 1981.

PROCESS DESCRIPTIONS

Epoxy resins are produced using a two stage reaction for liquid epoxy resins and either the Taffy process or the advancement process for solid resins. The process used depends upon the end product and properties desired. Typically, liquid epoxy resins have lower molecular weights than the solid resins.

TABLE 52.   U.S. MARKETS FOR EPOXY RESINS

Market	Thousand Metric Tons in 1981
Bonding and Adhesives	8
Flooring, Paving, Aggregates	8
Protective Coatings	
Appliance Finishes	3
Auto Primers	7
Can and Drum Coatings	18
Pipe Coatings	5
Plant Maintenance	13
Other (including trade sales)	16
Reinforced Uses	
Electrical Laminates	11
Filament Winding	10
Other	5
Tooling, Casting, Molding	10
Export	18
Other	13
TOTAL	145

Source: Modern Plastics, January 1982.

Epoxy resins are primarily the diglycidyl ethers of bisphenol-A (over 90 percent). Therefore, only the reaction chemistry and processing parameters used for these resins are discussed in this section. Other commercial epoxy resins produced are listed in Table 53. Epoxy resins are manufactured from epichlorohydrin and bisphenol-A via a condensation reaction which is usually carried to very low polymer molecular weights. A curing agent is typically added to produce crosslinking during the curing reaction, forming a tough, thermosetting solid. Epichlorohydrin and bisphenol-A first combine to form a chlorohydrin intermediate which, in the presence of sodium hydroxide, eliminates water and sodium chloride. This series of reactions is shown below:

Epichlorohydrin   Bisphenol-A

$$\text{ClCH}_2\text{CH-CH}_2 \ + \ \text{HO-}\langle\text{aryl}\rangle\text{-C(CH}_3)_2\text{-}\langle\text{aryl}\rangle\text{-OH} \longrightarrow$$

Chlorohydrin Intermediate

$$\text{HO-}\langle\text{aryl}\rangle\text{-C(CH}_3)_2\text{-}\langle\text{aryl}\rangle\text{-O-CH}_2\text{CHCH}_2\text{Cl} \ + \ \text{NaOH} \longrightarrow$$

$$\text{HO-}\langle\text{aryl}\rangle\text{-C(CH}_3)_2\text{-}\langle\text{aryl}\rangle\text{-O-CH}_2\text{CH-CH}_2 \ + \ \text{NaCl} \ + \ \text{H}_2\text{O}$$

The epoxy intermediate adds additional epichlorohydrin in a similar manner to produce the diglycidyl ether of bisphenol-A (DGEBA), with water and salt as by-products:

$$\text{HO-}\langle\text{aryl}\rangle\text{-C(CH}_3)_2\text{-}\langle\text{aryl}\rangle\text{-O-CH}_2\text{CH-CH}_2 \ + \ \text{ClCH}_2\text{CH-CH}_2 \longrightarrow$$

$$\text{ClCH}_2\text{CHCH}_2\text{-O-}\langle\text{aryl}\rangle\text{-C(CH}_3)_2\text{-}\langle\text{aryl}\rangle\text{-O-CH}_2\text{CH-CH}_2 \ + \ \text{NaOH} \longrightarrow$$

$$\text{CH}_2\text{-CHCH}_2\text{-O-}\langle\text{aryl}\rangle\text{-C(CH}_3)_2\text{-}\langle\text{aryl}\rangle\text{-O-CH}_2\text{CH-CH}_2 \ + \ \text{NaCl} \ + \ \text{H}_2\text{O}$$

TABLE 53.   OTHER COMMERCIAL EPOXY RESINS

Hydroxyl Compound Used	Epoxy Resin Obtained
4,4-isopropylidenebis(2,6-di-bromophenol)(tetrabromo-bisphenol A)	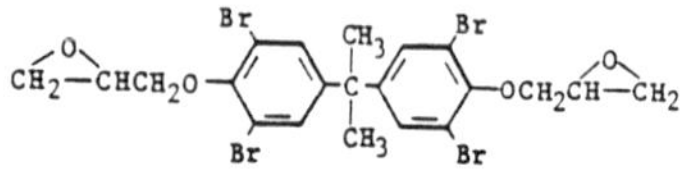
Resorcinol	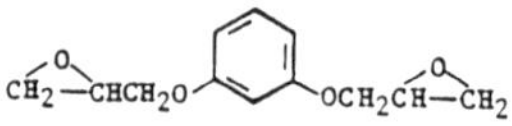
Phenol-formaldehyde novolak	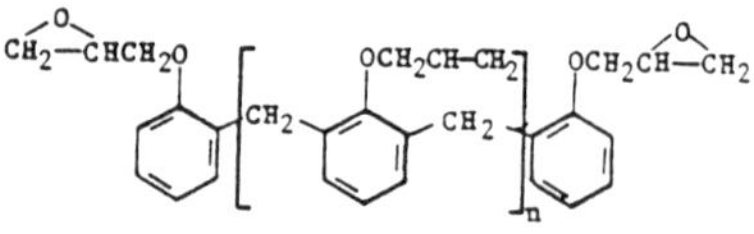
o-cresol-formaldehyde novolak	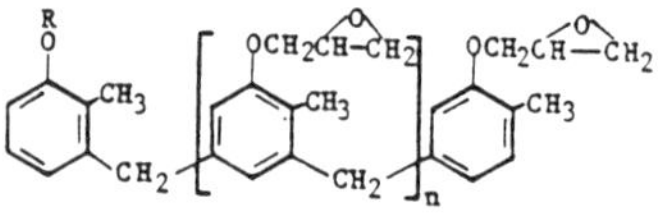
p-aminophenol	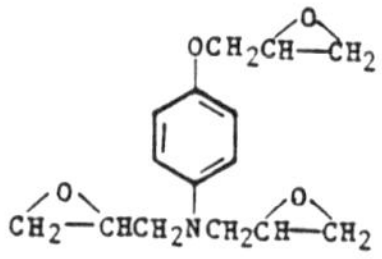
1,1,2,2-tetra(p-hydroxyphenyl)ethane	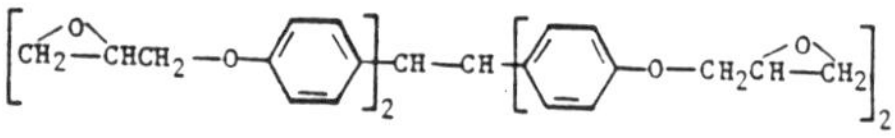

(continued)

TABLE 53 (continued)

Hydroxyl Compound Used	Epoxy Resin Obtained
1,4-butanediol	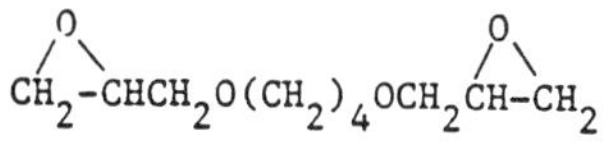 $CH_2-CHCH_2O(CH_2)_4OCH_2CH-CH_2$ (with terminal epoxide oxygens)
Glycerol	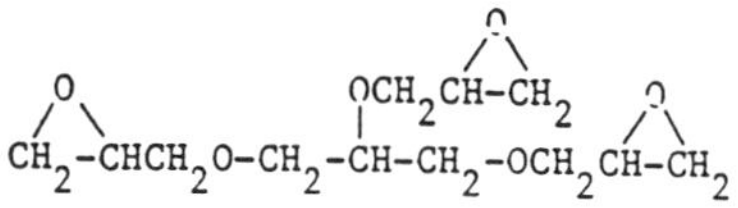 $CH_2-CHCH_2O-CH_2-CH-CH_2-OCH_2CH-CH_2$ with OCH_2CH-CH_2 branch (epoxide groups)
Poly(oxypropylene)glycol	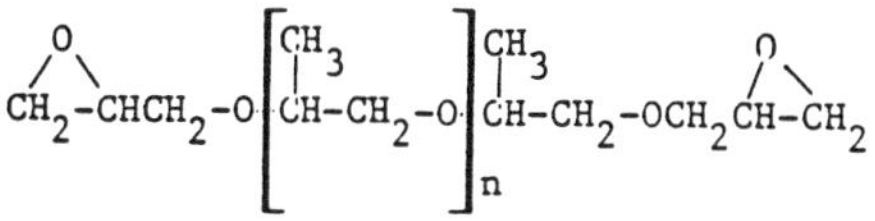 $CH_2-CHCH_2-O[CH(CH_3)-CH_2-O]CH(CH_3)-CH_2-O[CH(CH_3)-CH_2-OCH_2CH-CH_2]_n$
Inoleic dimer acid (one structure shown)	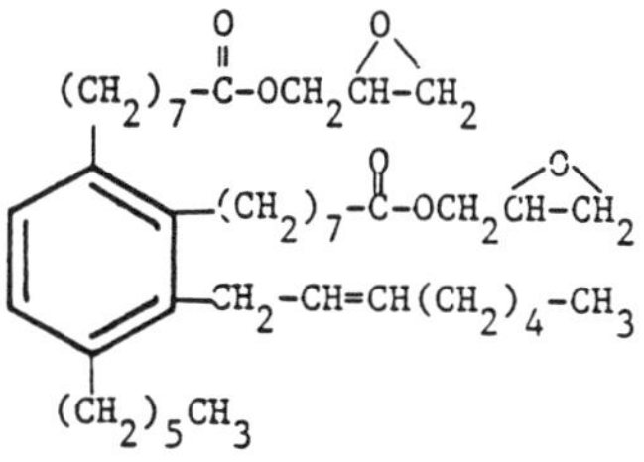 Aromatic ring with substituents: $(CH_2)_7-C(=O)-OCH_2CH-CH_2$ (epoxide); $(CH_2)_7-C(=O)-OCH_2CH-CH_2$ (epoxide); $CH_2-CH=CH(CH_2)_4-CH_3$; $(CH_2)_5CH_3$
1,1,3-tris(p-hydroxy-phenyl)propane	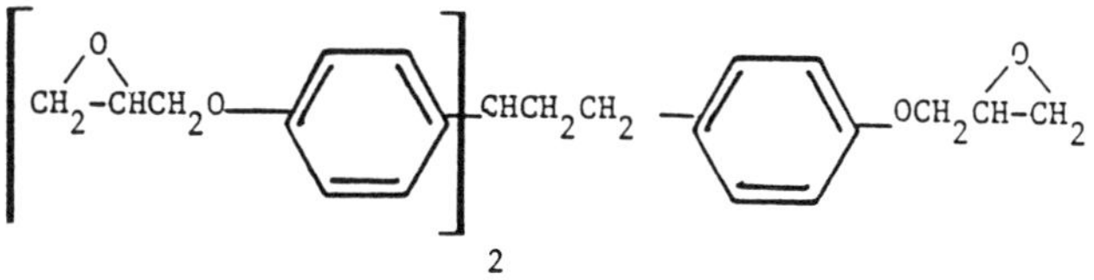 $[CH_2-CHCH_2O\!-\!C_6H_4\!-]_2 CHCH_2CH_2-C_6H_4-OCH_2CH-CH_2$ (epoxide groups)

TYPICAL COMMERCIAL PRODUCTS OBTAINED BY EPOXIDATION
WITH PERACETIC ACID

Epoxidized polybutadiene

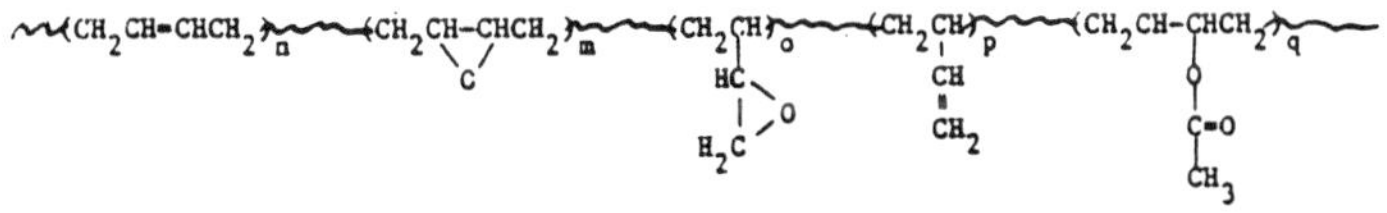

$\sim(CH_2CH=CHCH_2)_n \sim (CH_2CH-CHCH_2)_m \sim (CH_2CH)_o \sim (CH_2CH)_p \sim (CH_2CH-CHCH_2)_q \sim$

with pendant groups including epoxide ($CH-CH_2$ / HC, H_2C with O), vinyl ($CH=CH_2$), and acetate ($O-C(=O)-CH_3$)

(continued)

TABLE 53 (continued)

Vinylcyclohexene dioxide

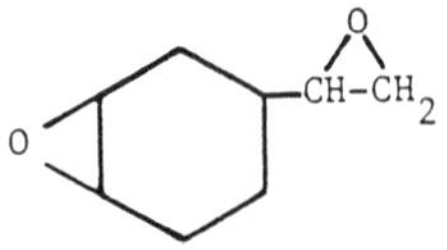

3,4-epoxycyclohexylmethyl-
    3,4-epoxycyclohexane
    carboxylate

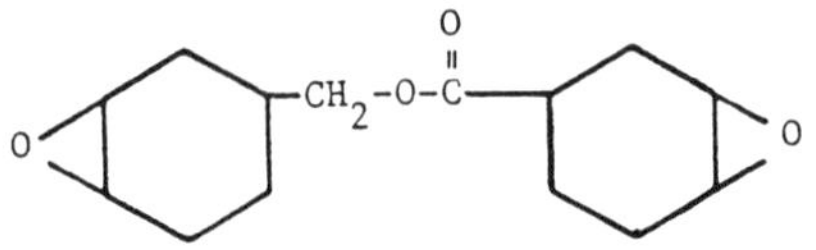

3-(3,4-expoycyclohexane)-8,9-
    epoxy-2,4-dioxaspiro[5,5]-
    undecane

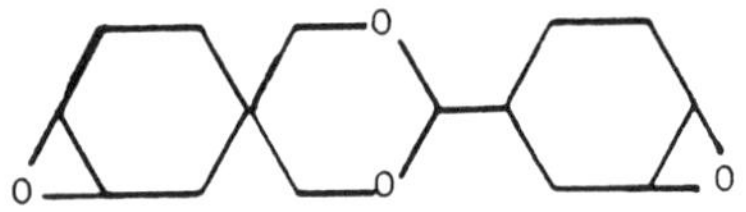

bis(2,3-epoxycyclopentyl)ether

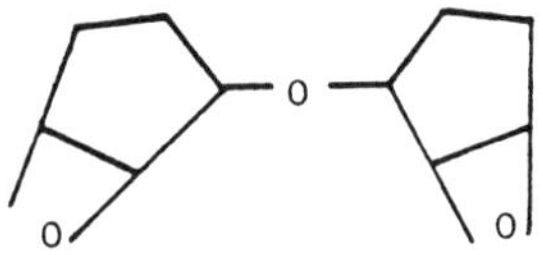

bis(3,4-epoxy-6-methylcyclo-
    hexylmethyl)adipate

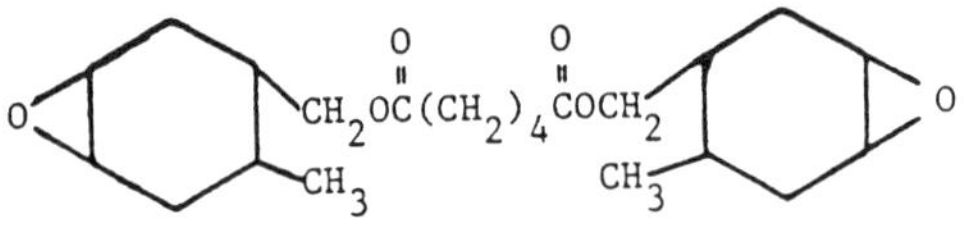

Source: Encyclopedia of Polymer Science and Technology.

This reaction continues until the epoxy groups have combined with all of the available bisphenol-A. A general structure used to represent both high and low molecular weight epoxy resins is shown below.

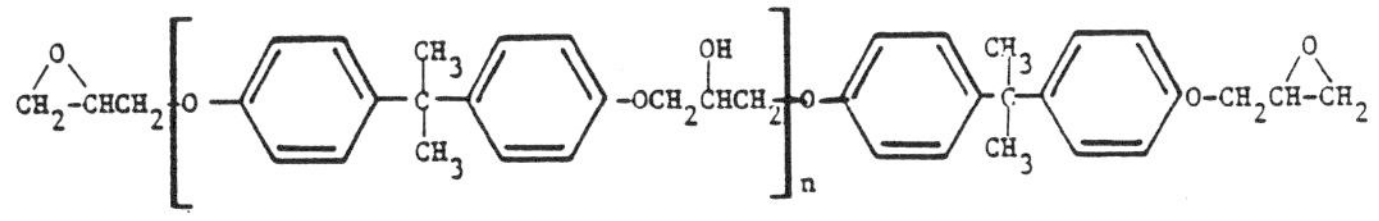

$$CH_2\text{-}CHCH_2 \cdot O \cdot \text{—} \cdots \text{—} \Big[O\text{—}\phi\text{—}\underset{CH_3}{\overset{CH_3}{C}}\text{—}\phi\text{—}OCH_2\underset{OH}{CH}CH_2 \Big]_n O\text{—}\phi\text{—}\underset{CH_3}{\overset{CH_3}{C}}\text{—}\phi\text{—}O\text{-}CH_2CH\text{-}CH_2$$

Tables 54 and 55 list typical input materials and operating parameters for epoxy resin production processes. Other input materials, including diluents, hardeners, fillers, and accelerators, are listed in Table 56.

<u>Liquid Resin Production</u>

The production of low molecular weight epoxy resins is accomplished by a two stage reaction. Epichlorohydrin and bisphenol-A are added to an agitated reactor in a 10:1 ratio. The condensation reaction is then started by adding a 40 percent aqueous sodium hydroxide solution to the reactor and heated to the desired temperature. The epichlorohydrin is separated from the water by condensation and returned to the reactor while the water is sent to wastewater treatment. After the sodium hydroxide addition is completed, the excess epichlorohydrin is recovered using distillation at reduced pressures. The resulting epoxy resin is cooled and a solvent is added to dissolve the resin and carry the salt out of the reactor. A hot water wash removes the brine and the solvent is removed after the resin is filtered and washed. The resin may be dried using heating under vacuum or thin film evaporation.[219] Figure 16 illustrates this batch process. A typical recipe for production of epoxy resins via the two-stage liquid process follows:[116, 219]

Material	Parts by Weight
Epichlorohydrin	46,000
Bisphenol-A	11,400
NaOH (40% aqueous solution) (dehydrochlorinator)	9,500
Methyl Isobutyl Ketone (solvent)	19,055
Water (polymer wash agent)	57,165

<u>Taffy Process</u>

The Taffy process is typically used to produce medium molecular weight epoxy resins. It is very similar to the liquid epoxy resin process since epichlorohydrin and bisphenol A are reacted to produce the epoxy resins. The molecular weight of the product is determined by the ratio of epichloro-hydrin to bisphenol-A. A smaller excess of epichlorohydrin than that used

TABLE 54.   TYPICAL INPUT MATERIALS TO EPOXY RESIN PRODUCTION PROCESSES

Process	Epichloro-hydrin	Bisphenol A	Liquid Epoxy Resin	Aqueous NaOH	Solvent
Liquid	X	X		X	X
Taffy	X	X		X	
Advancement		X	X	X	

TABLE 55.   TYPICAL OPERATING PARAMETERS FOR EPOXY RESIN PRODUCTION PROCESSES

Process	Temperature	Reaction Time	Yield
Liquid	99 - 119°C	3 1/2 hours	96.9%
Taffy	100°C[a]	2 - 3 hours[a]	>90%[a]
Advancement	100°C[a]	2 - 3 hours[a]	>90%[a]

[a]Estimate

Sources: Encyclopedia of Chemical Technology, 3rd Edition.
Encyclopedia of Polymer Science and Technology.

TABLE 56.   INPUT MATERIALS USED IN EPOXY RESIN MANUFACTURE
(IN ADDITION TO BISPHENOL-A AND EPICHLOROHYDRIN)

Function	Compound

Diluents
  Nonreactive
	α-methylnaphthalene
	bisphenols
	chlorinated phenol
	dibutyl phthalate
	dichlorohydrin
	4,4-dimethyl-5-hydroxymethylmetha-
	  dioxane
	dimethyl phthalate
	dioctyl phthalate
	ethylene glycol ethers
	glycol ether-esters
	hydrocarbon oils
	indene
	low molecular weight polystyrene
	methyl methacrylate
	monobasic aliphatic acids
	phenols
	  3-(pentadecyl) phenol
	pine oil
	styrene
	xylene

Epoxy-Containing
Reactive
  Monoepoxy
	allyl glycidyl ether
	α-pinene oxide
	butyl glycidyl ether
	cresyl glycidyl ether
	cyclohexene vinyl monoxide
	$(C_{12-14}H_{22-26}O_3)$glycidyl
	  ester of tert-carboxylic acid
	dipentene monoxide
	1,2-epoxy-4-vinylcyclohexane
	2,3-epoxyoctane (15% 1,2-isomer)
	glycidyl methacrylate
	4-methylphenyl glycidyl ether
	octylene oxide
	olefin oxides
	p-butyl phenyl glycidyl ether
	3-(pentadecyl)phenol glycidyl ether
	phenyl glycidyl ether
	styrene oxide

(continued)

TABLE 56 (continued)

Function	Compound
Epoxy	bis(2,3-epoxycyclopentyl) ether
	butanediol diglycidyl ether
	butadiene dioxide
	diethylene glycol diglycidyl ether
	diglycidyl ether
	diglycidyl ether of resorcinol
	2,6-diglycidyl phenyl glycidyl ether
	dimethylpentane dioxide
	divinylbenzene dioxide
	3,4-epoxy-6-methylcyclohexyl-methyl-3,4-epoxy-6-methylcyclohexane carboxylate
	2-glycidyl phenyl glycidyl ether
	limonene dioxide
	vinylcyclohexene dioxide
Non-epoxy Containing Reactive Diluents	acrylonitrile
	d-limonene
	α-caprolactam
	lactones
	butyrolactone
	phenols
	short chain polyols
	styrene
	tertiary amines combined with fatty polyamides
	unsaturated dienes
Colorants	
White	antimony oxide
	titanium dioxide
Black	carbon black
	hytherm black (Patent Chemicals)
Red	cadmium red
	National fast red (National Aniline)
	toluidene red
Brown	brown iron oxide

(continued)

TABLE 56 (continued)

Function	Compound
Yellow	cadmalith golden (Chemical and Pigment Co.) Calco condensation yellow BTC (American Cyanamid) Hansa yellow
Blue	Calco condensation blue (American Cyanamid) Phthalocyanine blue
Green	Calco condensation green AY (American Cyanamid) Phthalocyanine green
Gray	aluminum powder
Fillers	aluminum aluminum powder #44 (Metals Disintegration Co.) aluminum silicate antimony oxide asbestos barium titanate barytes calcium carbonate calcium sulfate carbon ceramic zircon colloidal silica copper dimethyl dioctadecylammonium bentonite ferric oxide flint powder glass graphite hydrated alumina iron kryolite limestone lithium aluminum silicate magnesium aluminum silicates magnesium silicate

(continued)

TABLE 56 (continued)

Function	Compound
Fillers (continued)	marble methane diamine mica molybdenum disulfide polypropylene Portland cement quartz silica silicone carbide silver steel tabular aluminum T-60 (ALCOA) talc wollastonite zinc oxide zirconium silicate
Solvent	toluene
Flame Retardant	cured, chlorinated polyester with antimony trioxide
Surface Lubrication	graphite molybdenum disulfide

Cure Accelerator	carboxyterminated polyesters
	dimerized and trimerized fatty acids
	morpholine salt of p-toluenesulfonic acid
	organic acids
	  maleic acid
	  oxalic acid
	  phthalic acid
	phenol
	phosphoric acid
	tertiary amines
	  benzyldimethylamine (BDMA)
	  tris(dimethylaminomethyl)phenol (DMP-30)
	zinc oxide

(continued)

TABLE 56 (continued)

Function	Compound
Curing Agents Amine Hardeners	amide derivative of 4-aminomethyl-1,8-diaminooctane amine adduct of o-hydroxylbenzoic acid aminoethylpiperazine amino resins 2-amino alkyl esters of polyhydric polyphenols aromatic bis(amino-imide) compounds aromatic polyamines benzyldimethylamine bisanthranilates of bis(hydroxyalkyl) ureas bisanthranilates of polyethylated carboxylic acid bisanthranilates of polyoxyethylated urea bis[4-amino-3-aminomethyl-piperidyl-(1)]alkanes 3,9-bis(2-hydrazidoethyl)-2,4,8,10-tetraoxaspiro[5.5]undecane carboxylic acid metal salt/amine complexes compounds derived from mannich bases cycloaliphatic epoxy and imidazoles diaminodiphenylsulfone 1,4-diamino butanes 1,3-diaminopropanes dicyandiamide dicyandiamide/imidazoline derivative diethanolamine diethylaminopropylamine diethylenetriamide diethylethanolamine 2,5-dimethyl-2,5-hexanediamine 2-ethyl-4-methylimidazole guanamines hydrogenated polynitrile mixture imidazoles imidazole-isocyanuric acid adducts macrocyclic polyamines

(continued)

TABLE 56 (continued)

Function	Compound

Curing Agents
  Amine Hardeners
  (continued)
	mixed amine-phenol products
	m-phenylenediamine
	n-alkyl substituted polyamine
	N-substituted piperazine adduct
	  cocuring agent
	oligomeric poly(ethylene
	  piperazine)
	phenolic amines
	phenol-polyamide-carbonyl compounds
	phenylenediamine
	phenylindane diamines
	phenylurea
	piperidine
	polyalkylene polyether polyol
	polyamines
	polyamides from polymerized fatty
	  acids
	polyamine-aryl imide groups
	polycyclic polyamines
	polyepoxide
	polyoxyalkylene polyamine
	pyridine
	pyrrolidone-5-carboxylic acid/
	  metal/amine complexs
	sterically hindered ureas
	  triethylenetetramine
	trimethlyamine
	2,4,6-tris(dimethylaminomethyl)
	  phenol

Anhydride and Acid Hardeners	aluminum chloride aromatic aminocarboxylic acids aryloxydianhydrides bisphenol anhydride carboxylic acid anhydrides chlorendic anhydride cyanoacetic acid derivatives dimerized fatty acids dodecylsuccinic anhydride ferric chloride hexahydrophthalic anhydride

(continued)

TABLE 56 (continued)

Function	Compound

Anhydride and Acid Hardeners (continued)	hydrophthalic anhydride
	imidyl phthalic anhydride
	latent Lewis acid catalyst system
	  BF_3 adducts
	  BF_3-monoethylamine
	  stannic chloride
	maleic acid
	maleic anhydride
	nadic methyl anhydride
	oxalic acid
	phthalic acid
	phthalic anhydrides
	polyester polycarboxylic acid
	salicylic acid and methyliminobis-
	  propylamine
	tetrahydrophthalic anhydride
	trimellitic acid ester
	trimerized fatty acids

Sulfur Containing Compounds	bis(fluoroalkylsulfonyl) methane
	  salt
	dialkyl hydroxyaryl sulfonium salts
	  and organic oxidant
	fluoroaliphatic sulfonyl substi-
	  tuted ethylenes
	methylol derivatives of polythiols
	polyether thiourea compounds
	polymercaptans

Other Catalysts	chromium III chelates
	diaryliodonium and copper chelate
	diaryliodonium salt, copper salt,
	  and reducing agent
	hydroxy functional organophosphate
	  ester
	inorganic bases
	Lewis bases
	onium salts and reducing agents
	stabilized iodonium/copper salt
	  compositions

Resinous Modifiers	acrylic resins amino resins butadiene-acrylonitrile resins

(continued)

TABLE 56 (continued)

Function	Compound
Resinous Modifiers (continued)	coal-tar pitch furfural resins isocyanates nylons petroleum-derived bitumens phenolic resins polyester resins polysulfide polymers polyurethane resins polyvinyl chloride silicone resins vinyl resins vinyl cyclohexane
Reinforcements	aramide boron carbon glass fibers graphite Kevlar® metal fibers
Thixotropic Filler	colloidal silica
Internal Release Agent	zinc stearate
Acidic Reactants	hydrogen peroxide peracetic acid

Sources: *Encyclopedia of Chemical Technology*, 3rd Edition.
Encyclopedia of Polymer Science and Technology.
Epoxy Resin Technology, Paul F. Bruins, editor, 1968.
Henry Lee and Kris Neville, *Handbook of Epoxy Resins*, 1967.
Modern Plastics Encyclopedia, 1981-1982.

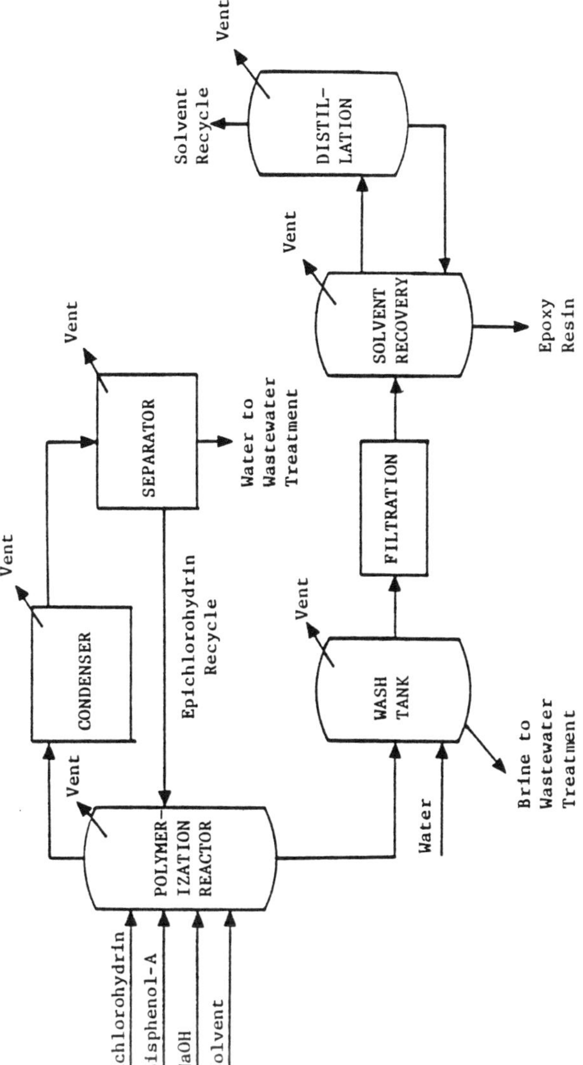

Figure 16. Low molecular weight liquid epoxy resin production process.

Source: W. G. Potter, Epoxide Resins, 1970.

for liquid resin production enables a medium molecular weight product
(usually 1 to 4 repeating groups) to be formed. Production of solid epoxy
resins via the taffy process typically have a recipe as follows:[244A]

Material	Parts by Weight
Bisphenol-A	228
Epichlorohydrin	145
NaOH (10% aqueous solution) (dehydrochlorinator)	750

The highly viscous product is actually a mixture of an alkaline-brine
solution and a water-resin emulsion. The epoxy resin is recovered by
separation of these phases, washing the resin with water, and removing the
water under vacuum. This process is presented in Figure 17.

Advancement Process

The advancement process, also referred to as the fusion method, uses
liquid epoxy resin as a starting material and extends the chain by adding
bisphenol A. In contrast to the liquid epoxy resin and Taffy processes,
this method uses a catalyst to accomplish the reaction. Catalysts are
considered to be proprietary, although a variety is available to provide
special requirements, such as resins free from resin branching.[249] The
molecular weight of the resulting resin is determined by the ratio of excess
liquid epoxy resin to bisphenol A. No by-products are generated by this
process, and it produces resins with only even numbers of repeating units.
[249] This process is depicted in Figure 18. A typical production recipe
for solid epoxy resin via the advancement process was not found in the
literature.

Energy Requirements

No data on the energy requirements for epoxy resin processing were
found in the literature consulted.

ENVIRONMENTAL AND INDUSTRIAL HEALTH CONSIDERATIONS

Cured epoxy resins are toxicologically inert compounds which are FDA
approved for coating the interiors of beverage cans.[63] However, the use
of curing agents, many of which are skin sensitizers, does increase the
toxic effect of uncured epoxy resins. Methyl isobutyl ketone, a by-product
from some epoxy resin processes, has been listed as hazardous, as well as
the following input materials: acrylonitrile, asbestos, di-n-butyl
phthalate, dimethyl phthalate, di-n-octyl phthalate, maleic anhydride,
methyl methacrylate, phenol, phthalic anhydride, pyridine, toluene, and
xylene.

Although the advancement process does not generate any wastes during
the chain extention reaction, the liquid epoxy resin used as an input to
this process generates wastewater during its production. Therefore, all of

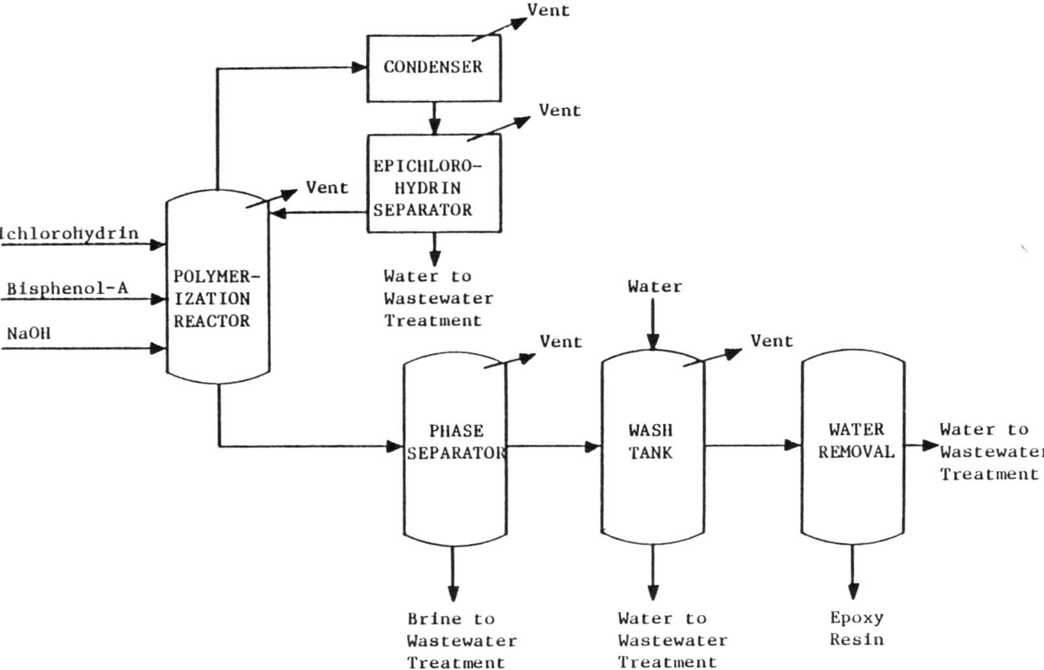

Figure 17.   Medium molecular weight epoxy resin production using the taffy process.

Source: Encyclopedia of Chemical Technology, 3rd Edition.

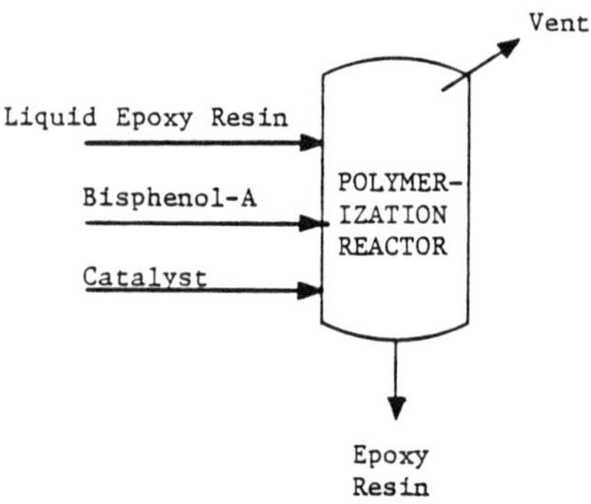

Figure 18.   High molecular weight epoxy resin using the advancement process.

Source: Encyclopedia of Chemical Technology, 3rd Edition.

these processes generate wastewater from polymer washing and brine removal
in addition to routine cleaning water.

High conversions in these processes (>90 percent) indicates that
sources upstream of and including the polymerization reactor are suspect as
the major VOC emissions contributors. The liquid epoxy resin production
process has added VOC emissions due to the use and recovery of the solvent.

Worker Distribution and Emissions Release Points

We have estimated worker distribution for epoxy resin production by
correlating major equipment manhour requirements with the process flow
diagrams in Figures 16 through 18. Estimates for the three major production
processes are shown in Table 57.

There are no major air emission point sources associated with epoxy
resin production. However, fugitive emissions from process vents and leaks
may present a significant environmental impact and/or worker health threat.
The magnitude of these problems is a function of the stream constituents,
process operating parameters, engineering and administrative controls, and
maintenance programs.

Probable fugitive emission sources and contaminants are identified in
Table 58. Typically, toluene is the solvent used in the liquid resin
process. Available toxicity information for these chemicals is summarized
under "Health Effects."

Numerous additives (see Table 56) may be incorporated into epoxy
resins. Available toxicity information for additives is provided in IPPEU
Chapter 10b.

In addition to the fugitive process emissions listed in Table 58
employees may be exposed to sodium hydroxide added as a reaction initiator.
Sodium hydroxide is highly corrosive and even brief contact with skin or
mucous membranes can produce severe irritation and chemical burns.

Maintenance employees may be exposed to methyl isobutyl ketone (MIBK)
used to periodically clean polymerization reactors. MIBK is irritating to
the eyes and respiratory tract and produces headache, nausea and drowsiness
at high concentrations. MIBK is a defatting agent and prolonged skin contact may produce dryness or dermatitis.

Little data are available from which employee exposure potential may be
estimated. However, one source estimates hydrocarbon emissions from epoxy
adhesive production at 25 g/kg product.[286]

Health Effects

The manufacture of epoxy resin involves the use of over 200 input
materials and specialty chemicals. Sufficient data do not exist with which
to evaluate potential health effects for over half of these materials.

TABLE 57.   WORKER DISTRIBUTION ESTIMATES FOR EPOXY RESIN PRODUCTION

Process	Unit	Workers/Unit/8-hour Shift
Liquid Resin	Batch Reactor	1.0
	Condenser	0.125
	Separator	0.25
	Washing	0.25
	Filter	0.25
	Solvent Recovery	0.25
	Distillation	0.25
Taffy	Batch Reactor	1.0
	Condenser	0.125
	Separator	0.25
	Phase Separator	0.25
	Washing	0.25
	Evaporator	0.25
Advancement	Batch Reactor	1.0

TABLE 58. SOURCES OF FUGITIVE EMISSIONS FROM EPOXY RESIN MANUFACTURE

Source	Constituent	Process		
		Liquid	Taffy	Advancement
Reactor Vent	Epichlorohydrin	X	X	
	Bisphenol A	X	X	X
	Solvent	X		
Condenser Vent	Epichlorohydrin	X	X	
Epichlorohydrin Separator Vent	Epichlorohydrin	X	X	
Phase Separator Vent	Epichlorohydrin		*	
Polymer Wash Tank	Solvent	X		
	Epichlorohydrin	*	*	
Solvent Recovery Vent	Solvent	X		
Distillation Column Vent	Solvent	X		

*Only trace amounts of epichlorohydrin are expected in these streams due to the removal upstream in the epichlorohydrin separation step.

Those that do have a significant history of animal test data and/or epidemiological data include one known human carcinogen--the filler asbestos--and
two suspected human carcinogens--acrylonitrile, a diluent; and polyvinyl
chloride, a resinous modifier. Four known animal carcinogens are also used
by the industry: epichlorohydrin, a major raw material in the process; the
filler polypropylene; and two diluents, vinyl cyclohexene dioxide and 3,4-
epoxy-6-methylcyclohexylmethyl-3,4-epoxy-6-methylcyclohexane carboxylate.
Other chemicals that may pose a significant risk to the health of plant
employees because of high toxicity, mutagenicity, and/or teratogenicity
include three diluents--allyl glycidyl ether, diglycidyl ether, and styrene;
two curing agents--piperidene and maleic anhydride; phenol, which is used as
a cure accelerator; the solvent toluene; and the colorant antimony oxide.
The reported health effects of exposure to these substances are summarized
below.

Acrylonitrile, a suspected human carcinogen [107], causes irritation of
the eyes and nose, weakness, labored breathing, dizziness, impaired judgment, cyanosis, nausea, and convulsions in humans. Mutagenic, teratogenic,
and tumorigenic effects have been reported in the literature. The OSHA air
standard is 2 ppm (8 hour TWA) with a 10 ppm ceiling.[69]

Allyl Glycidyl Ether has exhibited tumorigenic and mutagenic potential
in animal tests, and is currently being tested for carcinogenesis by the
National Toxicology Program.[233] It produces severe local reactions following acute exposure, but systematic effects are considered moderate by
inhalation and slight by ingestion and skin absorption.[242] Symptoms of
exposure include central nervous system depression and pulmonary edema.[242]
OSHA has set the ceiling for exposure at 10 ppm (8 hour TWA).[67]

Antimony Oxide is a suspected carcinogen [5] which has also produced
positive results in tests for mutagenic effecs.[155] Animal testing of this
substance is limited, however, and no epidemiological data exist.

Asbestos is classified as a human carcinogen by the International
Agency for Research on Cancer [104] and by the American conference of
Governmental Industrial Hygienists.[5] It is currently undergoing additional tests by the National Toxicology Program.[233] Most deaths from
exposure to asbestos are due to lung canger.[104] Not only is the lung
cancer risk dose-related, but there is also an important enhancement of the
risk in those exposed to asbestos who also smoke cigarettes.[104] The OSHA
emergency air standard is 0.5 fiber >5 um/cc (8 hour TWA).

Diglycidyl Ether is regarded as a hazardous compound and persons should
be protected from all contact with liquid or its vapor.[5] It is highly
toxic via oral and inhalation routes and moderately toxic via the dermal
route.[243] Effects observed in man include skin burns that are slow to
heal and acute irritation of eyes and respiratory tract.[170] In laboratory
animals, central nervous system depression, liver damage, and mutagenic
activity have been reported.[233] OSHA has set a ceiling for exposure at
0.5 ppm.[67]

Epichlorohydrin is highly toxic by ingestion, inhalation, and skin absorption.[242] In acute poisoning, death may be caused by respiratory paralysis [242] chronic exposure can cause kidney injury.[149] Symptoms of chronic poisoning at concentrations lower than 20 ppm include fatigue, gastrointestinal pains, and chronic conjunctivitis.[134] Epichlorohydrin has also produced positive results in animal tests for carcinogenicity [102] and for mutagenicity and teratogenicity.[233] The OSHA air standard is 5 ppm (8 hour TWA).[67]

3,4-Epoxy-6-Methylcyclohexylmethyl-3,4-Epoxy-6-Methylcyclohexane Carboxylate produced positive results in animal tests for carcinogenicity.[102] Very little animal data and no human data exist, however, with which to evaluate the toxicity or other adverse effects of exposure to this chemical.[233]

Maleic Anhydride is highly toxic by ingestion and inhalation [242] and is a power irritant to the skin and eyes.[149] Inhalation can cause pulmonary edema [149]; other effects of exposure include conjunctivitis, corneal damage, cough, bronchitis, headache, abdominal pain, nausea, and vomiting. [238] Animal data also suggest that maleic anhydride is a tumorigenic agent.[233] The OSHA air standard is 0.25 ppm (8 hour TWA).[67]

Phenol is extremely toxic by ingestion, inhalation, and skin absorption.[85] Approximately half of all reported cases of acute phenol poisoning has resulted in death.[278] Chronic poisoning may also occur from industrial contact and has caused renal and hepatic damage.[149] Evidence indicates that phenol is also a tumorigenic, mutagenic, and teratogenic agent; however, a recent carcinogenesis bioassay by the National Cancer Institute produced negative results.[233] The OSHA air standard is 5 ppm (8 hour TWA).[67]

Piperidine, according to animal test data, is highly toxic by ingestion and skin absorption and moderately toxic by inhalation.[233] It also produces mutagneic and teratogenic effects in laboratory animals.[233] However, there is no record in the available literature of.the effects on humans of exposure to piperidine.[233]

Polypropylene has produced positive results in animal testing for carcinogenicity.[107] It poses a very low toxic hazard, however. For example, polypropylene medical devices have produced little or no irritant response when placed into connective or muscle tissues for short periods of time.[31] The propensity for diffusion of polypropylene from the material is so small that biological response cannot be detected.[31]

Polyvinyl Chloride is a suspected human carcinogen.[107] Although the finished foam resin may cause allergic dermatitis, it is generally free from health hazard unless comminuted or strongly heated.[3] It burns readily and gives off objectionable smoke.[285] Several cases of angiosarcoma of the liver have been detected among long-time workers in plants that make polyvinyl chloride from monomeric vinyl chloride.[85]

Styrene has been linked with increased rates of chromosomal aberrations
in persons exposed in an occupational setting.[107]   Animal test data
strongly support epidemiological evidence of its mutagenic potential.[233]
Styrene also produces tumors and affects reproductive fertility in laboratory animals.[233]   Toxic effects of exposure to styrene usually involve the
central nervous system.[233]   The OSHA air standard is 100 ppm (8 hour TWA)
with a ceiling of 200 ppm.[67]

Toluene exhibits tumorigenic, mutagenic, and teratogenic potential in
laboratory tests.[233]   Human exposure data indicate that toluene is also
very toxic.   Exposures at 100 ppm have produced psychotropic effects and
central nervous system effects have been observed at 200 ppm.[233]   Symptoms
of exposure include headache, nausea, vomiting, fatigue, vertigo, paresthe-
sias, anorexia, mental confusion, drowsiness, and loss of consciousness.
The OSHA air standard is 200 ppm (8 hour TWA) with a 300 ppm ceiling.[67]

Vinylcyclohexene Dioxide, a suspected carcinogen [104], is currently
undergoing additional tests for carcinogenesis by the National Toxicology
Program.[233]   It is a strong irritant to skin and tissue, but only moderately toxic by ingestion and skin absorption.[42]   Risk by inhalation is
considered to be slight.[4]   Tests with microorganisms and mammals demonstrate the mutagenic potential of this substance.[233]

Air Emissions

VOC emission sources for epoxy resin processing are summarized in Table
58 by process and constituent type.   All of the process sources listed are
vents, and may be controlled by:[285]

- Routing the vented stream to a flare to incinerate any remaining
  hydrocarbons; and/or

- Routing the vented stream to blowdown.

Other sources of VOC emissions include leaks from valves, flanges, cooling
towers, drains, pumps, and compressors.   Although some equipment modifications may be performed to reduce these emissions, such as enclosing the
atmospheric side of open-ended valves or covering open drains where explosive limits are not a problem, the best control may be achieved by instituting a regular inspection and maintenance program.

No particulate emissions are expected from the polymerization process.
However, some particulate emissions may result from storage and handling of
the raw materials or end products.   According to EPA estimates, the population exposed to epoxy resin emissions in a 100 km^2 area around an epoxy
resin plant is: 64 persons exposed to hydrocarbons and 2 persons exposed to
particulates.[286]

<u>Wastewater Sources</u>

There are several wastewater sources associated with the various epoxy resin production processes, as shown in Table 59. The advancement process appears to generate less wastewater than the other two processes since this process advances the molecular weight of an existing liquid epoxy resin.

Ranges of several wastewater parameters for wastewaters from epoxy resin production are shown below. Values for the wastewaters from various processes presented were not distinguished by process type by EPA for the purpose of establishing effluent limitations for the epoxy resin industry. [284]

Epoxy Resin Wastewater Characteristic	Unit/Metric Ton of Epoxy Resin
Production	$2.5 - 5.1$ m^3
BOD_5	$57 - 82$ kg
COD	$30 - 127$ kg
TSS	$5 - 24$ kg

<u>Solid Wastes</u>

The solid wastes generated by this process may be: sodium chloride, other by-products, unrecovered epichlorohydrin, substandard epoxy resin which cannot be blended, and epoxy resin loss due to spillage and reactor cleaning. Values for some of the solid wastes generated by some epoxy resin processes are listed below.[93]

Solid Waste Source	Constituents	Amount, kg/Metric Ton of Epoxy Resin Produced
Brine	Sodium Chloride	344
Reactor Cleaning	Methyl Isobutyl Ketone	45
Reactor Cleaning (continued)	Sodium Monohydrogen Phosphate	16.7
	Epichlorohydrin-Methaonl By-Product	14.8
Epichlorohydrin Separator	Epichlorohydrin	0.9

TABLE 59.   SOURCES OF WASTEWATER FROM EPOXY RESIN MANUFACTURE

Source	Process		
	Liquid	Taffy	Advancement
Condensation Reaction	X	X	
Resin Wash	X	X	
Routine Cleaning Water	X	X	X

Environmental Regulation

Effluent limitations guidelines have been set for the epoxy resin industry. BPT, BAT, and NSPS call for the pH of the effluent to fall between 6.0 and 9.0 (41 Federal Register 32587, August 4, 1976).

New source performance standards (proposed by EPA on January 5, 1981) for volatile organic carbon (VOC) fugitive emissions include:

- Safety/release valves must not release more than 200 ppm above background, except in emergency pressure releases, which should not last more than five days; and

- Leaks (which are defined as VOC emissions greater than 10,000 ppm) must be repaired within 15 days.

The following compounds have been listed as hazardous wastes (46 Federal Register 27476, May 20, 1981):

Acrylonitrile	— U009
Asbestos	— U013
Di-n-butyl phthalate	— U069
Dimethyl phthalate	— U102
Di-n-octyl phthalate	— U107
Maleic anhydride	— U147
Methyl isobutyl ketone	— U161
Phenol	— U188
Phthalic anhydride	— U190
Pyridine	— U196
Toluene	— U220
Xylene	— U239

Disposal of these compounds or epoxy resins containing residual amounts of these compounds must comply with the provisions set forth in the Resource Conservation and Recovery Act (RCRA).

8. Fluoropolymers

INTRODUCTION

Fluoropolymers, named for the inclusion of fluorine in the monomer, offer unique performance characteristics. These polymers survive even the most corrosive conditions. Although they are higher in price than other thermoplastics, fluoropolymers exhibit outstanding electrical properties, possess a low coefficient of friction, and maintain nonabrasive and self-lubricating surface qualities. These properties make fluoropolymers useful in chemical processing equipment, as insulation and jacketing for wire and cable, in nonstick home cookware, and as parts for industrial equipment.

There are two major commercial fluoropolymers: polytetrafluoroethylene (PTFE) and polychlorotrifluoroethylene (PCTFE).[35] Table A-8 in Appendix A lists typical properties for PTFE and PCTFE. Other fluoropolymers used in specialty applications include fluorinated ethylene-propylene (FEP), polyvinylidene fluoride (PVDF), poly(ethylene-tetrafluoroethylene) (PETFE), poly (ethylene-chlorotrifluoroethylene) (PECTFE), perfluoroalkoxy resin (PFA), and polyvinyl fluoride (PVF). Table A-9 in Appendix A compares density and elongation for these polymers.

PTFE and PCTFE production processes are presented for this analysis since they dominate the fluoropolymer market.[35] PTFE, produced by suspension or emulsion polymerization, is available in three forms: granular (suspension), fine powder (emulsion), and aqueous dispersion (emulsion). PCTFE is produced by bulk, suspension, and emulsion polymerization processes. In order to discuss the differences in polymer properties as they relate to the production method used, these processes will be presented as PTFE production processes and PCTFE production processes.

INDUSTRY DESCRIPTION

The fluoropolymers are typically divided into two categories: PTFE and PCTFE resins; and other resins. PTFE and PCTFE producers are listed in Table 60 while miscellaneous resin producers and their products are presented in Table 61. Allied Corporation and Minnesota Mining and Manufacturing are the PCTFE producers and they consider the plant capacity data to be proprietary. The fluoropolymer industry is comprised of six producers with eight plants located in seven states: Alabama, Kentucky, New Jersey (2 sites), New York, Pennsylvania, Virginia, and West Viriginia. Although total capacity for the industry is not published in the literature, the capacity of the PTFE segment is 11,400 metric tons.[56]

TABLE 60. U.S. PRODUCERS OF PTFE RESINS

Producer	Location	1982 Capacity Thousand Metric Tons
Allied Corporation Allied Fibers & Plastics Co.	Elizabeth, NJ	1.4
E. I. du Pont de Nemours & Co., Inc. Polymer Products Dept.	Buffalo, NY Parkersburg, WV	7.7
ICI Americas Inc. Performance Resins Division	Bayonne, NJ	2.3
		11.4

U.S. PRODUCER OF PCTFE RESINS

Producer	Location	
Allied Corporation Allied Fibers & Plastics Co. Nypel, Inc., subsidiary	Chesterfield, VA	N.A.
Minnesota Mining and Manufacturing Co. Commercial Chemicals Division	Decatur, AL	N.A.

N.A. - Not Available.

Source: Directory of Chemical Producers, 1982.

TABLE 61. U.S. PRODUCERS OF MISCELLANEOUS FLUOROPOLYMERS

Producer	Location	Resin Type
Allied Corporation Allied Fibers & Plastics Co.	Elizabeth, NJ	Poly(chlorotrifluoro- ethylene-vinylidene fluoride) Poly(ethylene-chloro- trifluoroethylene)
E. I. du Pont de Nemours & Co., Inc. Polymer Products Department	Buffalo, NY Parkersburg, WV	Polyvinyl fluoride Fluorinated ethylene- propylene Poly(ethylene-tetra- fluoroethylene) Polyfluoroalkoxy

Minnesota Mining and
  Manufacturing Co.
Commercial Chemicals
  Division
	Decatur, AL	Polychlorotrifluoro-
		ethylene and poly
		(chlorotrifluoro-
		ethylene-vinylidene
		fluoride)

Occidental Petroleum
  Corporation
Hooker Chemical
  Corporation, subsidiary
  Plastics & Chemicals
  Specialties Group
  PVC Resins Division
	Pottstown, PA	Poly(chlorotrifluoro-
		ethylene-vinyl
		chloride)

Pennwalt Corporation
Chemicals Group
  Inorganic Chemical
  Division
	Calvert City, KY	Polyvinylidene
		fluoride

Source: *Directory of Chemical Producers*, 1982.

Fluoropolymer production follows the trends of the U.S. economy. Nonstick cookware, chemical processing equipment, and insulation needs tend to exhibit the same upswings and downturns as the U.S. economy. Sales of granular PTFE have increased from 5,000 to 6,140 metric tons during the 1976 to 1980 time period with a peak in 1979 at 6,550 metric tons.[65] Current capacity for this industry is sufficient to accommodate any near term growth.

PRODUCTION AND END USE DATA

In 1980, granular PTFE sales were 6,140 metric tons.[65] Uses for fluoropolymers include:[19, 23, 72, 77, 145, 245, 294]

- Electrical applications, such as hookup and hookup-type wire for the military and aerospace industries, coaxial cable tapes, interconnecting wires for airframes, computer wire, electrical tape, electrical components, "spaghetti" tubing, and insulation and jacketing for high performance wire and cable;

- Mechanical applications, such as seals and piston rings, bearings, mechanical tapes, coated glass fabrics, and basic shapes for industrial equipment, farm machinery, and passenger vehicles;

- Chemical applications, such as hard and soft packing, overbraided hose liners, thread-sealant tapes, gaskets, fiber and filament, membranes, and instrument parts for chemical equipment;

- Cryogenic seals, gaskets, valve seats, and liners;

- Medical packaging and instrument parts for medical equipment;

- Low molecular weight inert sealants and lubricants for handling oxygen or other corrosive materials, gyroscope flotation fluids, plasticizers for thermoplastics, and greases for specialty lubrication;

- Coatings and linings for valves, pumps, pipe and fittings; and large tanks and process vessels;

- Molded filter housings, pump casings, impellers, and other fluid handling equipment; and

- General applications such as sockets; pins and connectors; coatings for home cookware, tools, and food processing equipment; flame-retardant laminates for aircraft interiors; bearing pads for bridges, pipe lines, and high-rise buildings; and packaging and production line conveyor equipment parts that must be self-lubricating and noncontaminating.

PROCESS DESCRIPTIONS

Fluoropolymers are produced by mass (bulk), suspension, or emulsion polymerization. Suspension and emulsion processes are used to make both PTFE and PCTFE; mass polymerization produces PCTFE. Each of these processes has various advantages and disadvantages when related to the properties of the polymer obtained.

PTFE Reaction Chemistry

PTFE is polymerized using a free-radical initiator (R•) in the presence of small amounts of oxygen. The initiator reacts with a tetrafluoroethylene (TFE) monomer to form a TFE radical, which in turn reacts with other TFE monomers to produce the PTFE chain. Termination is predominantly by radical combination. This series of reactions is shown below.

Initiation

$$F_2C = CF_2 + \overset{\bullet}{R} \longrightarrow R\text{-}CF_2\text{-}CF_2\bullet$$

Propagation

$$R\text{-}CF_2\text{-}CF_2\bullet + nCF_2{=}CF_2 \longrightarrow R\text{-}[CF_2\text{-}CF_2]_n\text{-}CF_2\text{-}CF_2\bullet$$

Termination by Radical Combination

$$R\text{-}CF_2\text{-}CF_2\bullet + R\text{-}[CF_2\text{-}CF_2]_n\text{-}CF_2\text{-}CF_2\bullet \longrightarrow R\text{-}[CF_2\text{-}CF_2]_{n+2}\text{-}R$$

The close packing of the fluorine atoms around the carbon backbone provides a protective shield; thus, the polymer is resistant to many corrosive environments.

PCTFE Reaction Chemistry

PCTFE, production similar to that for PTFE, uses a free-radical initiator (R•) and the chlorotrifluoroethylene (CTFE) monomer. The initator combines with the monomer to produce a CTFE radical which then reacts with other monomers to form the PCTFE chain. Termination occurs primarily by radical combination. This sequence of events is shown below.

Initiation

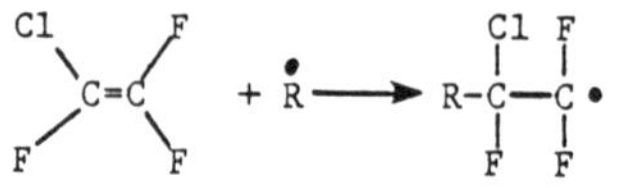

$$CClF{=}CF_2 + \overset{\bullet}{R} \longrightarrow R-\underset{\underset{F}{|}}{\overset{\overset{Cl}{|}}{C}}-\underset{\underset{F}{|}}{\overset{\overset{F}{|}}{C}}\bullet$$

Propagation

$$R-\underset{\underset{F}{|}}{\overset{\overset{Cl}{|}}{C}}-\underset{\underset{F}{|}}{\overset{\overset{F}{|}}{C}}\bullet + nCClF{=}CF_2 \longrightarrow R{-}\left[\underset{\underset{F}{|}}{\overset{\overset{Cl}{|}}{C}}-\underset{\underset{F}{|}}{\overset{\overset{F}{|}}{C}}\right]_n\underset{\underset{F}{|}}{\overset{\overset{Cl}{|}}{C}}-\underset{\underset{F}{|}}{\overset{\overset{F}{|}}{C}}\bullet$$

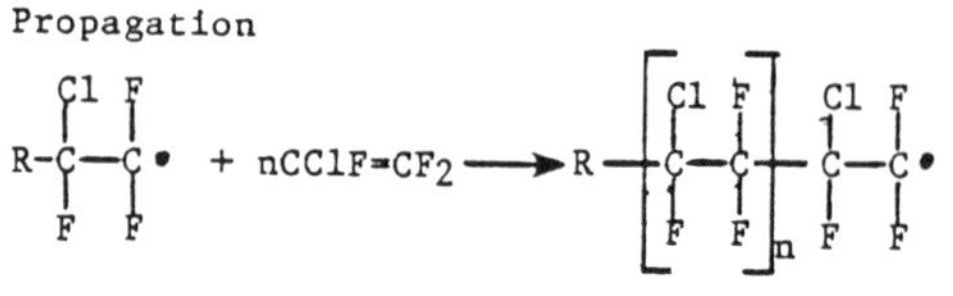

Termination by Radical Combination

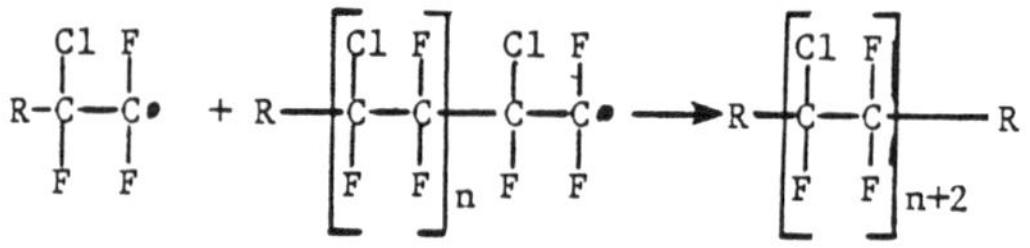

$$R-\underset{\underset{F}{|}}{\overset{\overset{Cl}{|}}{C}}-\underset{\underset{F}{|}}{\overset{\overset{F}{|}}{C}}\bullet + R{-}\left[\underset{\underset{F}{|}}{\overset{\overset{Cl}{|}}{C}}-\underset{\underset{F}{|}}{\overset{\overset{F}{|}}{C}}\right]_n\underset{\underset{F}{|}}{\overset{\overset{Cl}{|}}{C}}-\underset{\underset{F}{|}}{\overset{\overset{F}{|}}{C}}\bullet \longrightarrow R{-}\left[\underset{\underset{F}{|}}{\overset{\overset{Cl}{|}}{C}}-\underset{\underset{F}{|}}{\overset{\overset{F}{|}}{C}}\right]_{n+2}{-}R$$

Tables 62 and 63 list typical input materials and operating parameters for PTFE and PCTFE production. Other input materials, such as emulsifiers and protective colloids, are listed in Table 64.

PTFE Production Processes

PTFE is produced by emulsion and suspension polymerization processes. Emulsion polymerization produces two resin types: an aqueous dispersion and a fine powder. Granular PTFE is manufactured using the suspension polymerization process. A brief description of each process follows.

Emulsion Polymerization

The process used to form aqueous dispersions of PTFE is not a typical emulsion process. The TFE gas is introduced to a water-filled reactor where initiator, emulsifier, and any comonomer, if desired, are added. The dispersion formed must be stable enough to prevent premature coagulation; however, coagulation when the polymerization is completed is desired for the fine powder resin.

As shown in Figure 19, the polymer emulsion produced is typically stabilized and concentrated by evaporation if a dispersion is the desired end product. Figure 20 illustrates the steps taken to produce the PTFE fine powder. Both of these processes are performed batchwise. A typical production recipe was not found in the literature consulted.

Very little detail is available on the drying of the agglomerated PTFE resin. However, care must be taken to ensure the absence of shearing.[77]

TABLE 62. TYPICAL INPUT MATERALS TO PTFE AND PCTFE PRODUCTION PROCESSES

Product	Process	Water	Initiator	Monomer	Emulsifier
PTFE	Emulsion	X	X	X	X
	Suspension	X	X	X	
PCTFE	Emulsion	X	X	X	X
	Suspension	X	X	X	
	Mass		X	X	

TABLE 63. TYPICAL OPERATING PARAMETERS FOR PTFE AND PCTFE
          PRODUCTION PROCESSES

Product	Process	Temperature
PTFE		
Fine Powder	1,000 psi	55 – 240°C
Aqueous Dispersion	15 – 500 psi	0 – 95°C
PCTFE	50 – 150 psi	21 – 52°C

Sources: Encyclopedia of Chemical Technology, 3rd Edition.
          Encyclopedia of Polymer Science and Technology.
          Seymour S. Schwartz and Sidney H. Goodman, Plastics Materials and
              Processes, 1982.

TABLE 64.   INPUT MATERIALS AND SPECIALTY CHEMICALS
USED IN PTFE AND PCTFE MANUFACTURE

Function	Compound
Monomer	chlorotrifluoroethylene (CTFE) tetrafluoroethylene (TFE)
Comonomer	ethylene hexafluoropropylene isobutylene methyl methacrylate vinylidene fluoride
Initiator	disuccinic acid peroxide organic peroxides peroxydisulfates of ammonium, potassium, and sodium
Filler	asbestos bronze powder carbon graphite molybdenum disulfide polyimide powders poly(p-oxybenzoate)
Promoter	chloroform silver salts
Stabilization Aid	carboxylic acids chlorine cobalt trifluoride ozone
Emulsifier	C_{5-20} perfluorocarboxylic acid

Sources: *Encyclopedia of Chemical Technology*, 3rd Edition.
Encyclopedia of Polymer Science and Technology.
Fluoropolymers, Leo A. Wall, editor, 1972.
Seymour S. Schwartz and Sidney H. Goodman, *Plastics Materials and Processes*, 1982.

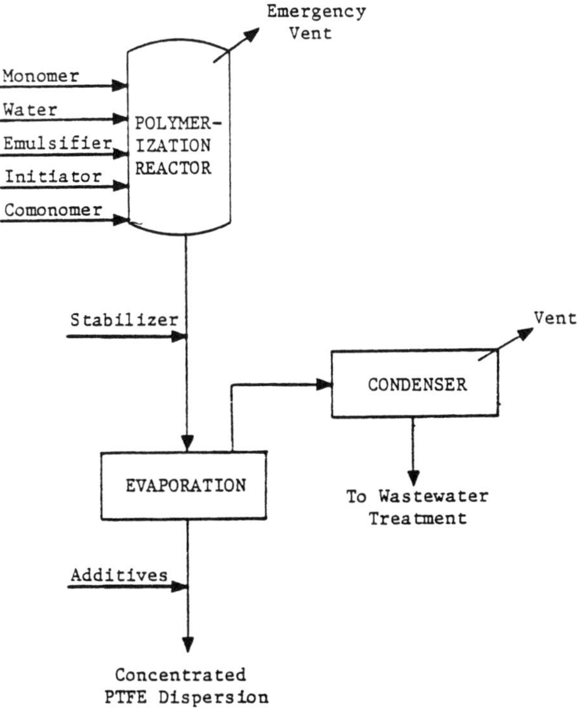

Figure 19.   Aqueous PTFE dispersion produced batchwise.

Sources: Encyclopedia of Chemical Technology, 3rd Edition.
           Seymour S. Schwartz and Sidney H. Goodman, Plastics Materials and
              Processes, 1982.

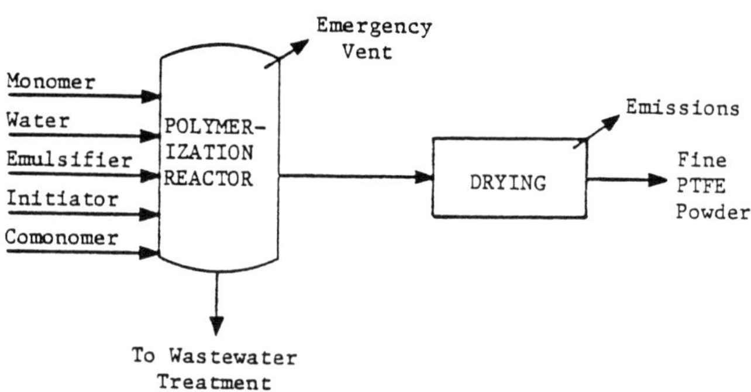

Figure 20.   Batch fine PTFE powder production.

Sources: Encyclopedia of Chemical Technology, 3rd Edition.
           Seymour S. Schwartz and Sidney H. Goodman, Plastics Materials and
           Processes, 1982.

Suspension Polymerization

Granular resins are manufactured using a batch suspension polymerization process. The TFE monomer and any comonomer are polymerized in an aqueous medium using an initiator. A dispersing agent may or may not be used. Vigorous agitation is used to keep the forming polymer in a partially coagulated state. When the batch reaches the desired conversion, the polymer is separated from the medium, shown in Figure 21, dried, and ground to the desired size. The resulting polymer has a higher void content than the fine powder. A typical production recipe was not found in the literature consulted.

PCTFE Production Processes

PCTFE is produced by mass, suspension, or emulsion polymerization. Each process exhibits advantages and disadvantages when compared to the other processes. Mass polymerization gives very pure resins since no suspension stabilizers or emulsifiers are used. However, the disadvantages include poor temperature control and low conversion. Suspension polymerization remedies the problems of temperature control and conversion, but produces a polymer of high melt viscosity for a given molecular weight which complicates resin fabrication. Overcoming the problems of both of these processes is emulsion polymerization. The major disadvantage of this final process is the presence of increased impurities in the resin produced. Ultimately, the end use of the product determines the process used.

Mass Polymerization

Mass polymerization of PCTFE uses only CTFE monomer, a comonomer (if desired), and an initiator. Mass polymerization gives products of high purity but suffers from poor temperature control, poor reproducibility, low conversion (35 to 40 percent), and long reaction times (seven days) at low temperatures (-49°C).

Monomer, initiator, and comonomer, if desired, are combined in a polymerization reactor. Shown in Figure 22, the reaction mass is discharged when the conversion reaches 35 to 40 percent. Unreacted monomer is removed and recycled while the polymer mass is extruded in a vacuum extruder. A typical production recipe was not found in the literature consulted.

Suspension Polymerization

Suspension polymerization of PCTFE overcomes some of the disadvantages of mass polymerization since the water used to suspend the polymer aids in temperature control and reproducibility, increases conversion, and reduces the reaction time. However, unless a comonomer (vinylidene fluoride) is included, it exhibits an unfavorable molecular weight—melt viscosity relationship. That is, the melt viscosity is higher at a given molecular weight than mass or emulsion polymerization resins, complicating resin processing.[19]

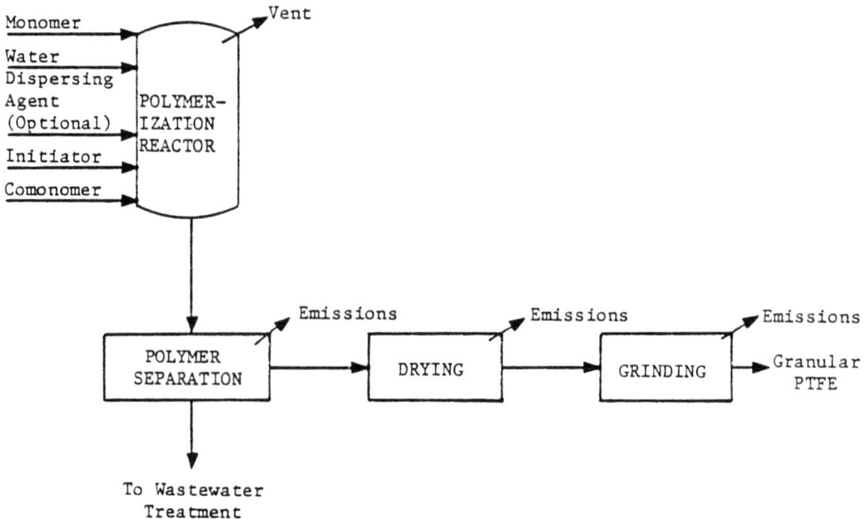

Figure 21.   Batch granular PTFE production.

Sources: Encyclopedia of Chemical Technology, 3rd Edition.
Seymour S. Schwartz and Sidney H. Goodman, Plastics Materials and Processes, 1982.

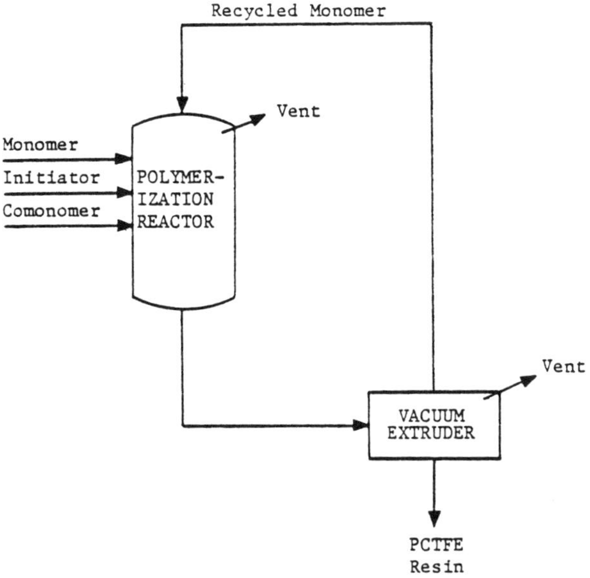

Figure 22.   Mass polymerization of PCTFE.

Source: Encyclopedia of Polymer Science and Technology.

The suspension polymerization process is very similar to that used in PTFE production. Shown in Figure 23, monomer, comonomer, initiator, and dispersing agent are fed to a water filled reactor. When the reaction is completed, the polymer is separated from the water and dried. A typical production recipe was not found in the literature consulted.

Emulsion Polymerization

Emulsion polymerization exhibits the same advantages as suspension polymerization as well as overcoming the unfavorable molecular weight-melt viscosity relationship. However, this produces a polymer with more impurities than either of the other two PCTFE types. Offsetting this problem is the best reproducibility of the three processes and the most thermally stable resin.

Emulsion polymerization of PCTFE is similar to the process used for PTFE. As shown in Figure 24, CTFE monomer, any desired comonomer, emulsifier, and initiator are added to a water filled polymerization reactor. A typical recipe for producing PCTFE by emulsion polymerization is shown below:[72]

Material	Parts by Weight
Chlorotrifluoroethylene (monomer)	
Deionized Water	200
Perfluorooctanoic Acid (emulsifier)	1
Potassium Persulfate (initiator component)	1
Ferrous Sulfate-Heptahydrate (initiator component)	0.1
Sodium Sulfate (initiator component)	0.4

The details of polymer separation were not available in the literature sources consulted; however, the polymer particles are probably coagulated and dried.

Energy Requirements

No data on the energy requirements for fluoropolymer processing were found in the literature consulted.

ENVIRONMENTAL AND INDUSTRIAL HEALTH CONSIDERATIONS

PTFE and PCTFE are nontoxic compounds used in coatings for food processing equipment. Although the polymer is not considered to be toxic, the TFE monomer must be inhibited to prevent the hazards associated with violent polymerization. TFE is also flammable in air, with limits of 11 and 60 percent. The monomer forms explosive mixtures in air and oxygen, and can disproportionate in the absence of air with the violence of black-powder explosions.[77] Proper precautions and ventilation can control these hazards.

Heated polymers are suspected as possible health hazards.[77]

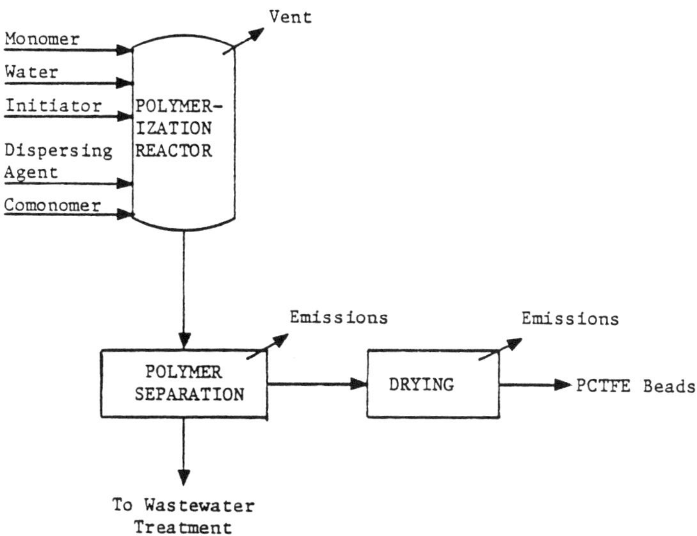

Figure 23.   Suspension polymerization of PCTFE.

Source: Encyclopedia of Polymer Science and Technology.

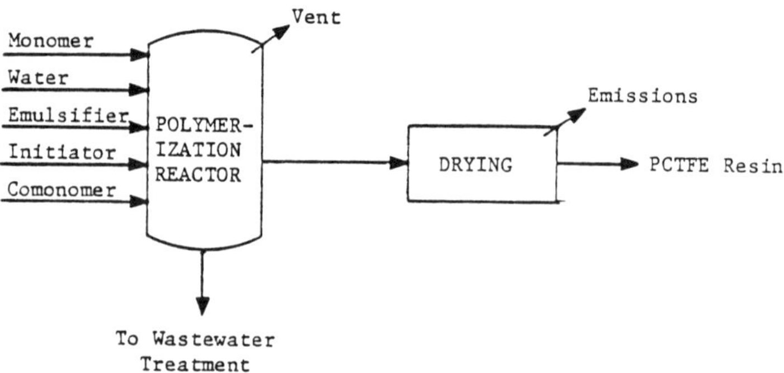

Figure 24.   Emulsion polymerization of PCTFE.

Source: Encyclopedia of Polymer Science and Technology.

Above 400°C, perfluoroisobutylene and hydrogen fluoride, toxic compounds, are released from PTFE in small amounts, though normal use does not involve these high temperatures. There have been reports of polymer fume fever in humans exposed to the unfinished product and thermodegration products.[107, 149] Exposure produces influenza-like symptoms, including chills, headaches, rigor-like shaking of limbs, mild respiratory discomfort and high fever.[107]

Only two compounds used in fluoropolymer production have been listed as hazardous under RCRA: asbestos, a filler, and chloroform, a polymerization promoter.

Mass polymerization of PCTFE is carried to 35 or 40 percent conversion. Therefore, it is expected that this process will have more VOC emissions downstream of the reactor than the other processes.

Worker Distribution and Emissions Release Points

Worker distribution estimates have been made by correlating major equipment manhour requirements with the process flow diagrams in Figures 19 through 24. Estimates for PTFE and PCTFE production processes are shown in Table 65.

No major point source air emissions are associated with fluoropolymer processing. The only air emissions are fugitive emissions from such sources as reactor vents, bead dryers, and extruders. The magnitude of the impact of these emissions on the environment and/or worker health depends upon process operating parameters, input materials, and engineering and administrative controls.

Table 66 shows sources of fugitive emissions from fluoropolymer manufacture. Employees may be exposed to vapor-phase monomer and comonomer (see Table 64) and polymer particulates which may contain residual fillers, stabilizers, and other additives. Available toxicity information for major contaminants is given below.

Health Effects

The manufacture of fluoropolymers involves the use of a confirmed human carcinogen, asbestos, and a suspected human carcinogen, chloroform. Asbestos is used as a filler and chloroform as a promoter in the process. Chlorine and ozone, both highly toxic substances, are used as stabilization aids. The reported health effects of exposure to these substances are summarized in the following paragraphs.

Asbestos is classified as a human carcinogen by the International Agency for Research on Cancer [104] and by the American Conference of Governmental Industrial Hygienists.[5] It is currently undergoing additional tests by the National Toxicology Program.[233] Many deaths from exposure to asbestos are due to lung cancer.[104] Not only is the lung cancer risk dose-related, but there is also an important enhancement of the

TABLE 65.   WORKER DISTRIBUTION ESTIMATES FOR FLUOROPOLYMER PRODUCTION

Process	Unit	Workers/Unit/8-hour Shift
PTFE:		
Aqueous Emulsion Polymerization	Batch Reactor	1.0
	Condenser	0.125
	Evaporator	0.25
Fine Powder Emulsion Polymerization	Batch Reactor	1.0
	Dryer	0.5
Suspension Polymerization	Batch Reactor	1.0
	Polymer Separator	0.25
	Dryer	0.5
	Grinding	0.25
PCTFE:		
Mass Polymerization	Batch Reactor	1.0
	Extruder	1.0
Suspension Polymerization	Batch Reactor	1.0
	Polymer Separation	0.25
	Dryer	0.5
Emulsion Polymerization	Batch Reactor	1.0
	Dryer	0.5

TABLE 66.   SOURCES OF FUGITIVE EMISSIONS FROM PTFE AND PCTFE PRODUCTION

| Source | Constituent | PTFE Processes | | PCTFE Processes | | |
		Emul-sion	Sus-pension	Mass	Sus-pension	Emul-sion
Reactor Vent	Monomer	X	X	X	X	X
	Comonomer	X	X	X	X	X
Evaporator Condenser Vent	Monomer	*a				
	Comonomer	*a				
Polymer Dryer Vent	Monomer	*b	*		*	*
	Comonomer	*b	*		*	*
	Particulates	X b	X		X	X
Polymer Separation	Monomer		*		*	
	Comonomer		*		*	
Polymer Grinding	Monomer		*			
	Comonomer		*			
	Particulates		X			
Vacuum Extruder	Monomer			X		
	Comonomer			X		

aAqueous dispersion resin only.

bFine powder resin only.

*Trace amounts of monomer and comonomer are expected due to high conversion
 levels and removal of unreacted compounds.

risk in those exposed to asbestos who also smoke cigarettes.[104]   The OSHA
emergency air standard is 0.5 fiber >5 um/cc (8 hour TWA).

Chlorine is highly toxic by inhalation and a strong local irritant.
[242]   Exposures to 3-6 ppm causes a stinging or burning sensation in the
eyes, nose, and throat, headache, sneezing and coughing, loss of voice, and
nose bleeding.[170]   Chronic exposures to concentrations on the order of 5
ppm have led to premature aging, bronchial disease, and predisposition to
tuberculosis.[170]   It has also caused mutations in human lymphocyte somatic
cells at 20 ppm.[36]   The OSHA air standard has been set at a ceiling for
exposure of 1 ppm.[69]

Chloroform produces tumorigenic, mutagenic, teratogenic, and carcinogenic effects in laboratory animals.[233]   It has been designated a suspected human carcinogen by the International Agency for Research on Cancer.
[108]   Chloroform is also moderately toxic and in high concentrations may
cause narcosis and death.[157]   Rapid death is attributable to cardiac
arrest, while delayed death has been associated with liver and kidney
damage.[108]   Symptoms of chloroform exposure at sublethal concentrations
include respiratory depression, dizziness, nausea, and intracranial pressure; after-effects include fatigue and headache.[5]   The OSHA air standard
is 50 ppm (8 hour TWA).[67]

Ozone produces tumorigenic, mutagenic, and teratogenic effects in
laboratory animals.[233]   It is also highly irritating to the respiratory
tract, skin, eyes, and mucous membranes.[243]   The primary site of acute
injury is the lung, which is characterized by pulmonary congestion and
hemorrhage, but there are indications in man that there are secondary sites
of reaction.[4]   When inhaled at concentrations not acutely injurious per
se, ozone may initiate, accelerate, or exacerbate respiratory tract disease
of bacterial or viral origin.[4]   The OSHA standard in air is 0.1 ppm (8
hour TWA).[67]

Air Emissions

Sources of fugitive emissions are summarized in Table 66 by process and
constituent type. These sources are primarly process vents and may be controlled by routing the vented stream to a flare or blowdown.[285]   Other
process VOC emissions result from leaks in valves, flanges, pumps and compressors. Although some equipment modifications are available to reduce
emissions, a regular inspection and maintenance program may be the best
control.

Sources of fugitive particulates are also listed in Table 66. These
emissions may be controlled by venting the stream to a baghouse or electrostatic precipitator for particulate removal.

Wastewater Sources

There are several sources of wastewater associated with fluoropolymer
processing, shown in Table 67. The major wastewater sources result from
emulsion and suspension polymerization which use water as a medium to aid in

TABLE 67.   SOURCES OF WASTEWATER FROM PTFE AND PCTFE PRODUCTION

Source	PTFE Processes		PCTFE Processes		
	Emulsion	Suspension	Mass	Suspension	Emulsion
Polymer Concentration	X[a]				
Polymer Isolation	X[b]				X
Polymer Separation		X		X	
Routine Cleaning Water	X	X	X	X	X

[a]Aqueous dispersion resin only.

[b]Fine powder resin only.

---

heat transfer. No data concerning wastewater production or characteristics were found in the literature consulted.

Solid Wastes

The solid wastes generated by this process are mostly fluoropolymers. Solid waste streams include: resin lost during spillage and cleaning, collected particulates, and substandard resin which cannot be blended.

Environmental Regulation

Effluent limitations guidelines have been set for only the PTFE resins portion of the fluoropolymer·industry. BPT, BAT, and NSPS call for the pH of the effluent to fall between 6.0 and 9.0 (41 Federal Register 32587, August 4, 1976).

New source performance standards (proposed by EPA on January 5, 1981) for volatile organic carbon (VOC) fugitive emissions include:

- Safety/release valves must not release more than 200 ppm above background, except in emergency pressure releases, which should not last more than five days; and

- Leaks (which are defined as VOC emissions greater than 10,000 ppm) must be repaired with 15 days.

Asbestos (a filler) and chloroform (a promoter), designated as U013 and U044, respectively, are listed as hazardous wastes (46 Federal Register 27476, May 20, 1981). All disposal of these compounds or fluoropolymers containing these compounds must comply with the provisions set forth in the Resource Conservation and Recovery Act (RCRA).

9. Phenolic Resins

Phenolic resins are heat resistant, chemical resistant, dimensionally and thermally stable, low cost resins. Their easy moldability, good electrical properties, and surface hardness make phenolic resins desirable in many applications, including use in the electrical, automotive, appliance, and consumer industry. Phenolic resins also have several industrial applications which include use in the sands used to make foundry shell molds and cores.

Phenolic resins are the condensation product of phenol and formaldehyde. There are two types of resins produced: single-stage and two-stage. Single-stage resins, which are more commonly called resols, are inherently thermosetting. Once the reaction is completed, the resin will crosslink without a catalyst at room temperature. Therefore, resols are stored at cool temperatures and exhibit a shorter shelf life than two-stage resins. Alkaline catalysis and a stoichiometric amount of formaldehyde and phenol characterize resol production.

Two-stage resins, called novolaks, are manufactured using acid catalysis and less than a stoichiometric amount of formaldehyde to react with the phenol present. The result is a brittle, thermoplastic solid which requires a catalyst, or curing agent, for cross-linking to occur.

Table A-10 in Appendix A presents typical properties of phenolic resin compounds according to specification. Molding compounds, as well as other phenolic resins, are formulated according to the specific end use. The properties of the resulting resins also depend upon the formulation used to manufacture them.

The two major phenolic resins are used in six different product areas: molding compounds, laminates, bonding resins for plywood, other bonding resins, foundry use, and coatings. Bonding and adhesive resins for plywood is the largest single market for phenolic resins, which captured over 29 percent of the sales for 1980.[65] Other phenolic resin uses include: coated and bonded abrasive production, fiber bonding for insulation, friction materials (including brake linings, clutch facings, and transmission bands), and molded parts for automotive and electrical uses.

INDUSTRY DESCRIPTION

In 1980, phenolic resins contributed 4.3 percent of the total sales for the plastics industry.[65] The phenolic resins industry consists of 50 producers which have 108 sites in 29 different states. One-fourth of the sites are located in the Great Lakes Region (EPA Region V); the remaining sites are distributed throughout the remaining states with New York and New Jersey, the southeast, the south-central, and the northwest regions each containing at least 10 percent of the sites. Major U.S. phenolic resin producers and their locations are listed in Table 68. The capacities of the plants listed are not available in the literature.

Phenolic resins are used in a wide variety of goods for consumers and industry. Adhesives, coatings, and laminates are used not only in the construction industry, but also in furniture and furnishings, appliances, and electrical products. The phenolic resin industry is affected by consumer spending and the automobile and housing industries. After growth to a production peak of 830,500 metric tons in 1978 from 609,100 metric tons in 1976, the 1981 level fell to 667,000 metric tons.[65, 143] The compound growth rate of phenolic resin production for 1975 to 1980 was 7.4 percent. [65]

PRODUCTION AND END USE DATA

In 1981, U.S. sales of phenolic resins totaled 667,000 metric tons. Phenolic resins are used in many products, which include:[22, 118, 264]

- Adhesives for plywood, particle board, fiberboard, wafer board, beams, and arches where durability in high humidity is required;

- Bonding for glass and rock wool fibers used in insulation, acoustical insulation, and carpet underlay;

- Decorative and industrial laminates for paper, cotton, or glass, including electronic circuit boards, gears, rods, bearings, tubes, furniture, wall paneling, and home and office furnishings;

- Foundry resins for synthetic, resin-bonded sand molds used to produce machine housings, automotive transmissions, cylinder heads, large metal parts, and intricate metal objects;

- Resins for bonding abrasives, including grinding wheels and snagging wheels, and for coated abrasives, including sandpaper, discs, and belts;

- Resins for friction materials used in automotive applications (including brake linings, clutch facings, and transmission bands) as well as for aircraft, train, and drilling rig parts;

- Molding compounds used for electrical sockets, switch gear, circuit breakers, automotive distributor caps, relays, brake pistons, coffee makers, utensil handles, and general appliance parts;

TABLE 68. U.S. PHENOLIC RESIN PRODUCERS

Producer	Location
American Cyanamid Co. Formica Corporation, subsidiary	Evendale, OH
American Hoechst Corporation Industrial Chemicals Division	Mount Holly, NC
AMETEK, Inc. Haveg Division	Wilmington, DE

Ashland Oil, Inc.
 Ashland Chemical Co., subsidiary
  Chemical Systems Division
  Foundry Products Division
	Columbus, OH
	Calumet City, IL
	Cleveland, OH

Baker International Corporation Magna Corporation, subsidiary	Houston, TX
The Bendix Corporation Friction Materials Division	Green Island, NY
Borden Inc. Borden Chemical Division Adhesives and Chemicals Division	Demopolis, AL Diboll, TX Fayetteville, NC Fremont, CA Rent, WA La Grande, OR Louisville, KY Missoula, MT Sheboygan, WI Springfield, OR
Brand-S Corporation Cascade Resins Division	Eugene, OR
Chargin Valley Co. Ltd. Nevamar Corporation, subsidiary	Odenton, MD
Clark Oil & Refinery Corporation Clark Chemical Corporation, subsidiary	Blue Island, IL
Core-Lube, Inc.	Danville, IL

(continued)

TABLE 68 (continued)

Producer	Location

CPC International Inc.
 CPC North American Division
  Industrial Diversified Unit
   Acme Resin Group
	Forest Park, IL

The Dexter Corporation Midland Division	Waukegan, IL

General Electric Co.
 Engineered Materials Group
  Electromaterials Business Department
 	
  Plastics Business Operations
	Coshocton, OH
	Schenectady, NY
	Pittsfield, MA

The P. D. George Company	St. Louis, MO
Georgia-Pacific Corporation Chemical Division	Albany, NY Columbus, OH Conway, NC Coos Bay, OR Crosett, AR Eugene, OR Louisville, MS Lufkin, TX Newark, OH Peachtree City, GA Port Wentworth, GA Richmond, CA Russellville, SC Taylorsville, MS Ukiah, CA Vienna, GA
Getty Oil Co. Chembond Corporation, subsidiary	Andalusia, AL Spokane, WA Springfield, OR Winnfield, LA
Gulf Oil Corporation Gulf Oil Chemicals Co. Industrial Chemicals Division	Alexandria, LA
Heresite-Saekaphen, Inc.	Manitowoc, WI

(continued)

TABLE 68 (continued)

Producer	Location
Hugh J.-Resins Co.	Long Beach, CA
Inland Steel Co. Inland Steel Container Co. Division	Alsip, IL
The Ironsides Co.	Columbus, OH
Koppers Co., Inc. Organic Materials Group	Bridgeville, PA
Lawter International, Inc.	Moundsville, AL
Libbey-Owens-Ford Co. LOF Plastic Products, subsidiary	Auburn, ME
Masonite Corporation Alpine Division	Gulfport, MS
Minnesota Mining and Manufacturing Co. Chemical Resources Division	Cordova, IL Cottage Grove, MN

Mobil Corporation
 Mobil Oil Corporation
  Mobil Chemical Co. Division
   Chemical Coatings Division
	Kankakee, IL
	Rochester, PA

Monogram Industries Inc. Spaulding Fibre Co., Inc. subsidiary	DeKalb, IL Tonawanda, NY
Monsanto Co. Monsanto Plastics and Resins Co.	Addyston, OH Chocolate Bayou, TX Eugene, OR Santa Clara, CA Springfield, MA
Niles Chemical Paint Co. Kordell Industries Division	Mishawaka, IN
The O'Brien Corporation The O'Brien Corporation-Southwestern Region	Houston, TX

(continued)

TABLE 68 (continued)

Producer	Location

Occidental Petroleum Corporation
 Hooker Chemical Corporation, subsidary
  Plastics and Chemical Specialties
   Group
    Durez Division
	Kenton, OH
	North Tonawanda, NY

Owens–Corning Fiberglass Corporation Resins and Coatings Division	Barrington, NJ Kansas City, KS Newark, OH Waxahachie, TX
Plastics Engineering Co.	Sheboygan, WI
Polymer Applications, Inc.	Tonawanda, NY
Polyrez Co., Inc.	Woodbury, NJ
Raybestos–Manhattan, Inc. Adhesivies Department	Stratford, CT
Reichhold Chemicals Inc.	Andover, MA Carteret, NJ Detroit, MI Kansas City, KS Moncure, NC South San Francisco, CA Tacoma, WA Tuscaloosa, AL White City, OR
Vacuum Division	Niagara Falls, NY
Rogers Corporation	Manchester, CT
Schenectady Chemicals Inc.	Oyster Creek, TX Rotterdam Junction, NY Schenectady, NY
The Sherwin–Williams Co. Chemicals Division	Fords, NJ
Simpson Timber Co. Oregon Overlay Division	Portland, OR

(continued)

TABLE 68 (continued)

Producer	Location

The Standard Oil Co. (Ohio)
 Sohio Industrial Products Co.,
  Division
  Don-Oliver Inc. Unit
	Niagara Falls, NY

Union Carbide Corporation Coatings Materials Division	Bound Brook, NJ Elk Grove, CA
United Technologies Corporation Inmont Corporation, subsidiary	Anaheim, CA Cincinnati, OH Detroit, MI
Valentine Sugars Inc. Valite Division	Lockport, LA
West Coast Adhesives Co.	Portland, OR
Westinghouse Electric Corporation Insulating Materials Division Micarta Division	Manor, PA Hampton, SC
Weyerhauser Co.	Longview, WA Marshfield, WI

Sources: Chemical Economics Handbook, updated annually, 1981 data.
Directory of Chemical Producers, 1982.

- Resins for coatings and adhesives which include automotive primers, can coatings, drum linings, anticorrosion paints, printing inks, wire enamels, and varnishes; and

- Foam for flower arrangement bases and decorative arts.

Table 69 lists phenolic resin consumption by market, while Table 70 presents major markets for phenolic resin molding compounds.

Novolaks exhibit greater molding latitude, better dimensional stability, and better long-term storage capabilities than resol resins. The addition of hexa as a curing agent, however, adds the possibility of ammonia emissions after the resin has been cured.

Resols are used in electrical, electronic, and appliance applications where ammonia outgassing, a problem with hexa addition in novolaks, might corrode metal or where odor may be objectionable. Resols offer better resistance to cracking where one side is wet and the other side is dry, as in steam iron handles, percolator bases, and dishwasher parts.

PROCESS DESCRIPTIONS

There are two types of phenolic resins produced: single-stage resins, called resols, and two-stage resins, called novolaks. Resols are produced using stoichiometric amounts of formaldehyde and phenol. The polymerization produces thermosetting chains which are not cross-linked. In fabrication of the desired end product, heat causes the resin to form an infusible cross-linked material. Novolaks are produced with only part of the formaldehyde stoichiometrically necessary for reaction with the phenol. The thermoplastic produced is a brittle solid at room temperature. Hexamethylenetetramine, called hexa, is added to provide the necessary crosslinking when heat is applied during product fabrication.

Resol Chemistry

Resol production uses an alkaline catalyst, such as sodium hydroxide, to initiate the reaction. A phenol anion is formed which then reacts with a formaldehyde molecule to yield a methylolphenol anion. This methylolphenol anion, in turn, reacts with a phenol molecule to produce a methylolphenol molecule and a phenol anion. Once the methylolphenol molecule is formed, it will react with other formaldehyde molecules until all the reactive sites are occupied. Once there is no free formaldehyde, the methylol groups on the phenol molecule combine with other methylol groups in a condensation reaction. These condensation reactions provide the structure necessary for crosslinking. Upon heating during the processing steps, the polymerization is completed when the infusible, cross-linked state is reached. This sequence of steps is shown below.

TABLE 69. 1981 CONSUMPTION OF PHENOLIC RESINS

Market	Thousand Metric Tons
Bonding and Adhesive Resins for:	
Coated and Bonded Abrasives	16
Fibrous and Granulated Wood	42
Friction Materials	15
Foundry and Shell Moldings	33
Insulation Materials	135
Laminating	
Building	15
Electrical/Electronics	10
Furniture	8
Other	5
Plywood	200
Molding Compounds	116
Export	12
Other	60
TOTAL	667

Source: _Modern Plastics_, January 1982.

TABLE 70. 1981 PHENOLIC RESIN MOLDING COMPOUND MARKETS

Market	Thousand Metric Tons
Appliances	17
Closures	8
Electrical/Electronics	55
Housewares	17
Industrial	4
Transporation	12
Other	3
TOTAL	116

Source: _Modern Plastics_, January 1982.

Initiation

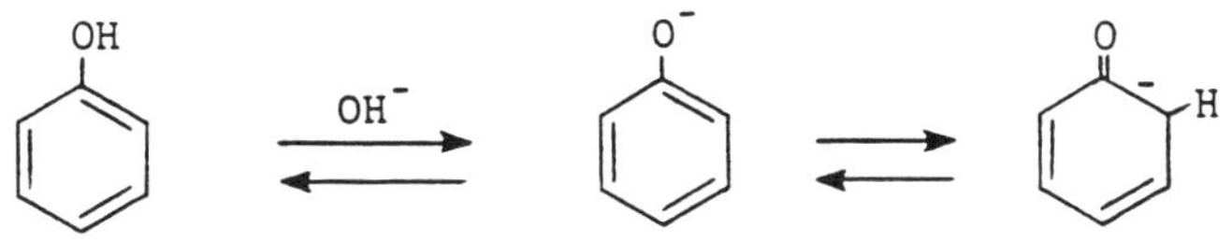

Intermediate Formation

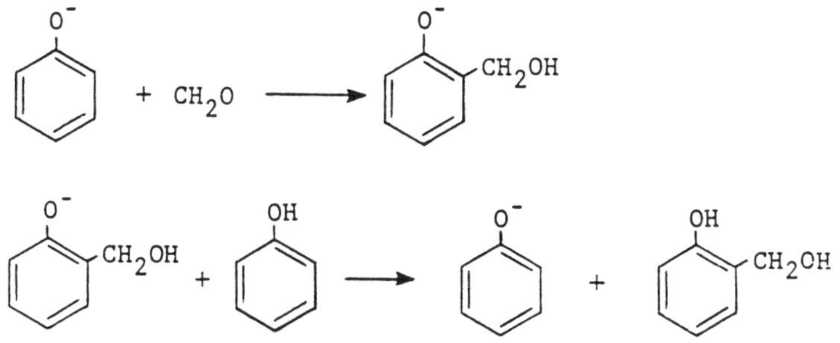

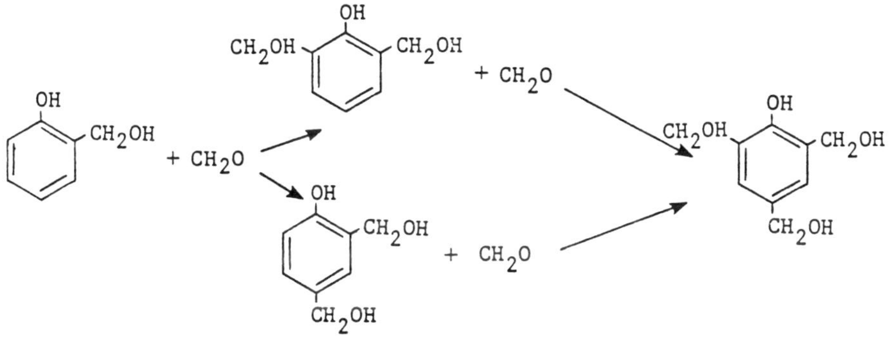

Condensation

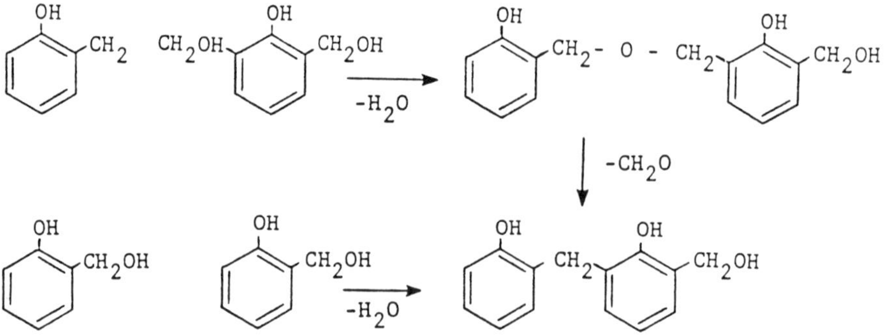

Cured State

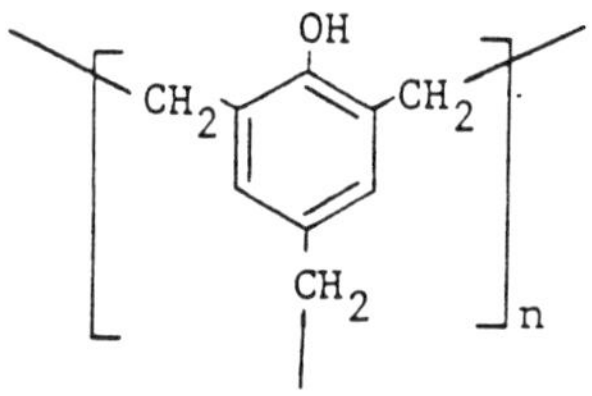

Novolak Chemistry

Novolaks are produced using an acid catalyst. This reaction is initiated when formaldehyde reacts with a hydrogen ion in solution to form a
protonated formaldehyde molecule. This reactive group then combines with
the phenol molecule at a position para to the OH group. This reaction also
regenerates a hydrogen ion. The hydrated carbonium ion and the alcohol (OH)
group on the cyclic ring may eliminate a molecule of water to form a reactive group. This group may then react with a phenol molecule to produce a
hydrogen ion and a dihydroxydiphenylmethane molecule. Hexamethylenetetramine (hexa) is added to the product before shipping. Upon the addition of
heat during product processing, the novolak structure will crosslink to form
an infusible solid with hexa providing the additional formaldehyde for the
linkages. This series of reactions is shown below.

Initiation

$$CH_2O + H^+ \longrightarrow CH_2OH^+$$

Intermediate Formation

Condensation

Curing Reaction

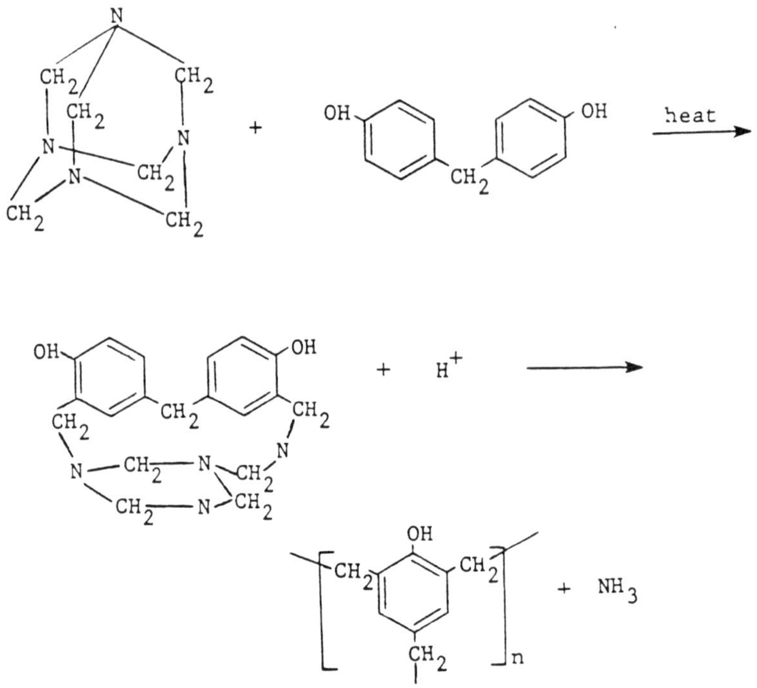

Tables 71 and 72 list typical input materials and operating parameters for the two types of phenolic resins produced. Other input materials, such as comonomers, fillers, and catalysts, are listed in Table 73.

Resols

Resol production, as presented in Figure 25, starts when phenol and formaldehyde are charged to a stirred reactor. An alkaline catalyst is added to adjust the pH of the solution to the reaction conditions desired. Vacuum removal of the water formed by the condensation reaction is performed at temperatures below 100°C to prevent over-reaction or gelation.

A typical recipe for resol production is shown below:[123]

Material	Parts by Weight
Phenol	94
Formaldehyde (37%)	123
Barium Hydroxide Hydrate (catalyst)	4.7

TABLE 71. TYPICAL INPUT MATERIALS TO PHENOLIC RESIN PRODUCTION
IN ADDITION TO FORMALDEHYDE AND PHENOL

Phenolic Resin Type	Alkaline Catalyst	Acid Catalyst	Fillers	Cure Catalyst
Resol	X		X	
Novolak		X	X	X

TABLE 72. TYPICAL OPERATING PARAMETERS FOR PHENOLIC RESIN PRODUCTION

Phenolic Resin Type	Temperature	Formaldehyde-Phenol Ratio	Reaction Time
Resol	80 to 100°C	1:1 to 1.5:1	1 to 3 hours
Novolak	85 to 90°C	0.75:1 to 0.95:1	3 to 6 hours

Source: _Encyclopedia of Chemical Technology_, 3rd Edition.

TABLE 73.   TYPICAL INPUT MATERIALS AND SPECIALTY CHEMICALS USED IN
PHENOLIC RESIN PRODUCTION IN ADDITION TO PHENOL AND
FORMALDEHYDE

Function	Compound
Curing Agent	Hexamethylenetetramine (hexa)
Acid Catalysts	Hydrochloric acid
	Oxalic acid
	Phosphoric acid
	p-toluenesulfonic acid
	Sulfuric acid
Alkaline Catalysts	Barium hydroxide
	Calcium hydroxide
	Organic amines
	Sodium carbonate
	Sodium hydroxide
Comonomers	Acetaldehyde
	Bisphenol-A
	Cresols
	Furfuraldehyde
	Nonylphenol
	Octylphenol
	p-phenylphenol
	p-tert-butylphenol
	Resorcinol
	t-butylphenol
	Xylenols
Fillers	Asbestos
	Cellulose
	Clay
	Cotton flock
	Glass
	Hydrated alumina
	Mica
	Organic fibers
	Wood flour

Sources: Encyclopedia of Chemical Technology, 3rd Edition.
Encyclopedia of Polymer Science and Technology.
Modern Plastics Encyclopedia, 1981-1982.

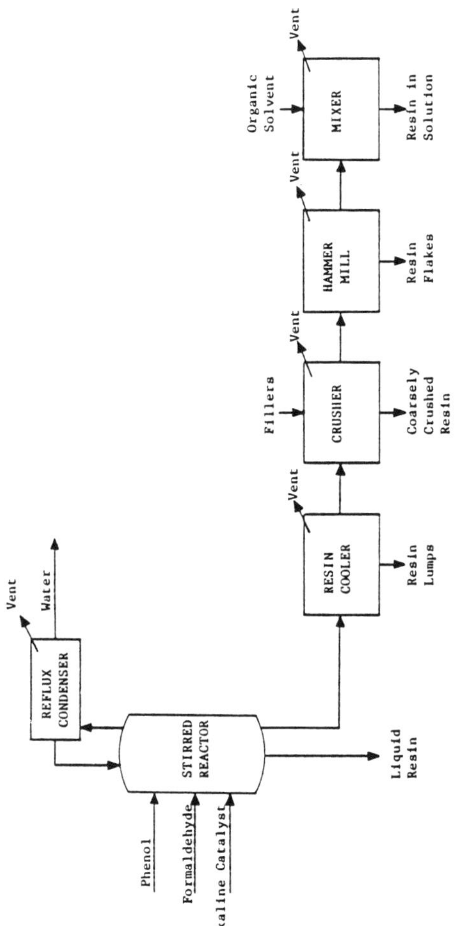

Figure 25.   Resol production process.

Sources: Encyclopedia of Chemical Technology, 3rd Edition. Encyclopedia of Polymer Science and Technology.

When the reaction has reached the desired end-point, usually over 95 percent conversion based on phenol, the contents are cooled by a rapid discharge to either a cooled stainless steel floor or a resin cooler.[22] Resin coolers are assemblies of water-cooled, baffle plates spaced approximately 4 cm apart.[22] The resin is cooled quickly in layers between the plates.

Resols may be produced as liquid resins (resins in solution or casting resins) or as solid resin lumps, solid resin flakes, or coarsely-crushed solid resin. Liquid resins require no dehydration step. They are fed from the reactor into drums or tank cars for shipping. The dried resin may be dissolved in an organic solvent, such as alcohol, and used as a varnish, coating, or prelamination treatment. Casting resins are transferred from the reactor to molds. Solid resins may be combined with fillers and are shipped in fiber drums, plastic bags, or multi-wall paper bags.

Resols have recently been produced in a dispersion. The addition of a protective colloid to the reactor allows the production of an in-situ dispersion. If a solid resin is desired, the spherical particles may be recovered from the liquid resin and dried via rapid sedimentation at the end of the reaction.[7] Protective colloids used include polyvinyl alcohol, natural gums, and cellulose derivatives.

Novolaks

As illustrated in Figure 26, novolaks are produced using phenol, formaldehyde, and an acid catalyst. A limiting amount of formaldehyde is used to prevent crosslinking when the phenol and formaldehyde are charged to a stirred reactor. The catalyst is added until the solution reaches the desired pH, the temperature of the solution is adjusted, and the water is refluxed to the reactor for the duration of the reaction period.

A typical recipe for novolak production is shown below:[251]

Material	Parts by Weight
Phenol	42
Formaldehyde	27
Wood Flour (filler)	52
Hexamethylenetetramine (curing agent)	3.7
Sulfuric Acid (catalyst)	0.1
Nigrosin (black dye)	1.3
Calcium Stearate (mold release agent)	0.7

When the reaction has proceeded to the desired end point, usually to a conversion greater than 95 percent, the water is no longer recycled to the reactor. Any remaining moisture is removed under a vacuum. When the water removal is complete, the resin is discharged to resin coolers, which may be shallow pans or cleaned floor areas.[22] If a flaked resin is desired, the molten resin is sent directly to a resin flaker. Resins used in solution are produced by the addition of solvent to the reactor, eliminating the dehydration step. Hexamethylenetetramine is added after all processing is completed to provide crosslinking upon heating.

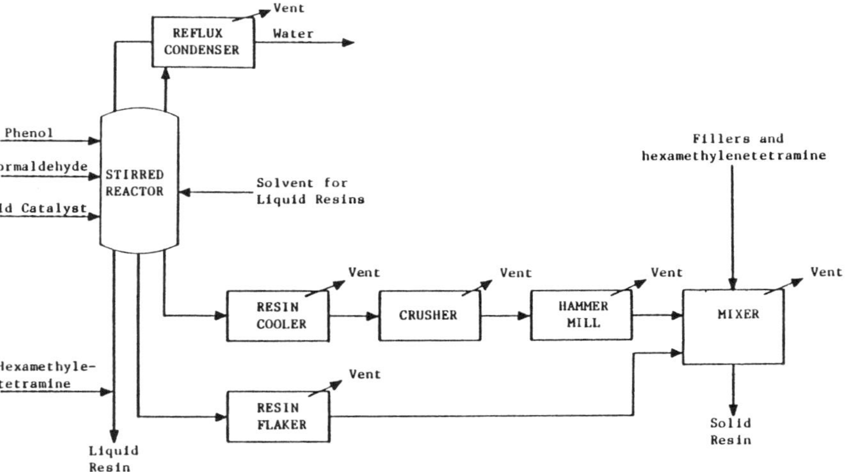

Figure 26.   Novolak production process.

Sources: Encyclopedia of Chemical Technology, 3rd Edition.
Encyclopedia of Polymer Science and Technology.

Energy Requirements

No data on the energy requirements for phenolic resin processing were found in the literature consulted.

ENVIRONMENTAL AND INDUSTRIAL HEALTH CONSIDERATIONS

Phenolic resins are considered to be nontoxic in the cured state. Uncured resins, however, contain small amounts of free phenol and formaldehyde. Therefore, care should be taken in the handling and storage of these resins. The manufacture of novolaks calls for a limiting amount of formaldehyde, and all phenolic resins are carried to over 95 percent conversion based on phenol. Most emissions of phenol and formaldehyde will occur in the reactor vent and reflux condenser vent; little emission of these substances is expected from downstream operations.

Worker Distribution and Emissions Release Points

Worker distribution estimates have been made by correlating major equipment manhour requirements with the process flow diagrams in Figures 25 and 26. Table 74 shows estimates for both resol and novolak production.

There are no major air emission point sources associated with the phenolic resin processes. Fugitive emissions from process sources as vents and leaks, however, may pose a significant environmental and/or worker health problem depending on stream constituents, process operating parameters, engineering and administrative controls, and maintenance programs.

Sources of fugitive emissions are listed in Table 75. Principal gaseous emissions of concern are phenol and formaldehyde. Cresols, bisphenol A, furfuraldehyde and acetaldehyde may be used as comonomers.

Phenol and formaldehyde handling is highly automated in most plants, thus minimizing employee exposure potential.[118]

The use of hexamethylenetetramine as a curing agent in novolak production presents the possibility of employee exposure to ammonia, a by-product of the curing reaction. Acid catalysts (sulfuric, hydrochloric or phosphoric acids) used in novolak production also may present a hazard.

Employees assigned to crushing, grinding and flaking operations may be overexposed to nuisance dust from polymer particulate emissions or to such additives as asbestos. Available toxicity information for contaminants associated with phenolic resin production is summarized in the following paragraphs.

Little data are available in the literature from which employee exposure potential may be estimated. However, one source gives the following emission factors for phenolic resin processing:[286]

- Grinding--16 g particulate matter/kg product
- Adhesives production--25 g hydrocarbon/kg product
- Molding--7-60 g hydrocarbon/kg product

TABLE 74. WORKER DISTRIBUTION ESTIMATES FOR PHENOLIC RESIN PRODUCTION

Process	Unit	Workers/Unit/8-hour Shift
Resol Production	Batch Reactor	1.0
	Reflux Condenser	0.125
	Resin Cooler	0.25
	Crusher	0.25
	Hammermill	0.25
	Batch Mixer	0.5
Novolak Production	Batch Reactor	1.0
	Reflux Condenser	0.125
	Resin Cooler	0.25
	Crusher	0.25
	Hammermill	0.25
	Flaker	0.25
	Batch Mixer	0.5

TABLE 75. SOURCES OF FUGITIVE EMISSIONS FROM PHENOLIC RESIN MANUFACTURE

Source	Constituent	Phenolic Resin	
		Resol	Novolak
Reactor Vent	Phenol	X	X
	Formaldehyde	X	X
Reflux Condenser Vent	Phenol	X	X
	Formaldehyde	X	X
Resin Cooler Vent	Phenol	*	*
	Formaldehyde	*	*
Resin Crusher Vent	Resin Particulates	X	X
Resin Flaker Vent	Resin Particulates		X
Hammermill Vent	Resin Particulates	X	X
Mixer Vent	Solvent	X	
	Resin Particulates		X

*Near complete reaction, combined with recycle of unreacted monomers to the reactor, reduces the phenol and formaldehyde emissions downstream to trace amounts.

Health Effects

Phenol and formaldehyde, two of the major raw materials in phenolic
resin production, pose a significant health risk if exposure occurs. The
comonomer resorcinol, and asbestos, which is used as a filler, also are
significant due to carcinogenic, mutagenic, teratogenic, or toxic properties. The reported health effects of exposure to these four substances
follow.

Asbestos is classified as a human carcinogen by the International
Agency for Research on Cancer [104] and by the American Conference of
Governmental Industrial Hygienists.[5] It is currently undergoing additional tests by the National Toxicology Program.[233] Many deaths from
exposure to asbestos are due to lung canger.[104] Not only is the lung
cancer risk dose-related, but there is also an important enhancement of the
risk in those exposed to asbestos who also smoke cigarettes.[104] The OSHA
emergency air standard is 0.5 fiber >5 um/cc (8 hour TWA).

Formaldehyde poses a high toxic hazard upon ingestion, inhalation, or
skin contact [242] with human effects reported at doses as low as 8 ppm.
[127] This suspected carcinogen is a potent mutagen and teratogen. The
OSHA air standard is 3 ppm (8 hour TWA) with a 5 ppm ceiling.[67]

Phenol is extremely toxic by ingestion, inhalation, and skin absorption.[85] Approximately half of all reported cases of acute phenol poisoning has resulted in death.[278] Chronic poisoning may also occur from
industrial contact and has caused renal and hepatic damage.[149] Evidence
indicates that phenol is also a tumorigenic, mutagenic, and teratogenic
agent; however, a recent carcinogenesis bioassay by the National Cancer
Institute produced negative results.[233] The OSHA air standard is 5 ppm (8
hour TWA).[67]

Resorcinol is a tumorigen and a mutagen according to animal test data
[233]; however, carcinogenic testing has produced indefinite results.[105]
The chemical is currently being tested for carcinogenesis under the National
Toxicology Program. Resorcinol is also highly toxic by ingestion. An oral
exposure of 29 mg/kg has caused death in humans.[52]

Air Emissions

Fugitive VOC emission sources are listed in Table 75 by resin type and
constituent. All of these sources are vents and may be controlled by:[285]

- Venting the stream to a flare to incinerate the hydrocarbons;
  and/or

- Venting the stream to blowdown.

Other VOC emissions may result from valves, relief valves, flanges, drains,
pumps, compressors, and cooling towers. Equipment modification may be used
to control these emissions, but a regular inspection and maintenance program
may effectively reduce most of the VOC emissions from these sources.

Fugitive particulate emissions are also summarized in Table 75. These also result from vents and may be controlled by:

- Venting the stream to a baghouse or electrostatic precipitator to remove particulates.

According to EPA estimates, the population exposed to phenolic resin emissions in a 100 km^2 area surrounding a phenolic resin plant is: 40 persons exposed to hydrocarbons and 22 persons exposed to particulates.[286]

Wastewater Sources

There are only two sources of wastewater associated with phenolic resin production, as shown in Table 76. The water generated from the condensation reaction used to produce the resin is the major wastewater stream. Routine cleaning water contributes a smaller amount to the wastewater produced.

Ranges of several parameters for wastewaters from phenolic resin processing are shown below. Values for the wastewater from the processes described were not distinguished by process type by EPA for the purpose of establishing effluent limitations for the phenolic resin industry.[284]

Phenolic Resin Wastewater Characteristics	Unit/Metric Ton of Phenolic Resin
Production	$1.67 - 24.03 \ m^3$
BOD5	$2 - 20.7$ kg
COD	$5 - 33.5$ kg
TSS	$0 - 30$ kg

Solid Wastes

The solid wastes generated by this process include substandard product which cannot be blended, collected particulates, and resin lost during routine cleaning and spillage. These wastes are comprised mainly of phenolic resins.

Environmental Regulation

Effluent limitations guidelines have been set for the phenolic resin industry. BPT, BAT, and NSPS call for the pH of the effluent to fall between 6.0 and 9.0 (41 Federal Register 32587, August 4, 1976).

New source performance standards proposed by EPA on January 5, 1981) for volatile organic carbon (VOC) fugitive emissions include:

- Safety/release valves must not release more than 200 ppm above background, except in emergency pressure releases, which should not last more than five days; and

TABLE 76.   SOURCES OF WASTEWATER FROM PHENOLIC RESIN MANUFACTURE

| | Phenolic Resin | |
Source	Resol	Novolak
Condensation Reaction Water	X	X
Routine Cleaning Water	X	X

- Leaks (which are defined as VOC emissions greater than 10,000 ppm) must be repaired within 15 days.

The following compounds are listed as hazardous wastes (46 Federal Register 27476, May 20, 1981):

Acetaldehyde	- U001
Asbestos	- U013
Cresols	- U052
Formaldehyde	- U122
Phenol	- U188
Resorcinol	- U201

All disposal of either of these compounds or phenolic resins which contain residual amounts of these compounds (i.e., uncured resins) must comply with the provisions set forth in the Resource Conservation and Recovery Act (RCRA).

10. Polyacetal

INTRODUCTION

Polyacetals, or acetal resins, are strong, hard, highly crystalline
thermoplastics. First commercially produced in the 1960's, these versatile
polymers possess a good balance of short- and long-term mechanical properties. Their high melt temperature, stiffness, and good processibility
(short-term properties) are combined with low friction, fatigue resistance,
corrosion resistance, and dimensional stability (long-term properties),
making them an ideal replacement for nonferrous metals in many applications.

Polyacetals are produced from the polymerization of formaldehyde. The
repeating oxymethylene structure, $-OCH_2-$, gives these resins a chemistry
similar to that of simple acetals. Acetal homopolymers consist of only this
repeating oxymethylene structure while acetal copolymers have an occasional
interruption in the oxymethylene chain for a comonomer unit, such as an
ethylene linkage. Trioxane, an oligomer of formaldehyde consisting of three
units in a cyclic structure, is used along with a cyclic comonomer to manufacture the acetal copolymer. Copolymers are more stable to thermal degradation, oxidation, and chemical attack due to the stabilizing nature of the
comonomer unit. Table A-11 in Appendix A lists some typical properties for
acetal homopolymers and copolymers.

Polyacetals are the first plastics with strength properties approaching
those of nonferrous metals.[247] These resins are rigid, tough, and resili-
ent but not brittle. Properties are retained over an extended time, even
during adverse temperature and humidity conditions or while being exposed to
most solvents. Polyacetals are denser than most plastics, yet they are
lighter than any of the die-casting metal alloys: 85 percent lighter than
brass, 80 percent lighter than zinc and iron, 45 percent lighter than
aluminum, and over 20 percent lighter than magnesium.[247]

INDUSTRY DESCRIPTION

Two producers make up the polyacetal industry. Celanese manufactures
various grades of acetal copolymers at their Bishop, Texas site while du
Pont produces acetal homopolymers at their Parkersburg, West Virginia plant.
These producers and their capacities are presented in Table 77.

Polyacetals are used primarily as metal replacements. Automotive,
plumbing, and machinery parts and consumer products are the largest markets.

TABLE 77.   U.S. PRODUCERS OF POLYACETALS

Producer	Location	1982 Capacity Thousand Metric Tons
Celanese Corp. Celanese Plastics and Specialties Co. Division Celanese Engineering Resins Div.	Bishop, TX	56.8
E.I. du Pont de Nemours & Co., Inc. Polymer Products Dept.	Parkersburg, WV	34.1
TOTAL		90.9

Source: <u>Directory of Chemical Producers</u>, 1982.

Polyacetal sales reached 48,000 metric tons in 1979, with a subsequent
decline in 1980 to 40,000 metric tons. A small increase in sales occurred
in 1981 (42,000 metric tons).[143] The polyacetal market is effected by the
general trend of the U.S. economy. However, this new plastic is also
experiencing growth in the number of applications for which it is used. The
current capacity (90.9 thousand metric tons) for polyacetal production is
sufficient for any growth in the near future.

PRODUCTION AND END USE DATA

In 1981, sales for polyacetals totaled 42,000 metric tons.[143] Uses
for acetal polymers include:[12, 173, 174, 247, 297]

- Transportation: automotive electrical switches, body hardware,
  safety-belt hardware, collapsible steering column hardware, fuel
  system components, and window handles;

- Machinery: specialized equipment for conveying and handling food,
  general purpose tabletop conveyor chain, mechanical couplings, small
  engine starter pulleys, pump impellers, and a variety of textile
  and agricultural machinery components;

- Plumbing and hardware: ball cocks, shower heads, faucets, pumps,
  sinks, water softeners, lawn sprinklers, irrigation systems, piping,
  and storage tanks;

- Appliances: gears, cams, springs, and bearings;

- Mill shapes: rods, slab, tubing, and sheet for the machining and
  stamping of commercial quantities of gears, bearings, wearstrips,
  thrust washers, and bushings;

- Packaging: aerosol bottles, valves, and containers; and

- Consumer products: pens, pencils, cigarette lighters, electric and
  hand razors, zippers, toys, and tape cartridge components, telephone
  parts, handles and other hardware items, meat hooks, milk pumps, and
  coffee spigots.

Table 78 lists major U.S. markets for polyacetal resins.

PROCESS DESCRIPTIONS

Polyacetals are produced according to the end product desired. Acetal
homopolymers are manufactured by the polymerization of formaldehyde while
acetal copolymers use trioxane to produce the desired resin. Copolymers are
more resistant to degradation caused by heat, oxidation, or solvent attack.
This resistance is gained, however, at the expense of impact strength.

TABLE 78. MAJOR U.S. MARKETS FOR POLYACETAL RESINS

Market	1981 Consumption Thousand Metric Tons
Appliances	3
Consumer Products	11
Electrical/Electronics	2
Industrial	8
Plumbing and Hardware	9
Sheet, Rod, and Tube	2
Transportation	6
Other	<u>1</u>
TOTAL	42

Source: *Modern Plastics*, January 1982.

<u>Acetal Homopolymer Reaction Chemistry</u>

The polymerization of formaldehyde is accomplished using an ionic initiator, A (either anionic or cationic) to start the reaction. Chain transfer agents, such as water, serve to terminate the growth of the poly-acetal chain. Chain transfer and reinitiation lead to more polymer chains than can be accounted for by the initiator concentration.[173] The polymerization reactions are shown below for the anionic case.

Initiation
$$A^- + CH_2O \longrightarrow ACH_2O^-$$

Propagation
$$ACH_2O^- + nCH_2O \longrightarrow A(CH_2O)_nCH_2O^-$$

Chain Transfer
$$A(CH_2O)_nCH_2O^- + H_2O \longrightarrow A(CH_2O)_{n+1}H + OH^-$$

Reinitiation
$$OH^- + CH_2O \longrightarrow HOCH_2O^-$$

A polyacetal chain-terminated with a hydroxyl end group tends to depolymerize due to the highly unstable hydroxyl group. The chain is stabilized by chemical modification, called end-capping. The hydroxyl group is replaced by a more stable group, commonly done by esterification with acetic anhydride.

<u>Acetal Copolymer Reaction Chemistry</u>

Like the reaction which produces acetal homopolymers, the acetal copolymer polymerization uses an ionic initiator (typically a cation, such as boron trifluoride). The initiator first reacts with any trioxane to open the rings, and no polymerization occurs until equilibrium with formaldehyde is reached. The chain then grows in a manner similar to the homopolymer, except several comonomer units are incorporated into the chain. The series of polymerization steps is shown below for the boron trifluoride initiator case.

<u>Initiation</u>

$$BF_3 + CH_2\underset{\diagdown O-CH_2 \diagup}{\overset{\diagup O-CH_2 \diagdown}{}}O \longrightarrow CH_2O + F_3B^- {-}O{-}CH_2{-}O{-}CH_2^+$$

<u>Propagation</u>

$$F_3B^- {-}O{-}CH_2{-}O{-}CH_2^+ + (n+1)CH_2O + mCH_2\overset{\overset{O}{\diagup \diagdown}}{{-}}CH_2 \longrightarrow F_3B{-}^-OCH_2OCH_2O[(CH_2)_2O]_m(CH_2O)_nCH_2^+$$

Chain Transfer

$$F_3B\text{-}^-OCH_2\text{-}OCH\text{-}O[(CH_2)_2O]_m(CH_2O)_nCH_2^+ + H_2O \longrightarrow F_3B\text{-}^-OCH_2\text{-}OCH_2\text{-}O[(CH_2)_2O]_m(CH_2O)_nCH_2OH + H^+$$

Polyacetal copolymers do not tend to depolymerize with hydroxyl end groups due to the oxyethylene groups randomly dispersed throughout the polymer.

Tables 79 and 80 list typical input materials and operating parameters for the two polyacetal production processes. Other input materials such as fillers and initiators are listed in Table 81.

Acetal Homopolymer Production

The polymerization of formaldehyde for acetal homopolymers consists of four basic steps: purification, polymerization, end-capping, and finishing. Each of these steps, shown in Figure 27, performs an essential function necessary in order to produce a useful commercial product.

Formaldehyde is typically supplied in a stable aqueous solution, about 40 percent of which is water, called formalin. Impurities present include methanol, formic acid, methyl formate, and carbon dioxide. Water, methanol, and formic acid retard polymerization and, therefore, must be removed. In order to accomplish the removal of impurities, partial polymerization is performed and the low molecular weight polymer is first precipitated with alkali, then washed with distilled water and vacuum dried. The dried formaldehyde polymer is pyrolyzed at 150 to 160°C; the formaldehyde which results is passed through a series of low temperature traps (-15°C) to remove any remaining impurities. This purified formaldehyde is necessary to produce the acetal homopolymer.

The purified formaldehyde is fed to a reactor which contains a dried inert solvent, such as heptane. A polymerization initiator, a chain transfer agent, and any stabilizers desired are then added. The polymerization reaction is complete when a 20 percent solids content is reached.[247] The polymer is filtered out of the solution, washed with solvent, and vacuum dried.

The resulting polymer now has predominantly hydroxyl end groups. The stability of the molecule is improved by end-capping the polymer chain by reacting the polymer with acetic anhydride, giving the chain acetate ester end groups which are more stable at elevated temperatures. This end-capping is achieved by reacting the polymer with an excess of acetic anhydride followed by filtration and washing.

TABLE 79.   TYPICAL INPUT MATERIALS TO POLYACETAL PRODUCTION PROCESSES

Process	Solvent	Formalde-hyde	Initiator	Comonomer
Homopolymer	X	X	X	
Copolymer	X	X	X	X

TABLE 80.   TYPICAL OPERATING PARAMETERS FOR POLYACETAL PRODUCTION PROCESSES

Process	Temperature	Reaction Time
Homopolymer	up to 119°C	*
Copolymer		
Polymerization	65–80°C	*
Stabilization	100–250°C	1–2 min.

*These processes are considered to be proprietary.

Sources: Encyclopedia of Chemical Technology, 3rd Edition.
           Encyclopedia of Polymer Science and Technology.

TABLE 81.   INPUT MATERIALS AND SPECIALTY CHEMICALS USED IN
POLYACETAL MANUFACTURE

Process	Compound
Monomer	formaldehyde
	trioxane
Comonomer	1,3-dioxolane
	ethylene oxide
Initiator	acetal perchlorate
	amides
	amidines
	amines
	ammonium salts
	arsines
	diazonium salts
	p-chlorophenyldiazonium
	hexafluoroarsenate
	p-chlorophenyldiazonium
	hexafluorophosphate
	Lewis acids
	boron trifluoride dibutyl etherate
	stannic chloride
	Lewis bases
	organometallic compounds
	phosphines
	phosphoric acid
	stibines
	sulfonium salts
	sulfuric acid
Chain Transfer Agent	acids
	alcohols
	water
Stabilizer	diphenylamine
	polyamides
	polyurethanes
	substituted ureas
Esterification Compounds	acetic anhydride
	amines
	soluble alkali metal salts
	sodium acetate

(continued)

TABLE 81 (continued)

Process	Compound
Antioxidant	α-conidendrol di-β-naphthyl-p-phenylenediamine
Property Enhancers	glass fibers molybdenum disulfide Teflon® fibers
UV Stabilizer	carbon black
Acid Acceptors	basic salts epoxides nitrogen compounds
Hydrolysis Solvents	water soluble alkanols isopropyl alcohol
Hydrolysis Initiator	ammonia basic hydroxides
Solvent	cyclohexane heptane hexane methylene chloride

Sources: *Encyclopedia of Chemical Technology*, 3rd Edition.

Encyclopedia of Polymer Science and Technology.

Seymour S. Schwartz and Sidney H. Goodman, *Plastics Materials and Processes*, 1982.

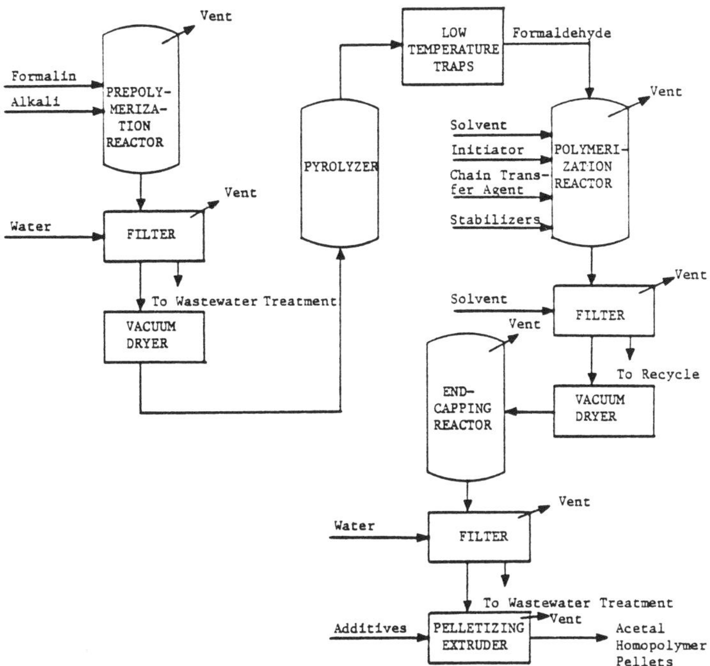

Figure 27.   Acetal homopolymer production process.

Sources: <u>Encyclopedia of Chemical Technology</u>, 3rd Edition.
Seymour S. Schwartz and Sidney H. Goodman, <u>Plastics Materials and
Processes</u>, 1982.

The end-capped polymer is then finished by adding the desired antioxidants and stabilizers. The formulated polymer is extruded into pellets
before shipping.

A typical recipe for the production of polyacetal via homopolymer
polymerization is as follows:[253]

Material	Parts by Weight
Formaldehyde (monomer)	1,000
Boron Trifluoride (catalyst)	1
Heptane (solvent)	500
Acetic Anhydride (esterification chain end compound)	800

Acetal Copolymer Production

Acetal copolymers are produced by the reaction of trioxane and a
comonomer, usually ethylene oxide. This process, shown in Figure 28, consists of four major steps: preparation and purification of trioxane,
copolymerization, stabilization, and finishing.

Trioxane is produced from formalin by introducing a strong mineral
acid, such as sulfuric acid, to the solution. Pure trioxane is subsequently
recovered by removing the water from the solution, typically by adding an
azeotropic solvent.

The copolymerization of the pure trioxane is typically performed in a
solvent, such as cyclohexane, hexane, or methylene chloride. The trioxane,
solvent, and small amounts of comonomer (0.01 to 15 percent is incorporated
to improve the thermal, oxidative, and solvent resistance of the polymer)
are combined in a reactor. As with formaldehyde polymerization, an initiator and chain transfer agent are also used. When the polymerization is
completed, the polymer is filtered and washed with solvent.

Acetal copolymers are stabilized by alkaline hydrolysis of the oxymethylene end groups of the polymer chain. More stable carbon-carbon linkages are unaffected. A water miscible organic solvent, such as isopropyl
alcohol, is used to hydrolyze the acetal copolymer with ammonia. After a
short contact time (one to two minutes), excess water is added to precipitate the polymer. The precipitated polymer is then washed and dried.

Further stabilization is achieved in the finishing step by the addition
of antioxidants and stabilizers. The formulated polymer is then extruded
into pellets before shipping.

A typical recipe for acetal copolymer production includes:[253]

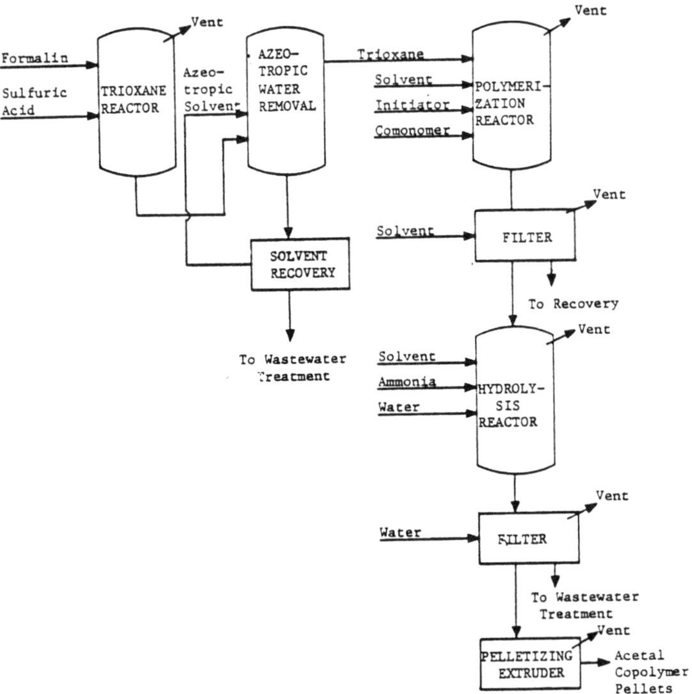

Figure 28.   Acetal copolymer production.

Sources: *Encyclopedia of Chemical Technology*, 3rd Edition.
Encyclopedia of Polymer Science and Technology.
Seymour S. Schwartz and Sidney H. Goodman, *Plastics Materials and Processes*, 1982.

Material	Parts by Weight
Trioxane (monomer)	900
Ethylene Oxide (comonomer)	0.44-77
Boron Fluoride Etherate Catalyst	0.03-0.90
Solvent	

The amount of solvent and other additives were not found in the literature.

Energy Requirements

The following energy requirements are listed for an unspecified polyacetal production process:[180]

Energy Required	Unit/Metric Ton of Product
Electricity	3.39×10^6 Joules
Steam	14.3 metric tons

ENVIRONMENTAL AND INDUSTRIAL HEALTH CONSIDERATIONS

Polyacetals are nontoxic resins used as pipe for potable water and con-
veyors for food. However, care must be used in handling both ethylene oxide
and formaldehyde since both can be toxic and improper handling may result in
in explosive mixtures with air.[173] Cyclohexane, ethylene oxide, and
formaldehyde are listed as hazardous materials.

Worker Distribution and Emissions Release Points

We have estimated worker distribution in polyacetal production processes by correlating major equipment manhour requirements with the process
flow diagrams in Figures 27 and 28. Estimates for homopolymer and copolymer
production are shown in Table 82.

No major air emission point sources are associated with polyacetal processing. However, fugitive emissions resulting from process sources and
leaks may pose significant environmental and worker health problems depending on stream constituents, process operating parameters, engineering and
administrative controls, and maintenance programs.

Sources of process fugitive emissions are shown in Table 83. Vapor-
phase emissions of concern include formaldehyde, ethylene oxide and such
solvents as cyclohexane, heptane, hexane and methylene chloride. Particulate emissions from pelletizing operations may result in elevated workplace
levels of nuisance dust.

Employees assigned to formaldehyde purification operations may be
exposed to formic acid, methyl alcohol, and methyl formate, all of which are
impurities removed from the formation feedstock. Workers in acetal copolymer production operations may contact sulfuric or other strong mineral acids
used in the trioxane formation reaction.

TABLE 82. WORKER DISTRIBUTION ESTIMATES FOR POLYACETAL PRODUCTION

Process	Unit	Workers/Unit/8-hour Shift
Homopolymer Production	Batch Reactor	1.0
	Vacuum Filter	0.25
	Vacuum Dryer	0.25
	Pyrolyzer	1.0
	Traps	0.125
	Extruder	1.0
Copolymer Production	Batch Reactor	1.0
	Vacuum Filter	0.25
	Extruder	0.5
	Azeotrope/Water Separation	0.25
	Solvent Recovery	0.25

TABLE 83. SOURCES OF FUGITIVE EMISSIONS FROM POLYACETAL MANUFACTURE

Source	Constituent	Process	
		Homopolymer	Copolymer
Prepolymerization Reactor Vent	Formaldehyde	X	
Polymerization Reactor Vent	Formaldehyde	X	X
	Solvent	X	X
	Trioxane		X
	Comonomer		X
Filters	Formaldehyde	*	*
	Particulates	X	X
	Solvent	*	*
	Trioxane		*
	Comonomer		*
End-Capping Reactor Vent	Formaldehyde	*	
	Solvent	*	
Pelletizing Extruder Vent	Formaldehyde	*	*
	Solvent	*	*
	Particulates	X	X
Trioxane Reactor Vent	Formaldehyde		X
	Trioxane		X
Hydrolysis Reactor Vent	Formaldehyde		*
	Solvent		*
	Ammonia		X

*Trace amounts of these compounds, if they are present in these streams, are expected due to high conversion and efficient removal steps upstream of these sources.

Available toxicity information regarding the above contaminants is given under "Health Effects" below and in Table 10.

Little data are available from which employee exposure potential may be estimated. However, one source estimates hydrocarbon emissions from poly-acetal molding at 1.5 g/kg product.[286]

Health Effects

The list of input materials and specialty chemicals used in the manu-facture of polyacetal include many generic, non-specific compounds which cannot be evaluated for potential health risk. The industry does use at least two highly toxic materials: the monomer formaldehyde and the comonomer ethylene oxide. The reported health effects associated with exposure to these two substances are summarized below.

Ethylene Oxide exposure of 500 ppm has been associated with convul-sions, gastrointestinal tract effects, and pulmonary system effects in humans.[58] Although testing for carcniogenic effects is inconclusive, ethylene oxide causes tumorigenic, metagenic, and teratogenic effects. The current OSHA air standard is 50 ppm (8 hour TWA).[67]

Formaldehyde poses a high toxic hazard upon ingestion, inhalation, or skin contact [242] with human effects reported at doses as low as 8 ppm. [127] This suspected carcinogen is a potent mutagen and teratogen. The OSHA air standard is 3 ppm (8 hour TWA) with a 5 ppm ceiling.[67]

Air Emissions

Sources of fugitive VOC emissions from polyacetal processing are sum-marized in Table 83 by process and constituent type. Most of these sources are vents, and controls which are applicable include:[285]

- Routing vent streams to a flare to incinerate the remaining hydrocarbons; and/or

- Routing vent streams to blowdown.

Other VOC emissions result from open sources such as filters, drains, and valves. Enclosing the atmospheric side of these sources is not feasible in this case due to the explosive limit of ethylene oxide. Leaks from flanges, relief valves, valves, pumps, and compressors also contribute to the VOC emissions for this process.

The one process source of fugitive particulates is also listed in Table 83. Again, due to the explosive limit of ethylene oxide, enclosing the filter is not an available control option. According to EPA estimates, the population exposed to air emissions in a 100 km^2 area surrounding a poly-acetal production facility are: 11 persons exposed to hydrocarbons and 2 persons exposed to particulates.[286]

Wastewater Sources

There are four sources of wastewater for polyacetal processing: formaldehyde purification, trioxane water removal, filtration, and routine cleaning water. If steam stripping is used for solvent recovery, an additional wastewater stream will be generated. These sources are listed in Table 84 by process. No data are available in the literature concerning wastewater parameters.

Solid Wastes

The solid wastes generated by this process are: polyacetal resin lost due to reactor cleaning and spillage; collected particulates; and substandard resin which cannot be blended.

Environmental Regulation

Effluent limitations guidelines have not been set for the polyacetal industry.

New source performance standards (proposed by EPA on January 5, 1981) for volatile organic carbon (VOC) fugitive emissions include:

- Safety/release valves must not release more than 200 ppm above background, except in emergency pressure releases, which should not last more than five days; and

- Leaks (which are defined as VOC emissions greater than 10,000 ppm) must be repaired within 15 days.

The following compounds have been listed as hazardous wastes (46 Federal Register 27476, May 20, 1981):

```
Cyclohexane    - U056
Ethylene oxide - U115
Formaldehyde   - U122
```

All disposal of these compounds or polyacetal resins containing residual amounts of these compounds must comply with the provisions set forth in the Resource Conservation and Recovery Act (RCRA).

TABLE 84. SOURCES OF WASTEWATER FROM POLYACETAL MANUFACTURE

Source	Homopolymer	Copolymer
Formaldehyde Purification (Prepolymerization)	X	
Filtration	X	X
Trioxane Water Removal (Solvent Recovery)		X
Routine Cleaning Water	X	X

11. Polyamide Resins

INTRODUCTION

Polyamide resins are condensation polymers which have a repeating amide group as an integral part of the linear chain structure. Flame resistance, chemical resistance, toughness, low coefficient of friction, stiffness, outstanding wear resistance, good electrical properties, and high temperature resistance are properties of polyamides which allow their use in a variety of applications.[246] The raw materials used in the manufacture of polyamide resins determine the physical and chemical properties of the product. However, additives may be used to provide heat and light stability, lower molecular weight, and increased mold release. Table A-12 in Appendix A lists typical properties of polyamide resins, nylon 6 and nylon 66.

Although polyamides are typically referred to as nylons, two distinct polyamide resins exist: linear nylon polymers and nonlinear non-nylon polymers. Non-nylon resins are classified as either reactive or nonreactive. Reactive resins are combined with epichlorohydrin to produce one type of nonreactive non-nylon resin. Approximately 75 percent of the nylon polyamide resins produced is made up of nylon 6 and nylon 66. Non-nylon polyamides contributed approximately 10 percent to the total polyamide market in 1981.[65]

Polyamide resins are used to produce extruded, injection molded, and cast parts as well as fibers. End products made from these resins include electrical grommets, bearings, gears, fasteners, valve components, rollers, tubing, pipe, film, monofilaments, wire straps, wire covering, electrical connectors, and sports equipment.

Two methods are used to produce polyamides: polyaddition and polycondensation. Nylon 6 is produced by initiating a ring opening reaction using ε-caprolactam and subsequent polyaddition. This reaction proceeds to a polymer/monomer equilibrium; therefore, careful control of the stoichiometry is not needed. Nylon 66 and non-nylon polyamides are manufactured using the polycondensation process. A carefully controlled reactor is necessary in order to provide the stoichiometric amount of the reactants used.

INDUSTRY DESCRIPTION

Thirty-two manufacturers comprise the polyamide industry, supporting 44 plants in 21 states. The largest concentration of plants is in the southeast (EPA Region IV), which contains 36 percent of the production locations.

The mideast (EPA Region III), New England states (EPA Region I), and Great
Lakes states (EPA Region V) contain 16, 14, and 14 percent of the production
sites, respectively. Table 85 lists nylon producers while Tables 86 and 87
list dimer-acid based and epichlorohydrin based non-nylon resins,
respectively.

Nylon resins are used in two forms: fibers and nonfibers. Fibers are
used to make tire cord, carpet filament, and textiles for garments and
furniture. Nonfibers are used in a variety of applications which require
high strength and high performance. Nylon sales increased 7 percent from
1980 to 1981, reaching 132,000 metric tons.[143] From 1976 to 1979, nylon
sales increased from 100,500 metric tons to a peak of 143,200 metric tons.
During the same period, production also increased from 112,300 metric tons
in 1976 to 148,600 metric tons in 1979. A decline in the 1980 production
brought the level back to 124,500 metric tons.[65]

Non-nylon resins are used in adhesives, coatings, and copolymers for
engineering applications. These resins reached a sales peak of 18,200
metric tons in 1977 when production also peaked at 18,600 metric tons. The
sales level increased from 1978 to 1979, however, to result in sales of
15,500 metric tons and production of 15,900 metric tons.[65]

Nylon resins are projected to exhibit the fastest growth, along with
thermoplastic polyesters, of the engineering plastics between now and 1990.
A 16 percent annual increase in nylon output for engineering applications is
forecast for this same period. The largest increase is expected for
reaction-injected-molded (RIM) and extrusion grades.[86]

PRODUCTION AND END USE DATA

In 1981, nylon sales totaled 132,000 metric tons.[143] Nylon has many
uses, which include:[159, 246, 247, 250, 267, 291]

- Production of synthetic fibers used in tires, carpets, stockings,
  and upholstery;

- Molding powders for valve seats, bearings, gears, cams, and machine
  parts;

- Automotive applications, including stone shields, emissions canis-
  ters, windshield wiper and speedometer gears, trim clips, wire
  jacketing, electrical connectors, dome light lenses, engine fan
  blades, radiator headers, brake fluid and power steering fluid
  reservoirs, valve covers, steering column housings, emission control
  valves, mirror housings, fender extensions, and license plate
  pockets;

- Electrical and electronic applications, including plugs, connectors,
  wiring devices, terminals, coil forms, cable ties, wire jacketing,
  antenna mounting devices, and mechanical components for use in com-
  puters, television sets, and other consumer-oriented products;

TABLE 85.   U.S. NYLON RESIN PRODUCERS

Producer	Location	Product Types	1982 Capacity Thousand Metric Tons
Adell Plastics, Inc.	Baltimore, MD	Nylon 6 and 66	9.1[a]
Allied Corp.			
Allied Fibers and Plastics Co.	Chesterfield, VA	Nylon 6	25.0
Nypel, Inc., subsidiary	Chesterfield, VA	Nylon 6 and 66	6.3[a]
American Grilon Inc.	Sumter, SC	Nylon 6 and 12	5.0
American Hoechst Corp.			
Plastics Division	Manchester, NH	Nylon 6	8.2
Badische Corp.	Freeport, TX	Nylon 6	6.8
Belding Heminway Co., Inc.			
Belding Chemical Industries, subsidiary	Grosvenordale, CT	Nylon 69	1.8
Bemis Co., Inc.			
Custom Resins Division	Henderson, KY	Nylon 6	9.1
Celanese Corp.			
Celanese Plastics & Specialties Co. Division			
Celanese Engineering Resins Division	Bishop, TX	Nylon 66	18.2
E. I. du Pont de Nemours & Co., Inc.			
Polymer Products Department	Parkersburg, WV	Nylon 66 and 612	75.0
The Firestone Tire and Rubber Co.			
Firestone Synthetic Fibers Co. Division	Hopewell, VA	Nylon 6	5.5
Monsanto Co.			
Monsanto Plastics & Resins Co.	Pensacola, FL	Nylon 66 and 69	31.8
Rilsan Corp.			
Rilsan Industries Inc., subsidiary	Birdsboro, PA	Nylon 11 and 12	5.0
Shakespeare Co.			
Monofilament Division	Columbia, SC	Nylon 6	1.8
Texapol Corp.	Bethlehem, PA	Nylon 6 and 66	—[b]
Wellman, Inc.			
Wellman Industries, Inc., subsidiary			
Man-Made Fiber Division	Johnsonville, SC	Nylon 6 and 66	3.6[a]
TOTAL			212.7

[a]These producers manufacture regenerated nylon resins from fiber waste.

[b]Company produces less than 2,275 kg/year of nylon resins.

Source: Directory of Chemical Producers, 1982.

TABLE 86. U.S. PRODUCERS OF DIMER ACID BASED NON-NYLON POLYAMIDE RESINS

Producer	Location	1982 Capacity Thousand Metric Tons
AZS Corp. AZ Products, Inc. Div.	Eaton Park, FL	5.9
Celanese Corp. Celanese Plastics & Specialties Co. Div. Celanese Specialty Resins Div.	Louisville, KY	0.9
Cooper Polymers, Inc.	Wilmington, MA	2.3
Crosby Chemicals, Inc.	Picayune, MS	1.8
Emhart Corp. USM Corp., sub. Bostik Div. Eastern Region	Middleton, MA	0.9
Henkel of America, Inc. Henkel Corp., sub.	Kankakee, IL	11.4
Lawter Int'l., Inc.	Moundsville, AL	1.4
Mobil Corp. Mobil Oil Corp. Mobil Chemical Co. Div. Chemical Coatings Div.	Edison, NJ	1.4
National Distillers and Chemical Corp. Emery Industries, Inc., sub.	Cincinnati, OH	4.5
The O'Brien Corp. The O'Brien Corp.- Southwestern Region	Houston, TX	0.5
Pacific Anchor Chem. Corp.	Los Angeles, CA	0.9

(continued)

TABLE 86 (continued)

Producer	Location	1982 Capacity Thousand Metric Tons
Reichhold Chemicals, Inc.	Andover, MA	2.3

Sun Chemical Corp.
  Chemicals Group
    Chemicals Div.
	Chester, SC	1.8

Union Camp Corp. Chemicals Products Div.	Dover, OH	4.5
TOTAL		40.5

Source: Directory of Chemical Producers, 1982.

TABLE 87. U.S. PRODUCERS OF EPICHLOROHYDRIN BASED POLYAMIDE RESINS

Producer	Location

Borden, Inc.
  Borden Chemical Div.
    Adhesives and Chemicals Div.
	Demopolis, AL
	Diboll, TX
	Fayetteville, NC
	Sheboygan, WI

Diamond Shamrock Corp.
  Int'l. Technology Unit
    Process Chemicals Div.
	Charlotte, NC

Georgia-Pacific Corp. Chemical Div.	Eugene, OR Newark, OH Peachtree City, GA Richmond, CA
Hercules, Inc.	Chicopee, MA Hattiesburg, MS Milwaukee, WI Portland, OR Savannah, GA

Source: Directory of Chemical Producers, 1982.

- Consumer goods, including roller skate sole plates, ski boots, and ice skate supports as well as bicycle wheels and racquetball racquets;

- Films for oven cooking bags and packaging for surgical instruments; and

- Other uses, including gun stocks, lighter fluid cases, printing plates, air conditioner hose, power tool housings and handles, brush bristles, fishing line, mallet heads, combs, self-sealing fuel tanks for military aircraft, medical devices, storm window and furniture parts, and beverage dispensing valves.

Table 88 presents major nylon markets for 1981.

Non-nylon polyamide resins are used in heat-seal adhesives, overprint varnishes, can side-seam cements, inks, thixotropic paints, thermosetting adhesives, potting compounds, casting compounds, solvent-based thermosetting coatings, solventless coatings, and plastic terrazos.[171]

PROCESS DESCRIPTIONS

Polyamide resins are produced by two basic processes: polyaddition and polycondensation. Monomeric ring structures, such as caprolactam, are polymerized by first opening the ring and then propagating the resulting polymer chain using the polyaddition process. The polycondensation process uses an acid and an amine to form the polyamide and by-product water.

Nylon 6 is the major product manufactured by the polyaddition process. Nylon 66 and non-nylon dimer acid based and epichlorohydrin based resins are produced via the polycondensation process.

Polyaddition Reaction Chemistry

Nylon 6, so named for the six carbon atoms in the caprolactam molecule, is produced by hydrolysis and polymerization of caprolactam. Water and 66 salts are used to open the ring structure. Once the ring is opened, the resulting chain may react with other opened ring chains or caprolactam rings to produce the polymer. This sequence of reactions is depicted below:[247]

<u>Hydrolysis of the Caprolactam Molecule (Ring Opening)</u>

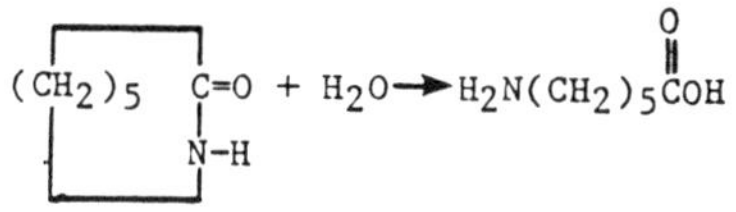

$$(CH_2)_5 \quad \overset{\displaystyle |}{\underset{\displaystyle N-H}{C=O}} + H_2O \longrightarrow H_2N(CH_2)_5\overset{\displaystyle O}{\overset{\|}{C}}OH$$

TABLE 88. MAJOR U.S. NYLON MARKETS

Market	1981 Thousand Metric Tons
Extrusion	
Filaments	11
Film	11
Sheet, Rod, Tube	5
Wire and Cable	4
Injection Molding	
Appliances	7
Consumer Products	20
Electrical/Electronic	14
Industrial	14
Transportation	28
Export	9
Other	9
TOTAL	132

Source: _Modern Plastics_, January 1982.

<u>Propagation of the Polymer Chain</u>

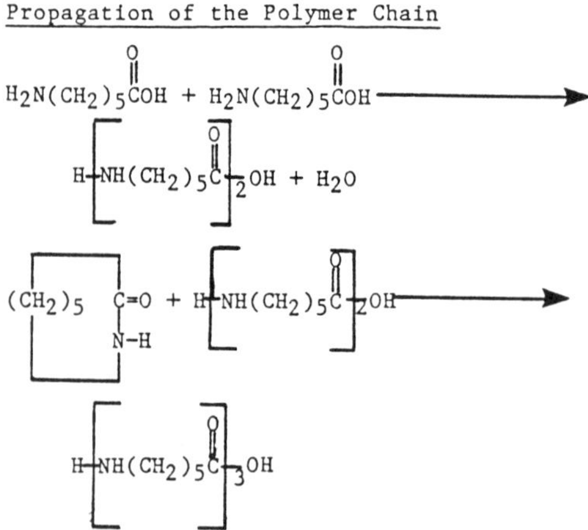

<u>Polycondensation Reaction Chemistry</u>

Nylon 66

Nylon 66 is produced by the reaction of adipic acid and hexamethylenediamine. Each of these reactants contains six carbon atoms, hence the 66
nomenclature. The dibasic acid groups on the adipic acid molecule combine
with the amine groups of the hexamethylenediamine molecule, resulting in the
elimination of a molecule of water after the formation of an intermediate
salt. An exact 1:1 molar ratio of reactants is necessary in order to form
the repeating group which characterizes nylon 66. The condensation reaction
is shown below.

$$HOC(CH_2)_4COH \ + \ H_2N(CH_2)_6NH_2 \longrightarrow$$

$$[^+H_3N(CH_2)_6NH_3{}^+]\,[^-OC(CH_2)_4CO^-] \longrightarrow$$

$$HOC(CH_2)_4CN(CH_2)_6NH_2 \ + \ H_2O$$
$$\overset{|}{H}$$

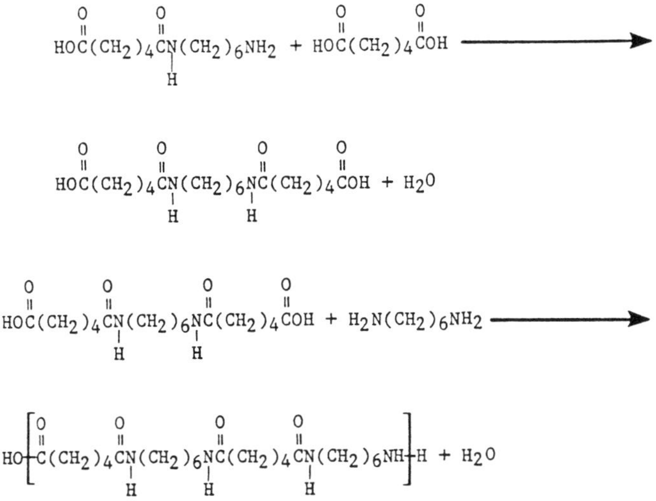

$$\text{HO}\overset{\text{O}}{\overset{\|}{\text{C}}}(\text{CH}_2)_4\overset{\text{O}}{\overset{\|}{\text{C}}}\underset{\text{H}}{\text{N}}(\text{CH}_2)_6\text{NH}_2 + \text{HO}\overset{\text{O}}{\overset{\|}{\text{C}}}(\text{CH}_2)_4\overset{\text{O}}{\overset{\|}{\text{C}}}\text{OH} \longrightarrow$$

$$\text{HO}\overset{\text{O}}{\overset{\|}{\text{C}}}(\text{CH}_2)_4\overset{\text{O}}{\overset{\|}{\text{C}}}\underset{\text{H}}{\text{N}}(\text{CH}_2)_6\underset{\text{H}}{\text{N}}\overset{\text{O}}{\overset{\|}{\text{C}}}(\text{CH}_2)_4\overset{\text{O}}{\overset{\|}{\text{C}}}\text{OH} + \text{H}_2\text{O}$$

$$\text{HO}\overset{\text{O}}{\overset{\|}{\text{C}}}(\text{CH}_2)_4\overset{\text{O}}{\overset{\|}{\text{C}}}\underset{\text{H}}{\text{N}}(\text{CH}_2)_6\underset{\text{H}}{\text{N}}\overset{\text{O}}{\overset{\|}{\text{C}}}(\text{CH}_2)_4\overset{\text{O}}{\overset{\|}{\text{C}}}\text{OH} + \text{H}_2\text{N}(\text{CH}_2)_6\text{NH}_2 \longrightarrow$$

$$\text{HO}\left[\overset{\text{O}}{\overset{\|}{\text{C}}}(\text{CH}_2)_4\overset{\text{O}}{\overset{\|}{\text{C}}}\underset{\text{H}}{\text{N}}(\text{CH}_2)_6\underset{\text{H}}{\text{N}}\overset{\text{O}}{\overset{\|}{\text{C}}}(\text{CH}_2)_4\overset{\text{O}}{\overset{\|}{\text{C}}}\underset{\text{H}}{\text{N}}(\text{CH}_2)_6\text{NH}\right]\text{H} + \text{H}_2\text{O}$$

Non-Nylon Dimer Based

Dimer acid based non-nylon polyamide resins use high molecular weight acids, such as the polymerization product of linoleic acid (9,11-octadecadienoic acid), and an amine, such as ethylenediamine, to produce the polymer chain. The condensation reaction is similar to the one for nylon 66 and is shown below.

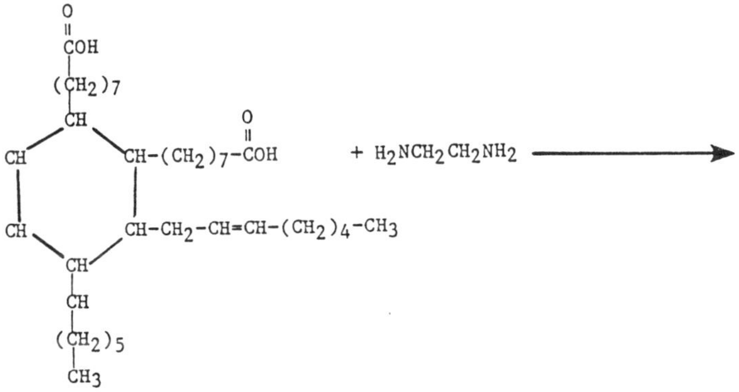

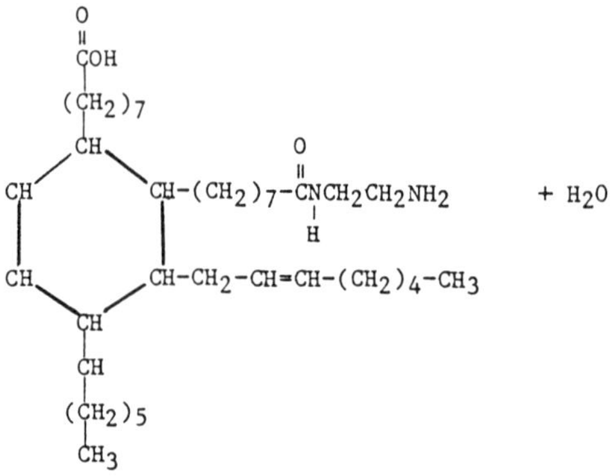

$$
\begin{array}{c}
O \\
\parallel \\
COH \\
| \\
(CH_2)_7 \\
| \\
CH \\
\end{array}
$$

(cyclohexane ring bearing the following substituents:)

- CH–(CH$_2$)$_7$–$\overset{\displaystyle O}{\overset{\displaystyle \parallel}{C}}$N(H)CH$_2CH_2NH_2$

- CH–CH$_2$–CH=CH–(CH$_2$)$_4$–CH$_3$

- CH
  | (CH$_2$)$_5$
  | CH$_3$

$+ \; H_2O$

Epichlorohydrin Based

Epichlorohydrin based polyamide resins are produced by the reaction of epichlorohydrin and a reactive polyamide resin. Adipic acid and diethylene-amine are commonly used as the raw materials for the reactive polyamide. This reaction is shown below.

<u>Reactive Polyamide Production</u>

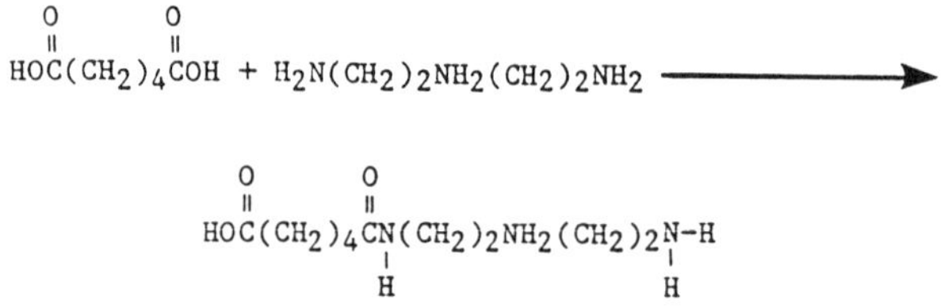

$$
\underset{}{HOC}(CH_2)_4\underset{}{COH} + H_2N(CH_2)_2NH_2(CH_2)_2NH_2 \longrightarrow
$$

(with carbonyl $\overset{O}{\overset{\parallel}{C}}$ groups)

$$
HOC(CH_2)_4\underset{H}{\underset{|}{C}}N(CH_2)_2NH_2(CH_2)_2\underset{H}{\underset{|}{N}}-H
$$

<u>Reaction with Epichlorohydrin</u>

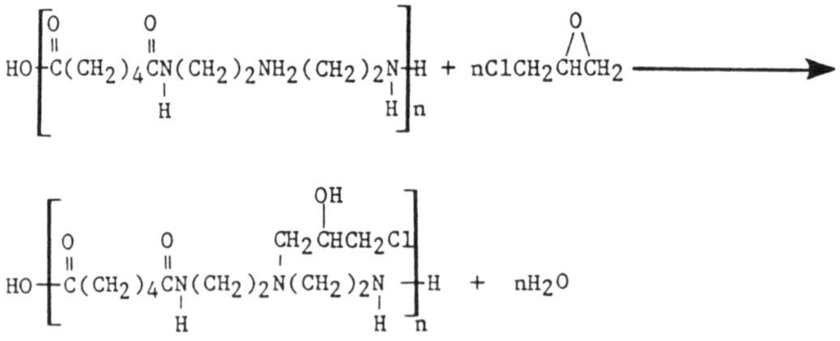

$$
HO\!\!\left[\,C(CH_2)_4\underset{H}{\underset{|}{C}}N(CH_2)_2NH_2(CH_2)_2\underset{H}{\underset{|}{N}}\right]_n\!\!H \; + \; nClCH_2\underset{}{CHCH_2} \longrightarrow
$$

(carbonyl $\overset{O}{\overset{\parallel}{C}}$ groups; epichlorohydrin epoxide $\overset{O}{\overset{/\backslash}{CHCH_2}}$)

$$
HO\!\!\left[\,C(CH_2)_4\underset{H}{\underset{|}{C}}N(CH_2)_2N(CH_2)_2\underset{H}{\underset{|}{N}}\right]_n\!\!H \; + \; nH_2O
$$

(with side chain $\overset{OH}{\overset{|}{CH_2}}CHCH_2Cl$)

Typical input materials and operating parameters for the polyaddition and polycondensation processes are presented in Tables 89 and 90, respectively. Table 91 lists input materials, including lubricants, fillers, UV light stabilizers, and colorants.

Polyaddition Process

The polyaddition process is based on the use of a compound which contains both an acid and an amine functional group. Nylon 6 is the major polymer produced using this method, depicted in Figure 29. Both batch and continuous processes are used for the polyaddition process. A typical production recipe for polyamides via the polyaddition process is given below:[125]

Material	Parts by Weight
Caprolactam (monomer)	90
Water (catalyst)	10
Acetic Acid (chain terminating agent)	

Caprolactam is fed to a polymerization reactor along with water, or another suitable catalyst, and a chain terminating agent (acetic acid). The temperature is raised to 250°C and the reaction starts by opening the caprolactam ring. The polymerization does not go to completion; instead, a monomer/polymer equilibrium is reached at around 85 to 90 percent conversion.[246]

When the reaction reaches the desired conversion, the reaction mixture is discharged from the reactor through a die to form a polymer rod. These rods are subsequently cut into pellets and dried. If the unreacted caprolactam monomer is not removed, it acts as a plasticizer. Caprolactam removal may be accomplished by feeding the polymer mixture from the reactor into a vacuum chamber where the monomer is flashed and recycled. Pelletization is then performed and the pellets dried to the desired moisture content.

Polycondensation Reaction

Polyamides made by the polycondensation reaction utilize a stoichiometric ratio of amine to acid of 1:1.[125] Therefore, careful control of losses from the reactor is necessary to maintain the correct reaction composition. Due to this constraint, batch processing is used for these reactions.

The polycondensation process used to manufacture nylon 66, shown in Figure 30, is actually a two-stage reaction. In the first stage, stoichiometric portions of hexamethylenediamine and adipic acid are mixed in water. The resulting intermediate salt, soluble in water, is decolored with activated carbon and concentrated to a 60 percent solution by evaporation.[246] If a chain terminating agent (acetic acid) is desired, it is added to the concentrated solution. A typical production recipe for this process was not found in the literature.

TABLE 89.   TYPICAL INPUT MATERIALS TO POLYAMIDE PRODUCTION PROCESSES

Process	Cyclic Monomer	Catalyst	Dibasic Acid	Amine	Reactive Polyamide Resin	Epichlorohydrin
Polyaddition	X	X				
Polycondensation						
Nylon 66			X	X		
Non-Nylon Dimer Based			X[a]	X		
Epichlorohydrin Based					X	X

---

[a]Higher molecular weight than those used for nylon 66 production.

TABLE 90.   TYPICAL OPERATING PARAMETERS FOR POLYAMIDE PRODUCTION PROCESSES

Process	Temperature	Pressure	Conversion
Polyaddition	240 – 280°C	1.8 MPa[a]	85 – 90%
Polycondensation	250 – 300°C	1.8 MPa	>99%

[a]Approximate.

Sources: Encyclopedia of Chemical Technology, 3rd Edition.
Encyclopedia of Polymer Science and Technology.

TABLE 91.   INPUT MATERIALS AND SPECIALTY CHEMICALS USED IN
POLYAMIDE RESIN MANUFACTURE

Function	Compound
Monomers	adipic acid diethylenetriamine ε-caprolactam epichlorohydrin ethylenediamine hexamethylenediamine high molecular weight fatty acid triethylenetetramine
Catalysts	phosphoric acid water
Molecular Weight Regulator	acetic acid
Fillers	asbestos glass fiber mica talc
Plasticizers	amides coumarone-indene resins resorcinol esters toluene sulfonamides
Comonomers	amino resins phenolic resins polyurethane polyvinyl acetate
Heat Stabilizers	combinations of cupric salts and halides of alkali metals
UV Light Stabilizer	carbon black copper compounds manganese compounds
Slip Enhancers	graphite metal stearates zinc stearate molybdenum disulfide polytetrafluoroethylene silicones
Delusterant	titanium dioxide

(continued)

TABLE 91 (continued)

Function	Compound
Antioxidants	aromatic amines
	organic phosphites
	phenolics
	phosphinates
Colorants	aluminum
	anthraquinone
	bronze
	cadmium mercury
	cadmium sulfide
	cadmium sulfoselenide
	carbon black
	ceramic black
	chrome cobalt aluminate
	chrome tin
	cobalt aluminate
	lithopone
	metallic oxides
	phthalocyanine blue
	phthalocyanine green
	titanium dioxide
	ultramarine blue
	ultramarine green
	ultramarine pink
	ultramarine red
	ultramarine violet
	zinc sulfide
Mold Release Agents	aliphatic monoalcohols
	aliphatic monoamines
	belenic acid
	esters of aliphatic monoalcohols
Nucleating Agents	aluminas
	cadmium acetate
	lead acetate
	magnesium oxide
	mercuric bromide
	mercuric chloride
	phenol phthalein
	silicas
	silver halides
	sodium isobutylphosphinate
	sodium phenyphosphinate

(continued)

Table 91 (continued)

Function	Compound
Fire Retardant	organic halide and a metallic oxide or halide

Sources: *Encyclopedia of Chemical Technology*, 3rd Edition.
Encyclopedia of Polymer Science and Technology.
Melvin I. Kohan, *Nylon Plastics*, 1973.
Modern Plastics Encyclopedia, 1981–1982.

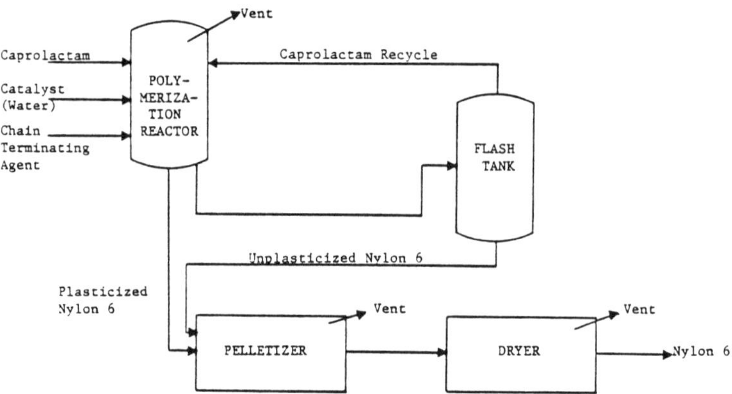

Figure 29. Polyamide production using the polyaddition process.

Source: <u>Encyclopedia of Polymer Science and Technology</u>.

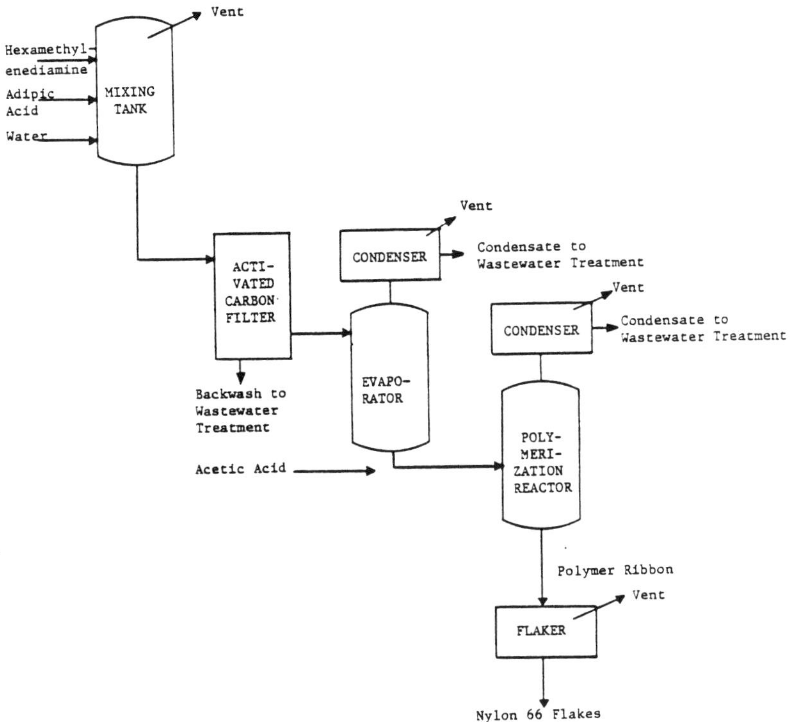

Figure 30. Polyamide resin production using the polycondensation process.

Source: Encyclopedia of Polymer Science and Technology.

The concentrated salt solution is then fed to the polymerization reactor where the second-stage of the reaction begins. Increased temperature and pressure are used to initiate the polymerization reaction. Water vapor is bled from the reactor to keep it at constant pressure. Diamine losses, which may be substantial due to low vapor pressure, are prevented by partially polymerizing the reactants to nylon 66 prepolymer. When the temperature of the batch reaches 275°C, the pressure is allowed to fall to atmospheric. The batch is held at 270°C and atmospheric pressure for half an hour to allow removal of the water vapor. If an extremely high conversion is desired, the batch is held under partial vacuum at the end of the polymerization.[246]

The polymer is extruded through the bottom of the reactor to form a ribbon which is flaked. Batches of flaked polymer are blended to minimize any variations from batch to batch.

Polycondensation reactions for non-nylon polyamides are much simpler than the nylon 66 process. The input materials are added to the polymerization reactor where the temperature and pressure is increased to the reaction conditions. Water evolved during the condensation reaction is vented as water vapor. Dimer acid based non-nylon resins are occasionally used in solution; the solvent is added to the reactor after the polymerization is completed. Epichlorohydrin resins are diluted to a 12.5 percent solids solution for use in the paper industry as coatings.

Energy Requirements
─────────────────

Data concerning the energy requirements for this process were not found in the literature consulted.

ENVIRONMENTAL AND INDUSTRIAL HEALTH CONSIDERATIONS

Polyamides are biologically inert compounds which have been approved by the FDA for use in food wrap. However, hexamethylenediamine and caprolactam are skin and tissue irritants. Care must be used in handling not only these materials but also nylon 6 plasticized with caprolactam. Two of the input materials to polyamide production, asbestos and lead acetate, are listed as hazardous wastes under RCRA.

Conversions of 90 percent or greater indicate that most of the emissions for these processes occur in or upstream of the reactor. Care is taken in the polycondensation process to reduce monomer loss since a stoichiometric amount of reactant is required to ensure the desired product. The polyaddition process removes and recycles caprolactam if an uplasticized product is desired. Therefore, although sources upstream of the reactor have relatively more emissions, the magnitude of these emissions is still small.

Worker Distribution and Emissions Release Points

Estimates of worker distribution have been made by correlating major equipment manhour requirements with the process flow diagrams in Figures 29 and 30. Table 92 shows estimates for both the polyaddition and polycondensation processes.

Only fugitive emissions from process sources and leaks are associated with polyamide production. Although there are no major point source air emissions, fugitive emissions may be a significant environmental and/or worker health problem. The magnitude of these emissions' impact is a function of stream constituents, process operating parameters, engineering and administrative controls, and maintenance programs.

Fugitive emissions and their principal sources are shown in Table 93. Additionally, employees may be exposed to monomers, comonomers, catalysts and other additives (see Table 91) during raw materials and final product handling and packaging.

Available toxicity information for contaminants of concern is discussed under "Health Effects" below and summarized in Table 10.

Little data are available from which employee exposure potential may be estimated. However, one source gives a hydrocarbon release rate of 15 g/kg product from nylon molding operations [286].

Health Effects

The manufacture of polyamide resin involves the use of one known human carcinogen and five suspected carcinogens. They are: the filler asbestos; lead acetate, a nucleating agent; polytetrafluoroethylene, which is used as a slip enhancer; polyurethane, a comonomer in the process; the monomer epichlorohydrin; and the colorant cadmium sulfide. Two nucleating agents, mercuric chloride and mercuric bromide, are considered extremely toxic and may also adversely affect the health of plant employees if exposure occurs. The following paragraphs summarize the reported health effects associated with exposure to these substances.

Asbestos is classified as a human carcinogen by the International Agency for Research on Cancer [104] and by the American Conference of Governmental Industrial Hygienists.[5] It is currently undergoing additional tests by the National Toxicology Program.[233] Most deaths from exposure to asbestos are due to lung cancer.[104] Not only is the lung cancer risk dose-related, but there is also an important enhancement of the risk in those exposed to asbestos who also smoke cigarettes.[104] The OSHA emergency air standard is 0.5 fiber >5 um/cc (8 hour TWA).

Cadmium Sulfide appears to be a mutagenic agent and has produced positive results in animal tests for carcinogenicity.[98] It is also highly

TABLE 92. WORKER DISTRIBUTION ESTIMATES FOR POLYAMIDE PRODUCTION

Process	Unit	Workers/Unit/8-hour Shift
Polyaddition	Batch or Continuous Reactor	1.0 or 0.5
	Flash Tank	0.125
	Pelletizer	0.5
	Dryer	0.5
Polycondensation	Batch Reactor	1.0
	Filter	0.25
	Evaporator	0.25
	Condenser	0.125
	Flaker	0.25

TABLE 93. SOURCES OF FUGITIVE EMISSIONS FROM POLYAMIDE MANUFACTURE

Source	Constituent	Process Polyaddition	Polycondensation
Polymerization Reactor Vent	Caprolactam	*	
	Amine Compound		*
	Acid Compound		*
	Epichlorohydrin		*
	Water	X	X
Mixing Tank Vent	Hexamethylenediamine		*
	Adipic Acid		*
	Water		X
Evaporator Vent	Hexamethylenediamine		*
	Adipic Acid		*
	Water		X
Pelletizer Vent	Caprolactam	*	
	Particulates	X	
Flaker Vent	Particulates		X
Dryer Vent	Particulates	X	

*Trace amounts of these compounds are expected, if they are present, due to the high conversion of the polyamide polymerization processes.

toxic by ingestion or inhalation following acute or chronic exposures.[242]
Sulfides of heavy metals are generally insoluble and usually have little
toxic action except through liberation of hydrogen sulfide.[242]   However,
inhalation of dust or fumes of cadmium compounds affects the respiratory
tract and may also involve kidney damage.[148]   Even brief exposures to high
concentrations may result in pulmonary edema and death.[148]   The OSHA air
standard for cadmium dust is 0.2 mg/m^3 (8 hour TWA) with a 0.6 mg/m^3
ceiling.

Epichlorohydrin is highly toxic by ingestion, inhalation, and skin
absorption.[242]   In acute poisoning, death may be caused by respiratory
paralysis [242]; chronic exposure can cause kidney injury.[149]   Symptoms of
chronic poisoning at concentrations lower than 20 ppm include fatigue,
gastrointestinal pains, and chronic conjunctivitis.[134]   Epichlorohydrin
has also produced positive results in animal tests for carcinogenicity [101]
and for mutagenicity and teratogenicity.[233]   The OSHA air standard is 5
ppm (8 hour TWA).[67]

Lead Acetate, a suspected human carcinogen [109], is also associated
with mutagenic and teratogenic effects in laboratory animals.[233]   The OSHA
air standard is 50 ug (Pb)/m^3 (8 hour TWA).[67]

Mercuric Bromide, as is the case for all forms of mercury, is highly
toxic if absorbed.[85]   It is capable of causing death or permanent injury
due to exposures of normal use [243] by inhalation, ingestion, and skin
absorption.[42]   The bromide ion is also toxic; the systematic effects of
the bromide ion are chiefly mental: drowsiness, irritability, ataxia,
vertigo, confusion, mania, hallucination, and coma.[85]   NIOSH has recommended an air standard in the work place for inorganic mercury of 0.05
mg(Hg)/m^3 (8 hour TWA).[158]   OSHA has set a ceiling air standard of 1
mg/10 m^3.

Mercuric Chloride is one of the most toxic forms of mercury.[85]
Fatalities have resulted from as little as 0.5 grams by mouth, although the
mean lethal dose in adults probably lies between 1 and 4 grams.[85]   Occupational poisoning due to mercury or its inorganic compounds is usually
associated with chronic exposure.[278]   Mercuric chloride is currently being
tested for carcinogenesis by the National Toxicology Program.[233]   NIOSH
has recommended an air standard in the work place for inorganic mercury of
0.05 mg(Hg)/m^3 (8 hour TWA).[158]   OSHA has set a ceiling air standard of
1 mg/10 m^3.

Polytetrafluoroethylene has produced positive results in animal tests
for carcinogenicity, although the data provide insufficient evidence to
assess the carcinogenic risk in humans.[107]   The finished compound is inert
under ordinary conditions, but there have been reports of polymer fume fever
in humans exposed to the unfinished product and thermodegration products.
[107, 149]   Exposure produces influenza-like symptoms, including chills,
headaches, rigor-like shaking of limbs, mild respiratory discomfort and high
fever.[107]

<u>Polyurethane</u>, in the form of foam, has produced positive results in animal tests for carcinogenicity.[107] However, very little animal data and no human data exist in the available literature with which to evaluate the potential for other adverse effects in humans following exposure to this substance.

<u>Air Emissions</u>

Sources of fugitive VOC emissions are listed in Table 93. All the process sources are vents which may be controlled by:[285]

- Routing the stream to a flare to incinerate the hydrocarbons present; and/or

- Routing the stream to blowdown.

One estimate of emissions from the mixing tank in nylon 66 production is 1.3 kg hexamethylenediamine per metric ton of product.[93] Other sources of fugitive emissions include valves, relief valves, flanges, pumps, compressors, and drains. Equipment modification (covering open drains or closing the atmospheric side of open-ended valves) may reduce these emissions; however, a regular inspection and maintenance program may be the most effective control.

Sources of fugitive particulates are also listed in Table 93. These emissions may be reduced by venting the evaporator, pelletizer, flaker, and dryer vents to either a baghouse or electrostatic precipitator to collect the particulates. According to EPA estimates, the population exposed to nylon resin emissions in a 100 km^2 area surrounding a nylon resin plant is: 27 persons exposed to hydrocarbons and two persons exposed to particulates.[286]

<u>Wastewater Sources</u>

The wastewater sources associated with polyamide production are listed in Table 94. Nylon 6 has only one source: routine cleaning water. Water vapor from the evaporator and reactor in nylon 66 production may be condensed although these streams contain very little amine or acid components.

Ranges of several wastewater parameters for wastewaters from nylon production are shown below. Values for the wastewater from the processes presented were not distinguished by process type by EPA for the purpose of establishing effluent limitations for the nylon segment of the polyamide industry.[284]

Nylon Wastewater Characteristic	Unit/Metric Ton of Nylon
Production	0 - 152.3
BOD$_5$	0.1 - 70
COD	0.2 - 300
TSS	0 - 8

TABLE 94. SOURCES OF WASTEWATER FROM POLYAMIDE PRODUCTION

	Process	
Source	Polyaddition	Polycondensation
Activated Carbon Filter Backwash		X
Evaporator Condensate		X
Reactor Condensate		X
Routine Cleaning Water	X	X

Solid Wastes

One solid waste may result from the polyaddition process: low molecular weight polymer from the flash tank. The magnitude of this waste is estimated to be 19 kg per metric ton of product.[93]

One solid waste also results from the polycondensation of nylon 66. The activated carbon filter used to purify the intermediate salt results in a solid waste containing 15 kg activated carbon and 3.7 kg trapped intermediate salt per metric ton of product.[93]

Reactor cleaning, product blending, particulate removal, and spillage contribute to the solid waste load. Substandard product which cannot be blended is also a solid waste for these processes.

Environmental Regulation

Effluent limitations guidelines have been set for the nylon 6, nylon 66, and nylon 612 segments of the polyamide industry. BPT, BAT, and NSPS call for the pH of the effluent to fall between 6.0 and 9.0 (41 Federal Register 32587, August, 4, 1976).

New source performance standards (proposed by EPA on January 5, 1981) for volatile organic carbon (VOC) fugitive emissions include:

- Safety/release valves must not release more than 200 ppm above background, except in emergency pressure releases, which should not last more than five days; and

- Leaks (which are defined as VOC emissions greater than 10,000 ppm) must be repaired within 15 days.

The following compounds have been listed as hazardous:

Asbestos   - U013
Lead acetate - U144

All disposal of these compounds and polyamide resins which contain residual amounts of these compounds must comply with the provisions set forth in the Resource Conservation and Recovery Act (RCRA).

12. Polybutylene

INTRODUCTION

Polybutylene refers to a group of polymers comprised primarily of
poly-1-butene. Commercial polybutylene resins are used for pipe and film.
These applications utilize the high temperature stability, chemical resistance, tear strength, puncture resistance, high impact strength, and moisture impermeability of the polymer. Polybutylene pipe is used for cold
water distribution and service; well piping; hot water service, which
includes residential and solar plumbing as well as underfloor and hot water
heating; industrial high-temperature and abrasive slurry lines; high-
temperature acid effluent lines; and chemical process lines. Polybutylene
film finds uses in heavy-duty shipping containers, food and meat packaging,
compression wraps, hot-fill containers, agricultural packaging, and industrial sheeting.[32]

Polybutylene is produced by the polymerization of 1-butene using a
Ziegler-Natta catalyst. Polybutylene has been identified as possessing four
separate solid forms: forms I, II, III, and I'. The structure for polybutylene is shown below.

$$\left[\begin{array}{c} CH_2-CH \\ | \\ CH_2-CH_3 \end{array}\right]_n$$

The commercial resin (form I) is one of four identified crystalline forms of
poly-1-butene. Poly-1-butene crystallizes into form II when the polymer is
heated above the melting point and resolidified. Form II transforms to form
I (via a solid-solid transformation) after about one week at ambient conditions. Poly-1-butene manufactured by solution polymerization of the monomer
and subsequent precipitation crystallizes into form III; gas-phase polymerization or low temperature solution polymerization of 1-butene results in
form I'. Both forms III and I' transform to form I upon heating.[121]

These solid-solid transformations cause shrinkage and warping in molded
parts since the elevated molding temperatures cause the polybutylene to
crystallize as form II, then transform to form I. The addition of propylene
as a comonomer prevents the formation of form II during molding, thus
eliminating the problem of shrinkage and warping.[124]

Polybutylene has a crystalline structure which is also spherical, giving the polymer excellent environmental stress crack resistance and recovery from elongation. Polybutylene samples demonstrate fast and complete recovery after 50 percent elongation, and 80 to 85 percent recovery after 100 percent elongation.[124] Table A-13 in Appendix A lists some typical properties of polybutylene.

INDUSTRY DESCRIPTION

At the present time, Shell Chemical Company is the only U.S. producer of polybutylene.[56, 124] The production facility, located in Taft, Louisiana, has a capacity of 22,300 metric tons. The total demand for polybutylene in 1980 was 7,300 metric tons. Therefore, sufficient capacity is available for growth in the polybutylene market in the near future.[53]

Polybutylene is a nontoxic compound which is used in packaging meat and other food items, transporting water, including potable water, and other piping and packaging applications. None of the chemicals which have been identified as inputs to polybutylene processing are known carcinogens. However, two substances used in polybutylene manufacture (1,1-dimethylhydrazine, a catalyst, and isotactic polypropylene, a transformation aid) have been identified as animal carcinogens and are therefore potential human carcinogens. One compound, 1,1-dimethylhydrazine, has been listed as hazardous under RCRA.

PRODUCTION AND END USE DATA

Polybutylene is still in the developmental stage of commercialization, with 1980 production slightly under 7,300 metric tons.[53] Polybutylene markets include:[32]

- Pipe for cold water service; wells; hot water service, including residential, solar, underfloor and hydronic heating plumbing; high-temperature and abrasive slurry lines; high-temperature acid effluent lines; and chemical process lines; and

- Film for heavy-duty shipping containers, food and meat packaging, compression wraps, hot-fill containers, agricultural packaging, and industrial sheeting.

PROCESS DESCRIPTION

Polybutylene is presently manufactured via mass polymerization using a Ziegler-Natta catalyst. The polymerization is initiated when 1-butene reacts with an active site on the catalyst to form a radical group. As other 1-butene molecules react with the 1-butene/catalyst radical, the polybutylene chain is propagated. The polymerization terminates when either the carbon-catalyst bond becomes weak in relation to the carbon-carbon bonds in the chain or a molecular weight regulating agent, such as hydrogen, reacts with the polymer chain. The polymerization reaction is depicted below.

Initiation

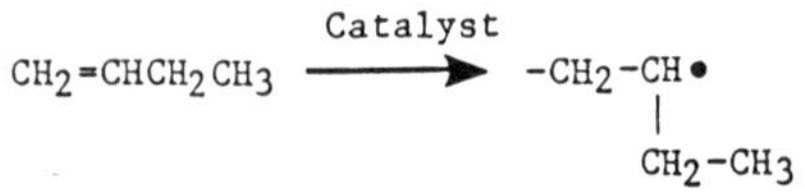

$$CH_2=CHCH_2CH_3 \xrightarrow{\text{Catalyst}} -CH_2-CH\bullet$$
$$|$$
$$CH_2-CH_3$$

Propagation

$$-CH_2-CH\bullet \ + \ nCH_2=CHCH_2CH_3 \longrightarrow$$
$$|$$
$$CH_2-CH_3$$

$$-\left[CH_2-CH\underset{\underset{CH_2-CH_3}{|}}{}\right]_n CH_2-CH\bullet$$
$$CH_2-CH_3$$

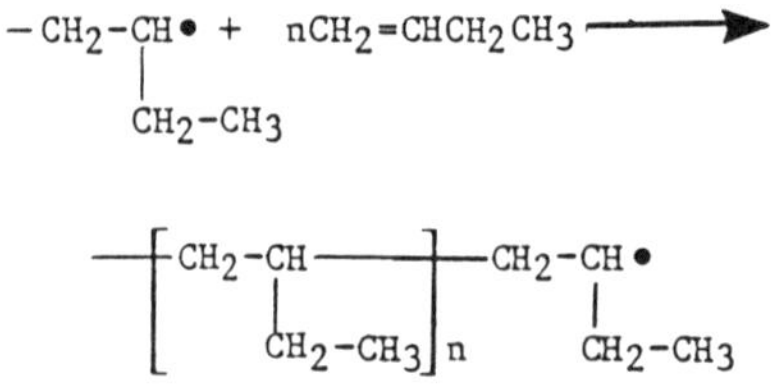

Termination by Elimination

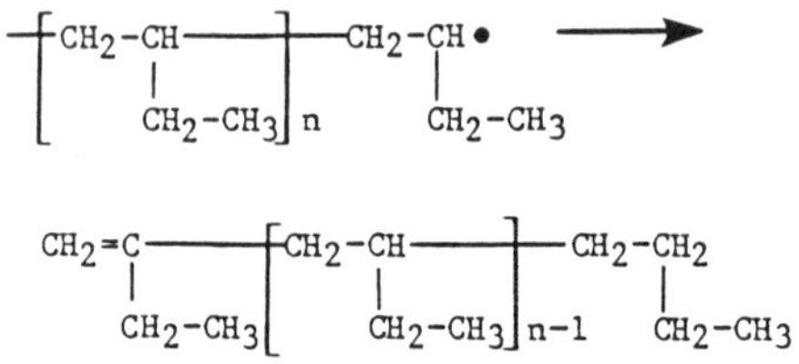

$$\left[CH_2-CH\underset{\underset{CH_2-CH_3}{|}}{}\right]_n CH_2-CH\bullet \longrightarrow$$
$$CH_2-CH_3$$

$$CH_2=C\underset{\underset{CH_2-CH_3}{|}}{}\left[CH_2-CH\underset{\underset{CH_2-CH_3}{|}}{}\right]_{n-1} CH_2-CH_2$$
$$CH_2-CH_3$$

Termination by Hydrogenation

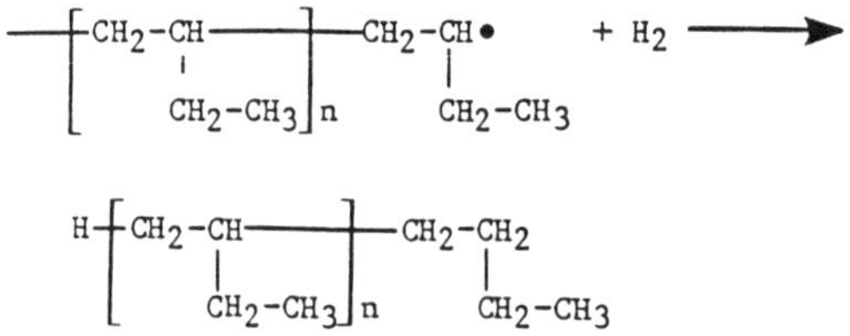

$$-\left[CH_2-CH\underset{\underset{CH_2-CH_3}{|}}{}\right]_n CH_2-CH\bullet \ + \ H_2 \longrightarrow$$
$$CH_2-CH_3$$

$$H-\left[CH_2-CH\underset{\underset{CH_2-CH_3}{|}}{}\right]_n CH_2-CH_2$$
$$CH_2-CH_3$$

Tables 95 and 96 list typical input materials and operating parameters for mass polymerization of polybutylene, respectively. The mass polymerization process, shown in Figure 31, uses high pressure to liquefy the 1-butene monomer. A comonomer, if desired, a molecular weight regulator (e.g., hydrogen), and the catalyst are added to the reactor. When the reaction has reached the desired conversion, the monomer-polymer mixture is fed to a wash tank.

TABLE 95. TYPICAL INPUT MATERIALS TO POLYBUTYLENE POLYMERIZATION

Function	Compound
Monomer	1-butene
Comonomer	1-pentene Propylene Ethylene
Catalyst	Titanium chloride Diethyl aluminum chloride Diethyl aluminum iodide 1,1-dimethylhydrazine Magnesium chloride Ethyl benzoate Triethyl aluminum
Molecular Weight Regulator	Hydrogen
Transformation Aids	Stearic acid Isotactic polypropylene

Sources: *Encyclopedia of Chemical Technology*, 3rd Edition.
          Modern Plastics Encyclopedia, 1981-1982.

TABLE 96. TYPICAL OPERATING PARAMETERS FOR POLYBUTYLENE POLYMERIZATION

Parameter[a]	Value
Temperature, °C	65
Pressure, Pa	800,000

[a]No other information on this process is available in the
literature.

Sources: *Encyclopedia of Chemical Technology*, 3rd Edition.
          Carl L. Yaws, *Physical Properties*, 1977.

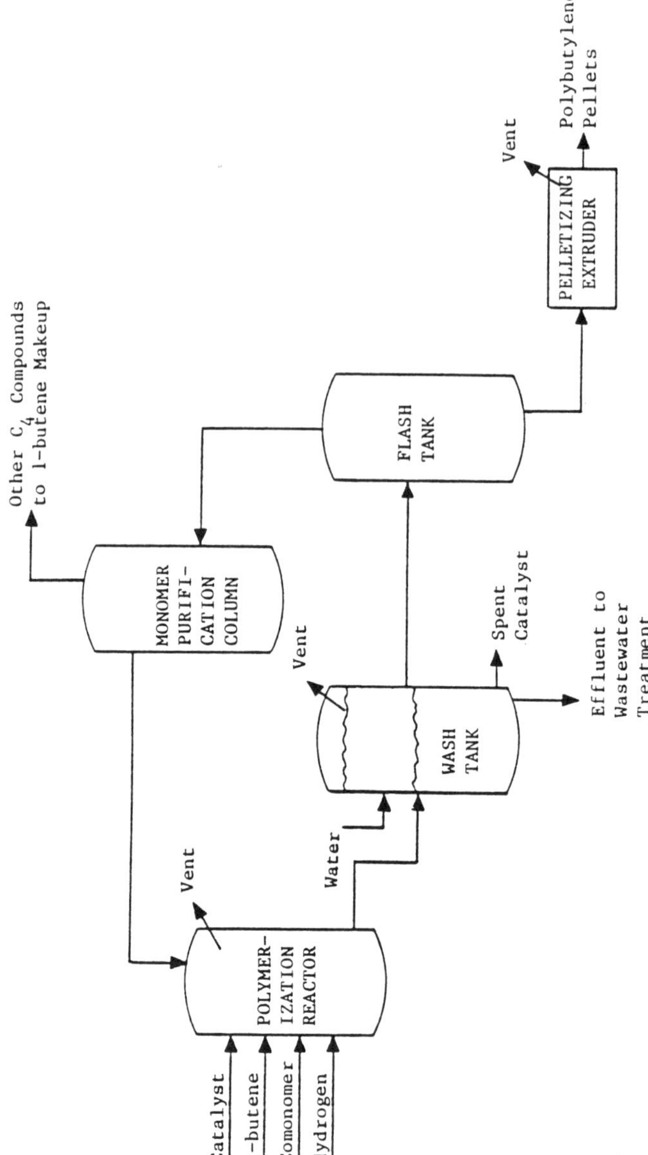

Figure 31. Mass polymerization of polybutylene.

Source: Encyclopedia of Chemical Technology, 3rd Edition.

Deoxygenated water is added to the wash tank to remove any catalyst
residues which are then either recycled and regenerated or sent to disposal.
The polymer melt is then fed to a flash tank and heated to facilitate
monomer removal. The separated monomers are purified while the devolatilzed
polymer is pelletized in an extruder.

The molecular weight of the polybutylene manufactured by this process
ranges from 250,000 to 650,000.[32] Polybutylene resins which are used for
film have antiblock and slip agents added to the polymer while it is in the
extruder.

A recipe for producing polybutylene by mass polymerization was not
found in the literature. However, a recipe for producing polybutylene by
solution polymerization, based on an Imperial Chemical Industries patent, is
shown below:[25]

Material	Quantity
1-Butene (monomer)	40 l
Titanium Tetrachloride (catalyst)	0.00931 kg
Triethyl Aluminum (catalyst)	0.0056 kg
Hexane (solvent)	0.15 l
Ethyl Alcohol (molecular weight regulator)	

Energy Requirements

No data giving energy requirements for polybutylene production were
found in the literature consulted.

ENVIRONMENTAL AND INDUSTRIAL HEALTH CONSIDERATIONS

Polybutylene is a nontoxic compound which is used in packaging meat and
other food items, transporting water, including potable water, and other
piping and packaging applications.[32] None of the chemicals which have
been identified as inputs to polybutylene processing are known carcinogens.
However, two substances used in polybutylene manufacture (1,1-dimethylhydra-
zine, a catalyst, and isotactic polypropylene, a transformation aid) have
been identified as animal carcinogens and are therefore potential human carcinogens. One compound, 1,1-dimethylhydrazine, has been listed as hazardous
under RCRA.

Worker Distribution and Emissions Release Points

Estimates of worker distribution have been made by correlating major
equipment manhour requirements with the process flow diagram in Figure 31.
Estimates are shown in Table 97.

Only fugitive emissions from process vents and leaks are present in
polybutylene processing. These fugitive emissions, however, may pose a
significant environmental and/or worker health problem based on the stream
constituents, process operating parameters, engineering and administrative
controls, and maintenance programs.

TABLE 97. WORKER DISTRIBUTION ESTIMATES FOR POLYBUTYLENE PRODUCTION

Unit	Workers/Unit/8-hour Shift
Batch Reactor	1.0
Monomer Purification Column	0.25
Washing	0.25
Flash Tank	0.125
Extrusion	0.5

Process sources of fugitive emissions are shown in Table 98. Neither
OSHA nor the American Conference of Governmental Industrial Hygienists
(ACGIH) have set standards or recommended exposure levels for 1-butene and
the other gaseous hydrocarbons potentially emitted from polybutylene production. However, polymer particulates from pelletizing may produce nuisance
dust levels which exceed the OSHA standard.

Propylene, ethylene and hydrogen are simple asphyxiants whose presence
in workplace air reduces the amount of available oxygen.

Available health effects information for substances which have yielded
positive animal carcinogenicity tests is given in the following paragraphs.
Other toxicity information is summarized in Table 10 and in IPPEU Chapter
10b.

Health Effects

Relatively little health effects data exist for the input materials to
polybutylene polymerization. Two substances, however, have tested positive
in animal tests for carcinogenicity and are therefore potential human carcinogens. They are 1,1-dimethylhydrazine, a catalyst, and the transformation aid, isotactic polypropylene. The reported health effects of exposure
to these substances are summarized below.

1,1-Dimethylhydrazine poses a high toxic hazard through ingestion,
inhalation, and skin absorption.[242] Although no epidemiological data are
available, it has produced positive results in animal testing for carcinogenic and mutagenic effects.[100] OSHA has set an air standard of 0.5 ppm
(8 hour TWA).[67]

Polypropylene has produced positive results in animal testing for
carcinogenicity.[107] It poses a very low toxic hazard, however. For
example, polypropylene medical devices have produced little or no irritant
responses when placed into connective or muscle tissues for short periods of
time.[31] The propensity for diffusion of polypropylene from the material
is so small that biological response cannot be detected.[31]

Air Emissions

Table 98 lists the sources of fugitive VOC emissions associated with
this process. These three sources are vents and may be controlled by:[285]

- Routing the stream to an incinerator to remove any remaining
  hydrocarbons; and/or

- Routing the stream to a blowdown.

Other sources of fugitive VOC emissions, such as flanges, valves, pumps,
compressors, and cooling towers, may be controlled by equipment modification. However, a regular inspection and maintenance program may be the best
method for reducing these emissions.

TABLE 98.   SOURCES OF FUGITIVE EMISSIONS FROM POLYBUTYLENE MANUFACTURE

Source	Constituent
Reactor Vent	1-butene Polybutylene Other C_4 monomers
Wash Tank Vent	1-butene Polybutylene Other C_4 monomers
Pelletizing Extruder Vent	1-butene and Polybutylene Particulates and Emissions

The particulate emissions associated with this process are also listed in Table 98. The extruder vent may be routed to a baghouse or electrostatic precipitator to collect particulates.

Wastewater Sources

Polybutylene manufacture generates two wastewater streams: the effluent from the wash tank and routine cleaning water. The effluent from the wash tank may contain a small amount of catalyst which is carried from the reactor to the wash tank with the polymer-monomer mixture. Polybutylene and 1-butene are virtually insoluble in water, and therefore are not found in the wastewater. There are no data in the literature which give wastewater characteristics for this process.

Solid Waste

The solid wastes generated during polybutylene production include substandard polymer which cannot be salvaged by blending and collected particulates. These wastes are not hazardous; therefore, their disposal should not pose an environmental problem.

Environmental Regulation

Effluent limitations guidelines for polybutylene have not been established.

New source performance standards (proposed by EPA on January 5, 1981) for volatile organic carbon (VOC) fugitive emissions include:

- Safety/release valves must not release more than 200 ppm above background, except in emergency pressure releases, which should not last more than five days; and

- Leaks (which are defined by VOC emissions greater than 10,000 ppm) must be repaired within 15 days.

One input to polybutylene product has been listed as hazardous (45 Federal Register 3312, May 19, 1980):

1,1-dimethylhydrazine - U098

All disposal of this material or polybutylene containing residual amounts of this must comply with the provisions of the Resource Conservation and Recovery Act (RCRA).

13. Polycarbonate

Polycarbonates, a special class of linear polyesters, result from the reaction of carbonic acid derivatives with aromatic, aliphatic, or mixed diols. Although many different polycarbonates exist, the polycarbonate of bisphenol-A (4,4'-dihydroxy-2,2'-diphenyl-propane) is the only resin of major commercial importance. Polycarbonates combine desirable product properties, high heat stability, toughness, and transparency with easy processability and low cost. Specialty resins contain small amounts of other polyhydric phenols in addition to bisphenol-A. Tetrabromobisphenol-A, tetramethylbisphenol-A, and 2,2-bis(4-hydroxyphenyl)-1,1-dichlorethylene constitute the major comonomers used.[75]

The most valuable properties of polycarbonates include: good mechanical properties, especially impact strength, over a wide temperature range; good resistance to high temperatures, even on long-term exposure; low water absorption; good dimensional stability; complete transparency; good weathering properties; physiological inertness; good resistance toward aqueous agents, oils, fats, aliphatic hydrocarbons, and alcohols; and self-extinguishing properties.[16] These properties make polycarbonates suitable as replacements for metals, glass, wood, and other thermoplastics. Table A-14 in Appendix A lists some typical polycarbonate properties.

The many differing uses of polycarbonates include electrical applications, appliances, instrumentation, communication and electronics, automotive, marine, aerospace, photography and films, light and optical technology, office equipment, various industrial equipment, food packaging, sporting goods, toys, medical appliances, household goods, and appliances.[16] Although most commercial polycarbonates are bisphenol-A based, pertinent details concerning the production processes are scarce. Direct reaction, transesterification, and interfacial polymerization reportedly work. Polymers produced by the transesterification process, first favored over direct reaction for its low cost, exhibit a narrow range of properties; therefore, use of this process has declined. Present production is probably performed by the interfacial polymerization process.[75]

INDUSTRY DESCRIPTION

Two producers comprise the polycarbonate industry: General Electric Company and Mobay Chemical Corporation. Production facilities are located

"""

in three states: Indiana, Texas, and West Virginia. Table 99 lists these
producers, their locations, and their 1982 capacities.

Polycarbonates, used as metal and glass replacements, have experienced
stable sales from 1979 to 1981. The small growth experienced during this
time frame (from 107,000 to 111,000 metric tons) concurs with the growth of
optic fiber usage. Current existing capacity of 193,200 metric tons is
sufficient to accomodate any industry growth in the near future.

PRODUCTION AND END USE DATA

In 1981, polycarbonate production totaled 111,000 metric tons.[143]
Polycarbonate applications include:[16, 30, 75, 247]

- Beer pitchers, mugs, and other high-volume non-electric housewares,
  such as tableware, flatware, coffee filters, trays, and baby
  bottles;

- Returnable bottles, particularly large milk and water bottles;

- Medical equipment, including biochemical laboratory apparatus,
  syringes, dental prostheses, and cages for rats and other small
  animals;

- Automotive markets, including lenses and bright trim extrusions,
  lamp housings, electrical components, decorative pieces, thin-wall
  mechanical parts, large exterior parts, instrument panels, and
  bumpers;

- Construction applications, including door and window components,
  drapery fixtures, furniture, and plumbing;

- Electrical and electronic products, including printers, copiers,
  other business machines, telephone connectors, circuit boards,
  wiring blocks, and other industrial components;

- Lighting applications, including diffusers and globes, housings, and
  other component parts;

- Mirrorized sheet for schools, off-highway equipment, and security
  facilities;

- Protective eyeware and helmets;

- Sheet for appliances, business machines, sound attenuation, aircraft
  interiors, transparent canopies for high speed military aircraft,
  solar collectors, and bullet-resistant parts;

- Electrical insulation for radios, television sets, X-ray equipment,
  and insulation of coils and wires;

TABLE 99. U.S. POLYCARBONATE PRODUCERS

Producer	Location	1982 Capacity Thousand Metric Tons
General Electric Co. Engineered Materials Group Plastics Business Operations	Mount Vernon, IN	120.5
Mobay Chemical Corp. Plastics & Coatings Division	Cedar Bayou, TX	59.1
	New Martinsville, WV	13.6
TOTAL		193.2

Source: Directory of Chemical Producers, 1982.

- Transparent covers for meter housings and transformers, street and industrial lamps, signal lamps, and housings for traffic signals;

- Implosion shields for television sets and housings for radio receivers and transmitters, hair dryers, electric shavers, and kitchen utensils;

- Cable coverings, cable casings, and battery cases;

- Engineering applications including frames for various metering and regulating apparatus, covers of various kinds where transparency is important, filter vessels, parts for pumps (such as rotors), and various other housing and pipe;

- Parts for fire extinguishers, protective masks, ski mobiles, hockey gear, and motorcycle shields;

- Photographic equipment including movie cameras and projectors as well as housings, cassettes, and reels;

- Drafting films, drafting tools, and ball point and fountain pens; and

- Break-resistant windows for buildings (gymnasiums, schools), buses, and trains.

Table 100 lists polycarbonate markets for 1981.

PROCESS DESCRIPTIONS

Most polycarbonates are produced by interfacial polymerization. Although other processes are reported in the literature, the most recent information indicates the predominance of the interfacial polymerization process. Transesterification processing gives polymers which exhibit a narrow range of properties; therefore, use of this process has declined in spite of the lower production cost.[75]

Polycarbonates of bisphenol-A produced by interfacial polymerization utilize an inert organic medium (methylene chloride) and an aqueous alkaline solution (sodium hydroxide) for the two-phase polymerization medium. The polymerization occurs in an emulsion at the interface of the two immiscible solvents. Phosgene is soluble in the chlorinated hydrocarbon, and bisphenol-A is soluble in the aqueous caustic solution. The two solvents are intimately mixed to form an emulsion. Crude polymer solubilizes into the hydrocarbon phase.[247]

During polymerization, bisphenol-A acts as a nucleophile and attacks the carbon atom of the phosgene. The chloride ion is displaced, and dissolves into the aqueous phase, as shown below:

TABLE 100. MAJOR POLYCARBONATE MARKETS

Market	1981 Thousand Metric Tons
Appliances	14
Communications and Electronics	22
Glazing	32
Lighting	4
Signs	3
Sports and Recreation	8
Transportation	13
Other	15
TOTAL	111

Source: _Modern Plastics_, January 1982.

In the presence of a catalyst or accelerator, these low molecular weight
polycarbonate building blocks react to form higher molecular weight polycarbonates. Accelerators enable the reaction to be carried out at room
temperature in a continuous process.

Molecular weight is regulated by adding a chain terminator such as a
monofunctional phenol to the monomer charge.

Table 101 lists typical input materials for this process. Additives are
used to give greater flame retardance, thermal stability, UV light stability, color stability, and easier mold release.[30] Operating parameters for
this process are considered to be proprietary.

Interfacial Polymerization

The interfacial polymerization process in Figure 32 presents a block
diagram which traces the flow of materials through the manufacture. Bisphenol-A and a small amount (1 to 3 percent) of monofunctional phenol are
dissolved or slurried in an aqueous sodium hydroxide solution. Methylene
chloride, a polymer solvent, is added along with a catalytic amount of
tertiary amine, an accelerator. Phosgene gas is dispersed in the stirred
mixture. The polycarbonate formed dissolves in the solvent while the sodium
chloride by-product remains dissolved in the aqueous phase. Polymerization
is complete when the concentration of phenol in the aqueous phase falls to
zero.

The two-phase reaction system is then separated and the organic phase,
containing the polymer, is washed with water and acid to remove any alkaline
residues. Recovery processes for isolating the polymer from the reaction
solvent are closely guarded trade secrets.[75] One proposed process uses
solvent stripping followed by drying and vacuum-vented extrusion. The
pellets produced are then ready for bagging and shipment.

TABLE 101.   TYPICAL INPUT MATERIALS FOR POLYCARBONATE PROCESSING

Function	Compound
Monomer	bisphenol-A phosgene
Comonomer	2,2-bis(4-hydroxyphenyl)-1,1- dichloroethylene tetrabromobisphenol-A tetramethylbisphenol-A
Initiator	potassium hydroxide sodium hydroxide water
Molecular Weight Regulator	monofunctional phenol
Color Enhancer	sodium dithionite
Solvent	methylene chloride

Accelerators	quarternary ammonium salts
	triethylbenzyl ammonium chloride
	tertiary amines
	   N,N-diethylcyclohexylamine
	   triethylamine

Acid Wash	hydrochloric acid phosphoric acid

Sources: _Encyclopedia of Chemical Technology_, 3rd Edition.
Encyclopedia of Polymer Science and Technology.

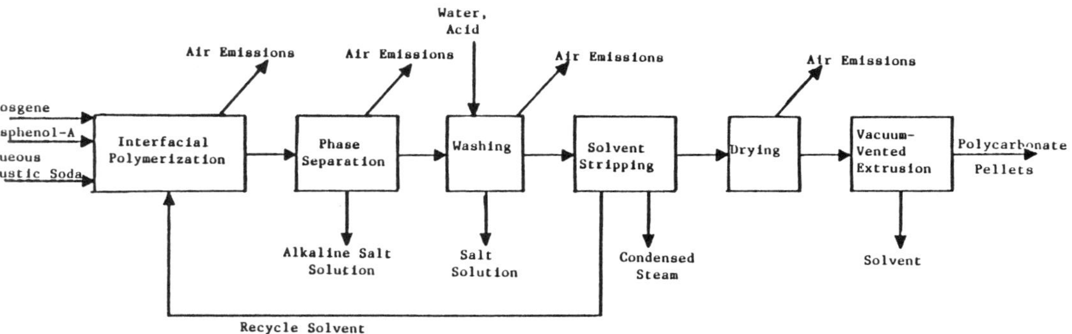

Figure 32. Block diagram for polycarbonate production (interfacial polymerization).

Source: Encyclopedia of Polymer Science and Technology.

A typical recipe for producing 100 parts of polycarbonate by the interfacial phosgenation process is shown below:[260]

Material	Parts by Weight
Bisphenol-A	92
Phosgene	42
P-tert-butyl Phenol (chain terminator)	1
Sodium Hydroxide	Sufficient to produce a 5-10% aqueous caustic solution
Methylene Chloride (solvent)	
Water	

Energy Requirements

The energy required for a polycarbonate process has been specified as:[181]

Energy Required	Unit/Metric Ton of Product
Steam	600 kg
Electricity	2.34×10^9 Joules

ENVIRONMENTAL AND INDUSTRIAL HEALTH CONSIDERATIONS

Polycarbonate resins are non-toxic compounds, approved by the FDA for food contact applications, which impart no odors or taste to foods. However, one of the input materials to the polycarbonate production process (phosgene) has been listed as a hazardous waste under RCRA.

Worker Distribution and Emissions Release Points

Worker distribution estimates for polycarbonate production are shown in Table 102. Estimates were made for the interfacial polymerization process outlined in Figure 32.

Probable air emission sources are listed in Table 103. No major air emission point sources are expected from this process. The only air emissions are fugitive emissions from such sources as reactor vents, dryer vents, and extruders. These may pose a significant environmental worker health impact depending on process operating conditions, input materials and degree of process control.

Health effects of three contaminants of concern are summarized below.

Potassium and sodium hydroxides, hydrochloric acid and phosphoric acid are highly irritating to the skin and mucous membranes. Brief contact to concentrated amounts of these substances can cause severe chemical burns.

TABLE 102.   WORKER DISTRIBUTION ESTIMATES FOR POLYCARBONATE PRODUCTION

Unit	Workers/Unit/8-hour Shift
Batch Reactor	1.0
Phase Separator	0.25
Washing	0.25
Solvent Stripping	0.25
Dryer	0.5
Extrusion	0.5

TABLE 103.   PROBABLE SOURCES OF AIR EMISSIONS FROM POLYCARBONATE PROCESSING

Reaction Step	Probable Source	Constituent
Interfacial Polymerization	Reactor Vent	Phosgene Bisphenol-A Methylene Chloride
Phase Separation	Separation Tank Vent	Methylene Chloride
Washing	Wash Tank Vent	Methylene Chloride
Solvent Stripping	Condenser Vent	Methylene Chloride
Drying	Dryer Vent	Methylene Chloride Particulates
Vacuum-Vented Extruder	Extruder Vent	Methylene Chloride Particulates

Little data are available from which employee exposure potential may be
estimated. However, one source estimates hydrocarbon emissions from polycarbonate molding operations at 32 g/kg product.[286]

Health Effects

Polycarbonate processing typically involves the use of less than 20
input materials and specialty chemicals. Most of these are only slightly to
moderately toxic. Three substances used by the industry are highly toxic,
however. They are: phosgene, one of the monomers; the accelerator triethylamine; and phenol, a molecular weight regulator. The paragraphs below
summarize the reported health effects of exposure to these three substances.

Phenol is extremely toxic by ingestion, inhalation, and skin absorption.[85] Approximately half of all reported cases of acute phenol poisoning has resutled in death.[278] Chronic poisoning may also occur from
industrial contact and has caused renal and hepatic damange.[149] Evidence
indicates that phenol is also a tumorigenic, mutagenic, and teratogenic
agent; however, a recent carcinogenesis bioassay by the National Cancer
Institute produced negative results.[233] The OSHA air standard is 5 ppm (8
hour TWA).[67]

Phosgene, a strong local irritant, is also extremely toxic by inhalation.[242] Concentrations greater than 0.25 mg/l of air (62 ppm) may be
fatal when breathed for one-half hour or more.[273] Three to 5 mg/l causes
death within a few minutes.[273] Any exposure is extremely dangerous
because there is no protective respiratory reflex to prevent deep inspira-
tion of phosgene.[88] It has been reported to have the ability to condition
the olfactory senses so that the characteristic odor may be detected only
briefly at the time of initial exposure.[238] The OSHA air standard is 0.1
ppm (8 hour TWA).[67]

Triethylamine is highly toxic by ingestion and inhalation and moderately toxic via dermal routes.[243] It is a strong irritant to tissue. At
least one animal study has shown that triethylamine produced mutagenic
effects in rats at the low dose of 1 mg/m^3.[96] The OSHA air standard is
25 ppm (8 hour TWA).[67]

Air Emissions

According to EPA estimates, the population exposed to air emissions in
a 100 km^2 area surrounding a polycarbonate production plant is: 41 persons exposed to hydrocarbons and 2 persons exposed to particulates.[286]
These sources may be controlled by venting VOC emission streams to a flare
or blowdown [285] and streams containing particulates to either an electrostatic precipitator or a baghouse.

Wastewater Sources

The major wastewater sources associated with this process include the
alkaline salt solution generated during phase separation, the salt solution

generated during washing, condensed steam from the solvent stripping process, and routine cleaning water. No data concerning either wastewater
production or wastewater characteristics were found in the literature
consulted.

Solid Waste

The major solid waste generated by this process is the by-product
sodium chloride, estimated at 760 kg of salt per metric ton of polycarbonate
produced.[93] Other solid wastes include polymer lost due to reactor cleaning and spillage, collected particulates, and substandard product which
cannot be blended.

Environmental Regulation

Effluent limitations guidelines have not been set for the polycarbonate
industry.

New source performance standards (proposed by EPA on January 5, 1981)
for volatile organic carbon (VOC) fugitive emissions include:

- Safety/release valves must not release more than 200 ppm above
  background, except in emergency pressure releases, which should not
  last more than five days; and

- Leaks (which are defined as VOC emissions greater than 10,000 ppm)
  must be repaired within 15 days.

Phosgene has been listed as a hazardous waste and designated as U095
(46 Federal Register 27476, May 20, 1981). Disposal of any polycarbonate
resin containing residual amounts of phosgene or any containerized phosgene
gas must comply with the provisions set forth in the Resource Conservation
and Recovery Act (RCRA).

14. Poly(Ester-Imide) and Poly(Ether-Imide) Resins

Poly(ester-imide) and poly(ether-imide) resins are condensation polymers which have a repeating imide group as an integral part of the chain structure. An imide group is shown below:

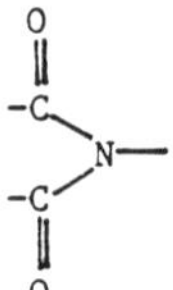

Poly(ester-imides) are characterized by repeating ester and imide linkages, while poly(ether-imides) contain ether and imide linkages.

High temperature resistance, good electrical properties, good wear and friction properties, chemical inertness, radiation and cryogenic temperature stability, and inherent nonflammability are properties of polyimides which allow their use in a variety of applications.[44] Modified polyimides have poorer thermal stabiity than the comparable aromatic polyimides; however, they offer greater versatility in processing. Typical properties of poly-(ether-imide) resins are presented in Table A-15 in Appendix A.

Poly(ester-imides) are used primarily for wire insulation.[248] Poly (ether-imides) are used for injection molding, extrusion, blow molding, and structural foam molding. Commercial use of poly(ether-imide) resins include circuit-breaker housings, under-the-hood automotive applications, microwave oven stirrer shafts, and integrated-circuit chip carriers.[150]

The polycondensation processes used to produce poly(ester-imide) and poly(ether-imide) resins involves the reaction, in solution, of an aromatic anhydride, a diamine, and a dianion, such as glycol or an alkali metal salt. Poly(ester-imide) resins can be produced by two processes, both of which start with trimellitic anhydride. The difference between the two processes is the reactant addition sequence. These processes, as well as the poly (ether-imide) resin production process, are described in more detail later in this section.

INDUSTRY DESCRIPTION

The producers of poly(ester-imide) and poly(ether-imide) resins are listed in Table 104.

Poly(ester-imide) resins are manufactured by three firms, representing three facilities in two states. Two facilities are located in New York State, and one facility is in Missouri. No production data were reported in the literature for these facilities.

Poly(ether-imide) resins are manufactured by two firms. General Electric began commercial production of poly(ether-imide) at their Mount Vernon, Indiana facility in 1982. G.E.'s production is reportedly between 2.3 and 4.5 thousand metric tons per year. G.E. is planning to expand production capacity at this facility.[150] LNP Corporation reports the availability of small quantities of poly(ether-imide) resins; however, the manufacturing location is not listed in the literature.[33]

PRODUCTION AND END USE DATA

The 1982 U.S. market for all types of polyimide resins is estimated to have been 2.5 thousand metric tons, valued at about $120 million. Modified polyimides, including poly(amide-imides), poly(ester-imides), and poly(ether-imides), comprised about 45 percent of the polyimide market. Modified polyimides have many uses, including:[44]

- Wire enamels and other coating applications, and
- Molded parts for the aerospace and hydraulic equipment industries.

Poly(ether-imide) is reported to be more competitive with engineering materials such as polyphenylene sulfide and polysulfone than with other aromatic polyimides. End uses for poly(ether-imide) resins being developed include fibers, film, sheet, and wire insulation.[71]

PROCESS DESCRIPTIONS

The polycondensation process used to produce poly(ester-imide) and poly(ether-imide) resins involves the reaction, in solution, of an aromatic anhydride, a diamine, and a dianion, such as glycol or an alkali metal salt. Not much information is available in the literature on the industrial processes used to manufacture either poly(ester-imide) or poly(ether-imide) resins. Little information is available on reaction mechanisms. Reaction chemistry, mainly available from patent literature, is discussed below for each resin.

Table 105 lists typical input materials for these resins.

<u>Poly(ester-imide) Resins</u>

Poly(ester-imide) resins can be produced by two processes. Both processes start with trimellitic anhydride (TMA). In one process, TMA is first reacted, in solution, with a diamine. The intermediate product is then

TABLE 104. U.S. POLY(ESTER-IMIDE) AND POLY(ETHER-IMIDE) RESIN PRODUCERS

Producer	Location	Product Type	1982 Capacity Thousand Metric Tons
General Electric Company			
Engineered Materials Group Electromaterials Business Department	Schenectady, NY	Poly(ester-imide)	N.R.
Plastics Business Operations	Mount Vernon, IN	Poly(ether-imide)	a
The P.D. George Company	St. Louis, MO	Poly(ester-imide)	N.R.
Schenectady Chemicals, Inc.	Schenectady, NY	Poly(ester-imide)	N.R.
LNP Corporation	Malvern, PA[b]	Poly(ether-imide)	N.R.

N.R. - Not reported.

[a]Modern Plastics reports a production range of 2.3 to 4.5 thousand metric tons per year.

[b]Plant location not reported; corporate headquarter location listed.

Sources: Directory of Chemical Producers, 1984.
Chemcyclopedia, 1985.

TABLE 105.   TYPICAL INPUT MATERIALS FOR POLY(ESTER-IMIDE)
             AND POLY(ETHER-IMIDE) RESINS

Anhydrides	Trimellitic Anhydride (TMA)
	3-Nitrophthalic Anhydride
	4-Nitrophthalic Anhydride
Arylenes	2,4'-dihydroxy diphenylmethane
	bis(2-hydroxyphenyl) methane
	2,2-bis(4-hydroxyphenyl) propane (Bisphenol A)
	1,2-bis(4-hydroxyphenyl) ethane
	4,4'-dihydroxydiphenyl ether
	4,4'-dihydroxybiphenyl
	Hydroquinone
	Phenylene
Glycols	1,2-ethylene glycol
	1,3-trimethylene glycol
	1,4-tetramethylene glycol
	1,5-pentamethylene glycol
	1,6-hexamethylene glycol
	3-methylpentane-1,5-diol
	1,10-decamethylene glycol
Solvents	Methanol
	Ethanol
	Isopropyl Alcohol
	Benzene
	Toluene
	Pentane
	Hexane
	Octane
	Cyclohexane
	Heptane
	Dimethylformamide
	N-methyl-2-pyrrolidone
	N,N-dimethylacetamide
	dimethyl sulfoxide
	M-cresol
	N-methylpyrrolidone
	Hexamethylphosphoramide
	Aroclor

(continued)

TABLE 105 (continued)

Diamines	Hexamethylene diamine
	Phenylenediamine
	Benzidene
	4,4'-Diaminodiphenyl ether
	3,3'-Dimethoxy-4,4'-diamine diphenyl
	4,4'-Diaminodiphenyl methane
	4,4'-Diaminodiphenyl sulfone
	3,5-Diaminobenzoic acid
	Durenediamine
Additives	Thionyl Chloride
	Tetrabutyl Titanate
	Zinc Acetate (catalyst)
	Antimony Trioxide (catalyst)
	Sodium Hydroxide
	Acetic Acid

Sources: Loncrini, "Aromatic Polyesterimides," Journal of Polymer Science:
           Part A-1, Vol. 4, pp. 1531-1541, 1966.
           Wirth, Joseph G., and Darrell R. Heath, U.S. Patent No. 3,838,097,
           September 24, 1974, assigned to General Electric Company.
           Ranney, Polyimide Manufacture 1971, Noyes Data Corporation,
           Park Ridge, NJ.

reacted with a glycol to produce the poly(ester-imide) resins. This
sequence of reactions is shown below:[230]

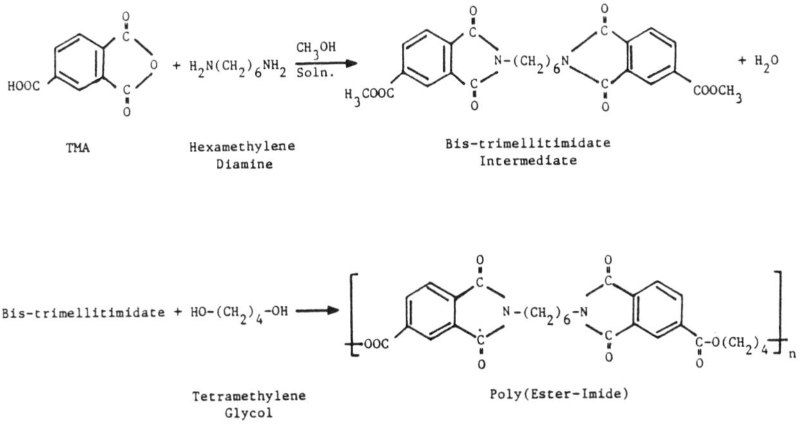

TMA

Hexamethylene
Diamine

Bis-trimellitimidate
Intermediate

Tetramethylene
Glycol

Poly(Ester-Imide)

In the second process, TMA is first reacted with an aromatic diacetoxy
compound to produce an aromatic bisesteranhydride intermediate. The intermediate is then reacted with an aromatic diamine in a polar solvent to give
a polyamic acid precursor. The precursor is then thermally converted to the
poly(ester-imide) resin. This sequence of reactions is shown below:[139]

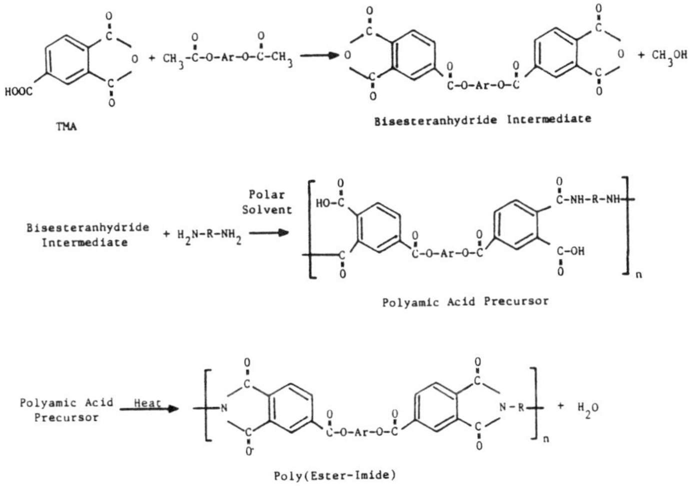

TMA

Bisesteranhydride Intermediate

Polyamic Acid Precursor

Poly(Ester-Imide)

The difference between these two processes is the order of adding reactants to TMA. In the first process, a diamine is added to the TMA-methanol solution before adding glycol. In this process, methanol is removed by distillation, and diimide diacid is recrystallized from dimethyl formamide. Bis-trimellitimidate intermediate is produced by reacting the diimide diacid with thionyl chloride in a methanol-toluene solution. Bis-trimellitimidate is converted to poly(ester-imide) by reacting with glycol and tetrabutyl titanate. A typical recipe for the production of poly(ester-imide) resin by this process is given below:[230]

Step	Material	Quantity
1	Trimellitic anhydride (TMA)	0.096 kg
	Methanol (solvent)	0.5 l
	Hexamethylene diamine	0.029 kg
2	N,N'-hexamethylene-bis- trimellitimide (diimide diacid)	0.047 kg
	Toluene (solvent)	0.2 l
	Methanol (solvent)	0.05 l
	Thionyl chloride	0.098 l

Step	Material	Parts by Weight
3	Bis-trimellitimidate	5
	1,4-tetramethylene glycol	2
	Tetrabutyl titanate	0.5

In the second process, the diacetoxy or glycol is reacted with the TMA solution to produce an intermediate before the diamine is introduced. A typical recipe for producing a bisesteranhydride intermediate is given below:[139]

Material	Quantity
Trimellitic anhydride (TMA)	0.192 kg
p-Phenylene diacetate	0.097 kg
Aroclor #1242 (solvent)	0.5 l
n-Heptane (solvent)	2 l
Diethylether	Washing

A recipe was not found for producing a poly(ester-imide) resin from this intermediate.

Although information was not found in the literature on industrial manufacturing processes for poly(ester-imide) resins, a process, based on engineering judgement, is shown in Figure 33.

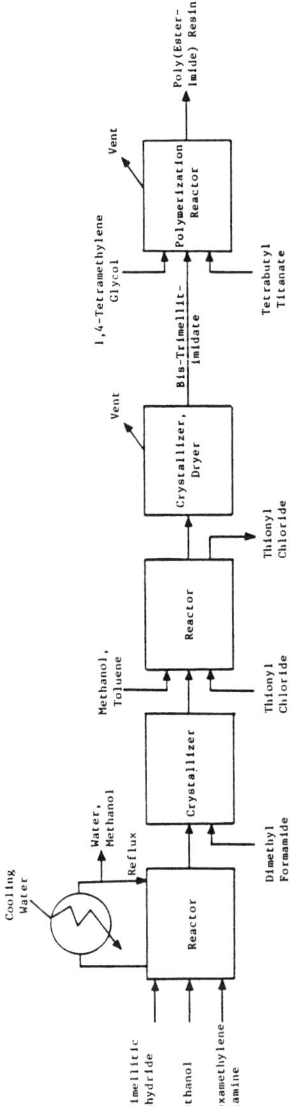

Figure 33. Poly(ester-imide) resin manufacturing process.

Poly(ether-imide) Resins

Poly(ether-imide) resins are produced by reacting a bis(nitrophthali-mide) with an alkali metal salt in a solvent. Bis(nitrophthalimide) is an intermediate produced by reacting a diamine with a nitro-substituted aromatic anhydride. This sequence of reactions is shown below:[298]

Diamine   Nitro-Substituted   Bis(Nitrophthalimide)
            Aromatic Anhydride

Alkali Metal   Poly(Ether-Imide)
Salt

A typical recipe for the production of poly(ether-imide) is shown below:[298]

Material	Quantity
Bisphenol A	2.2828×10^{-3} kg
Sodium hydroxide (50.3% solution)	1.5904×10^{-3} kg
Dimethylsulfoxide (solvent)	0.05 l
Benzene (solvent)	0.006 l
4,4'-Bis(3-nitrophthalimido) diphenylmethane	5.845×10^{-3} kg
Glacial acetic acid (quench reaction)	2×10^{-4} l
Methanol (precipitate product)	0.6 l

Although information was not found in the literature on industrial manufacturing processes for poly(ether-imide) resins, a process, based on engineering judgment, is shown in Figure 34.

Energy Requirements

Data concerning the energy requirements for this process were not found in the literature consulted.

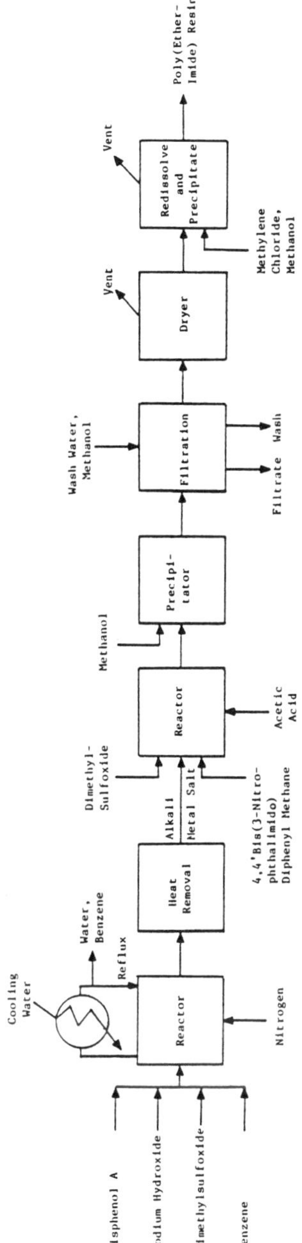

Figure 34.   Poly(ether-imide) resin manufacturing process.

ENVIRONMENTAL AND INDUSTRIAL HEALTH CONSIDERATIONS

Polyimide resins do not pose a significant health concern.[71] However, trimellitic anhydride and phthalic anhydride are reported as toxic [91], and care must be used in handling these materials. In addition, several of the solvents and diamines used in poly(ester-imide) and poly(ether-imide) resin manufacture are also regarded as toxic.

Worker Distribution and Emissions Release Points

Estimates of worker distribution have been made by correlating major equipment manhour requirements with the process flow diagrams in Figures 33 and 34. Table 106 shows estimates for both poly(ester-imide) and poly-(ether-imide) resin manufacturing processes.

There are no major point source air emissions associated with either poly(ester-imide) or poly(ether-imide) processing. The only air emissions are fugitive emissions from such sources as reactor vents, dryer vents, and condenser vents. The impact of these emissions depends on the process operating parameters, engineering and administrative controls, and input materials used.

Sources of fugitive emissions are shown in Table 107, based on engineering judgement. Solvents typically used in both processes include methanol, dimethyl formamide, dimethyl sulfoxide, benzene, and toluene. Available toxicity information for major process reactants and additives is summarized under "Health Effects" below.

No data were located in the literature from which worker exposure potential may be estimated.

Health Effects

The manufacture of poly(ester-imide) and poly(ether-imide) resins involves the use of several toxic materials. As discussed earlier, both trimellitic anhydride and phthalic anhydride are reported as toxic. In addition, several of the solvents, including methanol, benzene, toluene, dimethylformamide, and dimethylsulfoxide, and several of the diamines, including hexamethylene diamine, phenylenediamine, diaminodiphenyl methane, and diaminodiphenyl sulfone are reported as toxic compounds. Table 108 presents some additional information on these toxic compounds.

Air Emissions

Sources of fugitive VOC emissions are listed in Table 107. All the process sources are vents which may be controlled by:[285]

- Routing the stream to a flare to incinerate the hydrocarbons present; and/or

- Routing the stream to blowdown.

TABLE 106. WORKER DISTRIBUTION ESTIMATES FOR POLY(ESTER-IMIDE) AND POLY(ETHER-IMIDE) PRODUCTION

Process	Unit	Workers/Unit/8-Hour Shift
Poly(ester-imide)	Reactor	1.0
	Crystallizer	0.5
	Polymerization Reactor	1.0
Poly(ether-imide)	Reactor	1.0
	Heat Exchanger	0.25
	Filter	0.75
	Dryer	0.25
	Dissolver	0.75

TABLE 107.   SOURCES OF FUGITIVE EMISSIONS FROM POLY(ESTER-IMIDE) AND POLY(ETHER-IMIDE) MANUFACTURE

Source	Constituent	Process Poly(ester-imide)	Process Poly(ether-imide)
Precursor Reactor Vent	Trimellitic Anhydride		
	Methanol	X	
	Hexamethylene Diamine	*	
	Dimethyl Formamide	*	
	Thionyl Chloride		
	Toluene	*	
	Bisphenol A		
	Sodium Hydroxide		
	Dimethyl Sulfoxide		*
	Benzene		X
	Water	X	X
Crystallizer Vent	Bis-trimellitimidate	*	
	Water	X	
Precipitation, Filtration, Drying Vents	Alkali Metal Salt		
	Dimethyl Sulfoxide		*
	4,4'-Bis(3-Nitrophthalimido) Diphenyl Methane		
	Acetic Acid		
	Methanol		X
	Water		X
	Particulates		X
Polymerization Reactor Vent	Glycol	*	
	Tetrabutyl Titanate	*	
	Imidate	*	
	Polymer	*	
Dissolver and Crystallizer Vents	Methylene Chloride		*
	Methanol		*
	Intermediate		*
	Polymer		*

*Trace amounts of these compounds are expected, if they are present at all.

TABLE 108.   TOXIC COMPOUNDS USED IN POLY(ESTER-IMIDE) AND
             POLY(ETHER-IMIDE) MANUFACTURE

Input Material	Toxicity	Concern	OSHA Regulation
I. SOLVENTS			
Methanol	Eye, Human 5 ppm Effect: Irritation	Mutagen Teratogen Primary Irritant	200 ppm TWA
Benzene	Toxic Inhalation, Humans TC_{Lo}: 100 ppm/10Y-1 Effect: Carcinogen	Dangerous Fire Risk Ingestion Inhalation Skin Absorption Mutagen Teratogen Carcinogen Primary Irritant	10 ppm TWA
Toluene	Toxic Inhalation, Rat TC_{Lo}: 4000 ppm/4H	Primary Irritant Ingestion Inhalation Skin Absorption Dangerous Fire Risk Tumorigen Mutagen Teratogen	200 ppm TWA 300 ppm Ceiling

Dimethylformamide	Inhalation, Mouse
	LC_{50}: 9400 mg/m^3/
	  2H
		Tumorigen
		Teratogen
		Strong Tissue
		  Irritant
			10 ppm TWA
			(skin)

Dimethylsulfoxide	Oral, Rat LD_{50}: 17,500 mg/kg	Ingestion Mutagen	
II. DIAMINES			
Hexamethylenedia- mine	Toxic Oral, Rat LD_{50}: 750 mg/kg	Corrosive Ingestion Irritant	

(continued)

Table 108 (continued)

Input Material	Toxicity	Concern	OSHA Regulation
Phenylenediamine	Toxic Oral, Rat LD_{50}: 650 mg/kg	Inhalation Irritant to Skin Tumorigen Mutagen	
Diaminodiphenyl Methane	Subcutaneous, Rat LD_{50}: 3300 mg/kg	Causes Toxic Hepatitis	

Diaminodiphenyl Sulfone	Toxic
	Oral, Humans
	LD_{Lo}: 18 gm/kg/15
	  years
	Effect: Systemic
	Oral, Rat
	LD_{Lo}: 1000 mg/kg
		Positive Carcin-
		  ogen (Animal)
		Ingestion
		Tumorigen
		Mutagen	

Source: Industrial Process Profiles for Environmental Use. Chapter 10b-Plastics Additives, by Radian Corporation, July 1985.

No data were found in the literature on the quantity of air emissions from the polyimide manufacturing process.

Wastewater Sources

The wastewater sources associated with poly(ester-imide) and poly (ether-imide) resin production are listed in Table 109. Poly(ester-imide) has two sources of wastewater: routine cleaning water, and condensate from the reactor in which TMA is reacted with diamine. Poly(ether-imide) has three sources of wastewater: routine cleaning water, condensate from the reactor in which Bisphenol A is reacted with sodium hydroxide, and filtrate and wash water from the polymer recovery filter.

Data were not found on the quantity of wastewater generated by poly-imide manufacturing.

Solid Wastes

Only routine solid wastes are generated by the manufacture of poly (ester-imide) and poly(ether-imide) resins. These wastes include those from reactor cleaning, product blending, particulate removal, and spillage.

Environmental Regulation

Effluent limitations guidelines have not been set for the poly(ester-imide) and poly(ether-imide) resin manufacturing industry.

New source performance standards (proposed by EPA on January 5, 1981) for volatile organic carbon (VOC) fugitive emissions include:

- Safety/release valves must not release more than 200 ppm above background, except in emergency pressure releases, which should not last more than five days; and

- Leaks (which are defined as VOC emissions greater than 10,000 ppm) must be repaired within 15 days.

The following compounds have been listed as hazardous (46 _Federal Register_, May 20, 1981):

Benzene	– U019
Methanol	– U154
Phthalic Anhydride	– U190
Toluene	– U220

All disposal of the compounds and resins which contain residual amounts of these compounds must comply with the provisions set forth in the Resource Conservation and Recovery Act (RCRA).

TABLE 109.   SOURCES OF WASTEWATER FROM POLY(ESTER-IMIDE) AND POLY(ETHER-IMIDE) PRODUCTION

	Process	
Source	Poly(Ester-Imide)	Poly(Ether-Imide)
Precursor Reactor Condensate	X	X
Filtration Effluent and Wash		X
Routine Cleaning Water	X	X

15. Polyester Resins (Saturated)

INTRODUCTION

Two products dominate the saturated polyester market: polyethylene terephthalate (PET) and polybutylene terephthalate (PBT). PET is a stiff resin which exhibits higher impact and tensile strength when compared to PBT. PET is a relatively short chain polymer which crystallizes more slowly than PBT but exhibits greater dimensional stability at high temperatures. Good carbon dioxide barrier properties and the ability of PET bottles to withstand high internal pressures (100 psi) make these resins excellent containers for carbonated beverages. The PET containers used are also shatter resistant and light weight. Table A-16 in Appendix A presents some typical PET properties.

PBT is a crystalline thermoplastic whose rapid crystallization rate combined with its good mold flow results in very short molding cycles. PBT is a high-strength, rigid, dimensionally stable, creep-resistant material when filled. Mineral and mineral/glass fillers increase the strength and stiffness of PBT, as well as reducing the warpage which is experienced in glass reinforced resins. Unfilled resins are chemically resistant, and exhibit good tensile strength, toughness, and dimensional stability. Table A-17 in Appendix A lists some typical properties of PBT resins.

PBT is inherently resistant to most chemicals partly due to its crystalline nature. At moderate temperatures, the following organic solvents have little effect on PBT: hydrocarbons, chlorinated hydrocarbons, alcohols, ketones, esters, gasoline, and oils. Dilute bases and weak acids have little effect at room temperature. PBT parts are not harmed by soldering flux, acid, freon, cleaning solutions, and other substances used during cleaning, testing, and soldering.[9]

PET and PBT may be produced by either a batch or continuous process. The processes used to produce these two polymers are similar, with the major difference being the input materials.

INDUSTRY DESCRIPTION

The PET industry is made up of 12 producers with 19 sites located in six states: Virginia, West Virginia, Tennessee, North Carolina, South Carolina, and Alabama. Four producers constitute over 80 percent of the PET industry's capacity. DuPont commands 30 percent with Fiber Industry,

Eastman Kodak, and American Hoechst contributing 25, 16, and 10 percent, respectively. Table 110 lists PET resin producers and their 1982 capacities.

The PBT industry is much smaller with only three producers located in three different states. Table 111 presents these producers and their 1982 capacity.

PET and PBT resin usage is growing. In 1981, total sales of both PET and PBT resins grew from 480,000 metric tons in 1980 to 561,000 metric tons. PET resin production increased to 537,000 metric tons from 458,000 metric tons while PBT resin production grew by 2,000 metric tons to 24,000 metric tons.[143] PET fiber production increased from 1,809,000 metric tons in 1980 to 1,869,000 metric tons in 1981.[35] The current excess capacity in both the PET and PBT industries is sufficient to accommodate any industry growth in the near future.

PRODUCTION AND END USE DATA

In 1981, total PET production was 537,000 metric tons.[143] PET's many uses include:[80, 82]

- Carbonated soft drink containers;

- Noncarbonated beverage containers for fruit drinks, lemonade, and mineral water;

- Containers for mouthwash, edible oil, and chocolate syrup;

- Containers for foods, medicines, toiletries, cosmetics, and household chemicals;

- Strapping materials used for industrial banding applications, conveyor belting, filter fabrics, laundry bags, webbing, tire cords, sewing threads, and ropes;

- Upholstery fabrics and curtains as well as light weight, "easy-care," permanently pleatable articles of clothing;

- Ovenable paperboard that is used for cooking utensils in conventional and microwave ovens;

- Coatings for milk carton stock;

- Films, including magnetic tape, x-ray and photographic film, typewriter ribbons, and electrical insulation; and

- Packaging for boil-in-bags, retort pouches, packages for processed meats and cheeses, shrink films, and blister packs.

Table 112 lists 1981 PET consumption by market.

TABLE 110. U.S. POLYETHYLENE TEREPHTHALATE RESIN PRODUCERS

Producer and Location	1982 Capacity Thousand Metric Tons
Akzona, Inc.	
American Enka Co. Division	
Central, SC	25.0
Lowland, TN	48.2
Allied Corp.	
Allied Fibers and Plastics Co.	
Moncure, NC	38.6
American Hoechst Corp.	
Hoechst Fibers Industrial Division	
Spartanburg, SC	261.4
E. I. duPont de Nemours & Co., Inc.	
Textile Fibers Department	
Old Hickory, TN	250.0
Wilmington, NC	568.2
Eastman Kodak Co.	
Eastman Chemical Products, Inc.,	
subsidiary	
Carolina Eastman Co.	
Columbia, SC	204.6
Tennessee Eastman Co.	
Kingsport, TN	238.6
Fiber Industry, Inc.	
Florence, SC	
Greenville, SC	681.8
Salisbury, NC	
Shelby, NC	
The Firestone Tire & Rubber Co.	
Firestone Synthetic Fibers Co.	
Division	
Hopewell, VA	22.7
The Goodyear Tire & Rubber Co.	
Chemical Division	
Point Pleasant, WV	122.7

(continued)

TABLE 110 (continued)

Producer and Location	1982 Capacity Thousand Metric Tons
ICI Americas Inc. Films Division Hopewell, VA	36.4

Minnesota Mining and Manufacturing Co.
 Film and Allied Products Division
  Decatur, AL
  Greenville, SC
	25.0
	10.0

Monsanto Co. Monsanto Textiles Co. Decatur, AL	90.9
Rohm and Haas Co. Carodel Corp., subsidiary Fayetteville, NC	90.9
TOTAL	2,715.0

Source: *Directory of Chemical Producers*, 1982.

TABLE 111. U.S. POLYBUTYLENE TEREPHTHALATE RESIN PRODUCERS

Producer and Location	1982 Capacity Thousand Metric Tons
Celanese Corp. Celanese Plastics & Specialties Co. Division Celanese Engineering Resins Division Bishop, TX	31.8
GAF Corp. Chemical Products Calvert City, KY	4.5
General Electric Co. Engineered Materials Group Plastics Business Operations Mount Vernon, IN	59.1
TOTAL	95.4

Source: *Directory of Chemical Producers*, 1982.

TABLE 112. 1981 CONSUMPTION OF PET

Market	Thousand Metric Tons
Blow Molding	
Soft drink bottles	250
Other bottles (includes cosmetics, toiletries, pharmaceuticals, foods)	7
Extrusion	
Merchant film	130
Captive film	95
Coating for ovenable board	32
Sheeting for blister packs	11
Strapping	6
Injection Molding	6
Fibers	
Yarn and Monofilament	
Textiles	576
Industrial uses	159
Staple and tow	1,134
TOTAL	2,406

Sources: *Chemical Economics Handbook*, updated annually, 1981 data.
Modern Plastics, January 1982.

In 1981, PBT production totaled 24,000 metric tons.[143]   PBT has many uses, including:[9]

- Electrical/electronic applications, including connectors, switches, potentiometers, relays, TV tuner components, high voltage components, terminal boards, integrated circuit carriers, motor brush holders, end bells, and housings;

- Automotive applications from large exterior body components to windshield wiper structures, high-energy ignition caps and rotors, ignition coil caps, coil bobbins, fuel injection controls, sensors, connectors, window and door hardware, speedometer frames, and gears;

- Housewares, such as appliance housings for irons and toasters, handles, gasoline-powered tool housings, food processor blades, and hair dryer nozzles; and

- Industrial applications, including pump impellors, housings and support brackets, showerhead and faucet components, paint brush filaments, nonwoven fabric, irrigation valves and bodies, and water meter chambers and components.

PROCESS DESCRIPTIONS

In PET resin production, two routes may be used. Dimethyl terephthalate (DMT) and ethylene glycol may be reacted to yield a PET chain and methanol, or pure terephthalic acid (TPA) and ethylene glycol may be combined to form the PET resin and water. Both routes use a two-step esterification process. In the first step, the prepolymer bis(2-hydroxyethyl) terephthalate (BHET) is formed. In the second step, which is the same for both routes, the BHET monomer is polymerized by polycondensation to produce PET and ethylene glycol. These ester interchange reactions are shown below.

From DMT - Step 1

$$nCH_3OOC\text{-}\langle\ \rangle\text{-}COOCH_3 + 2nHOCH_2CH_2OH \xrightarrow{\text{Catalyst}} nHOCH_2CH_2O\overset{O}{\overset{\|}{C}}\text{-}\langle\ \rangle\text{-}\overset{O}{\overset{\|}{C}}OCH_2CH_2OH + 2nCH_3OH$$

BHET   Methanol

From TPA - Step 1

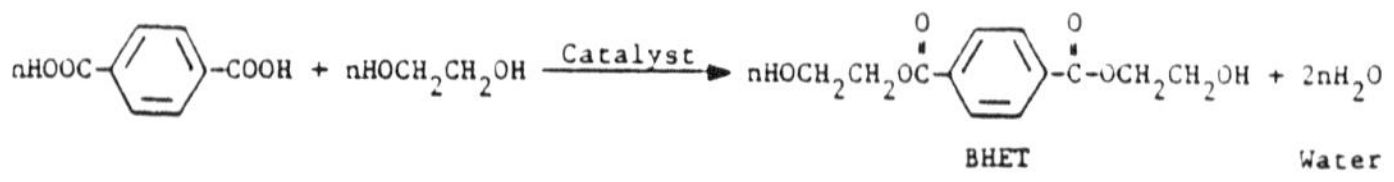

$$nHOOC-\!\!\bigcirc\!\!-COOH + nHOCH_2CH_2OH \xrightarrow{\text{Catalyst}} nHOCH_2CH_2OC\!\!\bigcirc\!\!C-OCH_2CH_2OH + 2nH_2O$$

BHET   Water

Step 2

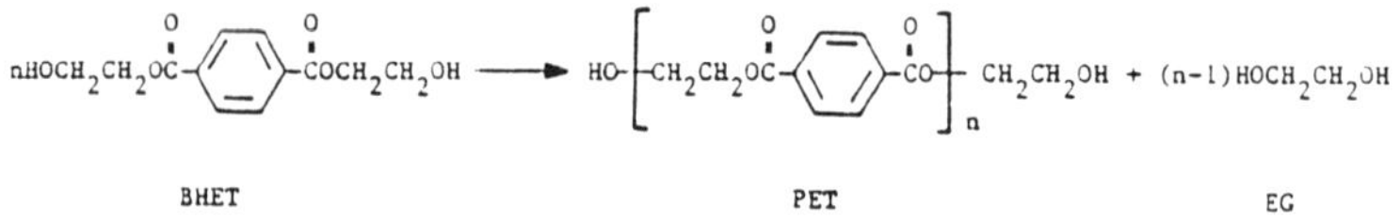

$$nHOCH_2CH_2OC\!\!\bigcirc\!\!COCH_2CH_2OH \longrightarrow HO\!\left[CH_2CH_2OC\!\!\bigcirc\!\!CO\right]_n\!CH_2CH_2OH + (n-1)HOCH_2CH_2OH$$

BHET   PET   EG

PBT Chemistry

In PBT resin production, 1,4-butanediol and DMT react to form the PBT chain and methanol. As with PET resin production, PBT esterification is also a two-step process, with bis(4-hydroxybutyl) terephthalate as the precursor. This ester exchange reaction is shown below.

$$nCH_3OOC-\!\!\bigcirc\!\!-COOCH_3 + 2nHOCH_2CH_2CH_2CH_2OH \xrightarrow{\text{Catalyst}} nHO(CH_2)_4-OC\!\!\bigcirc\!\!CO(CH_2)_4-OH + 2nCH_3OH$$

Bis(4-hydroxybutyl)terephthalate   Methanol

The condensation of the intermediate yields the polymer and 1,4-butanediol, as shown below.

$$HO-(CH_2)_4 \left(OOC-\!\!\!\bigcirc\!\!\!-COO(CH_2)_4 \right)_n OH + (n-1)HO-(CH_2)_4-OH$$

PBT   1,4-butanediol

Tables 113 and 114 list typical input materials and operating parameters for the PET/PBT process. Other input materials used in this process, including catalysts, fillers, and stabilizers or color improvers, are listed in Table 115.

PET Production

PET may be produced by either batch or continuous processes. As illustrated in Figure 35, ethylene glycol and a catalyst are mixed prior to being fed to an ester interchange reactor with either DMT or TPA. Typical process feeds using either DMT or TPA as monomer are as follows:[81]

Material	Parts by Weight
Dimethyl Terephthalate (DMT) (monomer)	1,460
Ethylene Glycol (monomer)	1,240
Catalyst	1
Additives	0.3-3
Terephthalic Acid (TPA) (monomer)	1,660
Ethylene Glycol (monomer)	1,240
Catalyst	1
Additives	0.3-3

The most frequently used ester-exchange catalyst is zinc acetate; however, numerous other catalysts for this polymerization process exist. Stabilizers and color improvers such as triaryl phosphites or phosphates are examples of additives used in the production of PET.

If DMT is used, methanol is separated overhead from the reflux stream in a distillation column. The bottoms stream, containing the partially reacted DMT/ethylene glycol mixture, is filtered before being fed to the polycondensation reactors. The excess ethylene glycol, added to facilitate methanol separation, is removed overhead and recycled to the mixing tank as the polymerization reaction proceeds under reduced pressures.

If pure TPA is used, water is separated overhead from the reflux stream in a distillation column. As depicted in Figure 36, the bottoms stream, which contains the partially esterified TPA/ethylene glycol mixture, is filtered before entering the polycondensation reactors. The excess ethylene

TABLE 113.   TYPICAL INPUT MATERIALS TO POLYETHYLENE TEREPHTHALATE AND POLYBUTYLENE TEREPHTHALATE PROCESSING

Resin	DMT	TPA	Ethylene Glycol	1,4-Butanediol	Catalyst
PET	X	X	X		X
PBT	X			X	X

TABLE 114.   TYPICAL OPERATING PARAMETERS FOR POLYETHYLENE TEREPHTHALATE AND POLYBUTYLENE TEREPHTHALATE PROCESSING

Resin	Ester-Interchange Reactor Temp.	Poly-condensation Reactor Temp.	Polycondensation Reaction Pressure
PET or PBT Batchwise	150 – 210°C	270 – 280°C	67.5 – 135.1 Pascals
Continuous PET	245°C	270°C[a]	2,030 – 3,380 Pascals[a]
		280 – 285°C[b]	67.5 – 135.1 Pascals[b]

[a]First stage.

[b]Second stage.

Source: Encyclopedia of Polymer Science and Technology.

TABLE 115. SPECIALTY CHEMICALS USED IN POLYETHYLENE TEREPHTHALATE
AND POLYBUTYLENE TEREPHTHALATE PRODUCTION

Function	Compound
Catalyst	antimony complexes
	antimony trioxide
	benzil
	benzophenone
	biacetyl
	calcium acetate
	calcium carbonate
	calcium oxide
	gallium compounds
	germanium compounds
	indium compounds
	lead oxide
	magnesium acetate
	magnesium carbonate
	magnesium oxide
	manganese acetate
	manganese carbonate
	manganese oxide
	tetramethylguanidine
	thallium compounds
	titanium complexes
	zinc acetate
	zinc carbonate
	zinc oxide
Stabilizer or Color Improver	triaryl phosphates
	triaryl phosphites
Deluster Agent	titanium dioxide
Thermal and Photostabilizer	phenolics
Carboxyl End Group Control	copper compounds
	diazomethane
	epoxides
	pyrocarbonates

Source: <u>Encyclopedia of Polymer Science and Technology</u>.

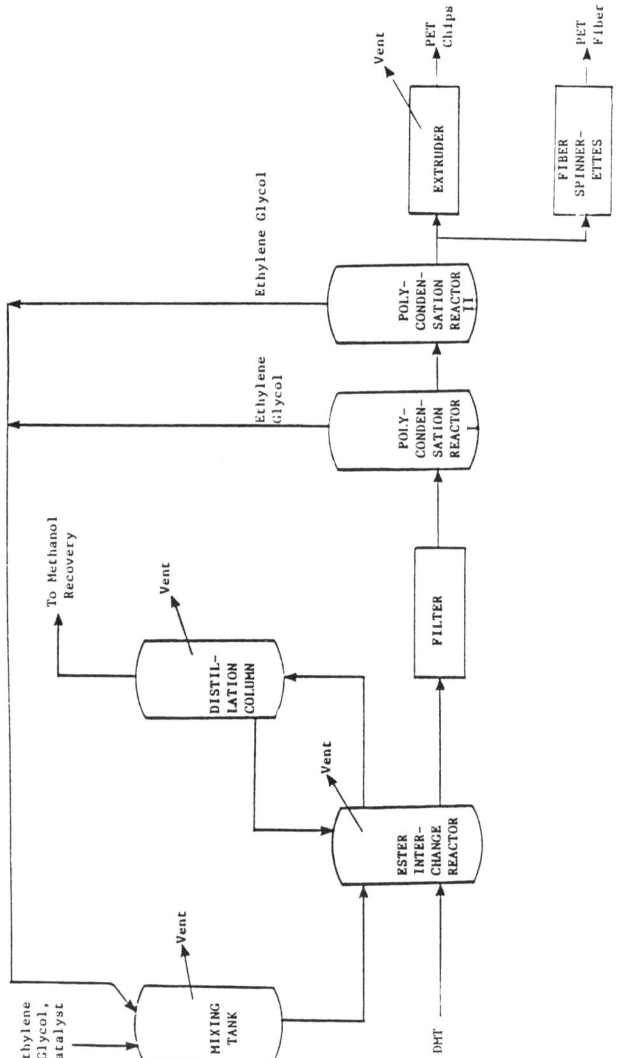

Figure 35.   PET production process using DMT.

Sources: Encyclopedia of Polymer Science and Technology.
Hydrocarbon Processing, November 1979 and 1981.

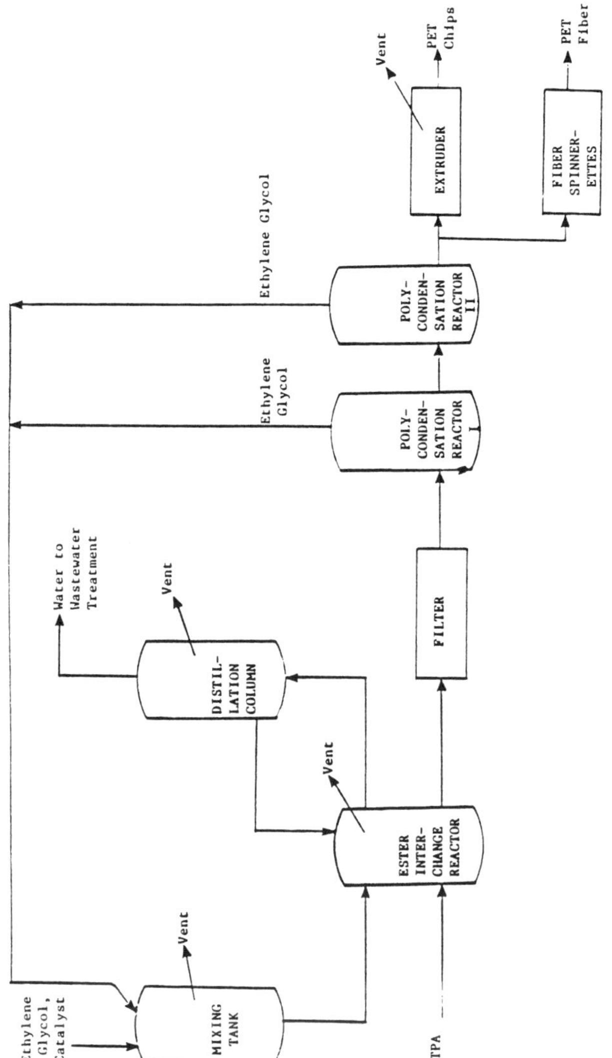

Figure 36. PET production process using TPA.

Sources: Encyclopedia of Polymer Science and Technology. Hydrocarbon Processing, November 1979 and 1981.

glycol used to promote water separation is removed overhead and recycled to the mixing tank. The polymerization reaction takes place under reduced pressures to aid in ethylene glycol removal and to shift the reaction equilibrium to the polymer side.

The PET resin is then either extruded or fed directly to fiber spinner-ettes when the reaction is completed. The excess ethylene glycol used speeds up the reaction rate. Residual acetaldehyde levels of less than 2.5 ppm are desired in bottle resins since acetaldehyde adversely effects beverage taste and resin color.[122]

PBT Production

PBT production is very similar to the process used to manufacture PET. As illustrated in Figure 37, 1,4-butanediol and a catalyst are combined in a mixing tank prior to entering the ester interchange reactor. DMT is then fed to the reactor to start the esterification. Methanol is separated from the overhead stream in a distillation column while the bottoms stream, containing the partially esterified DMT/1,4-butanediol mixture, is fed to the polycondensation reactors. Excess 1,4-butanediol, used to facilitate methanol separation, is removed and recycled to the mixing tank. The PBT mass is then extruded into PBT chips. The excess 1,4-butanediol used makes very high conversions possible for this process. As shown below, the recipe for PBT is very similar to that for PET [81]:

Material	Parts by Weight
Dimethyl Terephthalate (DMT) (monomer)	1,460
1,4-Butanediol (monomer)	1,800
Catalyst	1
Additives	0.3–3

Energy Requirements

No data listing the energy required for these processes were found in the literature consulted.

ENVIRONMENTAL AND INDUSTRIAL HEALTH CONSIDERATIONS

PET and PBT are considered to be nontoxic compounds. PET has been used as a food film wrap for over 20 years; therefore, any toxic effect would impact the use of this resin in the food and housewares industries. Similarly, PBT is used in many appliances and a toxic effect would also alter its usage.

Both PET and PBT processes achieve high conversion of raw materials due to the excess polyhydric alcohol used. This high conversion indicates that sources upstream of and including the ester interchange reactor contribute significantly more to the total VOC emissions than the sources downstream of the reactor. The polycondensation reactors are operated at reduced pressures, making them unlikely sources of VOC emissions.

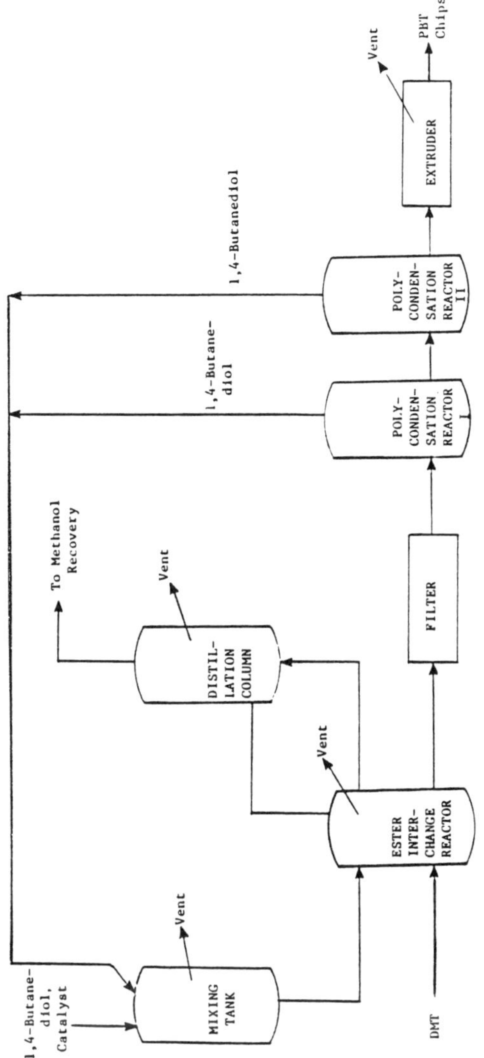

Figure 37.   PBT production process.

Source: Encyclopedia of Polymer Science and Technology.

Worker Distribution and Emissions Release Points

Worker distribution estimates for PET and PBT production will vary
because each can be made via either a continuous or batch process. We have
estimated worker distribution by correlating major equipment manhour
requirements with the process flow diagrams shown in Figures 35 through 37.
An estimate of 1.0 worker/unit/8-hour shift has been used for batch reactors
and mixers; a factor of 0.5 worker/unit/8-hour shift has been used for continuous reactors and mixers. Estimates are shown in Table 116.

Principal sources of employee exposure are the fugitive emission
sources listed in Table 119 and ether fugitive emissions from leaks in pump
and compressor seals, valves and drains. Health effects information for
process additives of particular concern due to their potential carcinogenicity or mutagenicity are discussed below. Available toxicity data for other
major input materials is summarized in Table 10.

Emission ranges from which potential for employee exposure to volatile
organic compounds estimated are shown in Tables 117 and 118 for PET production from TPA and DMT, respectively.

Health Effects

Relatively little health effects data exist for the input materials to
polyethylene terephthalate and polybutylene terephthalate production. Two
substances, however, have tested positive in animal tests for carcinogenicity and are therefore potential human carcinogens. They are antimony oxide,
a catalyst, and diazomethane, which is used for carboxy end group control.
The catalyst lead oxide is also a potential health risk due to its high
toxicity and tumorigenic and mutagenic potential. The reported health
effects of exposure to these three substances are summarized below.

__Antimony Oxide__ is a suspected carcinogen [5] which has also produced
positive results in tests for mutagenic effects.[155] Animal testing of
this substance is limited, however, and no epidemiological data exist.

__Diazomethane__ has produced positive results in animal tests for carcinogenicity.[101] It is also highly toxic by inhalation following acute or
chronic exposure.[242] Inhalation of diazomethane by man, depending on
degree of exposure, has caused chest pains, asthmatic symptoms, cough and
fever, fulminating pneumonia, moderate cyanosis, shock, and death.[101] The
OSHA air standard is 0.2 ppm (8 hour TWA).[67]

__Lead Oxide__ exhibits tumorigenic and mutagenic potential in animal
tests, but carcinogenesis tests have produced indefinite results.[99] In
addition, lead poisoning can occur through ingestion or inhalation of lead
oxide dust. The monoxide of lead is more toxic even than metallic lead or
other less soluble compounds.[88] The OSHA air standard is 50 ug (Pb)/m^3
(8 hour TWA).

TABLE 116. WORKER DISTRIBUTION ESTIMATES FOR SATURATED POLYESTER
RESIN PRODUCTION

Process	Unit	Workers/Unit/8-hour Shift
Polyethylene Terephthalate (DMT and TPA)	Batch or Continuous Reactor	1.0 or 0.5
	Batch or Continuous Mixer	1.0 or 0.5
	Distillation Column	0.25
	Vacuum Filter	0.25
	Extruder	1.0
	Spinerette	0.5
Polybutylene Terephthalate	Batch or Continuous Reactor	1.0 or 0.5
	Batch or Continuous Mixer	1.0 or 0.5
	Distillation Column	0.25
	Vacuum Filter	0.25
	Extruder	1.0

TABLE 117. CHARACTERISTICS OF VENT STREAMS FROM THE POLY(ETHYLENE TEREPHTHALATE) TPA PROCESS[a]

Process Section[b]	Stream	Name	Nature	Emission rate, kg VOC/Mg product	Temperature °C	Pressure psig	Composition, Wt.%
RMP	Methanol recovery	Esterifier Vent[c]	Continuous	0.04	38	Atm.	VOC N_2 H_2O
PR	Ethylene glycol removal	Polymerizer Reactors Cooling Water Tower	Continuous		25	0.2	VOC Air H_2O
		– with spent ethylene glycol spray condenser		0.315[d]			
		– without spent ethylene glycol spray condenser		9.5[e]			
		Total Emission Rate		----			
		– with EG spray condenser		0.355			
		– without EG spray condenser		9.5			

[a]Source of information: Industry correspondence.

[b]RMP = raw material preparation; PR = polymerization reaction.

[c]The potential emissions from this stream are sent to reactor cooling water tower when the spray condenser is not used.

[d]Based upon 0.02 kg VOC/Mg product from recovery equipment and 0.295 kg VOC/Mg product from cooling water tower.

[e]Includes potential emissions from Stream A. The actual emission rate will vary depending upon the number of polymerization reactors per line at a plant.

Source: Polymer Manufacturing Industry – Background Information for Proposed Standards, September 1983.

TABLE 118. CHARACTERISTICS OF VENT STREAMS FROM THE POLY(ETHYLENE TEREPHTHALATE) DMT PROCESS[a]

Process Section[b]	Stream	Name	Nature	Emission rate, kg VOC/Mg product	Temperature °C	Pressure psig	Composition, Wt.%
MR	Methanol recovery	Esterifier Vent (Methanol recovery vent)	Continuous	0.15	38	0	49 Methanol 51 Nitrogen
PR	Ethylene glycol removal	Polymerizer Reactors Cooling Water Tower	Continuous		25	0.2	VOC Air H_2O
		– with spent ethylene glycol spray condenser		0.355[c]			
		– without spent EG spray condenser		9.5[d]			
		Total Emission Rate		----			
		– with EG spray condenser		0.505			
		– without EG spray condenser		9.65			

[a]Source of information: Industry correspondence.
[b]MR = material recovery; PR = polymerization reaction.
[c]Assumed same as for the total TPA emission rate with EG spray condenser.
[d]Estimated to be the same as for the TPA process. All emissions from cooling water tower. The actual emission rate will vary depending upon the number of polymerization reactors per line at a plant.

Source: _Polymer Manufacturing Industry Background Information for Proposed Standards_, September 1983.

TABLE 119. SOURCES OF FUGITIVE EMISSIONS FROM POLYETHYLENE TEREPHTHALATE AND POLYBUTYLENE TEREPHTHALATE MANUFACTURE

Source	Constituent	Process		
		PET w/DMT	PET w/TPA	PBT
Mixing Tank Vent	Ethylene Glycol	X	X	
	1,4-Butanediol			X
Reactor Vent	Ethylene Glycol	X	X	
	1,4-Butanediol			X
	DMT	X		X
	TPA		X	
	Methanol	X		X
	Water		X	
Distillation Column Vent	Methanol	X		X
	Water		X	
	Ethylene Glycol	*	*	
	1,4-Butanediol			*
Extruder Vent	PET Emissions and Particulates	X	X	X

*Trace amounts of these compounds are expected due to the high conversion achieved in the ester interchange reactor.

Air Emissions

The sources of VOC fugitive emissions from these processes are listed in Table 119 according to process and constituent type. Controls for the process emissions, which are all vents, include routing the vented stream to a flare to incinerate the remaining hydrocarbons and routing the vented stream to blowdown.[285] Fugitive VOC emissions from leaks resulting from vents, flanges, open drains, pumps, and compressors may be controlled by equipment modification. However, the most effective control may be a regular inspection and maintenance program.

Sources of fugitive particulates are also listed in Table 119. These particulates may be collected by venting the stream to a baghouse or electrostatic precipitator.

Wastewater Sources

There are only two wastewater sources associated with PET production: water generated during the condensation reaction between TPA and ethylene glycol; and routine cleaning water. The TPA/ethylene glycol route generates water during reaction which, in turn, contributes to the wastewater load. However, this process also has reduced VOC emissions since the condensation by-product is water, not methanol.

PBT production has only one wastewater stream: routine cleaning water. No quantitative data are available on the wastewater streams generated by these processes.

Solid Wastes

The solid wastes generated by these processes are mostly PET and PBT. Solid waste streams include substandard resin which cannot be blended, resin removed during reactor cleaning, and resin lost during product blending or due to spillage. If the ethylene glycol is purified before recycle to the mixing tank, a solid waste stream consisting of solid ethylene glycol and polyethylene glycol is generated.[93]

Environmental Regulation

Effluent limitations guidelines have been set for the PET/PBT industry. BPT, BAT, and NSPS call for the pH of the effluent to fall between 6.0 and 9.0 (41 Federal Register 32587, 4 August 1976).

New source performance standards (proposed by EPA on 5 January 1981) for volatile organic carbon (VOC) fugitive emissions include:

- Safety/release valves must not release more than 200 ppm above background, except in emergency pressure releases, which should not last more than five days; and

- Leaks (which are defined as VOC emissions greater than 10,000 ppm) must be repaired within 15 days.

     The following compounds have been listed as hazardous (46 Federal
Register 27476, 20 May 1981):

     lead acetate - U144
     methanol   - U154

All disposal of these compounds or PET and PBT which contain any residual
amounts of these compounds must comply with the provisions set forth in the
Resource Conservation and Recovery Act (RCRA).

16. Polyester Resins (Unsaturated)

Unsaturated polyester resins are produced from the condensation reaction of glycols and dibasic acids, one component of which is unsaturated. These resins exhibit varying degrees of hardness, flexibility, and flexural strength based on the ratio of unsaturated to saturated acid, the glycol used, the crosslinking agent employed, and the concentration and type of acid used.

Unsaturated polyesters are used in a wide variety of applications including: marine products; automobile, truck, and bus parts; bathroom components and fixtures, general purpose tanks and pipes and corrosion resistant tanks and pipes; electrical generation, transmission, and distribution parts; consumer goods, such as major appliances and parts, recreational goods, and toys; gel coats; furniture and cultured marble castings; auto body coatings; and bonding and adhesive resins.[65]

The physical and chemical properties of unsaturated polyester resins are partially dependent upon the crosslinking agent used. Table A-18 in Appendix A lists typical crosslinking agents and the characteristics imparted. Table A-19 in Appendix A presents typical properties for unsaturated polyester resins.

One process is used to produce unsaturated polyester resins. The sequence used for the addition of the raw materials to the reactor determines the molecular structure of the resin. Unsaturation may be either randomly distributed along the polymer chain or concentrated in one area, for example at either the middle or the ends of the chain.[164]

INDUSTRY DESCRIPTION

The unsaturated polyester resin industry is comprised of 26 producers. These producers operate at 63 sites in 27 states. Over 20 pecent of the sites are located in California. The sites located in Ohio, Illinois, Indiana, and Pennsylvania contribute almost 30 percent of the total. The remaining plants are distributed through the rest of the states. Table 120 lists the producers of unsaturated polyester resins and plant capacities.

Unsaturated polyester resins are used in appliances, business equipment, construction, consumer goods, corrosion-resistant products,

TABLE 120.   U.S. PRODUCERS OF UNSATURATED POLYESTER RESINS

Producer	Location	Capacity as of July 1, 1980 in Thousand Metric Tons
The Alpha Corporation Alpha Resins Division	Collierville, TN Kathleen, FL Perris, CA	45
American Cyanamid Co. Industrial Chemicals Division	Wallingford, CT	*
Ashland Oil, Inc. Ashland Chemical Co., subsidiary Polyester Division	Ashtabula, OH Calumet City, IL Los Angeles, CA Philadelphia, PA	64
AZS Corporation AZ Products, Inc. Division Lancaster Chemical Corporation Division	Eaton Park, FL Newark, NJ	5
Barton Chemical Corporation	Chicago, IL	*
Beatrice Foods Co. Beatrice Chemical Division Farboil Co. Division	Baltimore, MD	*
Cargill, Inc. Chemical Products Division	Carpentersville, IL Forest Park, GA Lynwood, CA	23
Cook Industrial Coatings Co. Cook Paint and Varnish Co.	Detroit, MI Atlanta, GA Bethlehem, PA Miami, FL Milpitas, CA North Kansas City, MO	14

*All of these sources total 9,000 metric tons per year.

(continued)

TABLE 120 (continued)

Producer	Location	Capacity as of July 1, 1980 in Thousand Metric Tons
Emhart Corporation USM Corporation, subsidiary Bostik Division Eastern Region	Middleton, MA	*
The P. D. George Co.	St. Louis, MO	*
Hugh J.-Resins Co.	Long Beach, CA	*
ICI Americas Inc. Specialty Chemicals Division	New Castle, DE	11
Koppers Co., Inc. Organic Materials Group	Bridgeville, PA Cicero, IL Redwood City, CA Richmond, CA	59
Mobil Corporation Mobil Oil Corporation Mobil Chemical Co. Division Chemical Coatings Division	Rochester, PA	*
The O'Brien Corporation The O'Brien Corporation - Central Region	South Bend, IN	*
Owens-Corning Fiberglas Corporation Resins and Coatings Division	Anderson, SC Valparaiso, IN	54
Phillips Petroleum Co. Interplastic Corporation, subsidiary Commercial Resins Division	Jackson, MS Minneapolis, MN Pryor, OK	45

*All of these sources total 9,000 metric tons per year.

(continued)

TABLE 120 (continued)

Producer	Location	Capacity as of July 1, 1980 in Thousand Metric Tons
PPG Industries, Inc. Coatings and Resins Division	Circleville, OH Springdale, PA Torrance, CA	27
Reichhold Chemicals, Inc.	Azusa, CA Detroit, MI Elizabeth, NJ Houston, TX Jacksonville, FL Morris, IL South San Francisco, CA Tacoma, WA	159
H. H. Robertson Co. Freeman Chemical Corporation, subsidiary	Chatham, VA Marshall, TX Saukville, WI	45
Schenectady Chemicals Inc.	Schenectady, NY	*
SCM Corporation Glidden Coatings & Resins Division	Huron, OH Reading, PA San Francisco, CA	9
The Sherwin-Williams Co. Coatings Group	Oakland, CA	9
The Standard Oil Co. (Ohio) Vistron Corporation, subsidiary Chemicals Department Filon/Silmar Department	Covington, KY Hawthorne, CA	34

*All of these sources total 9,000 metric tons per year.

(continued)

TABLE 120 (continued)

Producer	Location	Capacity as of July 1, 1980 in Thousand Metric Tons
United States Steel Corporation U.S.S. Chemicals Division		136
	Bartow, FL	
	Colton, CA	
	Jacksonville, AR	
	Linden, NJ	
	Neville Island, PA	
	Swanton, OH	
TOTAL		748

*All of these sources total 9,000 metric tons per year.

Sources: Chemical Economics Handbook, updated annually, September 1980 Data.
Directory of Chemical Producers, 1982.

electrical equipment, marine products, and the transportation industry, which is the largest single category.[45] In 1981, unsaturated polyester resin production totaled 452,000 metric tons.[143] Since 1976, the sales of unsaturated polyester resins have risen from 436,000 metric tons to a peak of 545,000 metric tons in 1978. Sales declined in 1979 and 1980, but the 1981 production level was 21,000 metric tons above that of 1980. Present excess production capacity of 296,000 metric tons is sufficient to sustain growth in the near future.

PRODUCTION AND END USE DATA

In 1981, unsaturated polyester resin production totaled 452,000 metric tons.[143] These resins have many uses, including:[15, 143, 164]

- Coatings for wire insulation, cast products, and nonshiny furniture finishes;

- Coatings used to precoat molds or cured moldings;

- Parts for commercial and pleasure boats;

- Construction industry uses, including tub and shower stalls, sanitary ware, and cultured marble lavatories;

- Corrosion-resistant materials, such as pipe and filler elements for cooling towers;

- Molded electrical appliances, hand tools, and motor housings;

- Filament wound railway tank cars with a capacity of over 22,000 gallons, gasoline storage tanks, pipes for the oil industry, and rocket casings for military use;

- Fishing poles, golf clubs, and I-beams and channels for construction; and

- Automobile bodies, truck cabs, headlamp mountings, fender extensions, window frames, and hood scoops as well as recreational vehicle, airplane, camper, and trailer parts.

Table 121 lists unsaturated polyester resin uses while Table 122 presents markets for reinforced polyester resins.

PROCESS DESCRIPTION

Unsaturated polyester resins are manufactured from glycols and dibasic acids or anhydrides, with agents added to provide crosslinking. These resins are used either with or without reinforcing fillers in the areas of transportation, marine applications, construction, and electrical equipment. Only one process is used to manufacture unsaturated polyester resins. Various resin properties are achieved by changing the input materials to this process.

TABLE 121. 1981 CONSUMPTION OF UNSATURATED POLYESTER RESINS

Market	Thousand Metric Tons
Reinforced Polyester[a]	
Sheet, Flat and Corrugated	60
All Other	285
Surface Coatings	7
Export	5
Other (including cultured marble, body putty, furniture applications, buttons, etc.)	95
TOTAL	452

[a]Resin only.

Source: <u>Modern Plastics</u>, January 1982.

TABLE 122.   1981 CONSUMPTION OF REINFORCED UNSATURATED POLYESTER RESINS[a]

Market	Thousand Metric Tons
Aircraft/Aerospace (includes all military, commercial, and private aircraft)	12
Appliance/Business Equipment	38
Construction	138
Consumer (includes motor homes and recreational vehicles)	50
Corrosion-Resistant Products	120
Electrical	74
Marine	132
Transportation (includes passenger cars, trucks, buses, mass transit, etc.)	139
Other	37
TOTAL	740

[a]Tons include all reinforcements, fillers, etc.

Source: _Modern Plastics_, January 1982.

A typical recipe for unsaturated polyester production follows:[119]

Material	Parts by Weight
Phthalic Anhydride (monomer)	43
Maleic Anhydride (monomer)	19
Propylene Glycol (monomer)	38
Styrene (cross-linking agent)	70

  Unsaturated polyester resins are produced by first opening the dibasic acid in the presence of a catalyst. The newly generated polyester radical then reacts with other dibasic acid and glycol molecules to form the polyester chain. The crosslinking agent (M) is added to crosslink the chains. This series of reactions is shown below:

Initiation

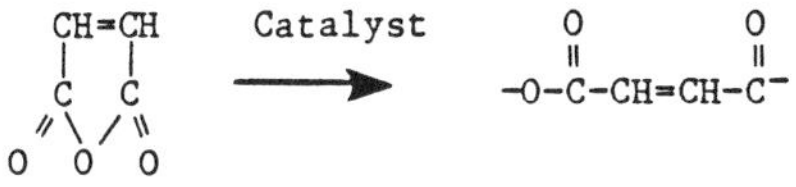

$$\underset{\underset{O}{\overset{C}{\diagdown}}\ \underset{O}{\diagup}\ \underset{O}{\overset{C}{\diagup\diagdown}}}{\overset{CH=CH}{\mid\quad\mid}} \qquad \xrightarrow{\text{Catalyst}} \qquad -O-\overset{O}{\overset{\|}{C}}-CH=CH-\overset{O}{\overset{\|}{C}}{}^-$$

Propagation

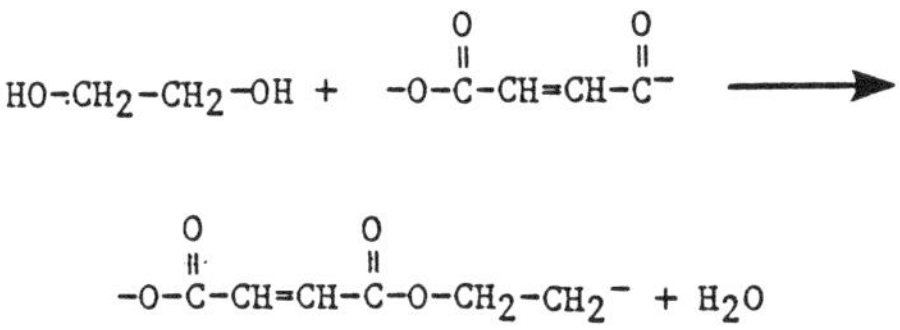

$$HO-CH_2-CH_2-OH \; + \; -O-\overset{O}{\overset{\|}{C}}-CH=CH-\overset{O}{\overset{\|}{C}}{}^- \longrightarrow$$

$$-O-\overset{O}{\overset{\|}{C}}-CH=CH-\overset{O}{\overset{\|}{C}}-O-CH_2-CH_2{}^- \; + \; H_2O$$

Crosslinking

$$\begin{array}{c}
\mid \\
-\!\!\!\!-\overset{O}{\overset{\|}{C}}-\overset{H}{\overset{\mid}{C}}-\overset{M}{\overset{\mid}{C}}-\overset{O}{\overset{\|}{C}}-O-\!\!\!\!- \\
\overset{\mid}{M}\ \ \overset{\mid}{R}
\end{array}$$

$$-R-O-\overset{O}{\overset{\|}{C}}-R'-\overset{O}{\overset{\|}{C}}-O-R-O-\overset{O}{\overset{\|}{C}}-\overset{H}{\overset{\mid}{C}}-\overset{\mid}{\underset{\underset{\mid}{M}\ \overset{\mid}{H}}{C}}-\overset{O}{\overset{\|}{C}}-O-$$

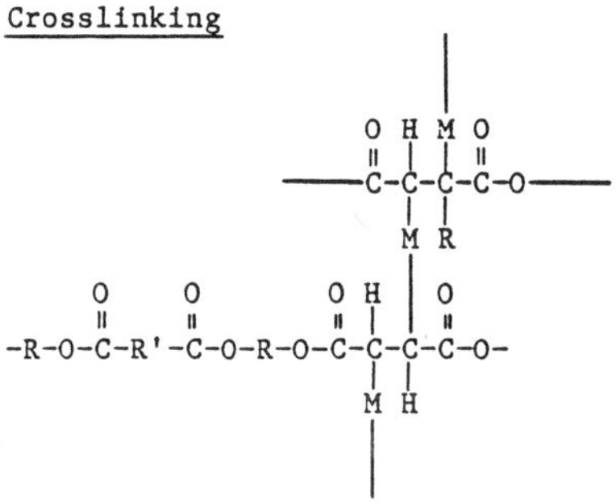

Tables 123 and 124 list typical input materials and operating parameters for unsaturated polyester resins.

TABLE 123.   INPUT MATERIALS AND SPECIALTY CHEMICALS USED IN
UNSATURATED POLYESTER RESIN PRODUCTION

Function	Chemical
Dibasic Acids or Anhydrides	adipic acid
	azelaic acid
	fumaric acid
	glutaric acid
	isophthalic acid
	isosebacic acid
	maleic anhydride
	α-methyl adipic acid
	phthalic anhydride
	pimelic acid
	sebacic acid
	succinic acid
	terephthalic acid
Glycols	Bisphenol-A
	1,4-butylene glycol
	2,3-butylene glycol
	cyclohexanediol
	diethylene glycol
	dipropylene glycol
	ethylene glycol
	propylene glycol
	triethylene glycol
Crosslinking Agents	acrylamide
	t-butyl styrene
	chlorostyrene
	diallyl fumarate
	diallyl phthalate
	divinyl benzene
	methyl acrylate
	methyl methacrylate
	α-methyl styrene
	styrene
	n-tert butyl acrylamide
	1,3,5-triacrylyl-hexahydro-s-triazine
	triallyl cyanurate
	triallyl(iso)cyanurate
	vinyltoluene

(continued)

TABLE 123 (continued)

Function	Chemical
Viscosity Aids	pyrogenic silica
	2-vinyl pyridine
Polymerization Inhibitors	benzaldehyde
	p-benzoquinone
	choranil
	copper
	dihydroxydiphenyl naphthaquinone
	2,5-diphenyl-p-benzoquinone
	ethylenebis(pyridinium chloride)
	hydroquinone
	hydroquinone monomethyl ether
	mono-t-butyl-hydroquinone
	nitrites
	oxygen
	phenyl trimethyl ammonium acetate
	phenyl trimethyl ammonium phosphate
	picric acid
	primary and secondary amines
	pyrogallol
	sulfur
	p-tert-butylcatechol
	toluhydroquinone
	toluquinone
	trimethylbenzyl ammonium acetate
	trimethylbenzyl ammonium bromide
	trimethylbenzyl ammonium chloride
Cure Agents	2-2'-azobisisobutyronitrile
	2-t-butylazo-2-cyano-4-methyl-pentane
	2-t-butylazo-2,4-dimethoxy-4-methylpentane
	2-(t-butylazo)isobutyronitrile
	1-cyano-1-(t-butylazo)cyclohexane

(continued)

TABLE 123 (continued)

Function	Chemical
Accelerators and Activators	aluminum compounds
	t-butyl peroxyacetate
	cobalt naphthenate
	cobalt octoate
	cobalt salts
	copper compounds
	cyclohexanone peroxide
	N,N-dimethylaniline
	2,5-dimethylhexyl-2,5-diperoxyocto-ate
	N,N-dimethyl-p-toluidine
	2,5-diperoxybenzoate
	iron compounds
	lauryl mercaptan
	manganese compounds
	sulfur compounds
	vanadyl acetylacetone with ketone peroxides or t-butyl hydroperoxide
Catalysts	acetal peroxide
	benzoyl peroxide
	bis(p-bromobenzoyl)peroxide
	t-butyl peroxybenzoate
	cumene hydroperoxide
	cyclohexanone peroxide
	2,4-dichlorobenzoyl peroxide
	diisobutylene ozonide
	diisopropylene ozonide
	2,5-dimethylhexane
	2,5-diperoxyoctoate
	lauryl(dodecyl)mercaptan
	lauroyl peroxide
	methyl ethyl ketone peroxide
	stannous oxalate in conjunction with sodium acetate or zinc acetate
	sulfuric acid
	tetrabutyl titanate
	tetrabutyl zirconate
	tetraoctyl titanate
	tetraoctyl zirconate
	p-toluenesulfonic acid

(continued)

TABLE 123 (continued)

Function	Chemical
Catalysts (continued)	t-butyl benzoyl peroxide
	t-butyl perbenzoate
	cetyl peroxide
	peroxycarbonates
Color Improvers	titanium dioxide
	triaryl phosphates
	triaryl phosphites
Flame Retardants	alkyl phosphates
	alkyl phosphites
	antimony oxide
	Chloran TM®
	(2,3-dicarboxy-5,8-endomethylene- 5,6,7,8,9,9-hexachloro-1,2,3,4, 4a,5,8,8a-octahydronaphthalene anhydride)
	chlorendic anhydride (2,3-dicarboxy- 1,4,5,6,7,7-hexachlorobicyclo [2.2.1]-5-heptene anhydride)
	dibromoneopentylene glycol
	dibromostyrene
	dichlorostyene
	halogen containing compounds chlorinated paraffin
	hydrated aluminum oxide
	monochlorostyrene
	tetrabromophthalic anhydride
	tetrachlorophthalic anhydride
	trichloroethyl phosphate
	triethyl phosphate
	vinyl phosphorus compounds
	zinc borate

Sources: Encyclopedia of Chemical Technology, 2nd Edition.
          Encyclopedia of Polymer Science and Technology.
          Modern Plastics Encyclopedia, 198-1982.

TABLE 124.   TYPICAL OPERATING PARAMETERS FOR UNSATURATED POLYESTER
RESIN PRODUCTION

	Temperature	Reaction Time
Polymerization	180–220°C	6–20 hours
Crosslinking Agent Addition	100–150°C	----
Polymerization of Anhydrides and Epoxides	95–125°C	3–8 hours
Post-Polymerization Heating for Resins from Epoxides	190–230°C	2–5 hours

Source: Encyclopedia of Chemical Technology, 2nd Edition.

As illustrated in Figure 38, the first step in the production of unsaturated polyester resins is polymerization. A catalyst, glycol, and dibasic acid or anhydride are fed to the reactor along with an inert gas. The inert gas provides an oxygen free atmosphere for the reaction since oxygen is an inhibitor. During polymerization, the water produced is removed overhead using a packed distillation column.[50] The unreacted glycol (the bottoms stream) is recycled to the reactor. When the reaction reaches the desired end point, the resin mixture is sent to a mixing tank where the polymer is dissolved in the crosslinking agent.[45] Other additives, including polymerization inhibitors, flame retardants, and pigments, may be added to the mixing tank. The formulated resin is then ready for shipping or storage. Cure catalysts are added just prior to the final use.

One method used to eliminate removing the water generated by the condensation of a glycol with a dibasic acid is to substitute an epoxide for the glycol. In addition to eliminating water removal, this combination of raw materials requires shorter reaction times and lower reaction temperatures than other unsaturated polyester resins; however, post-polymerization heating is required for this process to achieve sufficient crosslinking. [164]

<u>Energy Requirements</u>

No data listing the required energy for unsaturated polyester resin production were found in the literature consulted.

ENVIRONMENTAL AND INDUSTRIAL HEALTH CONSIDERATIONS

Unsaturated polyester resins are nontoxic compounds used in the construction, transportation, and consumer goods industries. Several of the chemicals which have been identified as inputs to unsaturated polyester resin processing have been listed as hazardous under RCRA: acrylamide, maleic anhydride, methyl ethyl ketone, methyl ethyl ketone peroxide, methyl methacrylate, and phthalic anhydride. Therefore, special handling of unsaturated polyester wastes which contain these compounds is necessary.

<u>Worker Distribution and Emissions Release Points</u>

Worker distribution estimates, made by correlating major equipment manhour requirements with the process flow diagram in Figure 38, are shown in Table 125.

No major air point source emissions are associated with unsaturated polyester resin production. However, fugitive and process emissions may contribute significantly to environmental and/or worker health problems. The magnitude of the impact is a function of stream constituents, process operating parameters, engineering and administrative controls, and maintenance programs.

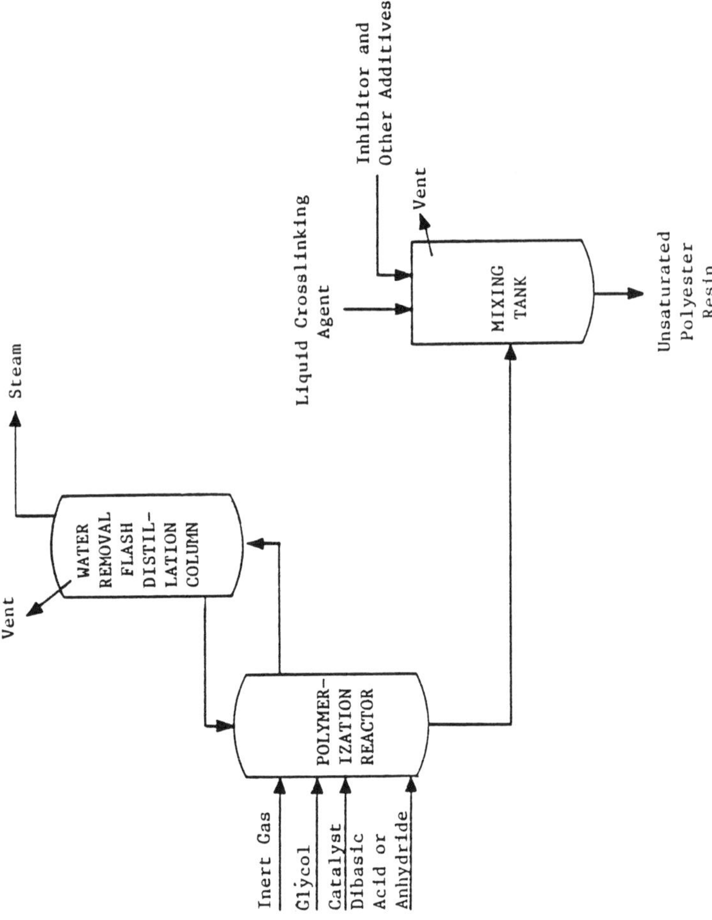

Figure 38. Unsaturated polyester resin production process.

Sources: Encyclopedia of Chemical Technology, 2nd Edition.
Cystic Research Center, Cystic Monograph No. 2, Polyester Handbook, 1977.

TABLE 125.   WORKER DISTRIBUTION ESTIMATES FOR UNSATURATED POLYESTER
RESIN PRODUCTION

Unit	Workers/Unit/8-hour Shift
Batch Reactor	1.0
Batch Mixer	1.0
Distillation Column	0.25

Table 126 shows process sources of fugitive emissions from unsaturated polyester resin manufacture. Employees also may be exposed to contaminants, fugitive emissions from pump and compressor seals, valves and drains.

The diversity of input materials used in unsaturated polyester production makes a complete discussion of exposure effects beyond the scope of this document. However, health effects of several compounds of concern are discussed in the paragraphs which follow. Additinal toxicity data for major input materials are given in Table 10.

Little data are available in the literature from which employee exposure potential may be estimated. However, one source gives an emission factor of 20 g/kg product for hydrocarbons generated in molding operations. [286]

Health Effects

Health effects data have not been published in the available literature for almost half of the input materials and specialty chemicals used in unsaturated polyester production processing. Although the industry uses many chemicals that have exhibited tumorigenic potential, only one, the flame retardant antimony oxide, has been classified as a suspected carcinogen. Several input materials are currently being tested for carcinogenesis by the National Toxicology Program.

Other chemicals are considered significant health risks due to high toxicity, mutagenic potential, and/or teratogenic potential. These include: maleic anhydride, a dibasic anhydride; three crosslinking agents--styrene, methyl acrylate, and acrylamide; four polymerization inhibitors--hydroquinone, benzoquinone, pyrogallol, and picric acid; methyl ethyl ketone peroxide and cumene hydroperoxide, which are used both as cure agents and catalysts; and N,N-dimethylaniline, an accelerator and activator.

The paragraphs which follow provide a brief synopsis of the reported health effects of exposure to these substances.

Acrylamide is a neurotoxin, causing symptoms of ataxia, hypersomnia, and vertigo.[170] Although the polymer is nontoxic, absorption of acrylamide through skin or dusts is associated with serious neuorological consequences.[87] Animal test data indicate that exposure to acylamide might also cause mutagenic and teratogenic effects. The OSHA air standard is 30 ug/m^3 (8 hour TWA).[67]

Antimony Oxide is a suspected carcinogen [5] which has also produced positive results in tests for mutagenic effects.[155] Animal testing of this substance is limited, however, and no epidemiological data exist.

Cumene Hydroperoxide is highly toxic after short exposures involving ingestion, inhalation, or skin absorption of relatively small quantities. [242] Animal test data indicate that the substance is also a tumorigenc and

TABLE 126. SOURCES OF FUGITIVE EMISSIONS FROM UNSATURATED
POLYESTER RESIN MANUFACTURE

Source	Constituent
Condenser Vent	Glycol Dibasic acid or anhydride
Mixing Tank Vent	Crosslinking agent Glycol Dibasic acid or anhydride

a mutagenic agent. Prolonged inhalation of vapors results in headache and
throat irritation, prolonged skin contact with contaminated clothing may
cause irritation and blistering.[278]

Hydroquinone has exhibited potential as a tumorigenic, mutagenic, and
teratogenic agent in animal tests.[233] Carcinogenesis tests have produced
indefinite results, but the substance is currently being tested by the
National Toxicology Program for carcinogenicity.[233] Hydroquinone is also
highly toxic. Fatal human doses have ranged from 5 to 12 grams, although
300 to 500 milligrams have been ingested daily for 3 to 5 months without ill
effects.[85] The OSHA air standard is 2 mg/m^3 (8 hour TWA).[67]

Maleic Anhydride is highly toxic by ingestion and inhalation [242] and
is a powerful irritant to the skin and eyes.[149] Inhalation can cause
pulmonary edema [149]; other effects of exposure include conjunctivitis,
corneal damage, cough, bronchitis, headache, abdominal pain, nausea, and
vomiting.[238] Animal data also suggest that maleic anhydride is a tumorigenic agent.[233] The OSHA air standard is 0.25 ppm (8 hour TWA).[67]

Methyl Acrylate is highly toxic by ingestion, inhalation, and skin
absorption upon acute exposure to relatively low doses.[242] The monomer is
very irritating to eyes, skin, and mucous membranes; lethargy and convulsions may occur if the vapor is inhaled in high concentrations.[149] Toxic
effects have been observed in humans at a concentration of 75 ppm.[52]
Methyl acrylate also produces tumors in laboratory animals, but sufficient
data do not exist to make a carcinogenic determination.[107] The OSHA air
standard is 10 ppm (8 hour TWA).[67]

Methyl Ethyl Ketone Peroxide has produced toxic effects in laboratory
animals by ingestion and inhalation of relatively low doses.[233] An oral
dose of 480 mg/kg has affected the gastrointestional tract in humans.[156]
There is some evidence of tumorigenic potential in the available literature
and the substance is currently being tested for carcinogenesis by the
National Toxicology Program.[233]

N,N-Dimethylaniline is highly toxic by ingestion, inhalation, and skin
absorption following acute exposure.[242] It can be absorbed through intact
skin to produce dangerous methemoglobinemia [85], and has been lethal to
humans by ingestion at a dose of 50 mg/kg.[156] N,N-Dimethylamiline has
exhibited tumorigenic potential in animal testing and is currently being
tested for carcinogensis by the National Toxicology Program.[233] OSHA has
set an air standard of 5 ppm (8 hour TWA).[67]

p-Benzoquinone is a tumorigenic and a mutagenic agent, but testing has
produced indefinite results for carcinogenicity.[105] The substance is
currently undergoing additional tests for carcinogenesis by standard
bioassay protocol under the sponsorship of the National Toxicology Program.
Benzoquinone is very toxic; the probable lethal oral dose for humans is 50
to 500 mg/kg.[85] The OSHA air standard is 0.01 ppm (8 hour TWA).[67]

Picric Acid is extremely toxic; the lethal dose for humans is less than
5 mg/kg.[85] Toxic effects, including headache, vertigo, nausea, vomiting
and diarrhea, yellow coloration of the skin and conjunctiva, gastroenteri-
tis, hemorrhagic nephritis, and acute hepatitis [5], may result from skin
contact inhalation, or ingestion of the dust of picric acid or its salts.
[278] Bacteria test data also provide evidence of mutagenic potential.[233]
The OSHA standard in air is 100 ug/m^3 (8 hour TWA).[67]

Pyrogallol is highly toxic by ingestion, inhalation, and skin absorption upon either acute or chronic exposure.[242] Because of its marked
reducing action, pyrogallol has tremendous affinity for the oxygen in the
blood; most of its toxic effects can be traced directly to its potent reducing action.[170] Toxic action invovles methemolobinemia, hemolysis, and
renal injury.[85] Delayed deaths from uremia have been reported.[85]
Pyrogallol has also produced tumorigenic and mutagenic effects in laboratory
animals.[233]

Styrene has been linked with increased rates of chromosomal aberrations
in persons exposed in an occupational setting.[107] Animal test data
strongly support epidemiological evidence of its mutagenic potential.[233]
Styrene also produces tumors and affects reproductive fertility in laboratory animals.[233] Toxic effects of exposure to styrene usually involve the
central nervous system.[233] The OSHA air standard is 100 ppm (8 hour TWA)
with a ceiling of 200 ppm.[67]

Air Emissions

Sources of fugitive VOC emissions are listed in Table 126. These two
sources may be controlled by venting to a flare or blowdown for hydrocarbon
removal.[285] Leaks from process equipment and other fugitive emissions
sources may be best controlled by a routine inspection and maintenance
program.

Wastewater Sources

There are only two sources of wastewater associated with unsaturated
polyester resin production: water from the condensation reaction and routine cleaning water. The water generated from the condensation reaction is
the major wastewater stream. Routine cleaning water contributes a smaller
amount to the wastewater produced. If the glycol is replaced by an epoxide,
only routine cleaning water is generated.

Ranges of several parameters for wastewaters from unsaturated polyester
resin processing are shown below. Values for the wastewater from the process described were not distinguished by polyester type by EPA for the purpose of establishing effluent limitations for the polyester resin industry.
[284]

Polyester Resin Wastewater Characteristics	Unit/Metric Ton of Unsaturated Polyester Resin
Production	$4.5\ m^3$
BOD_5	3 – 20 kg
COD	6 – 45 kg
TSS	0 – 12 kg

Solid Wastes

The solid wastes generated by this process include substandard product which cannot be blended and resin lost during routine cleaning and spillage. These wastes are comprised mainly of polyester resins.

Environmental Regulation

Effluent limitations guidelines have been set for the polyester resin industry. BPT, BAT, and NSPS call for the pH of the effluent to fall between 6.0 and 9.0 (41 Federal Register 32587, 4 August 1976).

New source performance standards (proposed by EPA on 5 January 1981) for volatile organic carbon (VOC) fugitive emissions include:

- Safety/release valves must not release more than 200 ppm above background, except in emergency pressure releases, which should not last more than five days; and

- Leaks which are defined as VOC emissions (greater than 10,000 ppm) must be repaired within 15 days.

The following compounds have been listed as hazardous wastes (46 Federal Register 27476, 20 May 1981):

Acrylamide	– U007
Maleic anhydride	– U147
Methyl ethyl ketone	– U159
Methyl ethyl ketone peroxide	– U160
Methyl methacrylate	– U162
Phthalic anhydride	– U190

All disposal of these compounds and unsaturated polyester resins which contain residuals of these compounds (i.e., uncured resins) must comply with the provisions set forth in the Resource Conservation and Recovery Act (RCRA).

17. Polyethylene—High Density (HDPE)

INTRODUCTION

Polyethylene is a lightweight, flexible, tough, chemical resistant polymer which exhibits outstanding electrical insulation properties. These characteristics combined with ease of fabrication and low production cost make polyethylene suitable for a wide variety of end products used in the packaging, housewares, construction, communications, and medical industries.

The empirical formula for polyethylene [$-(CH_2CH_2)-$] illustrates the repeating ethylene unit of the polymer. The molecular structure of polyethylene may be highly branched or basically linear depending upon the polymerization reaction conditions.

High density polyethylene (HDPE) is a very ordered linear structure with little branching. This structure makes the polymer highly crystalline, giving HDPE the following advantages over low density polyethylene (LDPE): increased stiffness, tensile strength, hardness, heat and chemical resistance, opacity, and barrier properties. However, the more crystalline nature of HDPE reduces impact strength and environmental stress-crack resistance.

This section presents the polymerization processes used to produce HDPE. Low density polyethylene (LDPE) and linear low density polyethylene (LLDPE) are discussed separately.

Table A-20 in Appendix A lists typical properties of high density polyethylene. HDPE is characterized by a density greater than 0.940 g/cm^3, with the usual density range from 0.955 to 0.970 g/cm^3. Some HDPE physical properties are determined by the density and the molecular weight of the polymer. Tables A-21 and A-22 in Appendix A present density and molecular weight dependent properties, respectively, for a range of HDPE resins. The end use of HDPE also requires some variation in physical properties. Table A-23 in Appendix A lists HDPE properties according to resin grade.

Two specialty grades of HDPE are produced: high molecular weight HDPE (HMW-HDPE) and ultra-high molecular weight HDPE (UHMWPE). HMW-HDPE is considered to be polyethylene with a molecular weight in the range of 200,000 to 500,000; UHMWPE is defined by a molecular weight of greater than 3

million. These polymers have chemical, physical, and electrical properties
very similar to those of HDPE. The differences lie in the unique abrasion
resistance, impact strength, and special processing characteristics due to
the higher melt viscosity of these polymers.[226] Typical properties of
UHMWPE are presented in Table A-24 in Appendix A for resins made using two
different catalysts and for one copolymer resin.

HDPE, due to its crystalline structure, is virtually insoluble in any
solvent at ambient temperature.[168] Chlorinated hydrocarbons cause swelling of the polymer below 100°C; at temperatures above 100°C, HDPE will
dissolve in these solvents. Polar compounds such as water, alcohols, acids,
esters, ketones, and nitriles have no effect on HDPE. Suitable HDPE solvents are listed in Table 127.

HDPE copolymers are formed by adding an α-olefin, such as 1-butene,
1-hexene, and propylene to the polymerization reactor. These compounds
increase branching along the HDPE chain and reduce the crystallinity of the
resin.

HDPE can be produced by particle form (slurry) polymerization, solution
polymerization, or low pressure gas phase polymerization. Fifty-eight
percent of the total United States installed capacity for HDPE uses the
particle form polymerization process with a Phillips catalyst. Thirty-two
percent of the capacity is comprised of either solution or particle form
polymerization with a Ziegler catalyst, while the remaining 10 percent
utilizes other processes, including gas phase polymerization.[169] Process
pressures are much lower than those for LDPE and are typcially less than
1.03×10^7 Pascals.

INDUSTRY DESCRIPTION

The HDPE industry is made of the 14 major chemical and petrochemical
producers listed in Table 128. These producers have 16 sites; 12 in Texas,
3 in Louisiana, and 1 in Iowa. The process technology used by these plants
is noted in Table 128. When available, the polymerization technology
(particle form or slurry, solution, and gas phase) is also listed.

HDPE is used primarily in consumer related industries including construction, packaging, and housewares. Due to the large number of products
manufactured from HDPE, the general trend of the economy and consumer spending affects the production and sales of HDPE resins. After a production
peak in 1979 of 2,277,000 metric tons, 1981 production (2,177,000 metric
tons) remains below the 1979 level but is still well above the production
level of 1,420,000 metric tons for 1976.[65, 143]

The emergence of high efficiency catalysts has resulted in a reduction
of processing steps for the three HDPE processes presented. These catalysts
provide low catalyst concentration in the recycle stream and the resulting
polymer. Therefore, catalyst deactivation using methanol and catalyst
removal from the polymer (polymer deashing) are eliminated. Simplifying the
process in this manner significantly reduces the amount of VOC emissions due
to methanol addition and catalyst removal.

TABLE 127.   SUITABLE SOLVENTS FOR HDPE

chlorocyclohexane

chloronaphthalene

decahydronaphthalene

dimethylcyclohexane

dimethylheptane

methylnaphthalene

n-hexyl chloride

octane

trichlorobenzene

xylene

Source: <u>Encyclopedia of Chemical Technology</u>, 3rd Edition.

TABLE 128.   U.S. PRODUCERS OF HDPE RESINS

Company	January 1, 1982 Capacity[a] Thousand Metric Tons	Process
Allied Corporation Fibers and Plastics Co. Baton Rouge, LA	300	Phillips
American Hoechst Corp. Plastics Division Bayport, TX	136	Hoechst
Atlantic Richfield Co. ARCO Polymers, Inc., subsidiary Port Arthur, TX	159	Solvay, USI
Chemplex (jointly owned by American Can Company and Getty Oil Co.) Clinton, IA	123	Phillips

Cities Service Co.[b]
 Chemicals and Minerals Group
  Petrochemicals Division
   Texas City, TX
	86	Monsanto

Dow Chemical USA[c] Freeport, TX Plaquemine, LA	54 150	Dow solution Dow slurry

E. I. duPont de Nemours and Co., Inc.
 Polymer Products Department
  Orange, TX
  Victoria, TX
	105
	102
		DuPont

Gulf Oil Corporation[d]
 Gulf Oil Chemicals Co.
  Plastics Division
   Orange, TX
	191	Phillips,
		Union Carbide

Hercules Inc. Lake Charles, LA	7	N/A[e]

N/A - Not Available.

(continued)

TABLE 128 (continued)

Company	January 1, 1982 Capacity[a] Thousand Metric Tons	Process
National Petrochemicals Corp. (jointly owned by National Distillers and Chemical Corporation and Owens Illinois, Inc.) La Porte, TX	284	Phillips, Solvay, USI
Phillips Petroleum Co. Plastics Division Pasadena, TX	682	Phillips
Soltex Polymer Corp. Deer Park, TX	341	Phillips, Solvay
Standard Oil Co. of Indiana AMOCO Chemicals Corp., subsidiary Chocolate Bayou, TX	173	AMOCO Gas Phase, Solvay, USI
Union Carbide Corporation[f] Plastics and Chemicals Division Seadrift, TX	91	Union Carbide Gas Phase
TOTAL	2,984	

---

[a]USS Chemicals and Texaco, Inc. have announced plans of a joint venture to construct an HDPE plant expected to have a capacity of approximately 272,000 metric tons per year at La Porte, TX by 1983. Conoco Inc. has announced plans to build a 195,000 metric tons per year plant by 1983 at Montagora County near Bay City, Texas. Exxon Corporation is also known to be interested in HDPE production although no plant construction is pending.

[b]Texas City plant is leased from Monsanto and will be operating until Cities Service completes the construction of a 204,000 metric ton HDPE plant at Bay City, TX by 1984.

(continued)

TABLE 128 (continued)

[c]An additional HDPE plant capacity of 68,000 metric tons per year is being used for the production of LLDPE at Freeport.

[d]Company has announced debottlenecking plans that will bring total capacity to 261,000 metric tons per year by 1982.

[e]Hercules produces Ultra-High Molecular Weight HDPE.

[f]Seadrift capacity is convertable between HDPE and LLDPE and additional capacity of 181,000 metric tons per year is being used to produce LLDPE in 1981.

Sources: Chemical Economics Handbook, updated annually, 1981 Data.
            Directory of Chemical Producers, 1982.

PRODUCTION AND END USE DATA

HDPE production is one of the largest in the plastics industry, following LDPE and polyvinyl chloride (PVC) in tons produced. In 1981, U.S. HDPE
production totaled 2,117,000 metric tons. The largest HDPE markets include:
[168]

- Wire and cable insulation;

- Bottles for liquid detergents, household bleach, and milk;

- Large drums and gas tanks;

- Cosmetic, medicine, and other consumer products. bottles;

- Containers including freezer boxes, fish crates, garbage barrels,
  fuel tanks, and water storage tanks;

- Pipe for transporting potable water, gas, acids, liquid hydrocarbons, oils, salt water, and other chemicals and solvents;

- Housewares;

- Toys;

- Institutional seating;

- Filament for cord, rope, ribbon, yarn, and gauze for surgical tissue
  reinforcement; and

- Film for boil-in-bag food packaging, hospital bags, heavy-duty
  bagging, drum liners, and merchant bags.

Table 129 presents HDPE consumption by end-use and major markets.

PROCESS DESCRIPTIONS

HDPE resins are manufactured by particle form (slurry), solution, or
gas phase polymerization processes. Particle form polymerization processes
using a Phillips catalyst contribute over half of the HDPE produced. One
third of the U.S. production is manufactured using either particle form or
solution polymerization processes and a Ziegler catalyst. The remaining
HDPE is produced from other methods, including gas phase polymerization.

The emergence of high yield catalysts has eliminated the need for
catalyst deactivation and catalyst removal from the polymer, called polymer
deashing. The simplification of the HDPE process by removing these steps
also reduces the amount of VOC emissions generated by this process.

HDPE is produced from ethylene using coordination polymerization.
During initiation, ethylene coordinates at the surface of the catalyst.
Propagation occurs by adding monomer in a similar manner. The polymeriza-
tion terminates either by monomer transfer or by hydrogenation.

TABLE 129. HDPE CONSUMPTION FOR 1982

Market	Thousand Metric Tons
Blow Molding	
Bottles	
Milk	239
Other Food	70
Household Chemicals	272
Pharmaceuticals/Cosmetics	66
Drums (15 gallon and larger)	27
Fuel Tanks (all types)	9
Tight-Head Pails	26
Toys	12
Housewares	5
Other Blow Molding	41
TOTAL BLOW MOLDING	767
Extrusion	
Coating	17
Film (12 mil and under)	
Merchandise Bags	65
Tee-Shirt Sacks	1
Trash Bags	4
Food Packaging (bags and box liners)	22
Deli Paper	5
Other	32
Pipe	
Corrugated	151
Water	28
Gas	24
Other	40
Sheet (over 12 mil)	29
Wire and Cable	47
Other Extrusion	13
TOTAL EXTRUSION	478
Injection Molding	
Industrial Containers	
Dairy Crates	18
Other Crates, Cases, Pallets	35
Pails	87

(continued)

TABLE 129 (continued)

Market	Thousand Metric Tons
Injection Molding (Continued)	
Consumer Packaging	
Milk Bottle Caps	9
Other Caps	16
Dairy Tubs	34
Ice-Cream Containers	20
Beverage-Bottle Bases	45
Other Food Containers	5
Paint Cans	10
Housewares	68
Toys	29
Other Injection Molding	93
TOTAL INJECTION MOLDING	469
Rotomolding	30
Export	200
Other (Chiefly resold resin and resin used for blending and compounding)	233
TOTAL	2,177

Source: Modern Plastics, January 1982.

Initiation

$$CH_2{=}CH_2 + Catalyst - R \longrightarrow Catalyst\text{-}CH_2\text{-}CH_2\text{-}R$$

Propagation

$$Catalyst\text{-}CH_2\text{-}CH_2\text{-}R + (n-1)CH_2{=}CH_2 \longrightarrow$$

$$Catalyst\text{---}(CH_2CH_2)_{\overline{n}}\text{-}R$$

Termination by Monomer Transfer

$$Catalyst\text{-}(CH_2CH_2)_{\overline{n}}R + CH_2{=}CH_2 \longrightarrow$$

$$Catalyst\text{-}CH_2\text{-}CH_3 + CH_2{=}CH\text{-}(CH_2\text{-}CH_2)_{\overline{n}}R$$

Termination by Hydrogenation

$$Catalyst\text{-}(CH_2CH_2)_{\overline{n}}\text{-}R + H_2 \longrightarrow Catalyst - H +$$

$$CH_3\text{-}(CH_2\text{-}CH_2)\text{-}R$$

Tables 130 and 131 list typical input materials and operating parameters for HDPE processes, respectively. Other input materials used in HDPE production such as catalysts, comonomers, molecular weight regulating agents, solvents, diluents, and specialty chemicals (e.g., antioxidants, UV light stabilizers, and crosslinking agents) are listed in Table 132.

Particle Form Polymerization

The particle form (or slurry) polymerization process is used to produce over half of the HDPE in the United States.[169] This process, depicted in Figure 39, consists of three major steps: polymerization, polymer separation, and pelletization. A typical recipe for HDPE production via particle form polymerization follows:[81]

Material	Parts by Weight
Ethylene (monomer)	1,025
Hydrogen (molecular weight regulator)	15
Hexane (diluent)	1
Chromium-Based Catalyst	<1

A diluent, typically hexane, is used to feed the chromium oxide catalyst suspended on silica gel to the reactor. The ethylene and comonomers are then continuously fed into the liquid-filled reactor, which may be either an autoclave or a loop reactor. Agitation is used to facilitate heat removal, prevent agglomeration of the polymer particles, and circulate the liquid diluent through the loop reactor.

The monomer-polymer slurry is continuously discharged from the reactor settling zone and fed into a flash tank. The diluent, unreacted ethylene,

TABLE 130. TYPICAL INPUT MATERIALS TO HDPE PROCESSES

Process	Monomer	Catalyst	Solvent or Diluent	Comonomer
Particle Form	X	X	X	X
Solution	X	X	X	X
Gas-Phase	X	X		X

TABLE 131. TYPICAL OPERATING PARAMETERS FOR HDPE PROCESSES

Process	Temperature	Pressure	Reaction Time
Particle Form	70–110°C	0.5–3×10^6 Pa	1 – 4 hours
Solution	150–250°C	2–4×10^6 Pa	several minutes
Gas Phase	70–110°C	2–3×10^6 Pa	--[a]

[a]No reaction times were given in the literature; however, several passes
are necessary to achieve the desired conversion.

Sources: Encyclopedia of Chemical Technology, 3rd Edition.
          Encyclopedia of Polymer Science and Technology.
          Modern Plastics Encyclopedia, 1981–1982.

TABLE 132. INPUT MATERIALS AND SPECIALTY CHEMICALS FOR
HDPE PROCESSES IN ADDITION TO MONOMER

Function	Compound	Process
Solvent	cyclohexane	solution
Diluent	butane hexane isobutane isopentane	particle form
Catalyst Phillips	chromium (VI) oxide supported on silica or silica-alumina gel	particle form solution gas phase
Ziegler	transition metal coupled with an alkyl metal compound	particle form solution gas phase
Standard Oil of Indiana (AMOCO)	molybdena (MoO_3) supported on an alumina gel	particle form solution
Molecular Weight Regulator	hydrogen	particle form
Comonomers	1-butene 1-hexene 1-octene propylene	particle form solution gas phase
Antioxidants	hindered phenols phosphates stearates	particle form solution gas phase
UV Light Stabilizers	benzophenones carbon black	particle form solution gas phase
Crosslinking Agents	benzoyl peroxide dicumyl peroxide	particle form solution gas phase

Sources: Encyclopedia of Chemical Technology, 3rd Edition.
Encyclopedia of Polymer Science and Technology.
Modern Plastics Encyclopedia, 1981-1982.

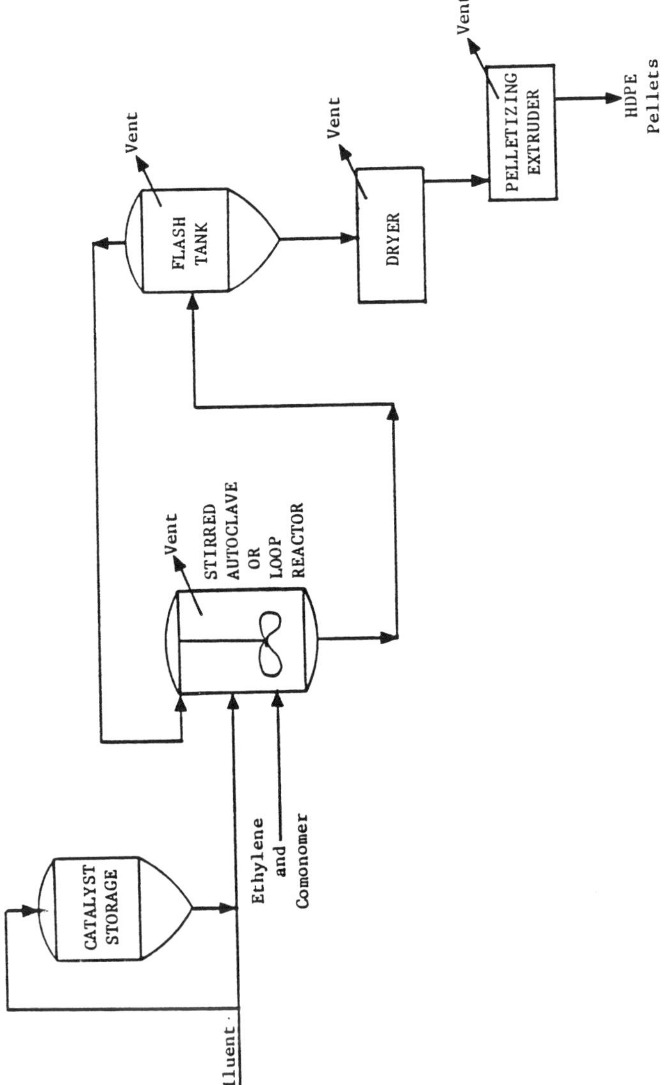

Figure 39. Particle form polymerization process.

Source: *Encyclopedia of Chemical Technology*, 3rd Edition.

and comonomer are returned to the reactor while the polyethylene particles are dried and pelletized before bagging.

HMW–HDPE and UHMWPE are manufactured using this process when a molecular weight inhibitor is not introduced. The average molecular weight of polyethylene from this process ranges from 250,000 to greater than 1,000,000.

Solution Polymerization

Solution polymerization is used for the production of homogeneous polymers of comparatively low molecular weight for injection molding and film extrusion. These polymers are more readily dissolved in hydrocarbons; in contrast to particle form polymerization, the HDPE produced by the solution process does not precipitate out of solution.

As depicted in Figure 40, ethylene and a comonomer (such as 1-hexene) are dissolved in the solvent and fed to the polymerization reactor. A high efficiency catalyst is fed separately into the reactor and the polymerization reaction is completed in several minutes. Information on a typical recipe for producing HDPE via solution polymerization was not found in the literature. The maximum polymer conversion is 35 to 40 percent; conversion is typically 18 to 25 percent in order to retain the polymer in solution. The high polymer production per kilogram of catalyst eliminates the need for the catalyst removal. The unreacted ethylene and solvent are removed in a flash tank and recycled to the reactor while the resulting polymer is dried and extruded into pellets.

Gas Phase Polymerization

A low pressure gas phase polymerization process is used to produce both LLDPE and HDPE. The density of the HDPE product from this process ranges from 0.940 to 0.970 g/cm^3, and the process is capable of producing polymers with densities in the range of 0.915 to 0.970 g/cm^3.[168]

As depicted in Figure 41, ethylene, comonomer, and a catalyst are fed to a fluidized bed reactor. Typical input to the reactor is as follows: [186, 187]

Material	Parts by Weight
Ethylene (monomer)	1,010
Hydrogen (molecular weight regulator)	4
Comonomer	20
Chromium-Based Catalyst	<1

Unreacted gases are recycled using a single-stage centrifugal compressor and are cooled in a heat exchanger before reentering the reactor. The granular polymer is discharged from the reactor intermittently. High catalyst conversion in the reactor eliminates the need for catalyst removal. Unreacted gases are separated and returned to the reactor and a nitrogen purge is used to remove any remaining hydrocarbons. This process, which requires no

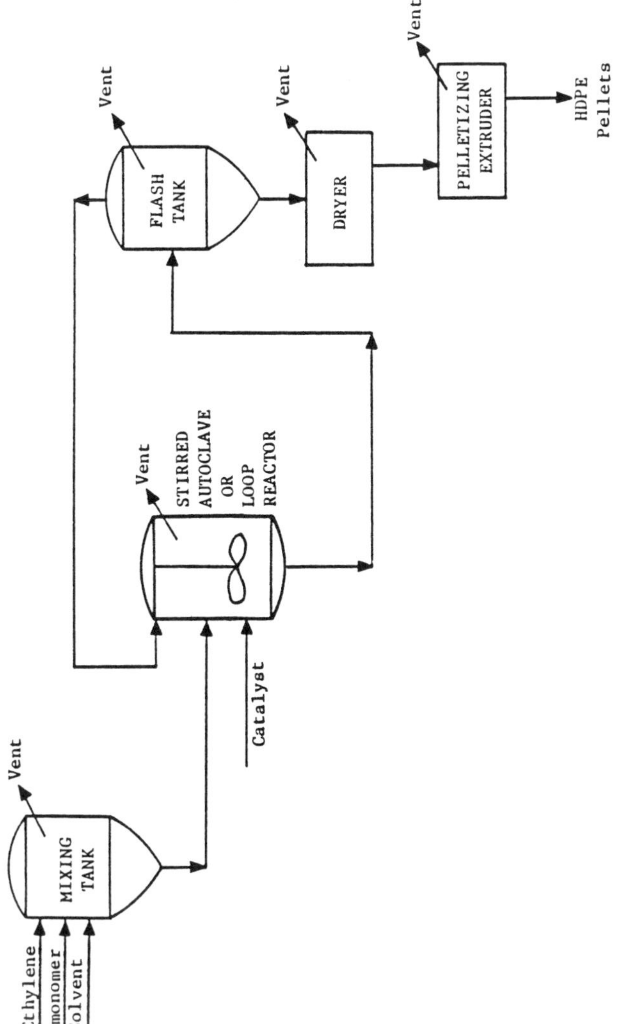

Figure 40.   Solution polymerization process.

Source: *Encyclopedia of Chemical Technology*, 3rd Edition.

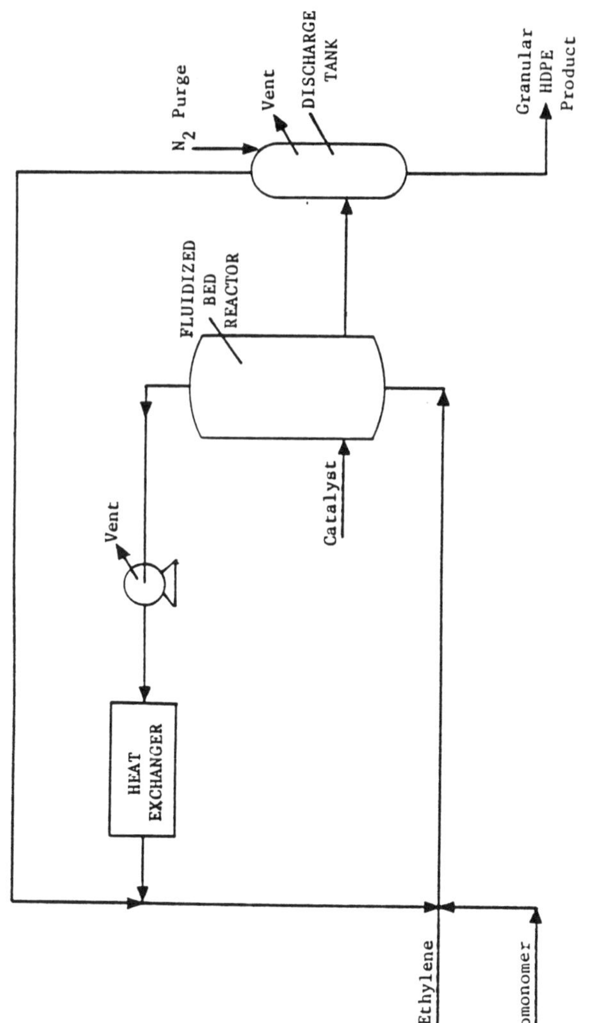

Figure 41.   Low pressure gas phase polymerization.

Source: Hydrocarbon Processing, November 1979.

solvent stripping or pelletizing, results in lower capital and operating costs when compared to particle form and solution polymerization.

<u>Energy Requirements</u>

The following data were found in the literature and listed according to the technologies designed below:[185, 186, 187]

Technology	Energy Required	Unit/Metric Ton of Product
Chemische Werke Huels AG	Electricity	1.58×10^9 Joules
Hoechst AG	Electricity	3.46×10^9 Joules
Naphthachimie	Electricity Steam	7.56×10^8 Joules 0.3 metric tons

ENVIRONMENTAL AND INDUSTRIAL HEALTH CONSIDERATIONS

HDPE, considered to be a nontoxic compound, is used in the packaging and medical industries as well as for pipe which transports potable water. These products affect a large percentage of the population and any toxic effect would have a significant impact on this industry.

High efficiency catalysts have eliminated the need for catalyst deactivation using methanol and polymer deashing. This simplification in processing also eliminates fugitive VOC emissions which are the result of methanol addition and the deashing step.

<u>Worker Distribution and Emissions Release Points</u>

Worker distribution estimates have been made for HDPE production via the three processes discussed previously. The estimates, shown in Table 133, were developed by correlating major equipment manhour requirements with the process flow diagrams in Figures 39 through 41.

No major air emission point sources are associated with the HDPE processes. However, fugitive emissions from such sources as reactor vents, storage tanks, polymer driers, and pelletizing extruder vents may pose significant environmental and/or worker health problems based on the constituents present in the stream, operating parameters for the process, engineering and administrative controls used, and maintenance programs.

Table 134 shows the principal process sources of fugitive emissions from HDPE manufacture. Ethylene, ethane, and several other possible input materials listed in Table 132 (hydrogen, butane, propylene), are simple asphyxiants, which pose a potential hazard because their presence limits the amount of oxygen available in the atmosphere. The nitrogen used to purge granular HDPE of residual hydrocarbon is also a simple asphyxiant.

TABLE 133.   WORKER DISTRIBUTION ESTIMATES FOR HDPE PRODUCTION

Process	Unit	Workers/Unit/8-hour Shift
Particle Form Polymerization	Storage	0.5
	Continuous Reactor	0.5
	Flash Tank	0.125
	Dryer	0.5
	Extruder	1.0
Solution Polymerization	Storage	0.5
	Continuous Reactor	0.5
	Flash Tank	0.125
	Dryer	0.5
	Extruder	1.0
Gas Phase Polymerization	Heat Exchanger	0.125
	Compressor	0.25
	Continuous Reactor	0.5
	Storage Tank	0.125

TABLE 134.   SOURCES OF FUGITIVE EMISSIONS FROM HDPE MANUFACTURE

		Process		
Source	Constituent	Particle Form	Solution	Gas Phase
Reactor Vents	Ethylene	X	X	X
	Ethane	X	X	X
	Solvent	X	X	
Flash Tank Vent	Ethylene	X	X	X
	Ethane	X	X	X
	Solvent	X	X	
Mixing Tank Vent	Ethylene		X	
	Solvent		X	
Polymer Drying Vent	Ethylene and Polyethylene Particulates and Emissions	X	X	
Pelletizing Extruder Vent	Ethylene and Polyethylene Particulates and Emissions	X	X	
Compressor Vent	Ethylene			X
	Ethane			X
Discharge Tank	Ethylene			X
	Ethane			X

Solvents and diluents used in HDPE production are central nervous system depressants. Overexposure may result in dizziness, headache, nausea, and narcosis.

Particulate emissions from drying and pelletizing operations may consist of polymer fines (nuisance dust) and specific additives which may be hazardous in themselves.

Available toxicity and standards information about contaminants potentially associated with HDPE production is summarized in Table 10.

Emission ranges from which potential for employee exposures to volatile organic compounds may be estimated are shown in Tables 135 through 137 for the particle form, solution, and gas phase processes, respectively.

Health Effects

Animal test data indicate that several input materials and specialty chemicals for HDPE processes are tumorigenic agents; however, only one, the catalyst chromium (VI) oxide, presents solid evidence of carcinogenicity. Others, including benzoyl peroxide, a crosslinking agent, and hexane, a diluent, exhibit teratogenic properties. In addition, the solvent cyclohexane has produced mutations in laboratory animals.

Chromium (VI) Oxide has produced positive results in animal tests for carcinogenicity [109] and there is extensive evidence of its mutagenic properties. Teratogenic effects have also been observed in laboratory animals. Chronic industrial exposures have led to severe liver damage and central nervous system involvement in humans; allergic reactions are common.[85] The American Conference of Governmental Industrial Hygienists (ACGIH) has recommended a Threshold Limit Value (TLV) of 0.05 mg/m^3 (8 hour TWA). The OSHA air standard is 0.5 mg/m^3 (8 hour TWA).

Air Emissions

Sources of fugitive VOC emissions from HDPE processing are summarized in Table 134 by process and constituent type. The major portion of the fugitive VOC emissions from HDPE processing originates in the polymerization section, which includes ethylene recycle and recompression. Polymerization section VOC emissions from vents are reported to contain 20 kg ethylene, 1 kg ethane, and 22 kg solvent, if used, per ton of HDPE produced.[93] VOC leak rates from valves and flanges are reported as 0.018 and 0.021 kg per hour per source, respectively, for liquids and 0.0059 and 0.0022 kg per hour per source, respectively, for gas service.[287] Control technologies for vents include venting the stream to a flare to incinerate any hydrocarbons and/or venting to blowdown.[285] Controls for valves, flanges, and other leak include equipment modification and a routine maintenance and inspection program.

TABLE 135. CHARACTERISTICS OF VENT STREAMS FROM THE HIGH DENSITY
POLYETHYLENE LIQUID PHASE PARTICLE FORM PROCESS

Stream Name	Nature	Emission rate, kg VOC/Mg product	Composition, Wt.%
Feed preparation	Intermittent	0.2	100.0 Ethylene
Dryer nitrogen blower	Continuous	0.06–0.4	0.3 Isobutane 99.7 Nitrogen
Continuous mixer	Continuous	0.006	0.6 Isobutane 99.4 Nitrogen
Recycle treaters	Continuous	12.7	61.0 Ethylene 18.0 Isobutane 20.0 Ethane 1.0 Hydrogen
Total Emission Rate		13.0–13.3	

Source: *Polymer Manufacturing Industry Background Information for Proposed Standards*, September 1983.

TABLE 136. CHARACTERISTICS OF VENT STREAMS FROM THE HIGH DENSITY
POLYETHYLENE LIQUID PHASE SOLUTION PROCESS

Stream Name	Nature	Emission rate, kg VOC/Mg product	Composition, % by volume
Tank vents	Continuous; cyclic	0.024	----
Cyclohexane treater vents	Intermittent	(to flare)	----
Catalyst preparation vent	Generally continuous	(to flare)	----
Compressor vents	Continuous and intermittent	(to flare)	----
Reactor vents	Generally intermittent	(to flare)	----
Separation vents	Some continuous and intermittent	(to flare)	----
Recycle ethylene treaters	Continuous	12.7	61.0 ethylene 1.0 hydrogen 18.0 isobutal 20.0 ethane
Recovery distillation	Continuous and intermittent	(to flare)	----
Extruder vents	Continuous	0.63	VOC[a] – 0.05 steam and air – 99.95
Stripper vents	Continuous	0.85	VOC[a] – 0.01 air – 99.99
All indicated streams to flare		14.8	VOC[b] air
Total Emission Rate		29.0	

[a]Cyclohexane.

[b]Includes ethylene, cyclohexane, and other VOC comonomers. Ethylene is emitted at about 4.35 kg VOC/Mg product; cyclohexane at about 8.7 kg VOC/Mg product; and other VOC at about 1.74 kg VOC/Mg product.

Source: *Polymer Manufacturing Industry Background Information for Proposed Standards*, September 1983.

TABLE 137.   CHARACTERISTICS OF VENT STREAMS FROM THE HIGH DENSITY
POLYETHYLENE GAS PHASE PROCESS

Stream Name	Nature	Emission rate, kg VOC/Mg product	Composition, Wt.%
Raw materials purification vent	Intermittent	0.04	0.5 VOC 99.5 N_2
Catalyst additive vent	Intermittent	0.01	35 VOC 65 N_2
Comonomer purification vent	Continuous	0.10	92 VOC 8 N_2
Catalyst dehydrator handling vent	Intermittent	0.004	2 VOC 98 Air
Vacuum pump vent	Intermittent	0.03	41 VOC 59 N_2
Recovery vessel vent	Intermittent	0.06	46 VOC 54 Air
Emergency reactor blowdown vent	Intermittent	0.05	83 VOC 17 N_2
Reactor purge vent	Intermittent	0.71	1-83 VOC 99-17 N_2
Product discharge vent	Continuous	22.3	35 VOC 65 N_2
Bin vent	Continuous	0.05	0.02 VOC 99.98 Air
Compressor seal (oil) vent	Continuous	0.01	50 VOC 50 N_2
Analyzer vents	Continuous	0.01	100 VOC
Total Emission Rate		23.37	

Source: Polymer Manufacturing Industry Background Information for Proposed
Standards, September 1983.

Sources of fugitive particulates are also listed in Table 134. Fugitive particulates may be collected by venting these streams to a baghouse or an electrostatic precipitator. According to EPA estimates, the population exposed to HDPE emissions in a 100 km^2 area surrounding an HDPE plant is: 21 persons exposed to hydrocarbons and 2 persons exposed to particulates. [286]

Wastewater Sources

HDPE processes generate only routine cleaning water. Several wastewater parameters from HDPE processes, shown below, are reported by EPA. These parameters were not separated according to process type for the purpose of establishing effluent limitations for the HDPE industry.[284]

Wastewater Characteristic	Unit/Metric Ton of HDPE
Production	0 - 30.87 m^3
BOD_5	0 - 1 kg
COD	0 - 2.4 kg
TSS	0 - 3.4 kg

The control treatment technologies common to HDPE facilities are: screening, chemical treatment, and aeration. This wastewater treatment system has been shown to effectively treat HDPE wastewaters.[284]

Solid Waste

The solid wastes generated during HDPE production include substandard polymer which cannot be salvaged by blending and low-molecular weight polyethylene waxes. The high efficiency catalysts have all but eliminated wax production.[168] These wastes are not hazardous; therefore, their disposal should not pose an environmental problem.

Environmental Regulation

Effluent limitations guidelines have been set for the polyethylene industry. BPT, BAT, and NSPS call for the pH of the effluent to fall between 6.0 and 9.0 (41 Federal Register 32587, August 4, 1976).

New source performance standards (proposed by EPA on January 5, 1981) for volatile organic carbon (VOC) fugitive emissions include:

- Safety/release valves must not release more than 200 ppm above background, except in emergency pressure releases, which should not last more than five days; and

- Leaks (which are defined as VOC emissions greater than 10,000 ppm) must be repaired with 15 days.

None of the input materials used in this process have been listed as hazardous (45 Federal Register 3312, May 19, 1980).

18. Polyethylene—Linear Low Density (LLDPE)

INTRODUCTION

Polyethylene is a lightweight, flexible, tough, chemical resistant polymer which exhibits outstanding electrical insulation properties. These characteristics combined with ease of fabrication and low production cost makes polyethylene suitable for a wide variety of end products used in the packaging, housewares, construction, communications, and medical industries.

The empirical formula for polyethylene $[-(CH_2CH_2)-]$ illustrates the repeating ethylene unit for the polymer. The molecular structure of polyethylene may be highly branched or basically linear depending upon the polymerization reaction conditions.

Linear low density polyethylene (LLDPE) has a linear structure with short side chains. These short side chains keep the polymer chain from forming a highly crystalline structure like high density polyethylene (HDPE). Short chain branching gives LLDPE the following advantages over low density polyethylene (LDPE): higher melting point, higher tensile strength, higher flexural modulus, better elongation, and better environmental stress crack resistance. This section presents the polymerization processes used to produce LLDPE. Low density polyethylene (LDPE) and high density polyethylene (HDPE) are discussed separately.

Table A-25 in Appendix A compares typical properties of LLDPE and LDPE resins. The LLDPE resins display higher viscosities at the same density due to increased crystallinity and the absence of long chain branching;[168] therefore, in order to use the same processing equipment as was used for LDPE, an LLDPE resin with a sufficiently high melt index must be selected.

Comonomers are added to the ethylene to control chain branching. Typically, 1-butene is used; however, 4-methyl-1-pentene, 1-hexene, and 1-octene are also used. The type of catalyst used determines which comonomer is selected since various catalysts display differing copolymerization activities.[147]

LLDPE is produced by particle form (slurry) polymerization, solution polymerization, or low pressure gas phase polymerization. Process pressures are much lower than those used for LDPE production, typically less than 1.03 x 10^7 Pascals. Low pressure gas phase polymerization requires no solvent

stripping which results in lower capital and operating costs than slurry or
solution polymerization.

INDUSTRY DESCRIPTION

The LLDPE industry is presently made up of two commercial resin producers: Dow Chemical, Freeport, Texas and Union Carbide, Seadrift, Texas.
[35] Although the present LLDPE production of 451,000 metric tons is
relatively small when compared to LDPE production, LLDPE is expected to
replace 70 percent of the LDPE resin and 20 percent of the HDPE resin
markets in the next 10 years.[35, 38]

PRODUCTION AND END USE DATA

LLDPE production is relatively small when compared to LDPE production.
As shown below, 363,700 metric tons were produced by Dow and Union Carbide
in 1980.[35]

| | 1980 Production |
Producer	Thousand Metric Tons
Dow	136,400
Union Carbide	227,300
Total	363,700

LLDPE markets and uses include:[41]

- Injection molded parts, including housewares, closures, and lids;

- Blow molded parts, including toys, bottles, and drum liners;

- Extruded wire and cable insulation;

- Extruded pipe and tubing;

- Extruded film used in food packaging applications, including ice
  bags, retail merchandise bags, and produce bags; and

- Extruded film used in nonfood applications, including trash bags,
  industrial liners, and garment bags.

Table 138 lists end-uses for LLDPE resins in 1981.

PROCESS DESCRIPTIONS

LLDPE resins are manufactured by particle form (slurry), solution, or
gas phase polymerization processes. There is no definitive information on
the process used by Dow. Union Carbide uses a low pressure gas phase
process to manufacture LLDPE. El Paso Polyolefins has built a demonstration
unit in Odessa, Texas which uses the particle form process to manufacture
LLDPE.[38]

TABLE 138.   LLDPE CONSUMPTION BY END-USE FOR 1981[a]

End-Use	Thousand Metric Tons
Blow Molding	1
Extrusion	
Film (12 mil and under)	266
Pipe and Conduit	30[b]
Sheet and Profile	2
Wire and Cable	42
Injection Molding	40
Rotomolding	20
Export	23
Other (blending resin and resold resin)	27
TOTAL	451

[a]Tonnage in this table is broken out from tonnage in the LDPE section, Table 147.

[b]Note discrepancy with pipe figures in LDPE Table 147.

Source: *Modern Plastics*, January 1982.

The emergence of high yield catalysts for the three LLDPE processes
presented has eliminated the need for catalyst deactivation and catalyst
removal from the polymer, called polymer deashing. The simplification of
the LLDPE process by removing these steps also reduces the amount of
fugitive VOC emissions generated by this process.

LLDPE is polymerized from ethylene using a catalyst with what can be
described as an anionic polymerization mechanism. During initiation,
ethylene is added to an organometallic active center, depicted as M^+R^-.
Propagation occurs by adding other ethylene molecules at the active organometallic center.

<u>Initiation</u>

$$M^+R^- + CH_2{=}CH_2 \longrightarrow M^+{-}^-CH_2{-}CH_2{-}R$$

<u>Propagation</u>

$$M^+{-}^-CH_2{-}CH_2{-}R + nCH_2{=}CH_2 \longrightarrow M^+{-}^-CH_2{-}CH_2{-}(CH_2CH_2)_{\overline{n}}R$$

The polymerization can be terminated by several mechanisms. Monomer transfer produces a polymer molecule containing an unsaturated bond, and an alkyl
group on the surface of the catalyst. A hydride ion may be transferred,
coupled with realkylation of the catalyst. Finally, a transfer reaction may
occur between an active polymer chain and a metal alkyl.

<u>Termination by Monomer Transfer</u>

$$M^+{-}^-CH_2{-}CH_2{-}(CH_2{-}CH_2)_{\overline{n}}R + CH_2{=}CH_2 \longrightarrow$$

$$M^+{-}^-CH_2{-}CH_3 + CH_2{=}CH{-}(CH_2{-}CH_2)_{\overline{n}}R$$

<u>Termination by Hydride Ion Transfer</u>

$$M^+{-}^-CH_2{-}CH_2{-}(CH_2{-}CH_2)_{\overline{n}}R \longrightarrow M^+{-}^-H + CH_2{=}CH{-}(CH_2{-}CH_2)_{\overline{n}}R$$

$$M^+{-}^-H + CH_2{=}CH_2 \longrightarrow M^+{-}^-CH_2{-}CH_3$$

<u>Termination by Transfer Reaction</u>

$$M^+{-}^-CH_2{-}CH_2{-}(CH_2{-}CH_2)_{\overline{n}}R + ZnR'_2 \longrightarrow$$

$$M^+{-}^-R' + R'{-}ZnCH_2{-}CH_2{-}(CH_2{-}CH_2)_{\overline{n}}R$$

Tables 139 and 140 list typical input materials and operating parameters for LLDPE processes, respectively. Other input materials used in
LLDPE production such as catalysts, comonomers, molecular weight regulating
agents, solvents, diluents, and specialty chemicals (e.g., antioxidants, UV
light stabilizers, and crosslinking agents) are listed in Table 141.

TABLE 139.   TYPICAL INPUT MATERIALS TO LLDPE PROCESSES

Process	Monomer	Catalyst	Solvent or Diluent	Comonomer
Particle Form	X	X	X	X
Solution	X	X	X	X
Gas-Phase	X	X		X

TABLE 140.   TYPICAL OPERATING PARAMETERS FOR LLDPE PROCESSES

Process	Temperature	Pressure	Reaction Time
Particle Form	55-70°C	1.5-2.9×10^6 Pa	1 - 2 hours
Solution	250°C	8×10^6 Pa	several minutes
Gas Phase	85-95°C	2×10^6 Pa	--[a]

[a]No reaction times were given in the literature; however, conversion is
approximately 2% per pass, and several passes are necessary to achieve the
desired conversion.

Sources: Chemical Engineering, May 17, 1982.
Encyclopedia of Chemical Technology, 3rd Edition.
Encyclopedia of Polymer Science and Technology.
Modern Plastics Encyclopedia, 1981-1982.

TABLE 141. INPUT MATERIALS AND SPECIALTY CHEMICALS FOR
LLDPE PROCESSES IN ADDITION TO MONOMER

Function	Compound	Process
Solvent	cyclohexane	solution
Diluent	butane hexane isobutane isopentane propane	particle form
Catalyst Phillips	chromium (VI) oxide supported on silica or silica-alumina gel	particle form solution gas phase
Ziegler	transition metal coupled with an alkyl metal compound	particle form solution gas phase
Standard Oil of Indiana (AMOCO)	molybdena (MoO$_3$) supported on an alumina gel	particle form solution
Molecular Weight Regulator	hydrogen	particle form
Comonomers	1-butene 1-hexene 1-octene propylene	particle form solution gas phase
Antioxidants	hindered phenols phosphates stearates	particle form solution gas phase
UV Light Stabilizers	benzophenones carbon black	particle form solution gas phase
Crosslinking Agents	benzoyl peroxide dicumyl peroxide	particle form solution gas phase

Sources: Chemical Engineering, May 17, 1982.
         Encyclopedia of Chemical Technology, 3rd Edition.
         Encyclopedia of Polymer Science and Technology.
         Modern Plastics Encyclopedia, 1981-1982.

Particle Form Polymerization

The particle form (or slurry) polymerization process, depicted in Figure 42, consists of three major steps: polymerization, polymer separation, and pelletization. A diluent, typically hexane, is used to feed the chromium oxide catalyst suspended on silica gel to the reactor. The ethylene and comonomers are then continuously fed into the liquid-filled reactor, which may be either an autoclave or a loop reactor. Agitation is used to facilitate heat removal, prevent agglomeration of the polymer particles, and circulate the liquid diluent through the loop reactor.

The following is a typical recipe for the production of LLDPE via particle form polymerization:[198]

Material	Parts by Weight
Ethylene and 1-Butene (monomer and comonomer)	18,680
Catalyst	1
Hydrogen (molecular weight regulator)	
Butane/Propane (diluent)	

The amounts of hydrogen and diluent were not found in the literature; however, they are expected to be on the order of 1 and 2 parts by weight, respectively.

The monomer-polymer slurry is continuously discharged from the reactor settling zone and fed into a flash tank. The diluent, unreacted ethylene, and comonomer are returned to the reactor while the polyethylene particles are dried and pelletized before bagging.

Solution Polymerization

Solution polymerization is used for the production of homogeneous polymers of comparitively low molecular weight for injection molding and film extrusion. These polymers are more readily dissolved in hydrocarbons; in contrast to particle form polymerization, the LLDPE produced by the solution process does not precipitate out of solution. Information regarding a typical production recipe for LLDPE by the solution polymerization process was not found in the literature.

As depicted in Figure 43, ethylene and a comonomer are dissolved in the solvent, typically cyclohexane, and fed to the polymerization reactor. A high efficiency catalyst is fed separately into the reactor and the polymerization reaction is completed in several minutes. The maximum polymer conversion is 35 to 40 percent; conversion is typically 18 to 25 percent in order to retain the polymer in solution. The high catalyst conversion eliminates the need for the catalyst removal. The polymer is either used in solution or the unreacted ethylene and solvent may be removed in a flash tank and recycled to the reactor while the resulting polymer is dried and extruded into pellets.

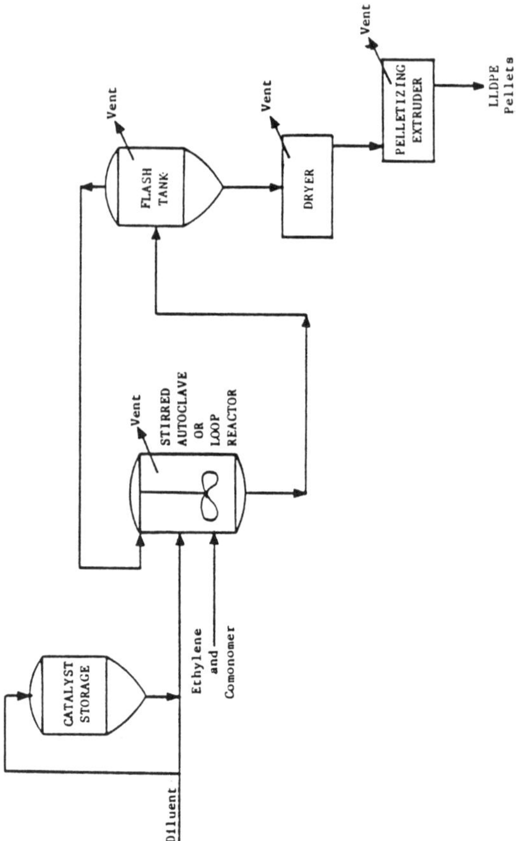

Figure 42. Particle form polymerization process.

Source: Encyclopedia of Chemical Technology, 3rd Edition.

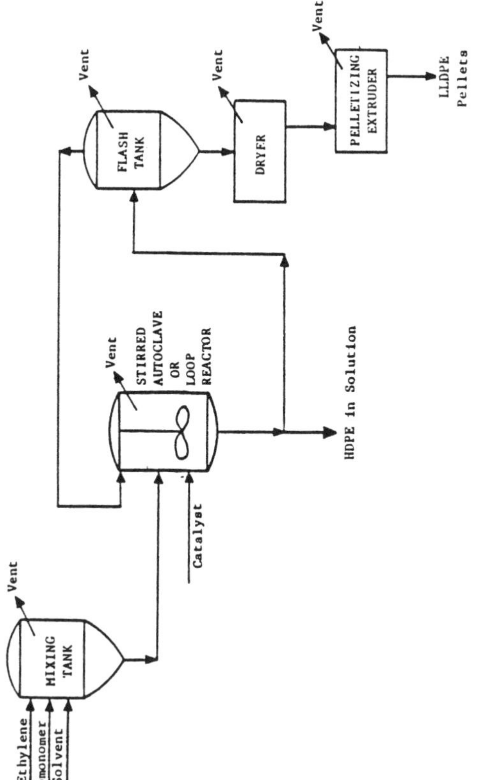

Figure 43. Solution polymerization process.

Source: Encyclopedia of Chemical Technology, 3rd Edition.

<u>Gas Phase Polymerization</u>

A low pressure gas phase polymerization process is used to produce both LLDPE and HDPE. The density of the LLDPE product from this process ranges from 0.915 to 0.940 g/cm^3, and the process is capable of producing polymers with densities in the range of 0.915 to 0.970 g/cm^3.[168]

As depicted in Figure 44, ethylene, comonomer, and a catalyst are fed to a fluidized bed reactor. Typical input materials and quantities are given below:[197]

Material	Parts by Weight
Ethylene and 1-Butene (monomer and comonomer)	1,020
Ziegler Catalyst	1
Hydrogen (molecular weight regulator)	

The amount of hydrogen was not reported, but is expected to approximately equal that of the catalyst. Unreacted gases are recycled using a single-stage centrifugal compressor and are cooled in a heat exchanger before reentering the reactor. The granular polymer is discharged from the reactor intermittently. High catalyst conversion in the reactor eliminates the need for catalyst removal. Unreacted gases are separated and returned to the reactor and a nitrogen purge is used to remove any remaining hydrocarbons. This process, which requires no solvent stripping or pelletizing, results in lower capital and operating costs when compared to particle form and solution polymerization.

<u>Energy Requirements</u>

No specific data for the energy required by LLDPE production were found in the literature consulted.

ENVIRONMENTAL AND INDUSTRIAL HEALTH CONSIDERATIONS

Like other polyethylenes, LLDPE is a nontoxic compound used in the packaging and consumer goods industries as well as for pipe which transports potable water. These products have the potential to affect a large percentage of the population and any toxic effect would have a significant impact on this industry.

High efficiency catalysts for the three LLDPE processes presented have eliminated the need for catalyst deactivation using methanol as well as polymer deashing. This simplification in processing also eliminates fugitive VOC emissions which are the result of methanol addition and the deashing step.

<u>Worker Distribution and Emissions Release Points</u>

Worker distribution estimates have been made for HDPE production via the three processes discussed previously. The estimates, shown in Table 142, were developed by correlating major equipment manhour requirements with the process flow diagrams in Figures 42 through 44.

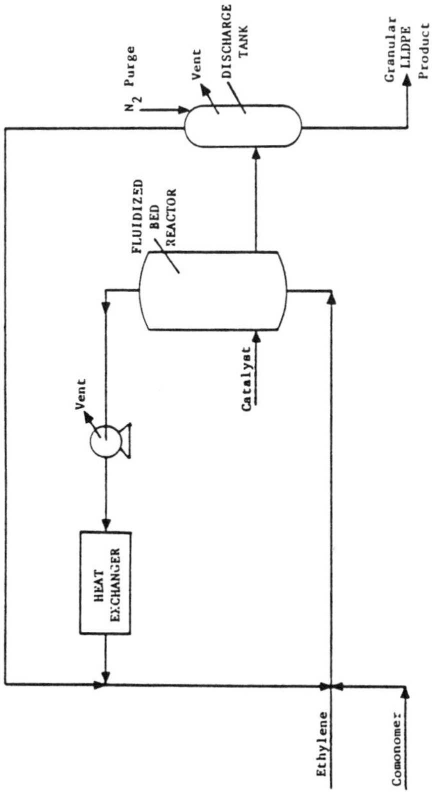

Figure 44. Low pressure gas phase polymerization.

Source: Hydrocarbon Processing, November 1979.

TABLE 142.   WORKER DISTRIBUTION ESTIMATES FOR LLDPE PRODUCTION

Process	Unit	Workers/Unit/8-hour Shift
Particle Form Polymerization	Storage	0.5
	Continuous Reactor	0.5
	Flash Tank	0.125
	Dryer	0.5
	Extruder	1.0
Solution Polymerization	Storage	0.5
	Continuous Reactor	0.5
	Flash Tank	0.125
	Dryer	0.5
	Extruder	1.0
Gas Phase Polymerization	Heat Exchanger	0.125
	Compressor	0.25
	Continuous Reactor	0.5
	Storage Tank	0.125

No major air emission point sources are associated with the LLDPE processes. However, fugitive emissions from such sources as reactor vents, storage tanks, polymer dryers, and pelletizing extruder vents may pose significant environmental and/or worker health problems based on the constituents present in the stream, operating parameters for the process, engineering and administrative controls used, and maintenance programs.

Table 143 shows the principal process sources of fugitive emissions from LLDPE manufacture. Ethylene, ethane, and several other possible input materials listed in Table 141 (hydrogen, butane, propylene), are simple asphyxiants, which pose a potential hazard because their presence limits the amount of oxygen available in the atmosphere. The nitrogen used to purge granular LLDPE of residual hydrocarbon is also a simple asphyxiant.

Solvents and diluents used in LLDPE production are central nervous system depressants. Overexposure may result in dizziness, headache, nausea, and narcosis.

Particulate emissions from drying and pelletizing operations may consist of polymer fines (nuisance dust) and specific additives which may be hazardous in themselves.

Available toxicity and standards information about contaminants potentially associated with LLDPE production is summarized in Table 10.

Health Effects

Animal test data indicate that several input materials and specialty chemicals for LLDPE processes are tumorigenic agents; however, only one, the catalyst chromium (VI) oxide, presents solid evidence of carcinogenicity. Others, including benzoyl peroxide, a crosslinking agent, and hexane, a diluent, exhibit teratogenic properties. In addition, the solvent cyclohexane has produced mutations in laboratory animals.

Chromium (VI) Oxide has produced positive results in animal tests for carcinogenicity [109] and there is extensive evidence of its mutagenic properties. Teratogenic effects have also been observed in laboratory animals. Chronic industrial exposures have led to severe liver damage and central nervous system involvement in humans; allergic reactions are common.[85] The American Conference of Governmental Industrial Hygienists (ACGIH) has recommended a Threshold Limit Value (TLV) of 0.05 mg/m^3 (8 hour TWA). The OSHA air standard is 0.5 mg/m^3 (8 hour TWA).

Air Emissions

Sources of VOC emissions are listed in Table 143 by process and constituent type. The major portion of the fugitive VOC emissions from LLDPE processing originates in the polymerization section, which includes ethylene recycle and recompression. These sources, all vents, may be controlled by routing the vented stream to a flare for the incineration of

TABLE 143.   SOURCES OF FUGITIVE EMISSIONS FROM LLDPE MANUFACTURE

Source	Constituent	Process — Particle Form	Solution	Gas Phase
Reactor Vents	Ethylene	X	X	X
	Ethane	X	X	X
	Solvent	X	X	
Flash Tank	Ethylene	X	X	X
	Ethane	X	X	X
	Solvent	X	X	
Mixing Tank	Ethylene		X	
	Solvent		X	
Polymer Drying Vent	Ethylene and Poly-ethylene Particulates and Emissions	X	X	
Pelletizing Extruder Vent	Ethylene and Poly-ethylene Particulates and Emissions	X	X	
Compressor Vent	Ethylene			X
	Ethane			X
Discharge Tank	Ethylene			X
	Ethane			X

hydrocarbons and/or routing the vented stream to blowdown.[285]   Other VOC
emissions may result from leaks associated with valves, flanges, pumps,
compressors, and drains.   Equipment modification as well as a routine
inspection and maintenance program are the available controls for these
sources.

Fugitive particulate sources are also listed in Table 143.   These
sources, also vents, may be controlled by venting the stream to a baghouse
or electrostatic precipitator for particulate collection.

Wastewater Sources

LLDPE processes generate only routine cleaning water.   No data
concerning LLDPE wastewater were found in the sources consulted.

Solid Waste

The solid wastes generated during LLDPE production include substandard
polymer which cannot be salvaged by blending and low-molecular weight
polyethylene waxes.   The high efficiency catalysts have all but eliminated
wax production.[38, 235]   These wastes are not hazardous; therefore, their
disposal should not pose an environmental problem.

Environmental Regulation

Effluent limitations guidelines have been set for the polyethylene
industry.   BPT, BAT, and NSPS call for the pH of the effluent to fall
between 6.0 and 9.0 (41 Federal Register 32587, August 4, 1976).

New source performance standards (proposed by EPA on January 5, 1981)
for volatile organic carbon (VOC) fugitive emissions include:

- Safety/release valves must not release more than 200 ppm above
  background, except in emergency pressure releases, which should not
  last more than five days; and

- Leaks (which are defined as VOC emissions greater than 10,000 ppm)
  must be repaired with 15 days.

None of the input materials used in LLDPE processing have been listed
as hazardous (45 Federal Register 3312, May 19, 1980).

19. Polyethylene—Low Density (LDPE)

INTRODUCTION

Polyethylene is a lightweight, flexible, tough, chemical resistant
polymer which exhibits outstanding electrical properties. These characteristics combined with ease of fabrication and low production cost make polyethylene suitable for a wide variety of end products used in the packaging,
housewares, construction, communications, and medical industries.

The empirical formula for polyethylene [$-(CH_2CH_2)-$] illustrates the
repeating ethylene unit of the polymer. The molecular structure of polyethylene may be highly branched or basically linear depending upon the
polymerization reaction conditions.

Low density polyethylene (LDPE) is a polymer with a branched chain
structure. The partially crystalline nature of the solid is due to the
structural symmetry of the molecules, which enables close packing of the
ordered regions of the chains known as crystallites. Complete crystallization is prevented since the highly branched regions of the polymer chain
will not pack closely, leaving unordered regions in the molecule. Linear
LDPE (LLDPE), on the other hand, has a very ordered structure with little
branching which resembles the structure of high density polyethylene (HDPE).

This section presents the polymerization processes used to produce
LDPE. Linear low density polyethylene (LLDPE) and high density polyethylene
(HDPE) are discussed separately.

Table A-26 in Appendix A lists typical properties of low density polyethylene. LDPE and LLDPE are characterized by a density of less than 0.94
g/cm^3, with the usual density ranging from 0.915 to 0.935 g/cm^3. LDPE
with densities between 0.926 and 0.940 g/cm^3 may be referred to as medium
density polyethylene. The properties for these polymers are also listed in
Table A-26 in Appendix A. LDPE is only slightly soluble in most organic and
inorganic solvents at temperatures below 40°C. Above 50°C, the solubility
increases in hydrocarbons and halogenated hydrocarbons; however, LDPE is
only slightly soluble in polar liquids such as alcohols, esters, amines, and
phenols. Animal, vegetable, and mineral oils are absorbed by LDPE.

LDPE is produced using high pressure mass polymerization, medium pressure solution polymerization, and low pressure gas phase polymerization.
Low pressure processes allow the production of LDPE or LLDPE at substantial

energy savings; however, most of the LDPE produced is manufactured by the high pressure processes since the low pressure technology is relatively new. LDPE copolymers are produced by adding butene, octene, hexene, or vinyl acetate to the polymerization reactor. These comonomers increase the branching of the LDPE structure.

INDUSTRY DESCRIPTION

The LDPE industry is composed of the 14 major chemical and petrochemical producers listed in Table 144. These producers have 20 sites in the United States: 13 in Texas, 3 in Louisiana, 2 in Illinois, and one each in Iowa and California; and one site in Puerto Rico. The type of reactor used is also listed in Table 144. High pressure processes are denoted by "tubular" or "autoclave" reactors while low pressure processes are designated as such.

LDPE is used primarily for consumer related goods including packaging, institutional products, and the electronic industry. Due to the large number of products manufactured from LDPE, the general trend of the economy and consumer spending affects the production and sales of LDPE resins. After a production peak in 1979 of 3,542,000 metric tons, 1981 production (3,417,000 metric tons) remains below the 1979 level, but is still well above the 1976 production level of 2,642,000 metric tons.[65, 143] Table 145 presents the LDPE capacities for plants more than 15 years old. These plant capacities may be permanently shut down or replaced by the new low pressure technology.[35] In the event that the entire 904,500 metric tons are permanently shut down, which represents 21 percent of the total LDPE capacity, additional LDPE plants would be needed in the near future.

PRODUCTION AND END USE DATA

LDPE production is the largest in the plastics industry. In 1981, U.S. LDPE production totaled 3,417,000 metric tons. The LDPE resin grades produced include: film, extrusion coating, injection molding, wire and cable coating, and specialty resins. Resin production is presented by LDPE producer in Table 146. The largest LDPE uses include:[168]

- High clarity extruded film;

- Coating for paper, metal foil, and other plastic films;

- Barrier coating;

- Pipe;

- Wire and cable coating;

- Injection molded items which require flexibility and toughness, such as housewares, pails, lids, packaging materials, furniture, automotive parts, crates, and toys;

- Blow molded items which require flexibility, such as squeeze bottles;

TABLE 144. U.S. PRODUCERS OF LDPE RESINS

Company	March 1, 1982 Capacity in Thousand Metric Tons	Type of Reactor
Atlantic Richfield Company ARCO Polymers, Inc. subsidiary Port Arthur, TX	182	Tubular
Chemplex Co. (jointly owned by American Can Co. and Getty Oil Co.) Clinton, IA	189	Autoclave
Cities Service Co. Chemicals and Minerals Group Petrochemicals Division Lake Charles, LA	330	Autoclave[a]
Dow Chemical USA Freeport, TX Plaquemine, LA	448 254	Tubular, auto-clave, and low pressure process[b]
E. I. duPont de Nemours & Co., Inc. Polymer Products Department Orange, TX Victoria, TX	211 109	Tubular and autoclave
Eastman Kodak Co. Eastman Chemical Products, Inc., subsidiary Texas Eastman Co. Longview, TX	166	Autoclave[c]
El Paso Natural Gas Co. El Paso Products Co., subsidiary El Paso Polyolefins Co. Rexene Co. Pasadena, TX Odessa, TX	182 68	Tubular
Exxon Corp. Exxon Chemical Co. Division Exxon Chemical Americas Baton Rouge, LA	300	Tubular and autoclave[d]

(continued)

TABLE 144 (continued)

Company	March 1, 1982 Capacity in Thousand Metric Tons	Type of Reactor
Gulf Oil Corporation		
Gulf Oil Chemical Co.		
Plastics Division		
Cedar Bayou, TX[e]	259	Autoclave
Orange, TX	130	special process
InterNorth, Inc.		
Northern Petrochemical Co.,		
subsidiary		
Polymers Division		
Morris, IL	295	Tubular
Mobil Corp.		
Mobil Oil Corp.		
Mobil Chemical Co. Division		
Petrochemicals Division		
Beaumont, TX	136	Tubular[f]
National Distillers and Chemical		
Corp.		
Chemicals Division		
U.S. Industrial Chemicals Co.		
Division		
Deerpark, TX	250	Tubular and
Tuscola, IL	75	autoclave
Phillips Petroleum Co.		
Plastics Division		
Pasadena, TX	---g	Low pressure process
Union Carbide Corp.		
Chemicals and Plastics Division		
Seadrift, TX[h]	545	Tubular and low pressure process
Taft, LA	273[i]	Low pressure
Union Carbide, Inc., subsidiary		
Penuelas, PR	141	Tubular
TOTAL[i]	4,543	

(continued)

TABLE 144 (continued)

[a]A 90,900 metric tons per year expansion came onstream in 1981.

[b]Company plans to add 90,900 metric tons of capacity for low pressure LDPE
at Plaquemine in 1982.

[c]Company plans to debottleneck and add new capacity for conventional LDPE
to give a total of 272,700 metric tons capacity at this location by 1983.

[d]Company is planning to add 18,200 metric tons of capacity at Baton Rouge
by 1983 and plans to build a new facility of 272,700 metric tons annual
capacity for low pressure LDPE at Mont Bellview, TX in 1982 or 1983.

[e]Company plans to add 227,300 metric tons capacity at Cedar Bayou.

[f]Company plans to expand its high pressue LDPE capacity to 227,300 metric
tons per year and construced 136,400 metric tons unit to produce low
pressure linear resin by 1983.

[g]The bulk of the plant's actual production is HDPE, but some LDPE resin (a
medium density pipe resin) is produced.

[h]Seadrift capacity includes 181,800 metric tons convertible HDPE/LLDPE
capacity which was used to make LLDPE in 1981. A portion of the capacity
at Taft, LA is also used for HDPE production.

[i]1981 production is reported as 3,417 metric tons, which includes 451,000
metric tons of LLDPE resins and 290,000 metric tons of ethylene-vinyl
acetate copolymer resins.

Sources: Chemical Economics Handbook, updated annually, 1982 data.
           Directory of Chemical Producers, 1982.
           Modern Plastics, January 1982.

TABLE 145.   LDPE PLANT CAPACITIES WHICH ARE MORE THAN 15 YEARS OLD

		Thousand Metric Tons/Year
ARCO	Port Arthur, TX	31.8
Dow	Freeport, TX	63.6
	Plaquemine, LA	36.4
DuPont	Orange, TX	136.4
	Victoria, TX	45.5
Eastman	Longview, TX	90.9
El Paso, Inc.	Odessa, TX	136.4
Gulf	Cedar Bayou, TX	90.9
	Orange, TX	90.9
National Distillers (USI)	Tuscola, IL	54.9
Union Carbide	Seadrift, TX	90.9
	Torrance, CA	36.4
		904.5

Source: Chemical Economics Handbook, updated annually, 1981 data.

TABLE 146.   1981 LDPE CONCENTRATION OF MAJOR GRADES BY PRODUCER

Company	Production in Thousand Metric Tons
ARCO	181
Chemplex	188
Cities Service	329
Dow	475
DuPont	320
Eastman	166
Exxon	299
Gulf	388
Mobil	136
Northern Petrochemical	294
El Paso	249
Union Carbide	573
USI	324

Source: Chemical Economics Handbook, updated annually, 1981 data.

- Powder coating used for textile, paper, flexible substrates, pipe, and metal drums, as well as automotive carpet backs;

- Rotational molded articles such as toys, storage tanks, and recreational vehicle tanks; and

- Extruded foam sheets and planks, which include the envelopes and bags used to protect delicate articles during shipping.

Tables 147 and 148 present LDPE consumption by end use and major markets for LDPE film, respectively.

PROCESS DESCRIPTIONS

LDPE resins are manufactured by high pressure mass polymerization, solution polymerization, or low pressure gas phase polymerization. Most of the LDPE produced is manufactured using the high pressure processes.

The emergence of high yield catalysts for solution and low pressure gas phase polymerization has eliminated the need for catalyst deactivation and catalyst removal from the polymer, called polymer deashing. The production of low molecular weight polyethylene wax is also reduced. The simplification of the LDPE process by removing or minimizing these steps reduces the amount of VOC emissions generated.[168, 184, 190, 193, 195, 240, 274]

LDPE is polymerized from ethylene using either a radical initiator in a high pressure mass polymerization process or a catalyst in a low pressure fluidized bed reactor. Solution polymerization processes may use either a catalyst or an initiator.

For the LDPE processes using a radical initiator, the ethylene reacts with the radical initiator (R) to form a radical group. As other ethylene molecules react with the ethylene radical, the polyethylene chain is propagated. The polymerization terminates when either two radical groups combine (radical combination), by disproportionation, or when the radical reacts with a molecular weight regulating agent, such as hydrogen. The polymerization reaction is depicted below:

<u>Initiation</u>

$$CH_2\!=\!CH_2 \;+\; R\!\bullet \;\longrightarrow\; \overset{\bullet}{C}H_2CH_2R$$

<u>Propagation</u>

$$\overset{\bullet}{C}H_2CH_2R \;+\; n(CH_2\!=\!CH_2) \;\longrightarrow\; \overset{\bullet}{C}H_2CH_2\!\!-\!\!(CH_2CH_2)_{\overline{n}}R$$

<u>Termination by Radical Combination</u>

$$\overset{\bullet}{C}H_2CH_2R \;+\; \overset{\bullet}{C}H_2CH_2\!\!-\!\!(CH_2CH_2)_{\overline{n}}\,R \;\longrightarrow$$

$$RCH_2CH_2CH_2CH_2\!\!-\!\!(CH_2CH_2)_{\overline{n}}\,R$$

TABLE 147. LDPE CONSUMPTION BY END-USE FOR 1981

End-Use	Thousand Metric Tons
Blow Molding	26
Extrusion	
Coating	251
Film (12 mil and under)	1,875
Pipe and Conduit	12
Sheet (over 12 mil)	6
Wire and Cable	155
Other Extrusion	29
Injection Molding	234
Rotomolding	25
Export	409
Other (chiefly resold resins and resin used for blending and compounding)	395
TOTAL	3,417

Source: Modern Plastics, January 1982.

TABLE 148.   LDPE FILM MARKET CONSUMPTION FOR 1981

Market	Thousand Metric Tons
Packaging, Food	
Baked Goods	118
Candy	15
Dairy	22
Frozen Food	42
Meat/Poultry/Seafood	61
Produce	59
Packaging, Retail Carryout Bags	
Tee-Shirt Sacks (both grocery and non-grocery applications)	18
Other Merchandise Bags (handle, drawstring, and other types)	30
Grocery Wetpack	15
Self-Service Bags (includes perforated rollstock - e.g. for produce)	22
Garment Bags (new-clothes bags and bags used by laundries or dry cleaners)	47
Packaging, Other	
Heavy-Duty Sacks (all-plastic sacks)	47
Industrial Liners (drum, bin, and box liners)	74
Rack and Counter Bags	79
Multiwall Sack Liners	15
Shrink Wrap, Pallet	30
Shrink Wrap, Other (for overwrapping or bundling)	82
Stretch Wrap	35
Textile	82
Packaging, Miscellaneous Food, Nonfood	210
TOTAL PACKAGING	1,103
Nonpackaging	
Agriculture	60
Diaper Backing	60
Household (food and storage bags and wraps)	49
Industrial Sheeting (construction and industrial rollstock)	93
Nonwoven Disposables	11
Trash Bags (for home, institutional, or industrial use)	529
Miscellaneous	49
TOTAL NONPACKAGING	851
TOTAL	1,954

(continued)

TABLE 148 (continued)

Market	Thousand Metric Tons
Less Tonnage From Resold Resin	-56
Less Tonnage From Downgaging[a]	-23
ADJUSTED TOTAL	1,875

[a]Assumes 15% downgaging factor with use of LLDPE.

Source: *Modern Plastics*, January 1982.

Termination by Disproportionation

$$R-(CH_2-CH_2)_n CH_2-\overset{\bullet}{C}H_2 \ + \ R'-(CH_2-CH_2)_m CH_2-\overset{\bullet}{C}H_2 \longrightarrow$$

$$R-(CH_2-CH_2)_n CH=CH_2 \ + \ R'-(CH_2 CH_2)_m CH_2-CH_3$$

Termination by Chain Transfer

$$R-(CH_2 CH_2)_n-CH_2\overset{\bullet}{C}H_2 \ + \ R'R'' \longrightarrow$$

$$R-(CH_2 CH_2)_n-CH_2 CH_2-R' \ + \ R''\bullet$$

The LDPE process may also be initiated by a Ziegler type catalyst. If a
catalyst is used, the ethylene reacts with an active site on the catalyst to
form a radical group. As other ethylene molecules react with the ethylene-
catalyst radical, the polyethylene chain is propagated. The polymerization
terminates when either the carbon-catalyst bond becomes weak in relation to
the carbon-carbon bonds in the chain or a molecular weight regulating agent,
such as hydrogen, reacts with the polymer chain. The polymerization reaction is depicted below:

Initiation

$$CH_2=CH_2 \ + \ Catalyst \longrightarrow Catalyst-CH_2-\overset{\bullet}{C}H_2$$

Propagation

$$Catalyst-CH_2-\overset{\bullet}{C}H_2 \ + \ nCH_2=CH_2 \longrightarrow$$

$$Catalyst-(CH_2 CH_2)_n-CH_2-\overset{\bullet}{C}H_2$$

Termination by Elimination

$$Catalyst-CH_2-CH_2-(CH_2 CH_2)_n-CH_2-\overset{\bullet}{C}H_2 \longrightarrow$$

$$Catalyst \ + \ CH_2=CH-(CH_2 CH_2)_n-CH_2-CH_3$$

Termination by Hydrogenation

$$Catalyst-(CH_2 CH_2)_n-CH_2-\overset{\bullet}{C}H_2 \ + \ H_2 \longrightarrow$$

$$Catalyst \ + \ H\bullet + \ H-(CH_2 CH_2)_n-CH=CH_2$$

Tables 149 and 150 list typical input materials and operating parameters for the LDPE processes, respectively. Other input materials used in
LDPE production such as radical initiators, comonomers, molecular weight
regulating agents, solvents, and specialty chemicals (e.g. antioxidants) are
listed in Table 151.

TABLE 149.   TYPICAL INPUT MATERIALS TO LDPE PROCESSES

Process	Monomer	Radical Initiator or Catalyst	Solvent	Comonomer
Mass	X	X		X
Solution	X	X	X	X
Gas-Phase	X	X		X

TABLE 150.   TYPICAL OPERATING PARAMETERS FOR LDPE PROCESSES

Process	Temperature	Pressure	Reaction Time	Conversion
Mass	150-350°C	6.9×10^6 - 3.4×10^8 Pa	20 sec-2 min	15-26%[a]
Solution	150-200°C	2.8 - 4.1×10^6 Pa	N/A	>97%
Gas Phase	100°C	6.9×10^5 - 2.1×10^6 Pa	--[b]	--[b]

[a]Autoclave - 8-20%.
 Tubular - 20-30%.

[b]No reaction times were given in the literature; however, several passes
are necessary to achieve the desired conversion.

N/A - Not Available.

Sources: Encyclopedia of Chemical Technology, 3rd Edition.
           Encyclopedia of Polymer Science and Technology.
           Hydrocarbon Processing, November 1979 and November 1981.
           Modern Plastics Encyclopedia, 1981-1982.

TABLE 151.   INPUT MATERIALS AND SPECIALTY CHEMICALS FOR
LDPE PROCESSES IN ADDITION TO MONOMER

Function	Compound	Process
Comonomer	Butene	Mass
	Hexene	Solution
	Octene	Gas Phase
	Vinyl Acetate	
Radical Initiator	Caprylyl peroxide	Mass
	Cumene hydroperoxide	Solution
	Decanoyl peroxide	
	Lauroyl peroxide	
	Oxygen	
	t-butyl hydroperoxide	
	t-butyl perbenzoate	
	t-butyl peroctoate	
	t-butyl peroxyacetate	
	t-butyl peroxyneodecanoate	
	t-butyl peroxypivalate	
Catalyst	Powdered chromium oxide	Solution
	Proprietary composition	Gas Phase
Solvent	Benzene	Solution
	Chlorobenzene	
Antioxidants	Dilauryl Thiodipropionate	Mass
	(DLTDP)	Solution
	Hindered phenols	Gas Phase
	Phosphates	
	Stearates	

Sources: Encyclopedia of Chemical Technology, 3rd Edition.
Encyclopedia of Polymer Science and Technology.
Modern Plastics Encyclopedia, 1981-1982.

High Pressure Mass Polymerization

High pressure mass polymerization has been used for LDPE production
since 1933.[168] As shown in Figure 45, the ethylene feed is first mixed
with ethylene from the low pressure recycle, compressed in a multistage
compressor, and mixed with the medium pressure recycle gas. The mixed
ethylene stream is then compressed in a second compressor to achieve the
desired pressure. The ethylene feed stream is mixed with the radical
initiator either before or after the second compression step and sent to the
reactor. A typical recipe for LDPE production via high-pressure mass
polymerization is as follows:[1, 192, 195]

Material	Parts by Weight
Ethylene	1,025
Initiator	2
Antioxidant	<1

The polymerization is performed in either a stirred autoclave or a
tubular reactor. Stirred autoclave reactors typically have length to
diameter ratios which start at 2:1 and may reach 20:1. Baffles divide the
autoclave reactor into separate reaction zones while tubular reactors have
three reaction zones: preheating, reaction, and cooling. Temperatures
increase from 150° to 250°C in the preheating zone to 300° to 350°C in the
reaction zone and fall back to 180° to 230°C in the cooling zone. Tube
diameters range from 7.6×10^{-3} to 7.6×10^{-2} meters and the reactors
typically have a length to diameter ratio which starts at 250:1 and may
reach 10,000:1.

Following polymerization, the ethylene/polyethylene gas stream is
separated in two steps. The first step separates unreacted gases, cools
them, and returns them to the secondary compressor. In the second step, the
low pressure separator removes the unreacted ethylene and routes it to a
cooler before returning it to the feed stream. The molten LDPE is fed to a
pelletizing extruder where antioxidants are added as well as slip, block,
and anti-block agents. The resulting LDPE pellets are then ready for
bagging.

Solution Polymerization

Solution polymerization can be used to produce polyethylene in a
density range from 0.92 to 0.96 g/cm^3. To manufacture LDPE, moderate
pressures and careful selection of the catalyst or radical initiator are
required. The polymerization, as depicted in Figure 46, occurs when
ethylene, solvent, and the initiator (either a catalyst or a free radical
initiator) are fed to the reactor. After the polymerization reaction
attains the desired conversion, the unreacted ethylene and the solvent are
removed from the polymer and recycled to the reactor while the LDPE polymer
is pelletized in an extruder.[184]

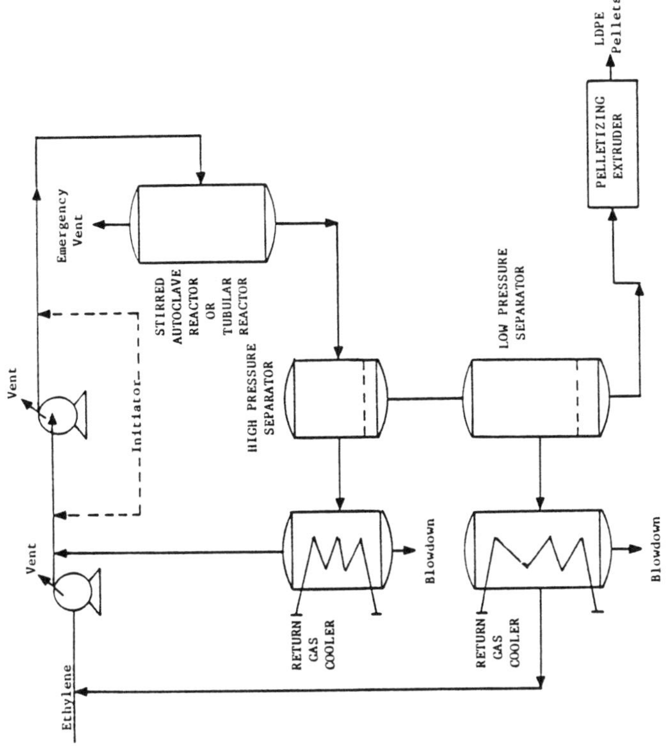

Figure 45. High pressure mass polymerization.

Source: Hydrocarbon Processing, 1981.

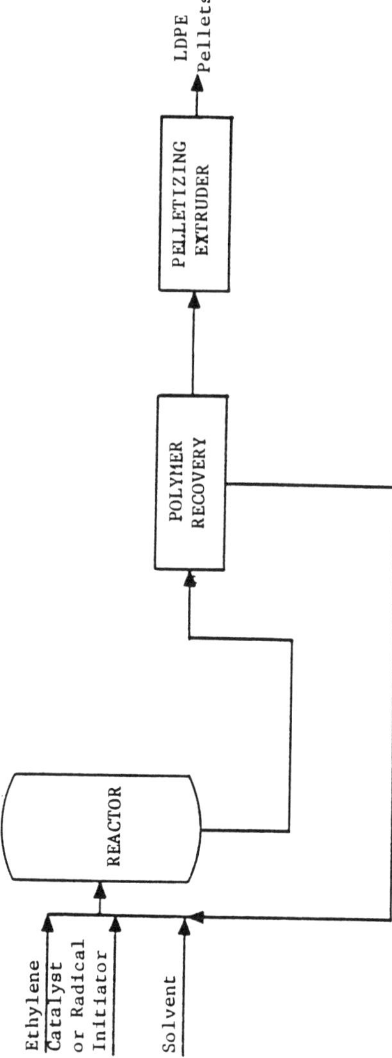

Figure 46.   Solution polymerization.

Source: Hydrocarbon Processing, 1981.

Typical input materials and quantities for solution polymerization are given
below:[193]

Material	Quantity
Ethylene (monomer)	918 kg
Organic Peroxide (initiator)	1.8 kg
Antioxidant	1.4 kg
Hydrocarbon Solvent	3.8 l

The major advantage of the solution polymerization process over high
pressure mass polymerization is the lower equipment cost. Lower pressures
also make the operating costs lower for solution polymerization when compared to mass polymerization. The elimination of the monomer stripping step
also helps reduce the capital and operating costs for this process.

Low Pressure Gas Phase Polymerization

Low pressure gas phase polymerization of LDPE is a relatively new technology. As depicted in Figure 47, ethylene, a comonomer, and catalyst are
fed to a fluidized bed reactor. The unreacted gases are rerouted to the
reactor through a single-stage centrifugal compressor followed by a heat
exchanger which cools the gases. The LDPE granular polymer leaves the reactor and enters a discharge tank. Nitrogen or another inert gas is used as a
purge to remove any remaining hydrocarbons since ethylene is explosive at
very low concentrations. The production recipe is the same as for LLDPE, as
follows:[197]

Material	Parts by Weight
Ethylene and 1-Butene (monomer and comonomer)	1,020
Catalyst	1
Hydrogen (molecular weight regulator)	

Low pressure gas phase polymerization manufactures LDPE with only half
the capital cost and one-fourth of the energy required for a traditional
high-pressure mass polymerization process.[34, 129] The process is simplified since no monomer stripping or pelletizing is required. Lower pressures
reduce the operating costs associated with the compressors.

Energy Requirements

The following energy requirements were classified by technology in the
literature consulted:[191, 192, 193, 194, 195, 196]

Technology	Energy Required	Unit/Metric Ton of Product
ANIC	Electricity	$2.88-3.06 \times 10^9$ Joules
	Steam (24 atm.)	0.15 metric tons
ARCO Technol- ogy, Inc.	Electricity	$4.16-5.03 \times 10^9$ Joules
	Steam (35 atm.)	0.6-1.0 metric tons

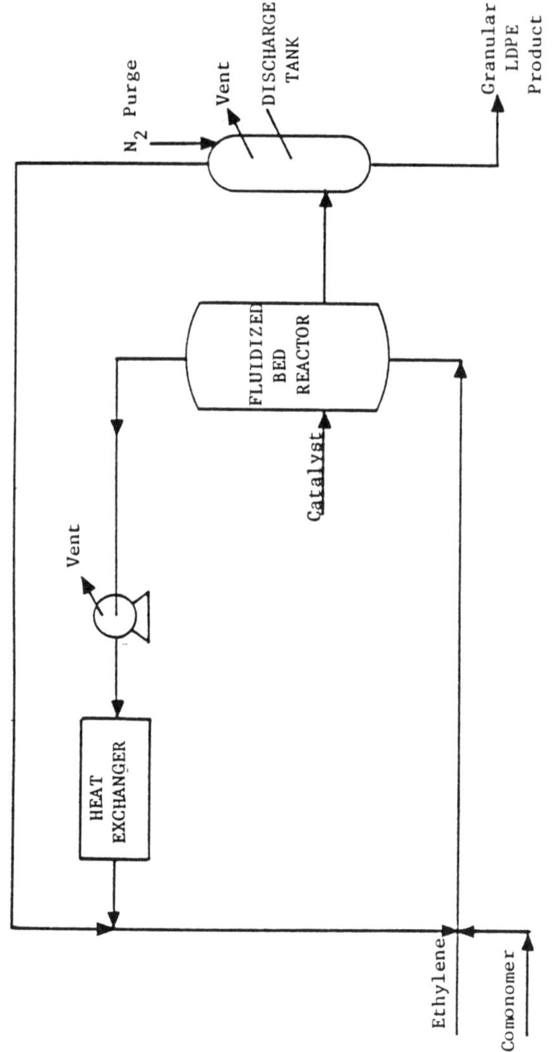

Figure 47. Low pressure gas phase polymerization.

Source: <u>Hydrocarbon Processing</u>, November 1979.

Technology	Energy Required	Unit/Metric Ton of Product
ATO Chimie	Electricity	3.78×10^9 Joules
	Steam	0.2 metric tons
El Paso Poly-olefins Co.	Electricity	3.46×10^9 Joules
	Steam	0.6 metric tons
Gulf Oil Chemicals Co.	Electricity	5.40×10^9 Joules
	Steam, low pressure (4 atm.)	0.5 metric tons
	Steam, high pressure (28 atm.)	0.125 metric tons
	Fuel for oil heater	5.28×10^8 Joules
Imhausen International Co.	Electricity	$2.83–3.89 \times 10^9$ Joules
	Steam (28 atm.)	0.12–0.20 metric tons

ENVIRONMENTAL AND INDUSTRIAL HEALTH CONSIDERATIONS

Polyethylene is a nontoxic compound used in the packaging and medical industries. Some evidence of tumorigenic activity in test animals has been observed; however, these results have not conclusively indicated the toxicity of polyethylene.[224] Benzene, a suspected carcinogen, and chromium oxide, a known animal carcinogen, are used in this process, as are the radical initiators cumene hydroperoxide and tert-butyl hydroperoxide.

The emergence of high efficiency catalysts and the use of radical initiators has resulted in a reduction of processing steps for the solution and low pressure gas phase polymerization processes. These catalysts provide low catalyst concentration in the recycle stream and the resulting polymer. Therefore, catalyst deactivation using methanol and catalyst removal from the polymer (polymer deashing) are eliminated. Simplifying the process in this manner significantly reduces the amount of VOC emissions due to methanol addition and catalyst removal.[168, 184, 190, 193, 195, 240, 274].

Worker Distribution and Emissions Release Points

Worker distribution estimates have been made by correlating major equipment manhour requirements with the process flow diagrams in Figures 45 through 47. Estimates for each polymerization process are given in Table 152.

No major air emission points are associated with LDPE production. Although there are no point source air emissions, fugitive and process sources may pose a significant environmental and/or worker health problem. The impact of these emissions depends on the stream constituents, process operating parameters, engineering and administrative controls, and maintenance program.

TABLE 152.   WORKER DISTRIBUTION ESTIMATES FOR LDPE PRODUCTION

Process	Unit	Workers/Unit/8-hour Shift
Mass Polymerization (High Pressure)	Compressor	0.25
	Continuous Reactor	0.5
	High/Low Pressure Separator	0.25
	Gas Cooler	0.25
	Extruder	1.0
Solution Polymerization	Continuous Reactor	0.5
	Polymer Recovery	0.25
	Extruder	1.0
Gas Phase Polymerization (Low Pressure)	Heat Exchanger	0.125
	Continuous Reactor	0.5
	Storage Tank	0.125
	Compressor	0.25

Process sources of fugitive emissions are listed in Table 153. Contaminants of major concern are ethylene and nitrogen, which are simple asphyxiants, chromium (VI) oxide, a human carcinogen, and benzene, a suspected human carcinogen. Particulate emissions from extruding operations may produce potentially hazardous airborne nuisance dust concentrations. Polymer additives (see Table 151) in elevated concentrations also may pose a hazard.

Health effects information for the contaminants associated with LDPE production are summarized below and in Table 10.

Emission ranges from which potential for employee exposure to volatile organic compounds may be estimated are given in Tables 154 and 155.

Health Effects

The input materials and specialty chemicals for LDPE processes include the suspected human carcinogen, benzene, and the known animal carcinogen, chromium oxide. In addition, the highly toxic cumene hydroperoxide and tert-butyl hydroperoxide, both radical initiators in the process, pose a significant risk to plant employees if exposure by ingestion, inhalation or skin absorption occurs. The following paragraphs provide a brief synopsis of the reported health effects of exposure to these substances.

Benzene is a suspected human carcinogen [101] which also exhibits mutagenic and teratogenic properties. It poses a moderate toxic hazard for acute exposures and a high hazard for chronic exposures through ingestion, inhalation, and skin absorption.[242] Effects on the central nervous system have been observed in humans upon exposure to concentrations of 100 ppm. [98] Intermittent exposure to 100 ppm over 10 years has been linked with the development of cancers in humans.[271] The OSHA standard in air is 10 ppm with a ceiling of 25 ppm.[67]

Chromium (VI) Oxide has produced positive results in animal tests for carcinogenicity [109] and there is extensive evidence of its mutagenic properties. Teratogenic effects have also been observed in laboratory animals. Chronic industrial exposures have led to severe liver damage and central nervous system involvement in humans; allergic reactions are common.[85] The American Conference of Governmental Industrial Hygienists (ACGIH) has recommended a Threshold Limit Value (TLV) of 0.05 mg/m^3 (8 hour TWA) for chromium VI compounds. The OSHA air standard is 0.5 mg/m^3 (8 hour TWA).

Cumene Hydroperoxide is highly toxic after short exposures involving ingestion, inhalation, or skin absorption of relatively small quantities. [242] Animal test data indicate that the substance is also a tumorigenic and a mutagenic agent. Prolonged inhalation of vapors results in headache and throat irritation, prolonged skin contact with contaminated clothing may cause irritation and blistering.[278]

TABLE 153.   SOURCES OF FUGITIVE EMISSIONS FROM LDPE MANUFACTURING PROCESSES

Source	Constituent	High Pressure Mass	Solution	Low Pressure Gas-Phase
Discharge Tank Vent	Ethylene			X
	Comonomer			X
Compressor Vent	Ethylene	X		X
	Comonomer	X		X
Extruder Vent	Ethylene and Poly-ethylene Particu-lates and Emissions	X	X	

TABLE 154.   CHARACTERISTICS OF VENT STREAMS FROM THE LOW DENSITY POLYETHYLENE HIGH-PRESSURE PROCESS

Stream Name	Nature	Emission rate, kg VOC/Mg product	Composition, Wt.%
Emergency reactor vent	Intermittent	0.08–4.65	95.3 Ethylene 1.0 Ethane 1.5 Propylene 2.2 Isopropanol
Combined dryer and storage bin vents	Continuous	0.71–6.01	0.2 Ethylene 99.8 Air
Emergency vent excluding reactor	Intermittent	0.2–7.65	100 Ethylene
Total Emission Rate		0.99–18.31	

Source: Polymer Manufacturing Industry Background Information for Proposed Standards, September 1983.

TABLE 155.   CHARACTERISTICS OF VENT STREAMS FROM THE
LOW DENSITY POLYETHYLENE LOW-PRESSURE PROCESS

Stream Name	Nature	Emission rate, kg VOC/Mg product	Composition, Wt.%
Raw materials purification vent	Intermittent	0.04	0.5 VOC 99.5 N_2
Catalyst additive vent	Intermittent	0.01	35 VOC 65 N_2
Comonomer purification vent	Continuous	0.10	92 VOC 8 N_2
Catalyst dehydrator handling vent	Intermittent	0.004	2 VOC 98 Air
Vacuum pump vent	Intermittent	0.03	41 VOC 59 N_2
Recovery vessel vent	Intermittent	0.06	46 VOC 54 Air
Emergency reactor blowdown vent	Intermittent	0.05	83 VOC 17 N_2
Reactor purge vent	Intermittent	0.71	1-83 VOC 99-17 N_2
Product discharge vent	Continuous	22.3	35 VOC 65 N_2
Bin vent	Continuous	0.05	0.02 VOC 99.98 Air
Compressor seal (oil) vent	Continuous	0.01	50 VOC 50 N_2
Analyzer vents	Continuous	0.01	100 VOC
Total Emission Rate		23.37	

Source: Polymer Manufacturing Industry Background Information for Proposed
Standards, September 1983.

Tert-Butyl Hydroperoxide has produced symptoms of severe depression, incoordination, cyanosis, and respiratory arrest in laboratory animals.[243] Skin contact is associated with severe local reactions. Limited animal test data also indicate mutagenic effecs. No epidemiological data are available, however, to evaluate the toxic hazard to humans.

Air Emissions

Sources of fugitive VOC emissions from LDPE processing are summarized in Table 153 by process and constituent type. Fugitive emissions of ethylene and comonomers are kept at very low levels for polyethylene production. Since ethylene is explosive in low concentrations, management practices prevent compressors and other pieces of equipment from developing significant leaks. Safety considerations dictate venting any air emissions outside the process area, as well as the use of Class 1, Group D electrical equipment and non-sparking maintenance tools. The major portion of the fugitive VOC emissions from LDPE processing originates in the polymerization section, which includes ethylene recycle and recompression. Polymerization section emissions from vents are typically flared and reported to contain 30 to 40 kg ethylene and 1 kg ethane per ton of LDPE produced.[93]

Sources of fugitive particulates are also listed in Table 153. The particulates may be collected by venting the stream to either a baghouse or electrostatic precipitator. According to EPA estimates, the population exposed to LDPE emissions in a 100 km^2 area surrounding an LDPE plant is: 30 persons exposed to hydrocarbons and 2 persons exposed to particulates.[286]

Wastewater Sources

LDPE processes generate only routine cleaning water. Several wastewater parameters from LDPE processes, shown below, are reported by EPA. These parameters were not separated according to process type for the purpose of establishing effluent limitations for the LDPE industry.[284]

Wastewater Characteristic	Unit/Metric Ton of LDPE
Production	0 – 41.72 m^3
BOD$_5$	0.2 – 4.4 kg
COD	0.2 – 54 kg
TSS	0 – 4.1 kg

The control treatment technologies common to LDPE facilities are: oil and water separator, equalization, aerated lagoon, clarification, and polishing. This wastewater treatment system has been shown to effectively treat LDPE wastewaters.[284]

Solid Waste

The solid wastes generated during LDPE production include substandard polymer which cannot be salvaged by blending and low-molecular weight polyethylene waxes. The high efficiency catalysts have reduced wax production, which is reported as 2.6 kg per metric ton LDPE.[93, 168] These wastes are not hazardous; therefore, their disposal should not pose an environmental problem.

Environmental Regulation

Effluent limitations guidelines have been set for the polyethylene industry. BPT, BAT, and NSPS call for the pH of the effluent to fall between 6.0 and 9.0 (41 Federal Register 32587, August 4, 1976).

New source performance standards (proposed by EPA on January 5, 1981) for volatile organic carbon (VOC) fugitive emissions include:

- Safety/release valves must not release more than 200 ppm above background, except in emergency pressure releases, which should not last more than five days; and

- Leaks (which are defined as VOC emissions greater than 10,000 ppm) must be repaired with 15 days.

Two input materials have been listed as hazardous (45 Federal Register 3312, May 19, 1980):

Benzene   − U019
Chlorobenzene − U037

Disposal of these materials and all LDPE resins containing residual amounts of these materials should comply with the provisions set forth in the Resource Conservation and Recovery Act (RCRA).

20. Polypropylene

Polypropylene (PP) is a stiff, heat resistant polymer. Its low specific gravity produces more finished parts per pound of polymer when compared to other thermoplastic resins. The moisture and chemical resistance of PP coupled with high tensile strength make the polymer an excellent choice for pipes and packaging. Although polypropylene homopolymers have poor impact strength at low temperatures, the addition of ethylene as a comonomer greatly improves the low temperature performance. Polypropylene production in 1981 totaled 1,774,000 metric tons.[143]

Polypropylene is produced by the polymerization of propylene using a catalyst to form the repeating structure shown below:

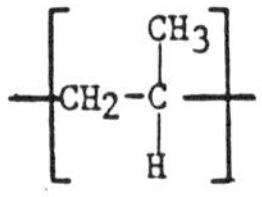

$$\left[-CH_2-\underset{\underset{H}{|}}{\overset{\overset{CH_3}{|}}{C}}- \right]$$

A highly crystalline form of polypropylene, called isotactic, is produced for commercial applications while the amorphous form, called atactic polypropylene, is an unwanted byproduct. The introduction of new catalysts has minimized the production of atactic PP, thereby eliminating the need for atactic polymer removal and deashing.

Table A-27 in Appendix A presents typical properties of polypropylene homopolymer, random copolymer (where the comonomer is copolymerized randomly with propylene), and block copolymer (where the comonomer is copolymerized with regular segments of PP). Random copolymers are formed by adding ethylene to the propylene monomer feed prior to polymerization reactor; block copolymers are manufactured by adding ethylene and propylene to a post--polymerization reactor which contains polypropylene.

Polypropylene may be manufactured to provide resins for a variety of applications. Table A-28 in Appendix A presents typical properties of various polypropylene grades.

INDUSTRY DESCRIPTION

Twelve chemical and petrochemical companies constitute the polypropylene industry. These producers have 16 sites; 11 are located in Texas while the other 5 sites are in Louisiana, New Jersey, Illinois, and West Virginia. Total annual polypropylene capacity for 1981 was 2,359,100 metric tons.[35, 56, 90, 142] Major U.S. PP producers are listed in Table 156 as well as their capacity, process type, and merchant source. Captive use constitutes 52 percent of the available capacity; the remaining 48 percent is available for production of PP for merchant use (external sales).

Polypropylene may be produced by solution, mass, or gas phase polymerization. New high yield catalysts used in any of these processes contribute significant energy savings and result in a reduction of the atactic PP concentration in the product.

Polypropylene is used in many segments of the U.S. economy. Therefore, general trends in consumer buying and the automotive industry are reflected in PP production. Polypropylene production in 1979, 1980, and 1981 was 1,746,000 metric tons, 1,658,000 metric tons, and 1,774,000 metric tons, respectively. The current excess PP capacity (33 percent) will accomodate any industry growth in the near future.

PRODUCTION AND END USE DATA

The polypropylene segment of the plastics and resins industry comprised 10 percent of the total industry production in 1980.[65] The largest markets for polypropylene are consumer oriented, with over half of the PP produced used for packaging, consumer and institutional goods, and furniture and furnishings. Polypropylene is used to manufacture:[46, 90]

- Syrup bottles;

- Packaging for detergents, shampoos, medicines, medical irrigation solutions, and a wide variety of over-the-counter drugs;

- Fenders, splash shields, automotive fan shrouds and heater ducts, replacement battery housings, and replacement for hard rubber original equipment;

- Primary and secondary carpet backing;

- Woven mats which separate the road surface from the road bed;

- Shipping sacks for potatoes, cabbage, and industrial products;

- Sandbags;

- Bags and overwrap film for soft goods;

- Electrical capacitors;

TABLE 156.   U.S. POLYPROPYLENE PRODUCERS

Producer	January 1982 Capacity (Metric Tons)	Process	Merchant Source
Atlantic Richfield Co. ARCO Polymers, Inc., subsidiary ARCO Chemical Company Division La Porte, TX	181,800	Mass	Captive
Eastman Kodak Co. Eastman Chemical Products, Inc., subsidiary Texas Eastman Co. Longview, TX	63,600	Solution	Captive
El Paso Natural Gas Co. El Paso Products, Co., subsidiary El Paso Polyolefins Co. Rexene Co. Pasadena, TX Odessa, TX	68,200 68,200	Mass, Gas Phase	Merchant Captive
Exxon Corp. Exxon Chemical Co. Division Exxon Chemical Co. USA Baytown, TX	181,800	Solution	Captive
Gulf Oil Corporation Gulf Oil Chemicals Co. Olefins & Derivatives Division Cedar Bayou, TX	181,800	Solution	Captive
Hercules Inc. Polymers Dept. Bayport, TX Lake Charles, LA	204,500 395,500	Solution	Merchant Merchant
InterNorth, Inc. Northern Petrochemical Co., subsidiary Polymers Division Morris, IL	90,900	Gas Phase	Captive

(continued)

TABLE 156 (continued)

Producer	January 1982 Capacity (Metric Tons)	Process	Merchant Source
Phillips Petroleum Co. Phillips Chemical Co., subsidiary Plastics Division Pasadena, TX	90,900	Mass	Captive
Shell Chemical Co. Norco, LA	136,400	Mass, Gas	Captive
Woodbury, NJ	136,400	Phase	Merchant
Soltex Polymer Corp. Deer Park, TX	90,900	Solution	Merchant
Standard Oil Co. (IN) AMOCO Chemicals Corp., subsidiary Chocolate Bayou, TX	234,100	Gas Phase, Solution	Captive
U.S. Steel Corp. USS Chemicals Division USS Novamont, Inc., subsidiary Kenova, WV	75,000	Solution	Merchant
La Porte, TX	159,100		Merchant
TOTAL	2,359,100		

Sources: Chemical Economics Handbook, updated annually, 1981 Data.
           Directory of Chemical Producers, 1982.
           Encyclopedia of Chemical Technology, 3rd Edition.
           Modern Plastics, May 1982.

- Aerosol covers and valves, tote boxes, beverage cases, fresh produce shipping containers, and thin sheet for food containers;

- Car bumpers;

- Large boxes and tanks for shipping products;

- Decorative ribbons, straps for automotive and nonautomotive packing, monofilaments for brushes, and yarns for textiles and cords; and

- Pipes for industrial waste systems, irrigation systems, and floor heating systems.

Table 157 lists polypropylene consumption according to major markets.

PROCESS DESCRIPTIONS

Polypropylene may be produced using solution, mass, or gas phase, or a combination of these polymerization processes. Of the twelve polypropylene producers listed in Table 156, six use solution polymerization. The remaining six producers utilize: mass polymerization (two producers), mass and gas phase polymerization (two producers), gas phase polymerization (one producer), and solution and gas phase polymerization (one producer). Solution polymerization contributes 57 percent of the total PP capacity, with mass, gas phase, mass/gas phase, and solution/gas phase adding the remaining 12, 4, 17, and 10 percent, respectively.

In polypropylene production, propylene monomer is polymerized to make the homopolymer. Random copolymers are manufactured when ethylene is fed along with propylene to the polymerization reactor; block copolymers are formed when ethylene and propylene are fed to a post-polymerization reactor which contains polypropylene. The reaction is initiated when a propylene molecule is added to an organometallic active center, depicted as M^+-R^-. Propagation occurs by adding monomer at the active organometallic center. The reaction terminates either by monomer transfer, metal alkyl transfer, or by hydride ion transfer, along with realkylation of the catalyst.

<u>Initiation</u>

$$M^+\text{-}R^- + CH_2\text{=}CH\text{-}CH_3 \longrightarrow M^+\text{-}^-CH_2\text{-}\underset{\underset{CH_3}{|}}{CH}\text{-}R$$

<u>Propagation</u>

$$M^+\text{-}^-CH_2\text{-}\underset{\underset{CH_3}{|}}{CH}\text{-}R + nCH_2\text{=}CH\text{-}CH_3 \longrightarrow M^+\text{-}^-CH_2\text{-}\underset{\underset{CH_3}{|}}{CH}\left[CH_2\text{-}\underset{\underset{CH_3}{|}}{CH}\right]_n R$$

TABLE 157.   MAJOR POLYPROPYLENE MARKETS

Market	1981 Metric Tons
Blow Molding	
Medical Containers	5,000
Consumer Packaging	28,000
TOTAL BLOW MOLDING	33,000
Extrusion	
Coating	5,000
Fibers andd Filaments	473,000
Film (up to 10 mil)	
Oriented	107,000
Unoriented	38,000
Pipe and Conduit	9,000
Sheet (over 10 mil)	17,000
Straws	15,000
Wire and Cable	5,000
Other Extrusion	11,000
TOTAL EXTRUSION	680,000
Injection Molding	
Appliances	43,000
Furniture	16,000
Housewares	86,000
Luggage and Cases	5,000
Medical	47,000
Packaging	
Closures	72,000
Containers and Lids	49,000
Toys and Novelties	41,000
Transportation	
Battery Cases	63,000
Other	63,000
Other Injection Molding	92,000
TOTAL INJECTION MOLDING	577,000
Export	282,000
Other (Chiefly resold material and material used for blending)	202,000
TOTAL	1,774,000

Source: *Modern Plastics*, January 1982.

Termination by Monomer Transfer

$$M^{+}\text{-}{}^{-}CH_2\text{-}\underset{CH_3}{CH}\text{-}\left[CH_2\text{-}\underset{CH_3}{CHR}\right]_n + CH_2{=}\underset{CH_3}{CH} \longrightarrow$$

$$M^{+}\text{-}{}^{-}CH_2\text{-}\underset{CH_3}{CH_2} + CH_2{=}\underset{CH_3}{C}\text{-}\left[CH_2\text{-}\underset{CH_3}{CH}\right]_n R$$

Termination by Metal Alkyl Transfer

$$M^{+}\text{-}{}^{-}CH_2\text{-}\underset{CH_3}{CH}\text{-}\left[CH_2\text{-}\underset{CH_3}{CH}\right]_n R + AlR'_3 \longrightarrow$$

$$M^{+}\text{-}{}^{-}R' + R'_2AlCH_2\text{-}\underset{CH_3}{CH}\text{-}\left[CH_2\text{-}\underset{CH_3}{CH}\right]_n R$$

Termination by Hydride Ion Transfer

$$M^{+}\text{-}{}^{-}CH_2\text{-}\underset{CH_3}{CH}\text{-}\left[CH_2\text{-}\underset{CH_3}{CH}\right]_n R \rightarrow M^{+}\text{-}{}^{-}H + CH_2{=}\underset{CH_3}{C}\text{-}\left[CH_2\text{-}\underset{CH_3}{CH}\right]_n R$$

$$M^{+}\text{-}{}^{-}H + CH_2{=}CH\text{-}CH_3 \rightarrow M\text{-}CH_2\text{-}CH_2\text{-}CH_3$$

The tertiary hydrogen atoms in the polypropylene chain are susceptible to attack by oxygen and UV radiation; therefore, antioxidants and light stabilizers are added to the pelletization step to ensure a stable polymer during storage and shipping.

Tables 158 and 159 present typical input materials and operation parameters for polypropylene production processes. Input materials, including specialty chemicals, used in polypropylene production are listed in Table 160.

<u>Solution Polymerization</u>

Solution polymerization processes constitute 63 percent of the U.S. polypropylene capacity. This process, as illustrated in Figures 48 and 49, uses a solvent, typically heptane, as the polymerization medium. Propylene, catalyst, and solvent are charged to the polymerization reactor. The resulting polymer slurry is fed to a flash tank where the unreacted propylene is removed and recycled to the reactor.

TABLE 158.   TYPICAL INPUT MATERIALS TO PP PRODUCTION
PROCESSES IN ADDITION TO MONOMER

Process	Input		
	Catalyst	Organic Solvent	Molecular Weight Regulating Agent
Solution	X	X	X
Mass	X		X
Gas Phase	X		X

TABLE 159.   TYPICAL OPERATING PARAMETERS FOR PP PRODUCTION PROCESSES

Process	Temperature	Pressure
Solution	50 – 90 °C	500,000 – 1,500,000 Pa
Mass	55 – 80 °C	2,700,000 – 4,100,000 Pa
Gas Phase	Proprietary	Proprietary

Sources: Chemical Engineering, April 20, 1981.
Encyclopedia of Chemical Technology, 3rd Edition.
Hydrocarbon Processing, November 1979, 1980, 1981.

TABLE 160.   INPUT MATERIALS AND SPECIALTY CHEMICALS USED IN
POLYPROPYLENE PRODUCTION (IN ADDITION TO MONOMER)

Function	Compound
Comonomer	Ethylene
Catalyst	
Ziegler	Titanium trichloride and aluminum diethylmonochloride[a]

Supported	Titanium tetrachloride supported on:
	  Alumina
	  Magnesium chloride
	  Magnesium hydroxide
	  Magnesium hydroxychloride
	  Silica

Solvent	Heptane
Catalyst Removal	Isopropyl alcohol Methanol
Molecular Weight Regulating Agent	Hydrogen
Antioxidants	Dilauryl thiodipropionate Distearyl pentaerythitol diphosphite Distearyl thiodipropionate 2,6-di-tert-butyl-4-methylphenol Octadecyl 3,5-di(tert-butyl-4-hydroxy) hydrocinnamate Sterically hindered phenols Tris(nonylphenyl)phospite
Quenchers (absorb excitation energy caused by UV radiation)	Nickel dibutyl dithiocarbamate
Sterically Hindered Amine Light Stabilizers (HALS) (capture radicals, decompose peroxides, act as quenchers)	Sterically hindered amines

[a]Second generation catalysts include Lewis bases (e.g., esters, ethers, amines, chlorides, and oxychlorides of phosphorus and phosphites).

Sources: *Encyclopedia of Chemical Technology*, 3rd Edition.
Hydrocarbon Processing, November 1977 and 1979.
Modern Plastics, November 1979.

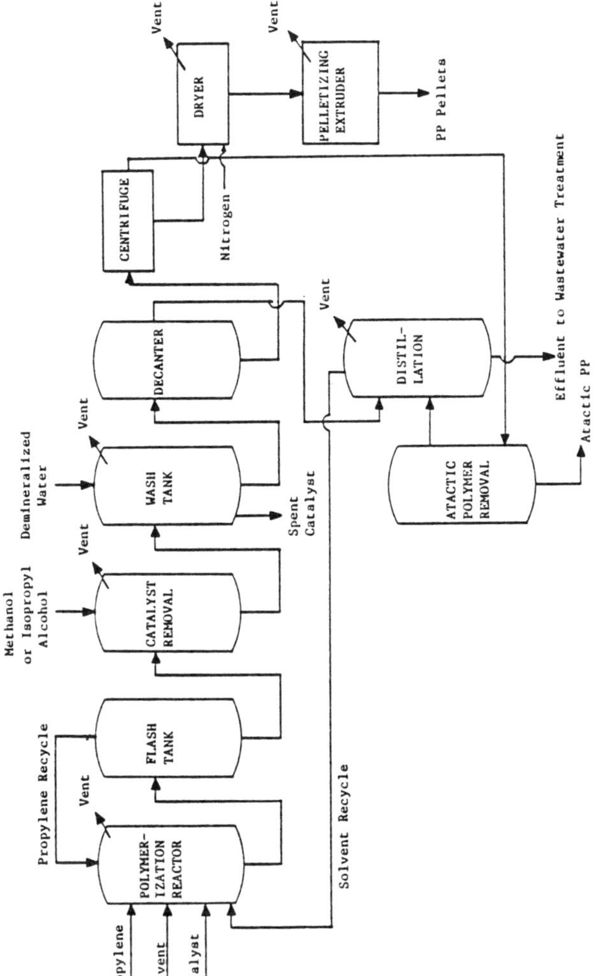

Figure 48. Solution polymerization of PP using a conventional catalyst.

Source: Encyclopedia of Chemical Technology, 3rd Edition.

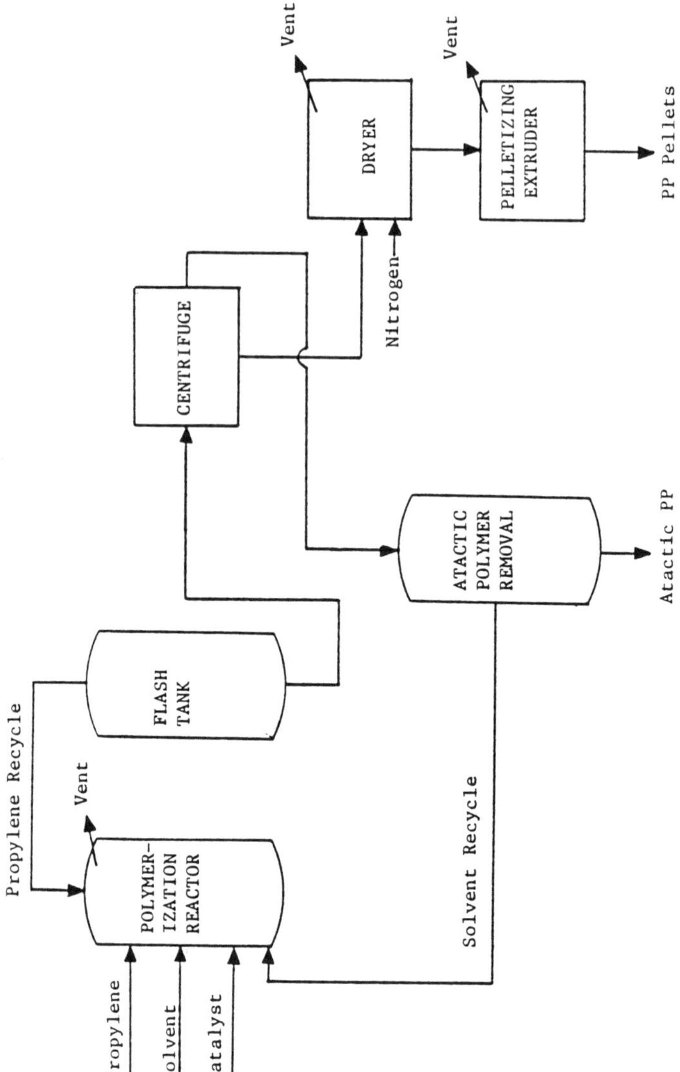

Figure 49. Solution polymerization of PP using a high yield catalyst.

Source: Encyclopedia of Chemical Technology, 3rd Edition.

Conventional Catalyst

A typical recipe for the production of polypropylene via solution polymerization using a conventional catalyst is as follows:[202]

Material	Parts by Weight
Propylene (monomer)	1,045
Titanium Catalyst	0.4
Alcohol	0.5
Hydrocarbon Solvent	8

In the conventional PP solution polymerization process, the polymer-solvent mixture is washed first with methanol (or isopropyl alcohol) to deactivate the catalyst, then with water to remove any remaining catalyst solids. The spent catalyst is either regenerated or sent to disposal. The washed slurry is decanted; the solids are centrifuged, dried in a nitrogen atmosphere, and pelletized in an extruder. The liquid mixture from the decanter is distilled so that the remaining solvent may be recycled to the reactor while the water is sent to wastewater treatment. The effluent from the centrifuge is heated to facilitate atactic polymer removal before the effluent is distilled.

High Yield Catalyst

The high yield catalysts reduce the complexity and energy consumption of the solution polymerization process. After any unreacted propylene is recycled to the reactor, the polymer-solvent mixture is centrifuged. The polymer is dried in a nitrogen atmosphere and pelletized while the effluent is heated to facilitate atactic polymer removal. The purified solvent is then recycled to the reactor. High polymer yield per pound of catalyst (typically 3,000 g/g for conventional catalysts and over 250,000 g/g for high yield catalysts) reduces the amount of catalyst in the polymer, eliminating the need for methanol addition for deactivation and subsequent washing. Since catalyst levels in the polymer are so small (ppb or less), the catalyst residues are not removed. Since methanol and water are not needed for catalyst removal, distillation of the centrifugate is also eliminated. This process simplification results in a reduction of process steam and electrical energy usage by 85 and 12 percent, respectively.[39]

Following are typical reactor inputs for polymerization of propylene using a high-yield catalyst:[202, 225]

Material	Parts by Weight
Propylene (monomer)	1,045
High-Yield Catalyst	0.004
Hydrocarbon Solvent	8

Mass Polymerization

Mass polymerization of polypropylene uses liquid propylene as the reaction medium. Propylene and a catalyst are fed into a loop or liquid pool reactor where the polymerization takes place. A typical production recipe for polypropylene via mass polymerization is shown below:[201]

Material	Parts by Weight
Propylene	1,010
Ziegler-Natta Supported Catalyst	0.14-0.20
Hydrocarbon (diluent)	

The amount of solvent was not found in the literature, but is expected to be less than 30 parts by weight.

As depicted in Figure 50, after the monomer-polymer mixture is removed from the reactor, methanol is added to facilitate catalyst removal. Spent catalyst residues are either recycled for regeneration or sent to disposal. Following catalyst removal, the mixture is washed with propylene and fed to a flash tank. The unreacted monomer is separated and recycled to the reactor while the polymer is dried in a nitrogen atmosphere and pelletized. If a high yield catalyst is used, the catalyst removal step is eliminated since the catalyst residues are left in the polymer and advantages similar to those presented earlier in this section are realized.

Gas Phase Polymerization

In gas phase polymerization, a highly active, highly stereospecific catalyst is used in order to minimize catalyst residues and atactic polymer in the product. As shown in Figure 51, polymerization occurs in an agitated reactor.

Propylene and the catalyst are fed to the polymerization reactor where the reaction proceeds to the desired conversion. The resulting gaseous polymer-monomer mixture is then fed to a flash tank where unreacted propylene is separated and recycled to the reactor. Catalyst residues are a very small part of the polymer mixture (ppb or less); therefore, they are not removed. The polymer is then dried in a nitrogen atmosphere and pelletized in an extruder; the dryer exhaust gases are recycled to the reactor to recover any unreacted propylene. A typical gas phase reactor production recipe for polypropylene was not found in the literature.

Energy Requirements

The following energy requirements are given for the various technologies listed below:[199, 203, 205]

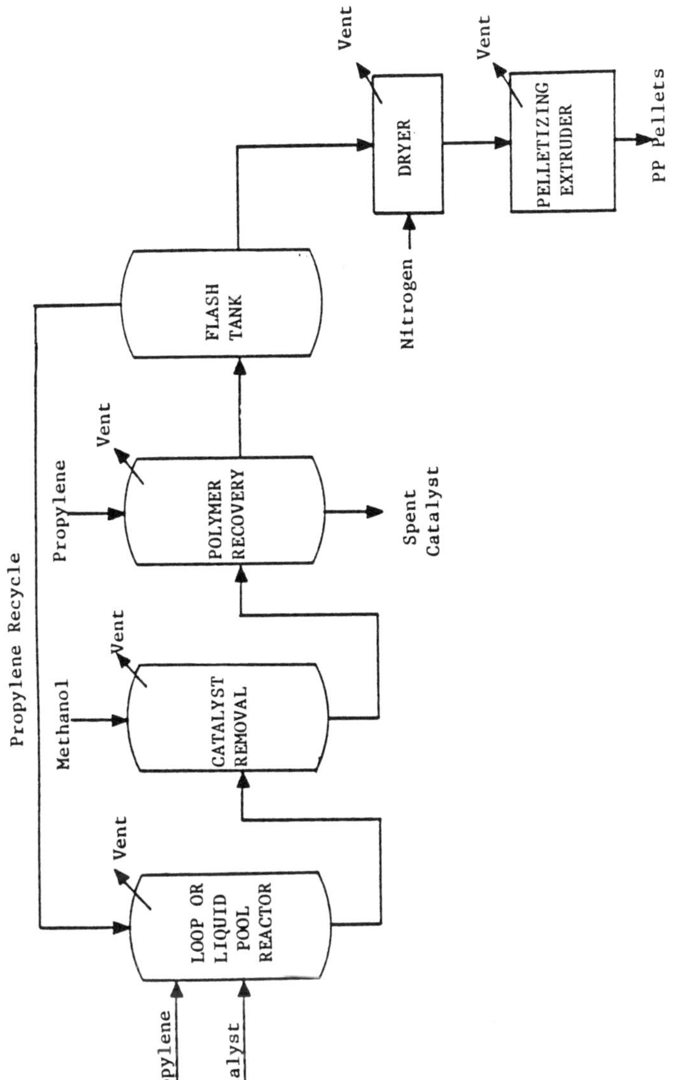

Figure 50. Mass polymerization of polypropylene using a conventional catalyst.

Source: Encyclopedia of Chemical Technology, 3rd Edition.

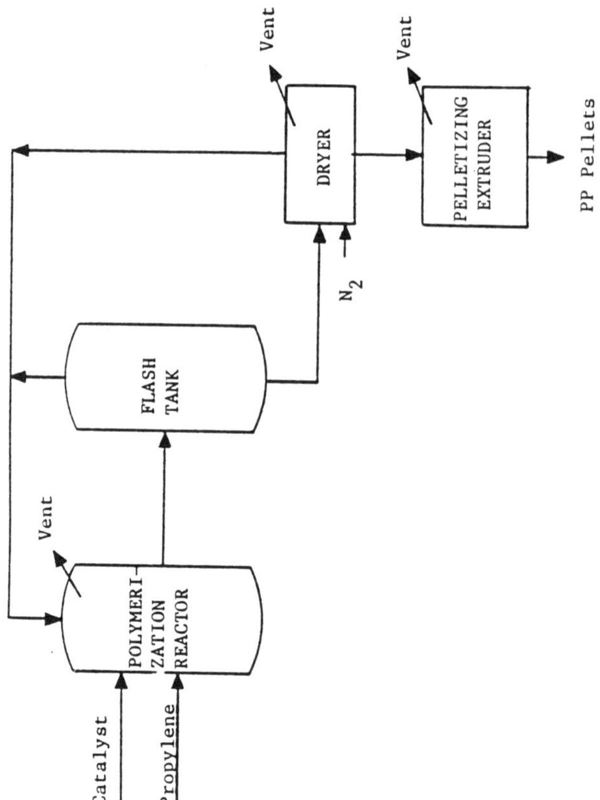

Figure 51. Gas phase polymerization.

Source: Encyclopedia of Chemical Technology, 3rd Edition.

Technology	Energy Required	Unit/Metric Ton Produced
El Pas Poly-olefins Co.	Electricity Steam	$1.69-1.98 \times 10^6$ Joules 0.5 metric ton
Mitsui Toatsu Chemicals, Inc.	Electricity Steam	1.87×10^9 Joules 2.6 metric tons
Sumitomo Chemi-cal Co., Ltd.	Electricity Steam	2.16×10^9 Joules 1.8 metric tons

ENVIRONMENTAL AND INDUSTRIAL HEALTH CONSIDERATIONS

Worker Distribution and Emissions Release Points

Worker distribution estimates have been made by correlating major equipment manhour requirements with the process flow diagrams shown in Figures 48 through 51. Estimates, shown in Table 161, have been made for solution polymerization using two catalyst types and for mass and gas phase polymerization.

Air emissions from process and fugitive sources comprise the emissions from PP production. Although there are no major point source air emissions, fugitive and process emissions may impose a significant environmental and/or worker health problem. The magnitude of this problem depends upon the stream components, process operating controls, and maintenance program.

Table 162 lists major sources of fugitive emissions from polypropylene production. Propylene, ethylene, and hydrogen are simple asphyxiants whose presence may reduce the available oxygen in the atmosphere. Heptane, iso-propanol and methanol, while not highly toxic, may be irritating to the skin and, in high concentrations, may cause central nervous system depression. Polymer and additive particulates may produce localized elevated nuisance dust concentrations.

Emission ranges from which potential employee exposure to volatile organic compounds may be estimated are given in Tables 163 and 164.

Health Effects

Several of the input materials and specialty chemicals used in poly-propylene production have exhibited tumorigenic, mutagenic, or teratogenic potential in animal tests. These include ethylene; methanol; isopropyl alcohol; 2,6-di-tert-butyl-4-methyl phenol; and nickel dibutyl dithiocarba-mate. Available data indicate, however, that these substances, except nickel dibutyl dithiocarbamate, are not highly toxic to humans upon expo-sures of short duration. Nickel dibutyl dithiocarbamate is considered very toxic by ingestion.[85]

Although polypropylene has produced positive results in animal testing for carcinogenicity [107], it poses a very low toxic hazard. For example,

TABLE 161.   WORKER DISTRIBUTION ESTIMATES FOR POLYPROPYLENE PRODUCTION

Process	Unit	Workers/Unit/8-hour Shift
Solution Polymerization (Conventional Catalyst)	Batch Reactor	1.0
	Flash Tank	0.125
	Catalyst Removal	0.25
	Washing	0.25
	Decanter	0.25
	Centrifuge	0.25
	Dryer	0.5
	Extruder	1.0
	Distillation	0.25
	Atactic Polymer Removal	0.25
Solution Polymerization (High Yield Catalyst)	Batch Reactor	1.0
	Flash Tank	0.125
	Centrifuge	0.25
	Dryer	0.5
	Extruder	1.0
	Atactic Polymer Removal	0.25
Mass Polymerization	Batch Reactor	1.0
	Catalyst Removal	0.25
	Polymer Recovery	0.25
	Flash Tank	0.125
	Dryer	0.5
	Extruder	1.0
Gas Phase Polymerization	Batch Reactor	1.0
	Flash Tank	0.125
	Dryer	0.5
	Extruder	1.0

TABLE 162. SOURCES OF FUGITIVE EMISSIONS FROM POLYPROPYLENE MANUFACTURE

Source	Constitutent	Solution	Mass	Gas Phase
Reactor Vent	Propylene	X	X	X
	Solvent	X		
Catalyst Removal[a]	Propylene	X	X	
	Alcohol	X	X	
Wash Tank Vent[a]	Propylene	X		
	Alcohol	X		
Polymer Recovery	Propylene		X	
Dryer Vent	Propylene and Poly-propylene Emissions and Particulates	X	X	X
	Solvent	X		
Pelletizing Extruder Vent	Propylene and Poly-propylene Emissions and Particulates	X	X	X
	Solvent	X		
Distillation Vent	Propylene	X		
	Methanol	X		
	Solvent	X		

[a]Not present in high yield catalyst processes.

TABLE 163. CHARACTERISTICS OF VENT STREAMS FROM THE POLYPROPYLENE SOLUTION PROCESS

Stream Name	Nature	Emission rate, kg VOC/Mg product	Composition[a,b]
Catalyst preparation	Continuous	0.07	$C_{10}HC$, isopropyl alcohol
Reactor vents	Continuous	4.07	C_3HC, $C_{10}HC$
Decanter vents	Continuous	11.49	C_3HC, $C_{10}HC$, isopropyl alcohol
Neutralizer vents	Continuous	1.82	C_3HC, C_4HC, $C_{10}HC$, isopropyl alcohol
Slurry vacuum/filter system vent	Continuous	7.93	$C_{10}HC$, isopropyl alcohol
Diluent separation and recovery	Continuous	8.72	$C_{10}HC$, isopropyl alcohol
Dryer vents	Continuous	0 to 0.6	Air and small amount of VOC
Extrusion/pelletizing vent	Continuous	2.0	25% HC
Total Emission Rate		36.79	

[a]Streams are diluted in 10 to 30 percent nitrogen.

[b]C_3HC - Propylene or any other hydrocarbon compound with three carbon atoms such as propane.
C_4HC - Butylene or any other hydrocarbon compound with four carbon atoms.
$C_{10}HC$ - A mixture of aliphatic hydrocarbons with 10-12 carbon atoms.

Source: _Polymer Manufacturing Industry Background Information for Proposed Standards_, September 1983.

TABLE 164. CHARACTERISTICS OF VENT STREAMS FROM THE POLYPROPYLENE
GAS PHASE PROCESS

Name	Nature	Emission rate, kg VOC/Mg product	Composition
Scrubber vent	Continuous	25	Propylene, Propane, Hexane
Reactor blow-down vent	Intermittent	11.5	Hydrogen, Propylene, Propane, Hexane
Total Emission Rate		36.5	

Source: Polymer Manufacturing Industry Background Information for Proposed
Standards, September 1983.

polypropylene medical devices have produced little or no irritant response when placed into connective or muscle tissues for short periods of time.[31] The propensity for diffusion of polypropylene from the material is so small that biological response cannot be detected.[31]

No data exist in the available literature with which to evaluate the potential health effects associated with the catalysts and most of the antioxidants used in the production of polypropylene.

Air Emissions

VOC fugitive emission sources for PP production are listed in Table 162. Emissions from the distillation column in solution polymerization are reported to be: ethane, 0.8 kg; propylene, 20 kg; propane, 0.2 kg; and alcohol and solvent, 10 kg per metric ton of polypropylene produced. Emissions from the dryers are reported to be 9 kg per metric ton of product.[93] The use of high yield catalysts reduces the number of processing steps, which in turn reduces the fugitive VOC emissions for the process. Other VOC emissions may result from leaks in process equipment including flanges, valves, pumps, compressors, and cooling towers. Although equipment modification may reduce these emissions, a routine inspection and maintenance program may be the most effective control.

Fugitive particulate sources are also listed in Table 162. These emissions may be controlled by venting these streams to baghouse or electrostatic precipitator to collect particulates. According to EPA estimates, the population exposed to PP emissions in a 100 km^2 area surrounding a PP plant is: 43 persons exposed to hydrocarbons and 2 persons exposed to particulates.[286]

Wastewater Sources

There are only two sources of wastewater associated with polypropylene processing: routine cleaning water and wash water from solution polymerization. Ranges of several wastewater parameters for wastewaters from polypropylene production are shown below. Values for the wastewater from the processes presented were not distinguished by process or resin type by EPA for the purpose of establishing effluent limitations for the polypropylene industry.[284]

Polypropylene Wastewater Characteristics	Unit/Metric Ton Polypropylene
Production	$2.50 - 66.75 \ m^3$
BOD_5	0 - 10 kg
COD	0 - 20 kg

The control treatment technologies common to polypropylene facilities are: screening, equalization, chemical treatment, activated sludge, and polishing pond. This wastewater treatment system has been shown to effectively treat PP wastewaters.[284]

Solid Waste

The solid wastes generated during polypropylene production include substandard resins which cannot be blended, collected particulates, atactic polymer, and catalyst residues. These wastes are not hazardous.

Environmental Regulation

Effluent limitations guidelines have been set for the polypropylene industry. BPT, BAT, and NSPS call for the pH of the effluent to fall between 6.0 and 9.0 (41 Federal Register 32587, August 4, 1976).

New source performance standards (proposed by EPA on January 5, 1981) for volatile organic carbon (VOC) fugitive emissions include:

- Safety/release valves must not release more than 200 ppm above background, except in emergency pressure releases, which should not last more than five days; and

- Leaks (which are defined as VOC emissions greater than 10,000 ppm) must be repaired with 15 days.

The only input to PP production which has been listed as hazardous is methanol, designated U154 (45 Federal Register 3312, May 19, 1980). Disposal of methanol and PP wastes containing residual amounts of methanol must comply with the criteria set forth in the Resource Conservation and Recovery Act (RCRA).

21. Polystyrene (General Purpose)

Polystyrene is a brittle thermoplastic which is clear, rigid, odor-free, and tasteless. Its physical properties include low specific gravity and good thermal stability. The ease with which polystyrene can be fabricated using heat contributes to the low cost of polystyrene products. Polystyrene has excellent thermal and electrical properties which render the polymer useful for manufacturing interior doors, lighting covers, air conditioner housings, and shutters. Other applications include flower pots, margarine tubs, caps, closures, disposable cutlery, toys, drinking cups, and egg cartons.

Polystyrene is produced by the polymerization of styrene by means of a free radical initiator. The formula

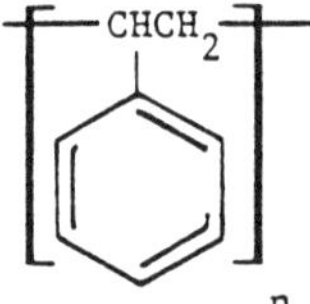

represents the repeating styrene group which provides the backbone for the polymer chain. Table A-29 in Appendix A presents typical physical properties of general purpose polystyrene.

General purpose polystyrene (GPPS) differs from impact (or rubber modified) polystyrene in that the general purpose resin is a pure homopolymer. The absence of comonomers or modifiers, such as rubber, make the polymer brittle. This section describes the manufacture of general purpose polystyrene; the section on impact (rubber modified) polystyrene presents the types of processes used in the manufacture of the copolymer. In 1980, general purpose polystyrene accounted for 49 percent of the polystyrene produced. Impact, or rubber modified, polystyrene accounted for the other 51 percent.[35]

General purpose polystyrene is produced by suspension polymerization, mass (or bulk) polymerization, or solution polymerization (also called continuous mass polymerization with solvents). Over half of the general purpose polystyrene produced is manufactured by the suspension polymerization process. Expandable bead polystyrene, used for cups and containers, is also produced by this method. Emulsion polymerization is not used for the production of general purpose polystyrene due to the detrimental effect large quantities of soap have on the clarity and electrical-insulation characteristics of the product. However, emulsion polymerization is used in the production of styrene copolymers (see the sections on acrylonitrile-butadiene-styrene polymers (ABS) and styrene-acrylonitrile polymers (SAN)).

INDUSTRY DESCRIPTION

The polystyrene industry is comprised of 18 producers of general purpose and/or impact polystyrene, three of which also produce styrene-butadiene latexes in addition to the polystyrene. These 18 producers have 44 sites, 17 of which are in Ohio, Michigan, Illinois, and Pennyslvania, and another 8 in California. The remaining plants are located in the east and southeast. Polystyrene producers are listed in Table 165 and styrene-butadiene latex producers are presented in Table 166.

Polystyrene is used in consumer related industries, such as construction, appliances, and packaging. Therefore, polystyrene consumption follows the trend set by housing starts and consumer spending. After a production peak of 1,820,000 metric tons in 1979, a drop to 1,490,900 metric tons occurred in 1980.[35, 64, 65] Most of the expansion planned is for expandable polystyrene. Excess capacity for impact polystyrene is sufficient to support industry growth for the near future. Therefore, no new plants for impact polystyrene are expected to be built.

PRODUCTION AND END USE DATA

In 1980, total polystyrene sales were 1,490,900 metric tons.[35] Polystyrene is used for packaging, toys, foam insulation, electrical and electronic uses, and serviceware, which includes the familiar cups. Table 167 presents the 1980 consumption of polystyrene. These products can be classified as either molding resins, extrusion resins, or expandable beads. General purpose polystyrene contributes 64 and 51 percent of the total polystyrene produced for molding and extrusion resins, respectively.[35]

PROCESS DESCRIPTIONS

General purpose polystyrene is produced by suspension, mass, or solution polymerization. According to one estimate, over half of the general purpose resin is manufactured using suspension polymerization,[17] as is the majority of the expandable bead polystyrene since the polymer is produced in a form which is readily used.[2] The remainder of the general purpose polystyrene production is performed by mass and solution polymerization. Suspension polymerization of general purpose polystyrene has excellent versatility for the production of numerous types of products. However, high-purity products are more difficult to obtain since water, suspension stabilizers, and unreacted initiators tend to be contaminants. The added

TABLE 165.   MAJOR U.S. PRODUCERS OF POLYSTYRENE

Producer	Capacity as of July 1, 1981 (Thousand Metric Tons)		
	General Purpose and Impact	Expandable Bead	Total
A&E Plastik Pak Co.			
A&E Plastics Division			
City of Industry, CA	15.9	0	15.9
American Hoechst Corp.			
Plastics Division			
Chesapeake, VA	118.2	0	118.2
Leominster, MA	54.5	0	54.5
Peru, IL	77.3	36.4	113.6
COMPANY TOTALS	250.0	36.4	286.4
American Petrofina Inc.			
Cosden Oil & Chemical Co.,			
subsidiary			
Big Spring, TX			0[a]
Calumet City, IL			77.3
Orange, CA			27.3
Windsor, NJ			54.4
Atlantic Richfield Co.			
ARCO Polymers, Inc., sub.			
Beaver Valley, PA	54.4	159.1[b]	213.6
BASF Wyandotte Corp.			
Polymers Groups			
Styropor Division			
Jamesburg, NJ	0	79.5[c]	79.5
Dow Chemical USA			
Allyn's Point, CT	N/A	N/A	86.4
Ironton, OH	N/A	N/A	86.4
Joliet, IL	N/A	N/A	63.6
Midland, MI	N/A	N/A	100.0
Torrance, CA	N/A	N/A	90.9[d]
COMPANY TOTALS	418.2	9.1	427.3

N/A — Not Available.

(continued)

TABLE 165 (continued)

| | Capacity as of July 1, 1981 (Thousand Metric Tons) | | |
Producer	General Purpose and Impact	Expandable Bead	Total
Carl Gordon Industries, Inc.			
Hammond Plastics			
Oxford Division			
Oxford, MA	45.5	0	45.5
Hammond Plastics Midwest, Inc.			
Owensboro, KY	22.7	0	22.7
COMPANY TOTALS	68.2	0	68.2
Gulf Oil Corporation			
Gulf Oil Chemicals Co.			
Plastics Division			
Marietta, OH	140.9e	0	140.9
Huntsman–Goodson Chemical			
Corp.			
Troy, OH	9.1	0	9.1
Kama Corp.			
Hazeltown, PA	11.4	0	11.4
Mobil Corp.			
Mobil Oil Corp.			
Mobil Chemical Co.,			
Division			
Petrochemicals Division			
Holyoke, MA	40.9	0	40.9
Joliet, IL	18.2	0	18.2f
Santa Anna, CA	29.5	0	29.5
COMPANY TOTALS	88.6	0	88.6
Monsanto Co.			
Monsanto Plastics and			
Resins Co.			
Addyston, OH	136.4	0	136.4
Decatur, AL	45.5	0	45.5
Long Beach, CA	22.7	0	22.7
Springfield, MA	136.4	0	136.4
COMPANY TOTALS	340.9	0	340.9

(continued)

TABLE 165 (continued)

Producer	Capacity as of July 1, 1981 (Thousand Metric Tons)		
	General Purpose and Impact	Expandable Bead	Total
Polysar Group			
Polysar Inc.			
Resins Division			
Copely, OH	54.5	0	54.5
Forest City, NC	18.2	0	18.2
Leominster, MA	54.5	0	54.5
COMPANY TOTALS	127.3	0	127.3
Shell Chemical Co.			
Belpre (Marietta), OH	136.4	0	136.4
Standard Oil Co. (IN)			
AMOCO Chemical Corp., sub.			
Joliet, IL	136.4	0	136.4
Torrance, CA	15.9	0	15.9
Willow Springs, IL	40.9	0	40.9
COMPANY TOTALS	193.2	0	193.2
Texstyrene Plastics, Inc.			
Fort Worth, TX	0	22.7	22.7
U.S. Steel Corp.			
USS Chemicals Division			
Haverhill, OH	9.18	0	9.1
Vititek Inc.			
Delano, CA	0	2.3	2.3
TOTAL	2,068.2	309.1[h]	2,377.3

---

[a]Will begin production of expandable PS at this location by the end of 1981; estimated capacity – 18,200 metric tons per year.

[b]Company intends to increase expandable PS capacity to a total of 227,300 metric tons per year.

(continued)

TABLE 165 (continued)

[c]Company will add expandable PS production capacity of 20,500 metric tons per year by 1983.

[d]One production train was put on standby in August 1981.

[e]Company will increase capacity to 200,000 metric tons per year by early 1983.

[f]Company capacity will be increased by 118,200 metric tons per year by early 1983.

[g]Facility was closed in September 1980 with the exception of this small amount of impact resin produced.

[h]Georgia Pacific Corp. plans to complete 29,500 metric tons per year of its expandable PS capacity in Painesville, CA by 1982.

Sources: Chemical Economics Handbook, updated annually, 1981 Data.
          Directory of Chemical Producers, 1981.

TABLE 166. POLYSTYRENE PRODUCERS WHICH ARE ALSO
            MAJOR STYRENE-BUTADIENE LATEX PRODUCERS

Dow Chemical USA
 Allyn's Point, CT
 Dalton, GA
 Freeport, TX
 Midland, MI
 Pittsburgh, CA

Polysar Group
 Polysar Inc.
  Polysar Latex
  Beaver Valley, PA
  Chattanooga, TN - 2 plants

U.S. Steel Corp.
 USS Chemicals Division
  Scotts Bluff, LA

Source: Chemical Economics Handbook, updated annually, 1981 Data.

TABLE 167.   CONSUMPTION OF POLYSTYRENE FOR 1980

	Thousand Metric Tons
Packaging	568.2
Housewares, furniture, and furnishing, consumer products	181.8
Electrical/electronic uses, appliances	168.2
Toys, sporting goods, recreational articles	140.9
Construction	136.4
Serviceware and flatware	109.1
Miscellaneous commercial and industrial molding	81.8
Other uses (medical/dental and other laboratory articles, combs, brushes, eyeglasses, signs, writing utensils, office and school supplies, novelties and miscellaneous uses)	104.5
	1,490.9

Source: _Chemical Economics Handbook_, updated annually, 1981 Data.

cost for the water, stabilizers, and wastewater treatment also make mass
polymerization a more favorable process as the operating costs of a
polystyrene plant increase.[2, 147] Due to the narrowing price increase
from styrene monomer to polymer, mass polymerization will probably become
the major process used for polystyrene production in the future.[147]
Tables 168 and 169 present typical input materials and operating parameters
for these processes. Specialty chemicals and other input materials are
listed in Table 170.

General purpose polystyrene is produced by the polymerization of
styrene. The reaction may be initiated with heat, but a free-radical initiator is typically used. The free-radical initiator (R) reacts with the
styrene monomer to start the polymerization reaction. As other styrene
monomers combine with the styrene radical, the polymer chain is propagated.
Polymerization terminates when the radical reacts with an electron accepting
group (R'), which is called chain transfer. Termination also occurs when
two radical groups react with each other, or by disproportionation. Typical
molecular weights for polystyrene range from 200,000 to 300,000.[17] The
polymerization steps are depicted below:

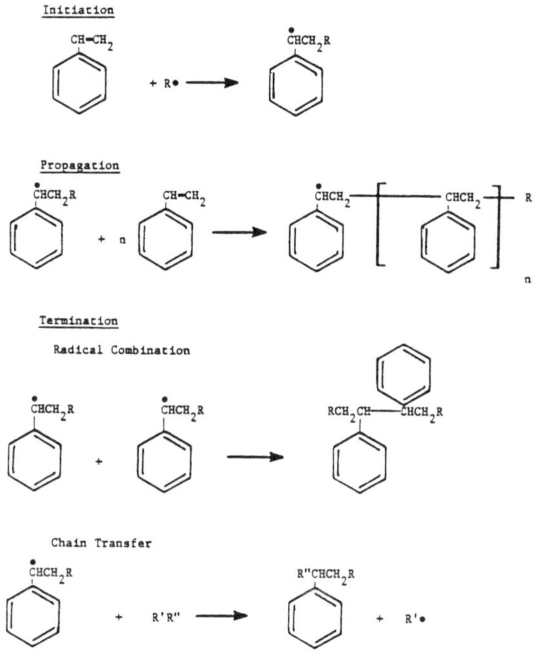

Initiation

Propagation

Termination

Radical Combination

Chain Transfer

TABLE 168. TYPICAL INPUT MATERIALS TO GENERAL PURPOSE POLYSTYRENE
PRODUCTION PROCESSES IN ADDITION TO MONOMER AND INITIATOR

Process	Water	Organic Solvent	Suspension Stabilizer	Blowing Agent[a]
Suspension	X		X	X
Mass				
Solution		X		

[a]For the production of expandable bead polystyrene.

TABLE 169. TYPICAL OPERATING PARAMETERS FOR GENERAL PURPOSE
POLYSTYRENE PRODUCTION PROCESSES

Process	Temperature	Pressure[a]	Reaction Time
Mass	80 - 200°C	Slightly reduced to 10-20 mm Hg	12-18 hours (batch) 2-8 hours (continuous)
Suspension	110-170°C	Reduced	5-9 hours
Solution	90-130°C	Atmospheric to 10-20 mm Hg	6-8 hours

[a]Including devolatilization step.

Sources: Lyle F. Albright, _Processes for Major Addition-Type Plastics and Their Monomers_, 1974.
Encyclopedia of Polymer Science and Technology, 1971.

TABLE 170.   INPUT MATERIALS USED IN GENERAL PURPOSE POLYSTYRENE
MANUFACTURE (IN ADDITION TO STYRENE MONOMER)

Free Radical Initiators

azobisisobutyronitrile
benzoyl peroxide
dimethylaniline[a]
di-tert-butyl peroxide
tert-butyl hydroperoxide

Chain Transfer Agents

alkyl mercaptans
 -methyl styrene dimer
carbon tetrabromide
carbon tetrachloride
ethylbenzene

Retardants and Inhibitors

benzoquinone
1,1-diphenyl-2-picrylhydrazyl
ferric salts
oxygen
sulfur
tert-butylcatechol
tert-butyl-pyrocatechol

Solvents

ethylbenzene

Suspension Stabilizers
(protective colloids)

alkaline earth phosphates,
   carbonates, and/or silicates
gelatin
magnesium silicates
methyl cellulose
pectin
polyacrylic acid and its salts
polyvinyl alcohol
polyvinyl pyrrolidone
starch
sulfonated polystyrene
zinc oxide

Flame Retardants

alkyl phosphates
antimony oxide
aryl phosphates
hydrated aluminum oxide

(continued)

TABLE 170 (continued)

Lubricant	mineral oil
Antioxidant	2,6-di(tert-butyl)-4-methyl-phenol

[a]Used only in conjunction with benzoyl peroxide at low temperatures.

Sources: Encyclopedia of Polymer Science and Technology, 1971.
Lyle F. Albright, Processes for Major Addition Type Plastics and Their Monomers, 1974.

Disproportionation

$$\overset{\bullet}{C}HCH_2R \quad + \quad \overset{\bullet}{C}HCH_2R' \quad \longrightarrow \quad CH_2CH_2R \quad + \quad CH\!=\!CHR'$$

Oxygen reacts with the free radical initiators used, thus making a peroxide radical shown below:

$$R\bullet + O_2 \longrightarrow R\overset{\bullet}{O}_2$$

which slows the reaction rate. Polymerization is performed in a nitrogen atmosphere to minimize this reaction.

Suspension Polymerization

Suspension polymerization of general purpose polystyrene uses water and a suspending agent to separate the monomer into droplets for the polymeriza- tion process. The major advantage of suspension polymerization as compared to other methods is the production of the polymer in small beads which are easily separated from the aqueous phase. Although the polymerization rate is slower, the cost is higher, and the contamination more severe for suspen- sion polymerization than mass or solution polymerization. The advantages of this process over mass or solution polymerization include: excellent heat transfer, no solvent recovery problems, fewer safety hazards, simpler temp- erature control, and narrower range particle size control possible. Another advantage of this process is the capability to produce both general purpose resins and expandable bead polystyrene in the same reaction train.

As illustrated in Figure 52, the styrene monomer is stripped of the inhibitor, which is added to prevent polymerization during shipping and storage, prior to entering the reactor. The stripper water is sent to wastewater treatment while the monomer is combined with a monomer soluble initiator and fed to an agitated reactor filled with water. A dispersing agent and a suspension stabilizer are added to maintain the monomer as aqueous phase droplets.

A typical recipe for suspension polymerization of GPPS includes the following:[2]

Material	Parts by Weight
Styrene (monomer)	7.27
Water	8.825
Additives	not specified

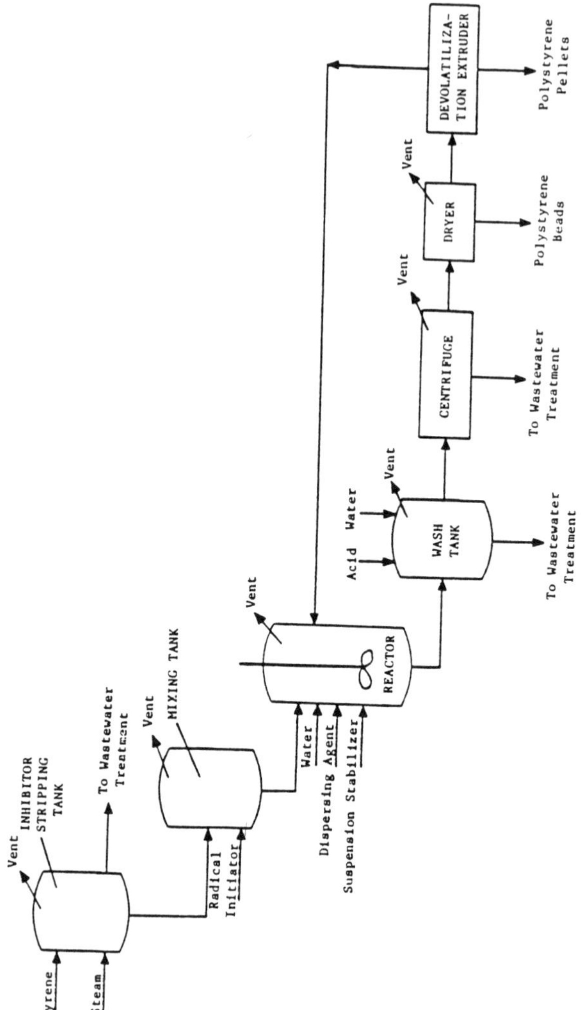

Figure 52.   Polystyrene production using the suspension polymerization process.

Source: Encyclopedia of Polymer Science and Technology.

If expandable polystyrene beads are desired, a blowing agent, such as
n-pentane, is also fed to the reactor where it is absorbed in the organic
phase and remains in the polymer beads. Expandable polystyrene bead
production is illustrated in Figure 53. The dried polymer beads typically
contain 5 to 7 percent pentane.[1]   When heated, these beads expand to form
molded products or, more recently, foam insulation.

After polymerization, the polymer beads are washed with acid to remove
any remaining initiators and suspension stabilizers. The wash water is sent
to wastewater treatment while the slurry of polymer beads is centrifuged.
The water removed during centrifugation is combined with the other wastewater while the beads are dried. The polymer beads are fed to a devolatilization extruder where any remaining monomer is removed during pelletization
and recycled to the reactor.

Typical water to monomer ratios range from 1:1 to 4:1 and the particle
size of the monomer ranges from 0.1 to 1 mm.[17] Final polymer volatile
concentrations as low as 0.5 percent can be produced using devolatilization
extrusion at reduced pressure, 10 to 20 mm Hg. Reactors are typically
aluminum, stainless steel, or stainless steel clad materials. Copper is not
used since it discolors the polymer.

Mass Polymerization

Mass polymerization is the simplest process by which polystyrene can be
made and produces a polymer of high clarity having good electrical-insulation properties. The major disadvantage of this process is the difficulty
in controlling the polymerization reaction due to its highly exothermic
nature and the increasing viscosity of the polymerizing mass. However, mass
polymerization of polystyrene results in little contamination of the polymer
since water and organic solvents are not used.

Mass polymerization, as depicted in Figure 54, consists of four basic
steps: prepolymerization, polymerization, devolatilization, and extrusion.
Prior to polymerization, the styrene monomer is stripped of the inhibitor
used to prevent polymerization during shipping. The styrene is fed to the
reactor while the stripper water is sent to wastewater treatment.

The equipment used in the prepolymerization step consists of a stirred
reactor with a reflux condenser operating under reduced pressure. This
enables control of the reaction temperature by removing a portion of the
vaporized monomer and returning the condensed liquid styrene to the reactor.
If peroxidic initiators are used, the reaction may be carried out at a lower
temperature (80°C) than that used for thermal initiation (130°C). The
polymer content of the resulting mixture varies from 25 to 35 percent.

From the prepolymerization section, the monomer-polymer mixture is fed
into a series of stirred reflux reaction kettles or reaction towers. The
temperature is progressively raised through the reaction zone to prevent
fall off in polymerization rate and to reduce the viscosity of the polymerizing mass. Typical final temperature in the polymerization section range
from 150 to 200°C.[2, 17] Materials of construction for the reactors are

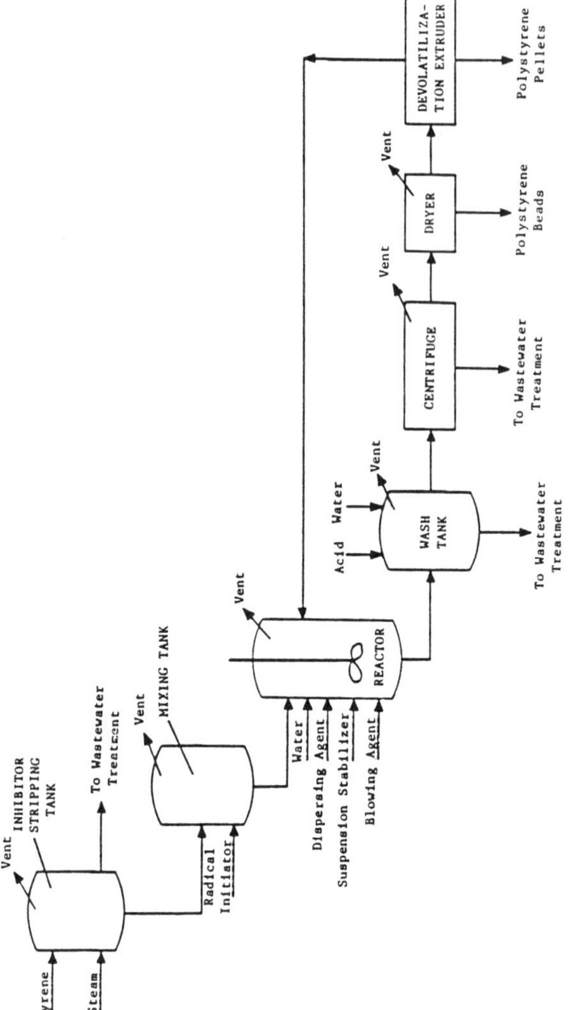

Figure 53. Expandable bead polystyrene production using the suspension polymerization process.

Source: Encyclopedia of Polymer Science and Technology.

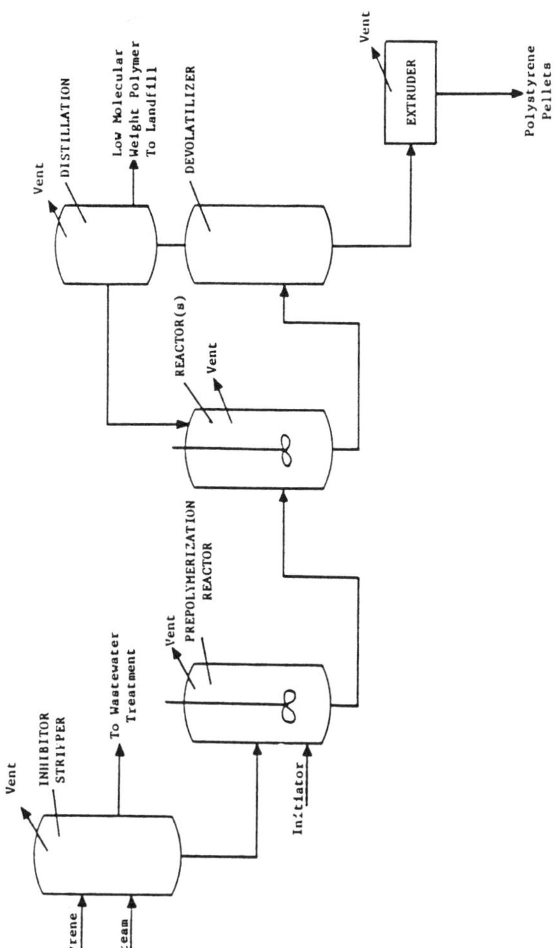

Figure 54. Polystyrene production via mass polymerization.

Sources: Lyle F. Albright, Processes for Major Addition Type Plastics and Their Manufacture, 1974. Encyclopedia of Polymer Science and Technology.

aluminum, stainless steel, or stainless-steel clad materials. Copper is not used since it causes discoloration of the polymer.

In order to remove unreacted monomer and low molecular weight polymers, the reaction mixture is then fed to a static devolatilizer where the volatiles are flashed off and distilled. The styrene monomer and oligomers are recycled while the low molecular weight polymers are sent to a landfill. The purified polymer is then extruded through a vented extruder and pelletized.

A typical recipe for mass polymerization of GPPS was not found in the literature. However, it was reported that the recipe for mass polymerization is similar to that for solution polymerization.

Solution Polymerization

Solution polymerization of polystyrene has several advantages over the mass polymerization process. The viscosity of the polymerization medium is lower, the temperature control is more precise, and the molecular weight of the polymer is lower. Although this process allows better heat dissipation, the solvent reacts with the polymer chain, resulting in a more contaminated product.

A typical recipe for solution polymerization of GPPS is shown below: [Albright]

Material	Parts by Weight
Styrene (monomer)	997
Ethylbenzene (solvent)	4 (makeup)
Additives	23.4
Blueing Dye	0.2

As illustrated in Figure 55, the styrene monomer is washed to remove the inhibitor added to prevent polymerization during shipping and storage. The wash water is sent to wastewater treatment while the styrene is diluted with ethylbenzene and fed to a series of agitated reactors with heat exchange zones. After polymerization, the unreacted styrene and solvent are removed from the polymer and recycled. The purified polymer is then extruded into strands and cut into pellets.

The ethylbenzene concentration of the incoming solution to the reactors ranges from 5 to 25 percent. Temperatures in the reactors start at 110 to 130°C in the first stage and increase to 105 to 170°C in the last stage. Devolatilization temperature, under reduced pressure, is usually 250°C. Small amounts of lubricant such as mineral oil (about 3 percent) may be added.[208] Reactors are aluminum, stainless steel, or stainless-steel clad materials. Copper is not used since it discolors the resulting polymer.

Energy Requirements

The energy required for polystyrene production is listed below by technology:[208, 209, 211]

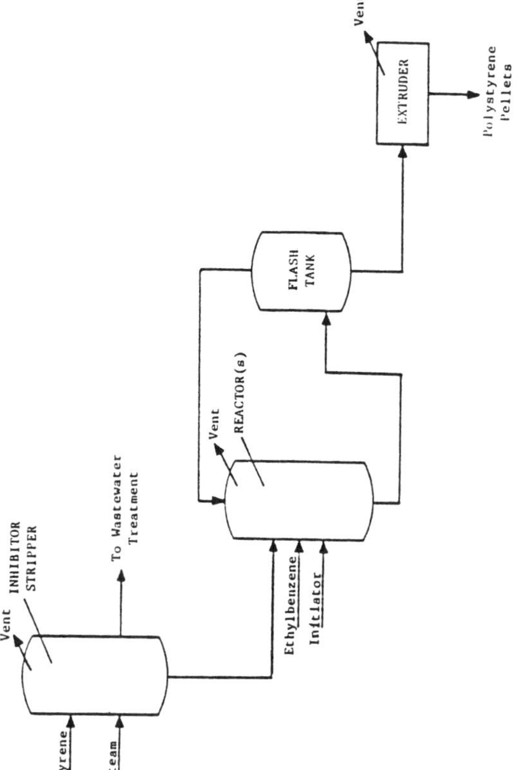

Figure 55. Polystyrene production via solution polymerization.

Source: Encyclopedia of Polymer Science and Technology.

| | | Unit/Metric |
Technology	Energy Required	Ton of Product
Cosden Technology,	Electricity	6.30×10^8 Joules
Inc.	Fuel	1.26×10^9 Joules
Gulf Oil Chemicals	Electricity	9.01×10^8 Joules
Co.	Steam	0.4 metric ton
Societe Chimique	Electricity	4.86×10^8 Joules
des Charbonnages	Net Fuel	4.19×10^8 Joules
(CdF)--The Badger		
Co., Inc.		

ENVIRONMENTAL CONSIDERATIONS

None of the chemicals which have been identified as inputs to polystyrene processing are known human carcinogens. However, there is some
evidence from tests conducted and reported to NIOSH of polystyrene and
styrene causing tumors in test animals. Although these results are not
conclusive, standards have been set for styrene and polystyrene exposure.
Three other input materials to polystyrene processing are suspected human or
known animal carcinogens: carbon tetrachloride, a chain transfer agent;
polyvinyl alcohol, a suspension stabilizer; and antimony oxide, a flame
retardant. Other highly toxic inputs include tert-butyl hydroperoxide, a
free radical initiator, and benzoquinone, a retardant.

One environmentally significant portion of polystyrene processing is
the devolatilization step. Residual styrene monomer in the polymer is
typically reduced to less than one percent, thereby reducing the exposure
risk and rendering the polymer usable for food applications such as meat
trays, egg cartons, and wrap for non-fatty foods.

Worker Distribution and Emissions Release Points

We have estimated worker distribution for polystyrene production by
correlating major equipment manhour requirements with the process flow
diagrams in Figures 52 through 55. Estimates are given in Table 171.

No major emission point sources are associated with polystyrene production. Although there are no major air emission point sources, fugitive
emissions may pose a significant environmental and/or worker health problem
depending on the components in the stream, operating conditions for the
process, engineering and administrative controls, and maintenance programs.

Process sources of fugitive emissions are listed in Table 172. Typical
inhibitors, stabilizers and initiators are listed in Table 170. Ethylbenzene is the solvent generally used in solution polymerization. Particulate
emissions from centrifugation, drying and extrusion operation may contain
harmful additives and may produce elevated nuisance dust levels.

TABLE 171.   WORKER DISTRIBUTION ESTIMATES FOR GENERAL PURPOSE
             POLYSTYRENE PRODUCTION

Process	Unit	Workers/Unit/8-hour Shift
Suspension Polymerization	Steam Stripper	0.25
(Pellets and Beads)	Batch Reactor	1.0
	Batch Mixer	1.0
	Washing	0.25
	Centrifuge	0.25
	Dryer	0.5
	Extruder	1.0
Mass Polymerization	Steam Stripper	0.25
	Batch Reactor	1.0
	Distillation Column	0.25
	Devolatilizer	0.25
	Extruder	1.0
Solution Polymerization	Steam Stripper	0.25
	Batch Reactor	1.0
	Flash Tank	0.125
	Extruder	1.0

TABLE 172.   SOURCES OF FUGITIVE EMISSIONS FROM POLYSTYRENE MANUFACTURE

Source	Constituent	Process		
		Suspension	Mass	Solution
Inhibitor Stripper Vent	Inhibitor	X	X	X
	Styrene	X	X	
Mixing Tank Vent	Styrene	X		
	Initiator	X		
Reactor Vents	Styrene	X	X	X
	Initiators	X	X	X
	Suspension Stabilizers	X		
	Solvents			X
Wash Tank Vent	Styrene	X		
Centrifuge Vent	Polystyrene and Styrene Particulates and Emissions	X		
Distillation Column Vent	Styrene		X	
Polymer Bead Drying Vent	Polystyrene and Styrene Particulates and Emissions	X		
Extruder Vent	Polystyrene and Styrene Particulates and Emissions		X	X

Health effects information about input materials of particular concern is provided in the paragraphs below. Table 10 summarized available toxicity information for major input materials.

Ranges of volatile organic emissions from which employee exposure potential to volatile organic compounds may be estimated are shown in Table 173 for expandable bead production via suspension polymerization and Table 174 for mass and solution polymerization.

Health Effects

The input materials used in general purpose polystyrene manufacture include three suspected human or known animal carcinogens. They are carbon tetrachloride, used as a chain transfer agent; polyvinyl alcohol, a suspension stabilizer; and the the flame retardant, antimony oxide. Other input materials that may pose a significant health risk to plant employees due to high toxicities are tert-butyl hydroperoxide, a free radical initiator, and the retardant benzoquinone. The reported health effects of exposure to each of these substances are summarized below.

Antimony Oxide is a suspected carcinogen [5] which has also produced positive results in tests for mutagenic effects.[156] Animal testing of this substance is limited, however, and no epidemiological data exist.

Carbon Tetrachloride is a suspected human carcinogen [108] which is also extremely toxic upon inhalation or ingestion of small quantities.[241] The toxic hazard posed by skin absorption is slight for acute exposures, but high for chronic exposures.[241] The toxicity of carbon tetrachloride appears to be primarily due to its fat solvent action which destroys the selective permeability of tissue membranes and allows escape of certain essential substances such as pyridine nucleotides.[78] The OSHA air standard is 10 ppm (8 hour TWA) with a 25 ppm ceiling.[67]

p-Benzoquinone is a tumorigenic and a mutagenic agent, but testing has produced indefinite results for carcinogenicity.[105] The substance is currently undergoing additional tests for carcinogenesis by standard bioassay protocol under the sponsorhip of the National Toxicology Program. Benzoquinone is very toxic; the probable and lethal oral dose for humans is 50 to 500 mg/kg.[85] The OSHA air standard is 0.1 ppm (8 hour TWA).[67]

Styrene has been linked with increased rates of chromosomal aberrations in persons exposed in an occupational setting.[107] Animal test data strongly support epidemiological evidence of its mutagenic potential.[233] Styrene also produces tumors and affects reproductive fertility in laboratory animals.[233] Toxic effects of exposure to styrene usually involve the central nervous system.[233] The OSHA air standard is 100 ppm (8 hour TWA) with a ceiling of 200 ppm.[67]

Tert-Butyl Hydroperoxide has produced symptoms of severe depression, incoordination, cyanosis, and respiratory arrest in laboratory animals.[243] Skin contact is associated with severe local reactions. Limited animal test

TABLE 173.   CHARACTERISTICS OF VENT STREAMS FROM EXPANDABLE POLYSTYRENE BEAD
PRODUCTION VIA SUSPENSION POLYMERIZATION

Stream Name	Nature	Emission rate, kg VOC/Mg product	Composition, Wt.%
Mix tank vents	Intermittent	0.123	0.75 Styrene 0.53 Acrylonitrile 98.72 Air
Reactor vents	Intermittent	1.14	58.8 VOC[a] 1.4 Styrene 39.8 Air
Holding tank vents	Intermittent	0.055	0.66 VOC[a] 99.34 Air
Wash tank vents	Continuous	0.024	0.35 VOC[a] 99.65 Air
Dryer vents	Continuous	2.89	0.066 − 0.27 VOC[a] 99.934 − 99.73 Air
Product improvement vents	Intermittent	----	0.022 VOC[a] 99.998 Air
Storage vents	Continuous	0.746	0.002 VOC[a] 99.998 Air
Total Emission Rate		4.98	

[a]Blowing agent.

Source: *Polymer Manufacturing Industry Background Information for Proposed
Standards*, September 1983.

TABLE 174.   CHARACTERISTICS OF VENT STREAMS FROM THE POLYSTYRENE
MASS OR SOLUTION POLYMERIZATION PROCESSES

Stream Name	Nature	Emission rate, kg VOC/Mg product	Composition, Wt.%
Monomer storage & feed dissolver	Continuous	0.09	100 Styrene
Reactor drum vent	Intermittent	0.12–1.35	6.9 Styrene 35.0 Water 58.1 Air
Devolatilizer condenser vent	Continuous	0.25–0.75	3 Styrene 97 Air
Extruder quench vent	Continuous	0.15–0.3	Styrene & steam
Total Emission Rate		0.61–2.5	

Source: Polymer Manufacturing Industry Background Information for Proposed
Standards, September 1983.

data also indicate mutagenic effects. No epidemiological data are available, however, to evaluate the toxic hazard to humans.

Air Emissions

The sources of VOC fugitive emissions for polystyrene production are listed in Table 172. All of these sources are vents. Control of emissions from these vents may include routing the stream to a flare to incinerate hydrocarbons or routing the stream to blowdown.[285] Other fugitive VOC emissions result from leaks in process equipment, such as flanges, valves, cooling towers, open drains, and pumps. Although control of these emissions may be accomplished by equipment modification, a routine inspection and maintenance program may be the best control.

The sources of fugitive particulates are also listed in Table 172. Control of these sources may be accomplished by venting these streams to either a baghouse or electrostatic precipitator for particulate collection. EPA estimates that only 30 people are affected by hydrocarbon emissions and 2 are affected by particulates in the 100 person per square km population surrounding a polystyrene plant.[286]

Wastewater Sources

There are several wastewater sources associated with the various polystyrene production processes as shown in Table 175. The major wastewater sources result from suspension polymerization which uses water to suspend the reaction mixture and wash the polymer. The aqueous phase of the polymerization mixture is removed from the polymer beads by centrifugation. This water, along with the polymer wash water, is reported to contain 20 kg of polystyrene per metric ton produced.[93] As Table 175 shows, all polystyrene processes generate water from monomer stripping and routine cleaning. Mass and solution polymerization have wastewater from only these two sources.

Ranges of several wastewater parameters for wastewaters from polystyrene production are shown below. Values for the wastewater from the three GPPS production processes and the processes presented in the IPS section were not distinguished by process or resin type by EPA for the purpose of establishing effluent limitations for the polystyrene industry.[284]

Polystyrene Wastewater Characteristics	Unit/Metric Ton of Polystyrene
Production	$0 - 141.8 \text{ m}^3$
BOD_5	$0 - 2.2$ kg
COD	$0 - 6.0$ kg
TSS	$0 - 8.4$ kg

TABLE 175. SOURCES OF WASTEWATER FROM POLYSTYRENE MANUFACTURE

Source	Suspension	Mass	Solution
Monomer Steam Stripping	X	X	X
Polymer Bead Washing	X		
Centrifugation	X		
Routine Cleaning Water	X	X	X

Solid Wastes

The solid wastes generated during polystyrene production include substandard resins or beads which cannot be blended and collected particulates, and low molecular weight polymers from mass polymerization. One report estimates losses from mass polymerization processes to be 2 percent.[93] No information is available in the literature regarding the quantity of particulates collected from air emission control devices. These wastes are not hazardous; therefore, their disposal should not pose an environmental problem.

Environmental Regulation

Effluent limitations guidelines have been set for the polystyrene industry. BPT, BAT, and NSPS call for the pH of the effluent to fall between 6.0 and 9.0 (41 Federal Register 32587, August 4, 1976).

New source performance standards (proposed by EPA on January 5, 1981) for volatile organic carbon (VOC) fugitive emissions include:

- Safety/release valves must not release more than 200 ppm above background, except in emergency pressure releases, which should not last more than five days; and

- Leaks (which are defined as VOC emissions greater than 10,000 ppm) must be repaired within 15 days.

Carbon tetrachloride, a chain transfer agent, has been listed as a hazardous waste and designated U211 (45 Federal Register 3312, May 19, 1980). Disposal of carbon tetrachloride or polystyrene resins which contain residual amounts of this compound must comply with the guidelines set forth in the Resource Conservation and Recovery Act (RCRA).

22. Polystyrene (Impact)

Impact, or rubber modified, polystyrene (IPS) is a translucent to opaque white polymer which is resistant to wear and exhibits high impact strength. IPS has less gloss and clarity than general purpose polystyrene (GPPS), but exhibits increased impact strength when compared to GPPS and displays the rigidity and ease of fabrication of the general purpose resin.[146] The production of IPS is achieved by the addition of either polybutadiene or styrene-butadiene rubber to the styrene monomer. The rubber remains as discrete particles dispersed in the polymer after polymerization.[17] Although the impact strength of the modified polymer is improved, the tensile strength and hardness are lowered, and the softening point is reduced by the addition of the rubber. This section describes the manufacture of rubber modified polystyrene; the section on general purpose polystyrene presents the types of processes used in the manufacture of the homopolymer.

Tables A-30 and A-31 in Appendix A include some typical properties for impact polystyrene. As these tables show, IPS has low specific gravity and good thermal stability. IPS is usually classified according to rubber content as: medium impact, with 3 to 10 percent rubber; high impact, with 5 to 10 percent rubber; and super impact, with up to 25 percent rubber.

The ease with which polystyrene can be fabricated using heat makes polystyrene products low in cost. Its excellent thermal and electrical properties render polystyrene useful for manufacturing interior doors, lighting covers, air conditioning housing, and shutters. Other applications include flower pots, margarine tubs, caps, closures, disposable cutlery, drain pipes, toys, drinking cups, and egg cartons. In 1980, impact polystyrene accounted for 51 percent of the total polystyrene produced. General purpose polystyrene and expandable beads captured the remaining 49 percent. [35]

Rubber modified polystyrene is produced via mass polymerization, suspension polymerization, or solution polymerization. The rubber can be added by: blending the rubber latex with the polystyrene latex followed by coagulation and drying; mechanically milling the dry rubber with the dry polystyrene; and grafting of preformed unsaturated rubber with styrene in mass, suspension, or solution polymerization processes. The copolymerization

process is often referred to as graft polymerization and is the most widely
used since it produces superior products with less rubber.[17]

INDUSTRY DESCRIPTION

The polystyrene industry is comprised of 18 producers of general purpose and/or impact polystyrene, three of which also produce styrene-butadiene latexes in addition to the polystyrene. Published information does not
disclose which producers manufacture impact polystyrene. These 18 producers
have 44 sites, 17 of which are in Ohio, Michigan, Illinois, and
Pennyslvania, and another 8 in California. The remaining plants are located
in the east and southeast. Polystyrene producers are listed in Table 176
and styrene-butadiene latex producers are presented in Table 177.

Polystyrene is used in consumer related industries, such as construction, appliances, and packaging. Therefore, polystyrene consumption follows
the trend set by housing starts and consumer spending. After a production
peak of 1,820,000 metric tons in 1979, a drop to 1,490,900 metric tons
occurred in 1980.[35, 65] Most of the expansion planned is for expandable
polystyrene. Excess capacity for impact polystyrene is sufficient to support industry growth for the near future. Therefore, no new plants for
impact polystyrene are expected to be built.

PRODUCTION AND END USE DATA

In 1980, total polystyrene sales were 1,490,900 metric tons.[35] Polystyrene is used for packaging, toys, foam insulation, electrical and electronic uses, and serviceware, which includes the familiar cups. Table 178
presents the 1980 consumption of polystyrene. These products can be classified as either molding resins, extrusion resins, or expandable beads.
Impact polystyrene contributes 64 and 51 percent of the total polystyrene
produced for molding and extrusion resins, respectively.[35]

PROCESS DESCRIPTIONS

Impact polystyrene is produced by suspension, mass, or solution polymerization. According to one estimate, over half of the impact polystyrene
produced is manufactured using a suspension process where the prepolymerization is carried to 20 to 40 percent completion in a bulk reactor and the
remaining conversion is performed by suspension polymerization.[17] The
remainder of the impact polystyrene production is performed by mass or solution polymerization. Suspension polymerization as a method of manufacturing
impact polystyrene offers excellent versatility for the production of
numerous types of products; however, high-purity products are more difficult
to obtain since water, suspension stabilizers, and unreacted initiators are
contaminants. The added cost for the addition of stabilizers and wastewater
treatment also make mass polymerization a more favorable process.[2, 147]
Due to the narrowing price increase from styrene monomer to polymer, mass
polymerization will probably become the major process used.[147]

Although the rubber can be incorporated into the polystyrene by
mechanical means, polymerization of the rubber with the polystyrene (often

TABLE 176. MAJOR U.S. PRODUCERS OF POLYSTYRENE

Producer	Capacity as of July 1, 1981 (Thousand Metric Tons)		
	General Purpose and Impact	Expandable Bead	Total
A&E Plastik Pak Co.			
A&E Plastics Division			
City of Industry, CA	15.9	0	15.9
American Hoechst Corp.			
Plastics Division			
Chesapeake, VA	118.2	0	118.2
Leominster, MA	54.5	0	54.5
Peru, IL	77.3	36.4	113.6
COMPANY TOTALS	250.0	36.4	286.4
American Petrofina Inc.			
Cosden Oil & Chemical Co.,			
subsidiary			
Big Spring, TX			0[a]
Calumet City, IL			77.3
Orange, CA			27.3
Windsor, NJ			54.4
Atlantic Richfield Co.			
ARCO Polymers, Inc., sub.			
Beaver Valley, PA	54.4	159.1[b]	213.6
BASF Wyandotte Corp.			
Polymers Groups			
Styropor Division			
Jamesburg, NJ	0	79.5[c]	79.5
Dow Chemical USA			
Allyn's Point, CT	N/A	N/A	86.4
Ironton, OH	N/A	N/A	86.4
Joliet, IL	N/A	N/A	63.6
Midland, MI	N/A	N/A	100.0
Torrance, CA	N/A	N/A	90.9[d]
COMPANY TOTALS	418.2	9.1	427.3

N/A — Not Available

(continued)

TABLE 176 (continued)

Producer	Capacity as of July 1, 1981 (Thousand Metric Tons)		
	General Purpose and Impact	Expandable Bead	Total
Carl Gordon Industries, Inc.			
Hammond Plastics			
Oxford Division			
Oxford, MA	45.5	0	45.5
Hammond Plastics Midwest, Inc.			
Owensboro, KY	22.7	0	22.7
COMPANY TOTALS	68.2	0	68.2
Gulf Oil Corporation			
Gulf Oil Chemicals Co.			
Plastics Division			
Marietta, OH	140.9e	0	140.9
Huntsman-Goodson Chemical			
Corp.			
Troy, OH	9.1	0	9.1
Kama Corp.			
Hazeltown, PA	11.4	0	11.4
Mobil Corp.			
Mobil Oil Corp.			
Mobil Chemical Co.,			
Division			
Petrochemicals Division			
Holyoke, MA	40.9	0	40.9
Joliet, IL	18.2	0	18.2f
Santa Anna, CA	29.5	0	29.5
COMPANY TOTALS	88.6	0	88.6
Monsanto Co.			
Monsanto Plastics and			
Resins Co.			
Addyston, OH	136.4	0	136.4
Decatur, AL	45.5	0	45.5
Long Beach, CA	22.7	0	22.7
Springfield, MA	136.4	0	136.4
COMPANY TOTALS	340.9	0	340.9

(continued)

TABLE 176 (continued)

Producer	Capacity as of July 1, 1981 (Thousand Metric Tons)		
	General Purpose and Impact	Expandable Bead	Total
Polysar Group			
Polysar Inc.			
Resins Division			
Copely, OH	54.5	0	54.5
Forest City, NC	18.2	0	18.2
Leominster, MA	54.5	0	54.5
COMPANY TOTALS	127.3	0	127.3
Shell Chemical Co.			
Belpre (Marietta), OH	136.4	0	136.4
Standard Oil Co. (IN)			
AMOCO Chemical Corp., sub.			
Joliet, IL	136.4	0	136.4
Torrance, CA	15.9	0	15.9
Willow Springs, IL	40.9	0	40.9
COMPANY TOTALS	193.2	0	193.2
Texstyrene Plastics, Inc.			
Fort Worth, TX	0	22.7	22.7
U.S. Steel Corp.			
USS Chemicals Division			
Haverhill, OH	9.18	0	9.1
Vititek Inc.			
Delano, CA	0	2.3	2.3
TOTAL	2,068.2	309.1[h]	2,377.3

[a]Will begin production of expandable PS at this location by the end of 1981; estimated capacity – 18,200 metric tons per year.

[b]Company intends to increase expandable PS capacity to a total of 227,300 metric tons per year.

(continued)

TABLE 176 (continued)

[c]Company will add expandable PS production capacity of 20,500 metric tons per year by 1983.

[d]One production train was put on standby in August 1981.

[e]Company will increase capacity to 200,000 metric tons per year by early 1983.

[f]Company capacity will be increased by 118,200 metric tons per year by early 1983.

[g]Facility was closed in September 1980 with the exception of this small amount of impact resin produced.

[h]Georgia Pacific Corp. plans to complete 29,500 metric tons per year of its expandable PS capacity in Painesville, CA by 1982.

Sources: Chemical Economics Handbook, updated annually, 1981 Data.
Directory of Chemical Producers, 1981.

TABLE 177.   POLYSTYRENE PRODUCERS WHICH ARE ALSO
MAJOR STYRENE-BUTADIENE LATEX PRODUCERS

Dow Chemical USA
 Allyn's Point, CT
 Dalton, GA
 Freeport, TX
 Midland, MI
 Pittsburgh, CA

Polysar Group
 Polysar Inc.
  Polysar Latex
  Beaver Valley, PA
  Chattanooga, TN - 2 plants

U.S. Steel Corp.
 USS Chemicals Division
  Scotts Bluff, LA

Source: Chemical Economics Handbook, updated annually, 1981 Data.

TABLE 178.   CONSUMPTION OF POLYSTYRENE FOR 1980

	Thousand Metric Tons
Packaging	568.2
Housewares, furniture, and furnishing, consumer products	181.8
Electrical/electronic uses, appliances	168.2
Toys, sporting goods, recreational articles	140.9
Construction	136.4
Serviceware and flatware	109.1
Miscellaneous commercial and industrial molding	81.8
Other uses (medical/dental and other laboratory articles, combs, brushes, eyeglasses, signs, writing utensils, office and school supplies, novelties and miscellaneous uses)	104.5
	1,490.9

Source: Chemical Economics Handbook, updated annually, 1981 Data.

referred to as graft polymerization) is the typical process. Copolymerization of the rubber and the styrene produces more homogeneous products with
less rubber.[17] Tables 179 and 180 present typical input materials and
operating parameters for these processes. Specialty chemicals used in this
process are listed in Table 181.

Impact polystyrene is produced by the polymerization of styrene in the
presence of rubber. The reaction may be initiated by heat, but a radical
initiator is typically used. The polymerization of impact polystyrene differs from polymerization of general purpose polystyrene since the unsaturated rubber molecule also becomes a radical group which then reacts with
the styrene monomers and polymers. The free-radical initiator (R) reacts
with the rubber molecule, typically polybutadiene, to start the polymerization reaction. As styrene monomers and polymers combine with the polybutadiene radicals, the polymer chain is propagated. Polymerization terminates
when the radical reacts with an electron accepting group (R'), which is
called chain transfer. Termination also occurs when two radical groups
react with each other, or by disproportionation. Molecular weights of commercial impact polystyrene grades range from 54,000 to 416,000.[147] The
polymerization steps are depicted below:

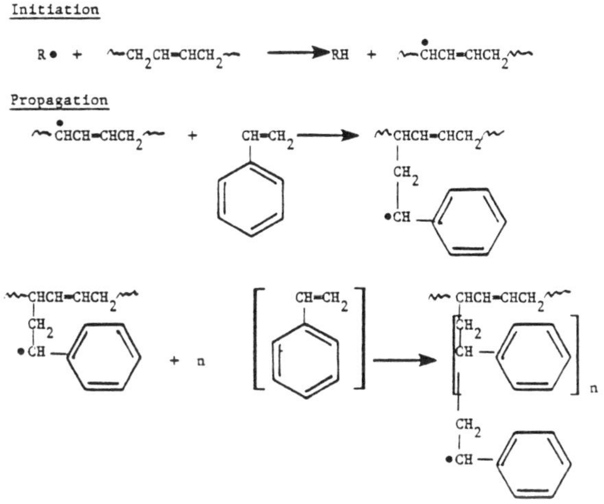

Initiation

R• + ~CH₂CH=CHCH₂~ ⟶ RH + ~ĊHCH=CHCH₂~

Propagation

~ĊHCH=CHCH₂~ + CH=CH₂(C₆H₅) ⟶ ~CHCH=CHCH₂~ with pendant CH₂–•CH–C₆H₅ group

~CHCH=CHCH₂~ (with CH₂–•CH–C₆H₅ group) + n [CH=CH₂(C₆H₅)] ⟶ ~CHCH=CHCH₂~ [with –CH₂–CH(C₆H₅)–]ₙ pendant group

TABLE 179.   TYPICAL INPUT MATERIALS TO IMPACT POLYSTYRENE PRODUCTION PROCESSES IN ADDITION TO STYRENE MONOMER AND RUBBER

Process	Water	Initiator	Organic Solvent	Suspension Stabilizers
Suspension	X	X		X
Mass		X		
Solution		X	X	

TABLE 180.   TYPICAL OPERATING PARAMETERS FOR IMPACT POLYSTYRENE PRODUCTION PROCESSES

Process	Temperature	Pressure[a]	Reaction Time
Suspension	110–170 °C	Reduced	5–9 hours
Mass	80 – 200 °C	Slightly reduced to 10–20 mm Hg	12–18 hours (batch) 2–8 hours (continuous)
Solution	90–130 °C	Atmospheric to 10–20 mm Hg	6–8 hours

[a]Including devolatilization step.

Sources: Lyle F. Albright, _Processes for Major Addition-Type Plastics and Their Monomers_, 1974.
Encyclopedia of Polymer Science and Technology.

TABLE 181.   INPUT MATERIALS USED IN IMPACT POLYSTYRENE MANUFACTURE
(IN ADDITION TO STYRENE MONOMER AND RUBBER)

Free Radical Initiators	azobisisobutyronitrile benzoyl peroxide dimethylaniline[a] di-tert-butyl peroxide tert-butyl hydroperoxide
Chain Transfer Agents	alkyl mercaptans -methyl styrene dimer carbon tetrabromide carbon tetrachloride ethylbenzene
Retardants and Inhibitors	tert-butyl-pyrocatechol benzoquinone sulfur ferric salts 1,1-diphenyl-2-picrylhydrazyl oxygen tert-butylcatechol
Solvents	ethylbenzene
Suspension Stabilizers (protective colloids)	alkaline earth phosphates, carbonates, and/or silicates gelatin magnesium silicates methyl cellulose pectin polyacrylic acid and its salts polyvinyl alcohol polyvinyl pyrrolidone starch sulfonated polystyrene zinc oxide
Flame Retardants	alkyl phosphates antimony oxide aryl phosphates hydrated aluminum oxide

(continued)

TABLE 181 (continued)

Lubricant   mineral oil

Antioxidant   2,6-di(tert-butyl)-4-methyl-
                                         phenol

[a]Used only in conjunction with benzoyl peroxide at low temperatures.

Sources: Lyle F. Albright, _Processes for Major Addition Type Plastics and Their Monomers_, 1974.
          Encyclopedia of Polymer Science and Technology.

<u>Termination</u>

Radical Combination

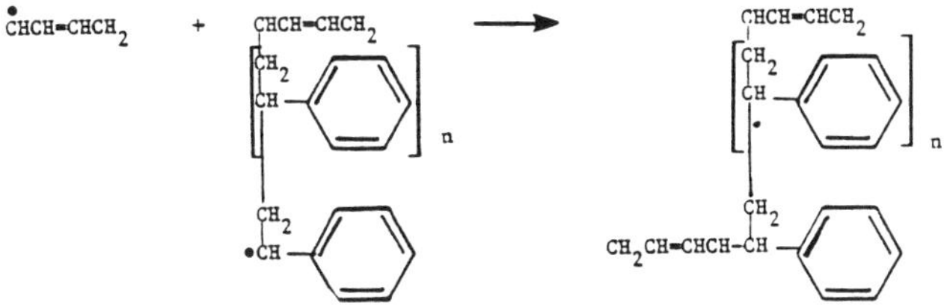

Chain Transfer

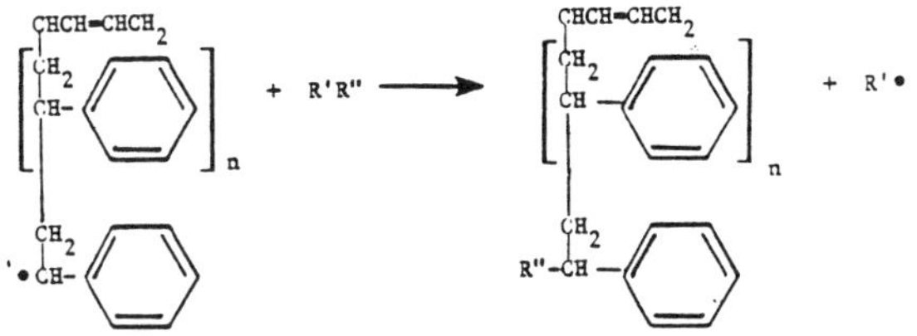

<u>Disproportionation</u>:

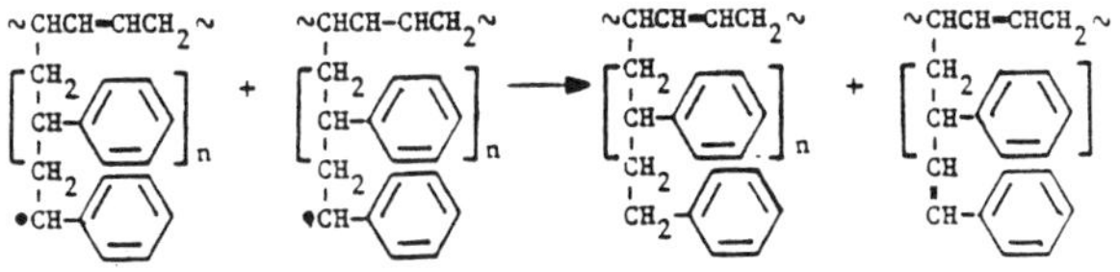

Oxygen reacts with the free radical initiators used, thus making a peroxide radical shown below:

$$R\bullet + O_2 \longrightarrow R\overset{\bullet}{O}_2$$

This slows the reaction rate. Polymerization is performed in a nitrogen atmosphere to minimize this reaction.

Suspension Polymerization

Suspension polymerization of impact polystyrene uses water and a suspending agent to separate the monomer into droplets for the polymerization process. The major advantage of this process as compared to other methods is the production of the polymer in small beads which may be easily separated. Although the polymerization rate is slower, the cost higher, and the contamination more severe for suspension polymerization than mass or solution polymerization, the advantages of this process include: excellent heat transfer, no solvent recovery problems, fewer safety hazards, simpler temperature control, and narrower range particle size control possible.

As illustrated in Figure 56, the styrene monomer is stripped to remove the inhibitor added to prevent polymerization during shipping and storage. The stripper water is sent to wastewater treatment while the styrene is combined with a monomer-soluble free radical initiator. The resulting mixture is fed to an agitated reactor filled with water where a dispersing agent and a suspension stabilizer (which maintain the styrene mixture as aqueous phase droplets) are added with the rubber. After the droplets have polymerized, the polymer beads are washed with acid to remove any residuals or contaminants. The wash water is sent to wastewater treatment while the wet polymer beads are centrifuged and dried. The water removed during centrifugation is combined with the other wastewater produced by this process. Any remaining unreacted styrene compounds and contaminants are removed from the polystyrene in the devolatilization extruder.

A typical recipe for suspension polymerization of IPS is given below for both the prepolymerization and polymerization steps:[2]

Prepolymerization – Quantities of each material are unknown.

Styrene-Rubber Mixture (6% rubber) (monomer)
Di(tert-butyl)peroxide (initiator)
Antioxidant
C_{12} Mercaptan (molecular weight regulator)
Refined Hydrocarbon Oil (lubricant and plasticizer)

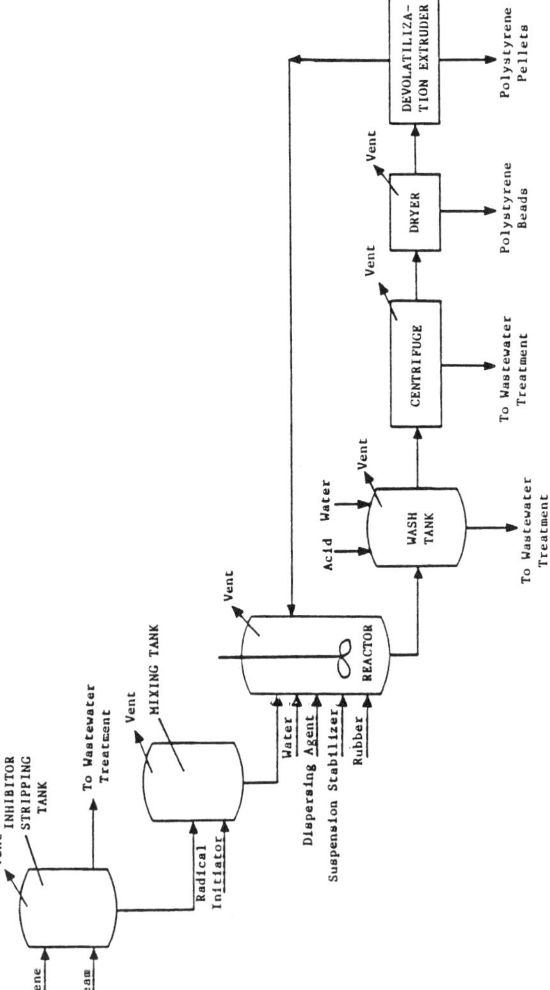

Figure 56.   Polystyrene production using the suspension polymerization process.

Source: Encyclopedia of Polymer Science and Technology.

<u>Polymerization</u>

<u>Material</u>	<u>Parts by Weight</u>
Prepolymerized Solution	100
Water	200
Surfactant	0.25
Calcium Chloride (stabilizer)	0.15
Acrylic Acids and Other Acrylates (comonomer)	0.25

Typical water to monomer ratios range from 1:1 to 4:1 and the particle size of the monomer ranges from 0.1 to 1 mm.[110] Final polymer volatile concentrations as low as 0.5 percent can be produced using devolatilization extrusion at reduced pressure, typically 10 to 20 mm Hg.[18] Reactors are aluminum, stainless steel, or stainless steel clad materials. Copper is not used since it discolors the polymer.

Mass Polymerization

Mass (or bulk) polymerization is the simplest process by which impact polystyrene can be made. The polymer chain is modified with the rubber molecules to form a copolymer which has slightly reduced clarity and gloss when compared to general purpose grades but exhibits improved impact strength. The polymerization of polystyrene is difficult to control due to its highly exothermic nature, and mass polymerization is the most difficult of the processes used since the viscosity of the polymerizing mass increases as the conversion increases. However, mass polymerization of polystyrene results in little contamination of the polymer by water, solvent, or suspension stabilizers.

Mass polymerization, as depicted in Figure 57, consists of four basic steps: prepolymerization, polymerization, devolatilization, and extrusion. Prior to polymerization, the styrene monomer is stripped of the inhibitor used to prevent polymerization during shipping. The styrene is fed to the reactor while the stripper water is sent to wastewater treatment. A typical recipe for mass polymerization of IPS was not found in the literature; however, it was reported to be similar to that used in the prepolymerization reaction described above.

The equipment used in the prepolymerization step consists of a stirred reactor with a reflux condenser operating under reduced pressure. This enables control of the reaction temperature by removing a portion of the vaporized monomer and returning the condensed liquid styrene to the reactor. If peroxidic initiators are used, the reaction may be carried out at a lower temperature (80°C) than that used for thermal initiation (130°C). The polymer content of the resulting mixture varies from 25 to 35 percent.

From the prepolymerization section, the monomer-polymer mixture is fed into a series of stirred reflux reaction kettles or reaction towers. The temperature is progressively raised through the reaction zone to prevent

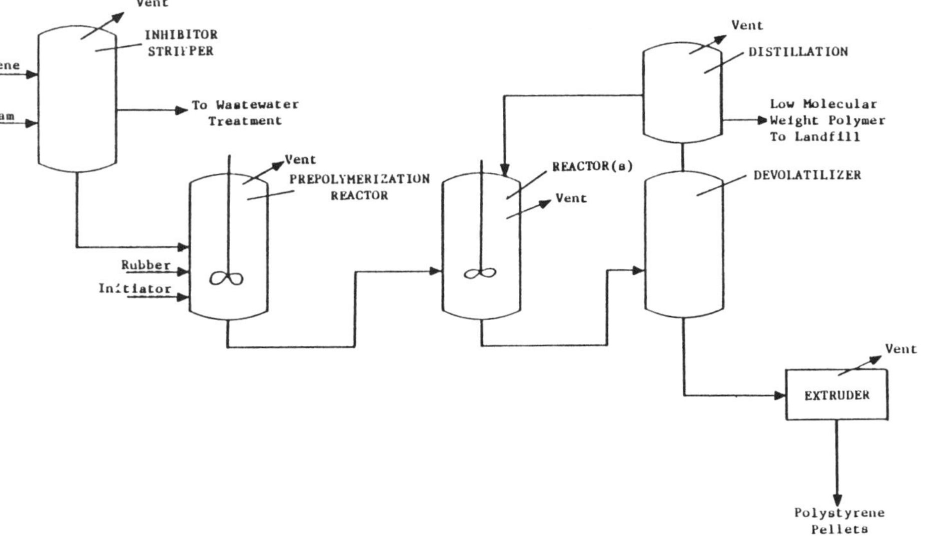

Figure 57.   Polystyrene production via mass polymerization.

Sources: Lyle F. Albright, Processes for Major Addition Type Plastics and Their Manufacture, 1974. Encyclopedia of Polymer Science and Technology.

fall off in polymerization rate and to reduce the viscosity of the polymerizing mass. Typical final temperatures in the polymerization section range
from 150 to 200°C.[2, 17] Materials of construction for the reactors are
aluminum, stainless steel, or stainless steel clad materials. Copper is not
used since it causes discoloration of the polymer.

In order to remove unreacted monomer and low molecular weight polymers,
the reaction mass is then fed to a static devolatilizer where the volatiles
are flashed off and distilled. The styrene monomer and oligomers are
recycled while the low molecular weight polymers are sent to a landfill. The
purified polymer is then extruded through a vented extruder and pelletized.

Solution Polymerization

Solution polymerization of impact polystyrene has several advantages
over the mass polymerization process. The viscosity of the polymerization
medium is lower, the temperature control is easier, and the molecular weight
of the polymer is lower. Although this process allows better heat dissipation, the solvent reacts with the polymer chain, resulting in more contamination of the product.

As illustrated in Figure 58, the styrene monomer is stripped to remove
the inhibitor added to prevent polymerization during shipping and storage.
The stripper water is sent to wastewater treatment while the styrene is
diluted with ethylbenzene and fed along with the rubber to a series of
agitated reactors, or a tower reactor, with heat exchange zones. After
polymerization, the unreacted styrene, rubber, and solvent are removed from
the polymer and recycled. The purified polymer is then extruded into
strands and cut into pellets.

The ethylbenzene concentration of the incoming solution to the reactors
ranges from 5 to 25 percent. Temperatures in the reactors start at 110 to
130°C in the first stage and increase to 150 to 170°C in the last stage.
Devolatilization temperature, under reduced pressure, is usually 250°C.[18]
Small amounts of a lubricant such as mineral oil (about 3 percent) may be
added.[208] Reactors are aluminum, stainless-steel, or stainless steel clad
materials. Copper is not used since it discolors the resulting polymer.

A typical recipe for solution polymerization of IPS is given below:[2]

Material	Parts by Weight
Styrene (monomer)	2,079
Polybutadiene (rubber)	121
Ethylbenzene (solvent)	220
Mineral Oil (lubricant and plasticizer)	small amount
2,6-Di(tert-butyl)-4-methyl phenol (antioxidant)	
α-methylstyrene (molecular weight regulator)	

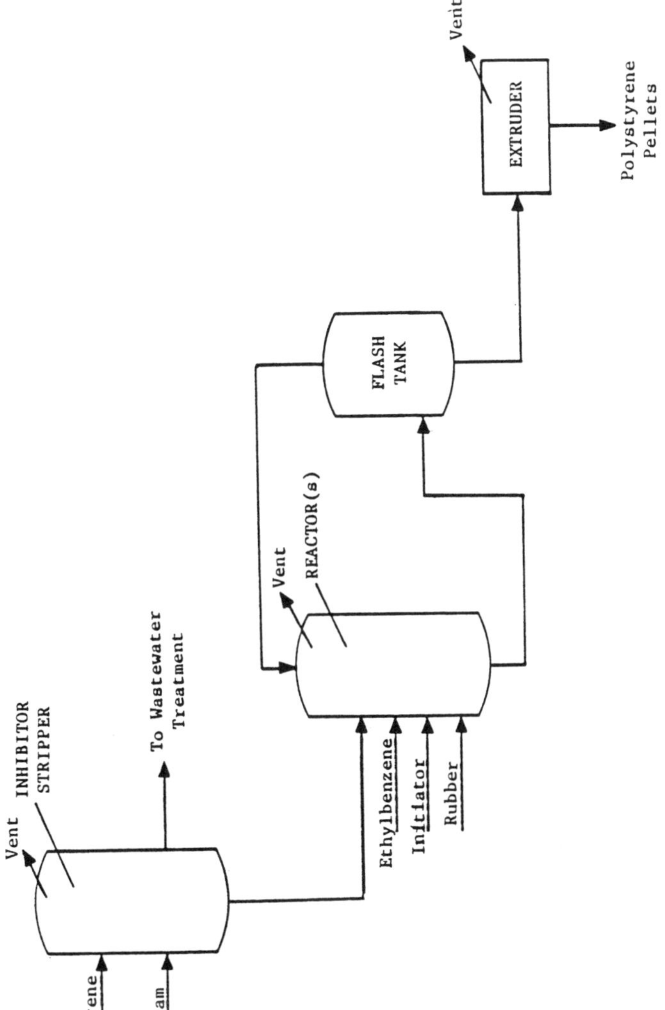

Figure 58. Polystyrene production via solution polymerization.

Source: <u>Encyclopedia of Polymer Science and Technology</u>.

<u>Energy Requirements</u>

The energy required for polystyrene production is listed below by technology:[208, 209, 211]

Technology	Energy Required	Unit/Metric Ton of Product
Cosden Technology, Inc.	Electricity Fuel	6.30×10^8 Joules 1.26×10^9 Joules
Gulf Oil Chemicals Co.	Electricity Steam	9.01×10^8 Joules 0.4 metric ton
Societe Chimique des Charbonnages (CdF)--The Badger Co., Inc.	Electricity Net Fuel	4.86×10^8 Joules 4.19×10^8 Joules

ENVIRONMENTAL AND INDUSTRIAL HEALTH CONSIDERATIONS

None of the chemicals which have been identified as inputs to polystyrene processing are known human carcinogens. However, there is some evidence from tests conducted and reported to NIOSH of polystyrene and styrene causing tumors in test animals.[232] Although these results are not conclusive, standards have been set for styrene and polystyrene exposure. Three other inputs to polystyrene processing are suspected human or known animal carcinogens: carbon tetrachloride, a chain transfer agent; polyvinyl alcohol, a suspension stabilizer; and antimony oxide, a flame retardant. Other highly toxic inputs include a free radical initiator, tert-butyl hydroperoxide, and a retardant, benzoquinone.

One environmentally significant portion of polystyrene processing is the devolatilization step. Residual monomer in the polymer is typically reduced to less than one percent, thereby reducing the exposure risk and rendering the polymer usable for food applications such as meat trays, egg cartons, and wrap for non-fatty foods.

<u>Worker Distribution and Emissions Release Points</u>

We have estimated worker distribution for polystyrene production by correlating major equipment manhour requirements with the process flow diagrams in Figures 56 through 58. Estimates are given in Table 182.

No major emission point sources are associated with polystyrene production. Although there are no major air emission point sources, fugitive emissions may pose a significant environmental and/or worker health problem depending on the components in the stream, operating conditions for the process, engineering and administrative controls, and maintenance programs.

TABLE 182.   WORKER DISTRIBUTION ESTIMATES FOR
IMPACT POLYSTYRENE PRODUCTION

Process	Unit	Workers/Unit/8-hour Shift
Suspension Polymerization	Steam Stripper	0.25
	Batch Reactor	1.0
	Batch Mixer	1.0
	Washing	0.25
	Centrifuge	0.25
	Dryer	0.5
	Extruder	1.0
Mass Polymerization	Steam Stripper	0.25
	Batch Reactor	1.0
	Distillation Column	0.25
	Devolatilizer	0.25
	Extruder	1.0
Solution Polymerization	Steam Stripper	0.25
	Batch Reactor	1.0
	Flash Tank	0.125
	Extruder	1.0

Process sources of fugitive emissions are listed in Table 183. Typical inhibitors, stabilizers and initiators are listed in Table 181. Ethylbenzene is the solvent generally used in solution polymerization. Particulate emissions from centrifugation, drying and extrusion operation may contain harmful additives and may produce elevated nuisance dust levels.

Health effects information about input materials of particular concern is provided in the paragraphs below. Table 10 summarized available toxicity information for major input materials.

Health Effects

The input materials used in impact polystyrene manufacture include three suspected human or known animal carcinogens. They are carbon tetrachloride, used as a chain transfer agent; polyvinyl alcohol, a suspension stabilizer; and the the flame retardant, antimony oxide. Other input materials that may pose a significant health risk to plant employees due to high toxicities are tert-butyl hydroperoxide, a free radical initiator, and the retardant benzoquinone. The reported health effects of exposure to each of these substances are briefly summarized below.

Antimony Oxide is a suspected carcinogen [5] which has also produced positive results in tests for mutagenic effects.[156] Animal testing of this substance is limited, however, and no epidemiological data exist.

Carbon Tetrachloride is a suspected human carcinogen [108] which is also extremely toxic upon inhalation or ingestion of small quantities.[241] The toxic hazard posed by skin absorption is slight for acute exposures, but high for chronic exposures.[241] The toxicity of carbon tetrachloride appears to be primarily due to its fat solvent action which destroys the selective permeability of tissue membranes and allows escape of certain essential substances such as pyridine nucleotides.[78] The OSHA air standard is 10 ppm (8 hour TWA) with a 25 ppm ceiling.[67]

p-Benzoquinone is a tumorigenic and a mutagenic agent, but testing has produced indefinite results for carcinogenicity.[105] The substance is currently undergoing additional tests for carcinogenesis by standard bioassay protocol under the sponsorhip of the National Toxicology Program. Benzoquinone is very toxic; the probable and lethal oral dose for humans is 50 to 500 mg/kg.[85] The OSHA air standard is 0.1 ppm (8 hour TWA).[67]

Styrene has been linked with increased rates of chromosomal aberrations in persons exposed in an occupational setting.[107] Animal test data strongly support epidemiological evidence of its mutagenic potential.[233] Styrene also produces tumors and affects reproductive fertility in laboratory animals.[233] Toxic effects of exposure to styrene usually involve the central nervous system.[233] The OSHA air standard is 100 ppm (8 hour TWA) with a ceiling of 200 ppm.[67]

TABLE 183.   SOURCES OF FUGITIVE EMISSIONS FROM POLYSTYRENE MANUFACTURE

Source	Constituent	Suspension	Mass	Solution
Inhibitor Stripping Vent	Styrene	X	X	X
	Inhibitor	X	X	X
Mixing Tank Vent	Styrene	X		
	Initiator	X		
Reactor Vents	Styrene	X	X	X
	Initiators	X	X	X
	Suspension Stabilizers	X		
	Solvents			X
Wash Tank Vent	Styrene	X		
Centrifuge Vent	Polystyrene, Styrene, and Rubber Particulates and Emissions	X		
Distillation Column Vent	Styrene	X		
Polymer Bead Drying	Polystyrene, Styrene, and Rubber Particulates and Emissions	X		
Extruder	Polystyrene, Styrene, and Rubber Particulates and Emissions		X	X

Tert-Butyl Hydroperoxide has produced symptoms of severe depression, incoordination, cyanosis, and respiratory arrest in laboratory animals.[243] Skin contact is associated with severe local reactions. Limited animal test data also indicate mutagenic effects. No epidemiological data are available; however, to evaluate the toxic hazard to humans.

Air Emissions

The sources of VOC fugitive emissions for polystyrene production are listed in Table 183. All of these sources are vents. Control of emissions from these vents may include routing the stream to a flare to incinerate hydrocarbons or routing the stream to blowdown.[285] Other fugitive VOC emissions result from leaks in process equipment, such as flanges, valves, cooling towers, open drains, and pumps. Although control of these emissions may be accomplished by equipment modification, a routine inspection and maintenance program may be the best control.

The sources of fugitive particulates are also listed in Table 183. Control of these sources may be accomplished by venting these streams to either a baghouse or electrostatic precipitator for particulate collection. EPA estimates that only 30 people are affected by hydrocarbon emissions and 2 are affected by particulates in the 100 person per square km population surrounding a polystyrene plant.[286]

Wastewater Sources

There are several sources of wastewater associated with the various polystyrene production processes as shown in Table 184. The major wastewater sources result from suspension polymerization which uses water to suspend the reaction mixture and wash the polymer. The aqueous phase of the polymerization mixture is removed from the polymer beads by centrifugation. This water, along with the polymer wash water, is reported to contain 20 kg of polystyrene per metric ton produced.[93] As Table 184 shows, all polystyrene processes generate water from monomer stripping and routine cleaning. Mass'and solution polymerization have wastewater from only these two sources.

Ranges of several wastewater parameters for wastewaters from polystyrene production are shown below. Values for the wastewater from the three IPS production processes and the processes presented in the GPPS section were not distinguished by process or resin type by EPA for the purpose of establishing effluent limitations for the polystyrene industry.[284]

Polystyrene Wastewater Characteristics	Unit/Metric Ton of Polystyrene
Production	$0 - 141.8 \ m^3$
BOD_5	$0 - 2.2$ kg
COD	$0 - 6.0$ kg
TSS	$0 - 8.4$ kg

TABLE 184.   SOURCES OF WASTEWATER FROM POLYSTYRENE MANUFACTURE

Source	Suspension	Mass	Solution
Monomer Steam Stripping	X	X	X
Polymer Bead Washing	X		
Centrifugation	X		
Routine Cleaning Water	X	X	X

Solid Wastes

The solid wastes generated during polystyrene production include substandard resins or beads which cannot be blended and collected particulates, and low molecular weight polymers from mass polymerization. One report estimates losses from mass polymerization processes to be 2 percent.[93] No information is available in the lieterature regarding the quantity of particles collected from air emission control devices. These wastes are not hazardous; therefore, their disposal should not pose an environmental problem.

Environmental Regulation

Effluent limitations guidelines have been set for the polystyrene industry. BPT, BAT, and NSPS call for the pH of the effluent to fall between 6.0 and 9.0 (41 Federal Register 32587, August 4, 1976).

New source performance standards (proposed by EPA on January 5, 1981) for volatile organic carbon (VOC) fugitive emissions include:

- Safety/release valves must not release more than 200 ppm above background, except in emergency pressure releases, which should not last more than five days; and

- Leaks (which are defined as VOC emissions greater than 10,000 ppm) must be repaired within 15 days.

Carbon tetrachloride, used as a chain transfer agent, has been tested as hazardous and designated U211 (45 Federal Register 3312, May 19, 1980). Disposal of carbon tetrachloride or polystyrene containing residual amounts of this compound must comply with the guidelines established in the Resource Conservation and Recovery Act (RCRA).

23. Polyurethane Foam

Polyurethane foams are cellular plastics produced by the reaction of a polyol and a polyisocyanate in the presence of a blowing agent, catalyst, and surfactant.[65] Many multifunctional alcohols (polyols) may be used to produce a wide range of products. Flexible and rigid polyurethane foams are used in furniture, bedding, seating, construction, refrigeration, and automotive parts.[217] Polyurethane foam constitutes 90 to 95 percent of the polyurethane produced.[179]

In addition to urethane groups ($-N-C-O-$), a typical urethane polymer may contain aliphatic and aromatic residuals from the polyol used as well as ester, ether, amide, or urea groups. These groups do not necessarily repeat in any regular order. The degree of polyol functionality, reactant molecular weight and type, isocyanate structure, catalysis, and reaction conditions all effect the resulting polymer's form and physical properties.[217] Typical properties for polyurethane foams of various formulations are listed in Table A-32 in Appendix A. In general, as the molecular weight of the polymer increases, the solubility decreases while a corresponding increase occurs in the tensile strength, melting point, elongation, elasticity, and glass-transition temperature of the polymer.[178, 179] The density of the foam is controlled by regulating the water content, i.e., low density foams require more water if the foam is water blown. Other blowing agents are mainly halocarbons, such as dichlordifluoromethane and methylene chloride.

Polyurethane foams are typically categorized as either flexible or rigid. Flexible foams are evaluated in terms of recovery from compression, or set, since they are used primarily in seating cushions, bedding and rug underlays. Good recovery from compression allows use of the product over a long period without deterioration in firmness or shape.

Rigid polyurethane foams have desirable properties which enable their use in construction and refrigeration applications. These properties include: excellent thermal insulation, especially when blown with fluorocarbons; a combination of high strength and light weight; good heat resistance; good energy absorption for use in sound deadening or vibration dampening; and excellent adhesion to wood, metal, glass, ceramic, and fiber surfaces.[178, 179] The use of polymeric isocyanates in the manufacture of rigid foam give improved heat and flame resistance.

Polyurethane foam is made by the prepolymer, quasi-prepolymer, and one-shot processes. Virtually all of the polyurethane foam produced is from the one-shot process; therefore, this process will be the only one presented.[113]

INDUSTRY DESCRIPTION

The polyurethane foam industry is comprised of 24 producers with 64 plants in 25 states. Three regions contain almost two-thirds of the plants: the southeastern states (EPA Region IV), the Great Lakes states (EPA Region V), and the southwestern states (EPA Region IX) contain 17, 22, and 27 percent of the plants, respectively. Polyurethane foam producers and their locations are listed in Table 185.

Polyurethane foam is used in the manufacture of consumer hard-goods, including furniture and appliances, and the construction industry. In 1981, U.S. polyurethane foam sales totaled 787,000 metric tons. Polyurethane flexible foam sales reached a peak in 1978 of 655,000 metric tons while rigid foam sales reached a peak in 1981 of 256,000 metric tons.[65] Total polyurethane foam sales peaked in 1978 at 891,000 metric tons. The current capacity, as seen from the 1978 production level, is adequate to enable industry growth in the near future.

PRODUCTION AND END USE DATA

In 1981, U.S. sales of polyurethane foam totaled 787,000 metric tons.[65] Polyurethane foam's many uses include:[178, 179, 217]

- Molded foam automotive parts, including safety pads, sun visors, arm rests, bucket seats, seat cushioning, instrument panel trim, floor mats, underlays, roof insulation, weather stripping, and air filters;

- Rigid foam for use in the aircraft industry, as insulation for household and commercial refrigerators, and in the flotation, transportation, and furniture industries which includes decorative parts, mirror frames, and chair shells;

- Flexible foam for mattresses, furniture cushions, carpet underlay;

- Bonding material for fabric;

- Insulation for refrigerated trucks; and

- Foam for use in the construction industry, including curtain-wall construction, preformed rigid panels, spray-applied wall construction, and roofing insulation.

Table 186 lists polyurethane foam consumption for 1981 according to foam type and market area.

TABLE 185. U.S. POLYURETHANE FOAM PRODUCERS

Producer	Location
Applied Plastics Co., Inc.	El Segundo, CA
E. R. Carpenter Co., Inc.	Conover, NC La Mirada, CA Richmond, VA Riverside, CA Russellville, KY Temple, TX Tupelo, MS
Cook Paint and Varnish Co.	Milpitas, CA North Kansas City, MO
Dow Chemical U.S.A.	Ironton, Ohio
The Firestone Tire & Rubber Co. Firestone Foam Products Division	Conover, NC Corry, PA Elkhart, IN Milan, TN Thomasville, GA
General Latex and Chemical Corp.	Ashland, OH Cambridge, MA Cucamonga, CA Dalton, GA
General Motors Corporation Inland Division	Dayton, OH

The BF Goodrich Co.
  BF Goodrich Engineered Products Group
    Fabricated Polymers Division
	Akron, OH

The Goodyear Tire & Rubber Co. Chemical Division	Bakersfield, CA Logan, OH Luckey, OH
Great Western Carpet Cushion Co., Inc.	Orange, CA

(continued)

TABLE 185 (continued)

Producer	Location

Gulf Oil Corporation
 Millmaster Onyx Group, subsidiary
  Apache Building Products Co.
   Division
	Belvidere, IL
	Jackson, MS
	Linden, NJ
	North Salt Lake, UT

Mobay Chemical Corporation Polyurethane Division	Santa Ana, CA
Olin Corporation Olin Chemicals Group	Fogelsville, PA
Plastic Management Corporation G.F.C. Foam Corporation, subsidiary	Bridgeview, IL Carlstadt, NJ East Rutherford, NJ Hazleton, PA
Reeves Bros. Inc. Curon Division	Corneluis, NC
H.H. Robertson Co. Freeman Chemical Corporation, subsidiary	Burlington, IA
Scott Paper Co. Foam Division	Eddystone, PA Fort Wayne, IN
Sheller—Globe Corporation	Tupelo, MS
D. Sobek Co.	Fremont, CA
Textron Inc. Industrial Product Group Burkart/Randall Division	Cairo, IL St. Louis, MO

(continued)

TABLE 185 (continued)

Producer	Location
United Foam Corporation	Boise, ID
	City of Commerce, CA
	Denver, CO
	Fresno, CA
	Hayward, CA
	Honolulu, HA
	National City, CA
	Phoenix, AZ
	Portland, OR
	Sacramento, CA
	Salt Lake City, UT
	Seattle, WA
United Merchants & Manufacturers, Inc.	
Glascoat Midwest Division	Elkhart, IN
Glascoat Southwest Division	Miami, FL
Thalco Division	Los Angeles, CA
The Upjohn Co.	
CPR Division	Columbus, OH
	Torrance, CA
Polymer Chemicals Division	La Porte, TX
Witco Chemical Corporation	
Isocyanate Products Division	Chicago, IL
	New Castle, DE

Source: *Directory of Chemical Producers*, 1982.

TABLE 186. U.S. CONSUMPTION OF POLYURETHANE FOAMS

Market	1981 Thousand Metric Tons
Flexible Foam	
Bedding	72
Furniture	200
Rug Underlay[a]	52
Transportaton	130
Other	77
FLEXIBLE FOAM TOTAL	531
RIGID FOAM	
Insulation	140
Furniture	10
Household and Commercial Refrigeration	50
Industrial Insulation	23
Transportation	22
Other	11
RIGID FOAM TOTAL	256
TOTAL	787

[a]Does not include 120,000 metric tons of rebonded underlay.

Source: Modern Plastics, January 1982.

PROCESS DESCRIPTION

Polyurethane foam is predominantly manufactured via the one-shot process. Although the prepolymer and quasi-prepolymer systems have been used, the one-shot process, which combines all of the input materials simultaneously to produce the foam, is the only commercial process now in use.[113, 178]

Polyurethane foam is produced by a polycondensation reaction between an isocyanate and a polyol. The reaction is influenced by the structure, functionality, and type and location of substituents of the isocyanate; structure of the polyol, presence of impurities or traces of acid in the isocyanate; and the reaction temperature, especially if it exceeds 100°C. Although choice of temperature, isocyanate, and polyol give some control over the reaction, the choice of catalysts exerts the most control.[178] The polyurethane condensation reaction can be very complex since the input materials and products may also act as catalysts for this process.

If no catalyst is used, the polycondensation reaction between an isocyanate and a polyol to form a urethane would appear as shown below:

$$R'NCO + ROH \rightleftarrows \left[\begin{array}{c} R'N=\overset{\cdots}{C}-O \\ \diagdown \overset{O}{\diagup} \\ H \quad R \end{array} \right]$$

$$\left[\begin{array}{c} R'N=\overset{\cdots}{C}-O \\ \diagdown \overset{O}{\diagup} \\ H \quad R \end{array} \right] \xrightarrow{ROH} \left[\begin{array}{c} H \diagdown \diagup R \\ O \\ \vdots \\ R'N-\overset{\cdots}{C}-O \\ \vdots \\ O \\ \diagup \diagdown \\ R \quad H \end{array} \right] \longrightarrow \overset{O}{\underset{\|}{R'NH-C-OR}} + ROH$$

If a basic catalyst is used, the reaction will proceed as shown below:

$$[R-N=C=O \rightleftarrows R-N=C^+-O^-] + Base \longrightarrow \left[\begin{array}{c} R-N=\overset{\cdots}{C}-O^- \\ \vdots \\ Base^+ \end{array} \right]$$

$$\left[\begin{array}{c} R-N=\overset{\cdots}{C}-O^- \\ \vdots \\ Base^+ \end{array} \right] \xrightarrow{R'OH} \left[\begin{array}{c} H \diagdown \overset{+}{\diagup} R' \\ O \\ \vdots \\ R-N^- -\overset{\cdots}{C}-O^- \\ \vdots \\ Base^+ \end{array} \right] \longrightarrow \overset{O}{\underset{\|}{RNH-C-OR'}} + Base$$

If an acid catalyst is used, the reaction proceeds as shown below:

$$R'NCO + H\text{-}Acid \longrightarrow [R'\text{-}N\text{=}C\text{-}O\cdots H\cdots Acid]$$

$$[R'\text{-}N\text{=}C\text{-}O\cdots H\cdots Acid] \xrightarrow{ROH} \left[\begin{array}{c} H \overset{+}{\diagdown\diagup} R \\ O \\ \vdots \\ R'N^-\text{-}C\text{-}O^- \cdots H \cdots Acid \end{array} \right] \longrightarrow$$

$$\overset{\displaystyle O}{\overset{\|}{R'NH\text{-}C\text{-}OR}} + H\text{-}Acid$$

The progresson of urethane to polyurethane foam is depicted below. The urethane molecule reacts with isocyanate groups to form the polyurethane chain. Carbon dioxide, which in turn acts as a blowing agent, is produced by reacting water with isocyanates. Trace amounts of free NH_2 and OH end groups are found in the polyurethane foam.

$$O\text{=}C\text{=}N\text{-}R\text{-}NH\overset{O}{\overset{\|}{\text{-}C}}\text{-}O\text{-}R'\text{-}O\overset{O}{\overset{\|}{\text{-}C}}\text{-}NH\text{-}R\text{-}N\text{=}C\text{=}O + 2R(NCO)_2 + H_2O \longrightarrow$$

$$R'\text{-}O\overset{O}{\overset{\|}{\text{-}C}}\text{-}NH\text{-}R\text{-}NH\overset{O}{\overset{\|}{\text{-}C}}\text{-}NH\text{-}R\text{-}NH\overset{O}{\overset{\|}{\text{-}C}}\text{-}NH\text{-}R\text{-}NH\overset{O}{\overset{\|}{\text{-}C}}\text{-}NH\text{-}R\text{-}R' + CO_2$$

The highly reactive $-N\text{=}C\text{=}O$ group which characterizes an isocyanate also leads to two important side reactions. One is the reaction with urea to form a biuret and the other is reaction with a urethane molecule to form an allophanate. These reactions are depicted below.

Urea   Biuret

$$RNCO + RNH\overset{O}{\overset{\|}{\text{-}C}}\text{-}NHR \longrightarrow R\underset{\underset{\displaystyle CONHR}{|}}{N}\text{-}CONHR$$

Urethane   Allophanate

$$RNCO + RHN\overset{O}{\overset{\|}{\text{-}C}}\text{-}OR' \longrightarrow R\underset{\underset{\displaystyle CONHR}{|}}{N}\text{-}COOR'$$

Tables 187 and 188 list typical input materials and operating parameters used for polyurethane foam manufacture.

Polyurethane foam is produced primarily by the one-shot process, which mixes all ingredients simultaneously. The polyol, polyisocyanate, catalyst, surfactant, and water (or other blowing agent) are combined in a reactor which is kept well agitated. Once the ingredients are uniformly mixed, represented by the cream time, the contents of the reactor are emptied. The reacting mixture which will gel to form the foam is poured onto a moving conveyor, inclined 5 to 10 degrees above the horizontal to keep the liquid from falling into the foam. The resulting slab is treated with heat or steam to cure the surface, then baked in a curing oven to cure the polymer. This process is illustrated in Figure 59.

Variations in the foam are achieved by altering the foam formulation. The use of polymeric isocyanates gives a wide range of obtainable foam properties which minimizes the need for compounding, a foam which can be cold cured, and a reduced risk of isocyanate emissions due to the lower vapor pressure of these compounds when compared to toluene diisocyanate (TDI).[178] TDI is used almost exclusively in polyurethane foam manufacture.[179] Choice of polyol also gives a variety of foam types. For example, encapsulating foams are obtained by using castor oil as the polyol.

A recipe to produce a typical furniture grade flexible polyurethane foam is given below:[277]

Material	Parts by Weight
Polyether Triol	100
Toluene Diisocyanate	50
Water	4
Stannous Octoate (catalyst)	1
Silicone Copolymers (surfactant)	1.5
Trichlorofluoromethane (blowing agent)	3

Polyurethane foam may be produced as slab stock, molded, sprayed, or foamed into a retaining cavity, such as a refrigerator housing. For a more detailed description of "foaming in place", see IPPEU Chapter 10a, Section 10.

Energy Requirements

No data were found in the literature consulted listing energy required for polyurethane foam production.

ENVIRONMENTAL AND INDUSTRIAL HEALTH CONSIDERATIONS

Several chemicals associated with the production of polyurethane foam have been listed as hazardous under RCRA: di-n-butyl phthalate, dichlorodifluoromethane, dimethyl phthalate, dioctyl phthalate, ethylenediamine, hydrocyanic acid, pyridine, resorcinol, toluene diisocyanate, trichlorofluoromethane, and tris(2,3-dibromopropyl) phosphate. Isocyanates are generally regarded to be skin sensitizers and irritants.

TABLE 187.   TYPICAL INPUT MATERIALS TO POLYURETHANE FOAM PRODUCTION

Function	Compound
Isocyanate	dianisidine diisocyanate
	2,5-dichlorophenyl isocyanate
	3,4-dichlorophenyl isocyanate
	4,4'-diphenylmethane diisocyanate
	ethyl isocyanate
	hexamethylene diisocyanate
	hydrogenated methylene diphenyl isocyanate
	isophorone diisocyanate
	m-chlorophenyl isocyanate
	m-xylene diisocyanate
	methyl isocyanate
	methylene diphenyl isocyanate (MDI)
	n-butyl isocyanate
	n-propyl isocyanate
	o-chlorophenyl isocyanate
	octadecyl isocyanate
	p-chlorophenyl isocyanate
	phenyl isocyanate
	tolidine diisocyanate
	toluene diisocyanate (TDI)
Polyols	poly(oxypropylene) adducts of glycerol
	poly(oxypropylene) adducts of 1,2,6-hexanetriol
	poly(oxypropylene) adducts of pentaerythritol
	poly(oxypropylene) adducts of sorbitol
	poly(oxypropylene) adducts of trimethylol propane
	poly(oxypropylene) glycols
	poly(oxypropylene-b-oxyethylene) adducts of ethylenediamine
	poly(oxypropylene-b-oxyethylene) adduct of trimethylolpropane
	poly(oxypropylene-b-oxyethylene) glycols
Polyesters	Polyesters made from:
	adipic acid
	1,3-butylene glycol
	1,4-butylene glycol
	caprolactone
	diethylene glycol
	ethylene glycol
	phthalic glycol
	propylene glycol

(continued)

TABLE 187 (continued)

Function	Compound

Polyethers	propylene oxide adducts of
	  α-methylglucoside
	  ethylenediamine
	  glycerol
	  pentaerythritol
	  sorbitol
	  sucrose
	  trimethylol propane

Catalysts Acid	boron trifluoride etherate hydrogen chloride
Tin	dibutyltin diacetate dibutyl tin salts used with low concentrations of the following antioxidants; catechol, resorcinol, t-butyl catechol, or tartaric acid dimethyltin dichloride di-n-butyltin dilaurate n-butyltin trichloride stannic chloride stannous chloride stannous octoate stannous oleate trimethyltin hydroxide tri-n-butyltin acetate tetra-n-butyltin
Amines and Other Organic Catalysts	benzyltrimethylammonium hydroxide bis(2-dimethyl aminoethyl) ether cobalt benzoate cobalt 2-ethylhexoate cobalt naphthenate cobalt octoate diethylcyclohexylamine dimethylaminoethyl piperazine 4-dimethylamine pyridine dimethylethanolamine lead benzoate lead 2-ethylhexoate lead oleate

(continued)

TABLE 187 (continued)

Function	Compound
Amines and Other Organic Catalysts (Continued)	lithium acetate
	manganese 2-ethylhexoate
	manganese linoresinate
	manganese naphthenate
	methylmorpholine
	N-aminoethyl piperazine
	N-ethylenediamine
	N-ethyl morpholine
	N-methyl morpholine
	N-tetramethylenediamine
	N-tetramethylenediamine-1,3-butane-diamine
	N,N-dimethylbenzylamine
	N,N-dimethylcyclohexylamine
	N,N-dimethylethanolamine
	N,N,N',N'-tetrakis(2-hydroxypropyl) ethylene diamine
	N,N,N',N'-tetramethyl butane diamine
	oxybis(dimethylaminoethane)
	potassium oleate
	pyridine
	sodium propionate
	tetramethyl guanidine
	triethylenediamine
	1,2,4-trimethylpiperazine
	zinc 2-ethylhexoate
	zinc naphthenate
	zirconium 2-ethylhexoate
	zirconium naphthenate
	zirconium toluene
Other	antimony pentachloride
	antimony trichloride
Blowing Agents	barium azodicarboxylate
	benzene sulfonyl hydrazide
	chlorofluorocarbons
	dichlorodifluoromethane
	trichlorofluoromethane
	diazoaminobenzene
	ethyl chloride
	methyl chloride
	methylene chloride

(continued)

TABLE 187 (continued)

Function	Compound
Flame Retardants	ammonium polyphosphate (Phosgard P/30®) chlorinated paraffins (with antimony trioxide) fine grade alumina trihydrate O,O-diethyl-N,N-bis-(2-hydroxyethyl) aminomethyl phosphonate tris(2-chloroethyl)phosphate tris(2,3-dibromopropyl)phosphate ("tris") tris(2,3-dichloropropyl)phosphate zinc borate
Surfactant	alkylsilicone-peroxyalkylene copolymers peroxyalkylene-polydimethyl siloxane organosilicones dimethyl silicone polymer
Cell Size Control Agent	mineral oil
Color Enhancer (Color Stabilizer)	triphenyl phosphite zinc dibutyl tricarbonate
UV Light Stabilizer	benzophenones
Clarifier	copper salts
Flow Control Agent	cellulose acetate butyrate
Gloss Control	colloidal silicas silica aerogels
Pigments	acetylene black aniline dyes antimony silico colors antimony trioxide benzidine toners cadmium colors carbon black chrome oxides

(continued)

TABLE 187 (continued)

Function	Compound
Pigments (Continued)	chromium colors
	dioxazine colors
	fluorscent zirconium dioxide
	hydroxyazo-metal chelate
	iron oxides
	Lake toners
	Lithol toners
	molybdenum colors
	nickel azo colors
	phthalocyanines
	Quinacridine colors
	strontium colors
	titanium dioxide
	titanium zirconates
	zirconium dioxides
Fillers	α-cellulose fiber
	aluminum silicates
	asbestine
	barium sulfate (barytes)
	barium zirconate
	barium zirconate silicate
	calcium carbonate
	calcium silicate
	calcium zirconate silicate
	cellulosics
	China clay (Kaolin)
	chopped glass fibers
	glass beads
	glass flake
	graphite
	hydrated alumina
	mangesium zirconium silicate
	mica, natural and synthetic
	nylon fiber
	pearlite
	polyester fiber
	polypropylene fiber
	silica, natural and treated
	talc
	vermiculite
	zirconium silicate
	zirconium spinel

(continued)

TABLE 187 (continued)

Function	Compound
Plasticizers	adipates
	didecyl adipate
	dioctyl adimpate
	aromatic oils
	chlorinated diphenyls and polyphenyls
	chlorinated waxes or paraffins
	phosphates
	cresyl diphenyl phosphate
	octyl diphenyl phosphate
	tributyl phosphate
	tricesyl phosphate
	trioctyl phosphate
	triphenyl phosophate
	tris(2-chloroethyl) phosphate
	tris(2-chlorphenyl) phosphate
	tris(2-chloropropyl) phosphate
	phthalates
	butyl/benzyl phthalate
	dibutyl phthalate
	dicapryl phthalate
	didecyl phthalate
	diisobutyl phthalate
	diisooctyl phthalate
	dimethyl phthalate
	dioctyl phthalate
Stabilizers	benzaldehyde
	hydroquinone
	pyrogallol
	resorcinol
	t-butyl catechol
Blocking Agents	
Low Temperature	acetyl acetone
	bisphenols
	dimethylamino methyl phenols
	ethyl acetoacetate
	ethyl malonate
	hydrocyanic acid
	novolaks

(continued)

TABLE 187 (continued)

Function	Compound
Medium Temperature	caprolactam
	imides
	monoethylaniline
	phenol mercaptans
	phenolic tertiary amines
	pyrocatechol
	quarternary amines
High Temperature	diphenyl amine
	hydroxy biphenyl
	isooctyl phenol
	phenol
	pyrrolidone

Sources: *Advances in Urethane Science and Technology*, K. C. Frisch, editor, 1971–1978.
Anna W. Crull, *Polyurethane and Other Foams*, 1979.
Anna W. Crull, *Polyurethane: Developments, Processes*, 1978.
E. N. Doyle, *The Development and Use of Polyurethane Products*, 1971.
Encyclopedia of Chemical Technology, 3rd Edition.
Encyclopedia of Polymer Science and Technology.
Polyurethane Technology, Paul F. Bruins, editor, 1969.

TABLE 188.   TYPICAL OPERATING PARAMETERS FOR POLYURETHANE FOAM PRODUCTION

Parameter	Value
Premix Temperature	18 - 22°C
Isocyanate Temperature	18 - 22°C
Cream Time[a]	10 sec
Gel Time	45 sec
Rise Time	90 sec

[a]Time required to achieve uniform mixing.

Source: _Polyurethane Technology_, Paul F. Bruins (ed), 1969.

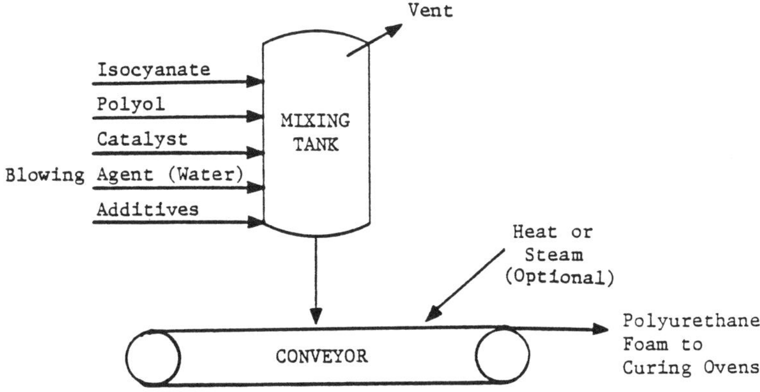

Figure 59.   Polyurethane foam production using the one-shot process.

Source: Encyclopedia of Polymer Science and Technology.

Tris(2,3-dibromopropyl) phosphate, tris(2-chloromethyl phosphate), antimony oxide, and dioctyl phthalate (DOP) are suspected human carcinogens. Other input materials posing a significant health risk include hydrocyanic acid, pyrocatechol, phenol, antimony trichloride, triethylamine, hydrcquinone, resorcinol, and methyl isocyanate.

Worker Distribution and Emissions Release Points

Worker distribution estimates for polyurethane production, derived by correlating major equipment manhour requirements with the process illustrated in Figure 59, are shown below:

Unit	Workers/Unit/8-hour Shift
Batch Reactor	1.0
Conveyor	0.5

No major air emission point sources are associated with the one-shot process. However, fugitive and process emissions result from the mixing tank vent and from the formation and cure of the foam. Control of the reactor vent stream may be accomplished by venting to a flare to incinerate the remaining hydrocarbons or to blowdown.[285] Control of fugitive emissions from valves, flanges, and pumps could result from equipment modification. A routine inspection and maintenance program may be the best control available.

Of major concern in polyurethane manufacture is employee exposure to toluene-2,4-diisocyanate, methylene bisphenyl isocyanate or other isocyanate compounds. The effects of acute and chronic exposure to these compounds are discussed in the "Health Effects" portion of this section.

Little data are available in the literature from which employee exposure potential may be estimated. However, one source gives an emission factor of 14-420 g hydrocarbon/kg product for polyurethane foam production [286].

Health Effects

The production of polyurethane foam involves the use of over 190 input materials, some of which are toxic, mutagenic, and/or teratogenic. Three flame retardants are suspected carcinogens: tris(2,3-dibromopropyl) phosphate, tris(2-chloroethyl) phosphate, and antimony trioxide. In addition to these, some of the input chemicals are known to produce tumors in laboratory animals and are currently being tested for carcinogenesis. Other chemicals which could pose a significant health risk to plant employees if exposure occurs include: three blocking agents-- hydrocyanic acid, pyrocatechol, and phenol; three catalysts--antimony pentachloride, antimony trichloride, and triethylamine; hydroquinone, which is used as a stabilizer; the antioxidant resorcinol, which is used in low concentrations; and methyl isocyanate. The

reported health effects of exposure to these substances are summarized
below.

Several organometallic compounds also are used in the production of
polyurethane foam. Included among these are five compounds of tin, which
are used as catalysts, for which OSHA has set an air standard of 0.1
mg(Sn)/m^3 (8 hour TWA). These are tributyltin acetate, trimethyltin
hydroxide, dimethyltin dichloride, dibutyltin diacetate, and tetrabutyltin.
OSHA has set an air standard of 2 mg(Sn)/m^3 (8 hour TWA) for stannous
chloride and stannic chloride.

Antimony Oxide is a suspected carcinogen [5] which has also produced
positive results in tests for mutagenic effects.[156] Animal testing of
this substance is limtied, however, and no epidemiological data exist.

Antimony Pentachloride is extremely toxic [85] and can cause systemic
damage upon acute exposure by ingestion or inhalation.[242] The fumes from
reaction with moisture are also toxic.[157] Animal test data suggest that
this substance, in addition to being highly toxic, is a mutagenic agent.
[233] The OSHA air standard is 0.5 mg(Sb)/m^3 (8 hour TWA).[67]

Antimony Trichloride, like antimony pentachloride, is extremely toxic.
[85] It is reported to irritate not only the upper respiratory tract, but
to produce slightly delayed abdominal pain and loss of appetite indicative
of effects over and above that of hydrochloric acid, believed to be the
major hydrolytic product from contact with moist tissue.[5] Toxic effects
have been reported in humans after inhalation of a dose equivalent to 73
mg/m^3.[20] The OSHA air standard is 0.5 mg/m^3 (8 hour TWA).[67]

Hydrocyanic Acid is highly toxic by ingestion, inhalation, and skin
absorption [242]; the probable oral lethal dose for humans is less than 5
mg/kg, or just a taste.[85] Inhalation of vapors causes toxic effects and
death within several minutes to several hours, depending upon the concentration.[83] Symptoms of poisoning include giddiness, hyperpnea, headache,
palpitation, cyanosis, and unconsciousness; asphyxial convulsions may
precede death.[83] The OSHA standard in air is 10 ppm (8 hour TWA).[67]

Hydroquinone has exhibited potential as a tumorigenic, mutagenic, and
teratogenic agent in animal tests.[233] Carcinogenesis tests have produced
indefinite results, but the substance is currently being tested by the
National Toxicology Program for carcinogenicity.[233] Hydroquinone is also
highly toxic. Fatal human doses have ranged from 5 to 12 grams, although
300 to 500 milligrams have been ingested daily for 3 to 5 months without ill
effects.[85] The OSHA air standard is 2 mg/m^3 (8 hour TWA).[67]

Methylene Bisphenyl Isocyanate is a strong irritant and may cause
systemic damage via skin absorption.[242] Animal data indicate that the
substance is also very toxic by ingestion and inhalation.[233] The OSHA air
standard is 0.02 ppm (8 hour TWA).[67]

Phenol is extremely toxic by ingestion, inhalation, and skin
absorption.[85] Approximately half of all reported cases of acute phenol

poisoning has resulted in death.[278] Chronic poisoning may also occur from
industrial contact and has caused renal and hepatic damage.[149] Evidence
indicates that phenol is also a tumorigenic, mutagenic, and teratogenic
agent; however, a recent carcinogenesis bioassay by the National Cancer
Institute produced negative results.[233] The OSHA air standard is 5 ppm
(TWA).[67]

Polyurethane, in the form of foam, has produced positive results in
animal tests for carcinogenicity.[107] However, very little animal data and
no human data exist in the available literature with which to evaluate the
potential for other adverse effects in humans following exposure to this
substance.

Pyrocatechol is very toxic, probably 1.5 to 3.0 times more toxic than
phenol, depending on the route of administration. [4] Symptoms of exposure
are similar to those associated with exposure to phenol (e.g., vasoconstric-
tion causing rise in blood pressure, blood dyscrasias, renal and hepatic
damage, convulsions), but convulsions are more frequent.[85] Pyrocatechol
exhibits tumorigenic and mutagenic potential, and is currently being tested
for carcinogenesis by the National Toxicology Program.[233]

Resorcinol is a tumorigen and a mutagen according to animal test data
[233]; however, carcinogenic testing has produced indefinite results.[105]
The chemical is currently being tested for carcinogenesis under the National
Toxicology Program. Resorcinol is also highly toxic by ingestion. An oral
exposure of 29 mg/kg has caused death in humans.[52]

Toluene-2,4-Diisocyanate (TDI) is a strong irritant of the eyes, mucous
membranes and skin. Acute exposure to TDI produces coughing, choking, chest
pain, nausea, and pulmonary edema. The importance of acute effects of TDI
overexposure is overshadowed by its ability as a potent respiratory tract
sensitizer. In sensitized individuals, severe asthmatic symptoms may occur
within minutes of exposure. The OSHA air standard is a ceiling value of
0.02 ppm. The American Conference of Governmental Industrial Hygienists
(ACGIH) has recommended a Threshold Limit Value (TLV) of 0.005 ppm (8 hour
TWA).

Triethylamine is highly toxic by ingestion and inhalation and
moderately toxic via dermal routes.[243] It is a strong irritant to tisse.
At least one animal study has shown that triethylamine produced mutagenic
effects in rats at the low dose of 1 mg/m^3.[96] The OSHA air standard is
25 ppm (8 hour TWA).[67]

Tri(2-Chloroethyl) Phosphate, a suspected carcinogen [6], is currently
undergoing additional tests for carcinogenesis by the National Toxicology
Program.[233] Very little animal and human toxicity data exist in the
available literature with which to evaluate toxic effects of acute or
chronic exposure to this substance.[233]

Tris(2,3-Dibromopropyl) Phosphate is classified as a suspected human
carcinogen by the International Agency for Research on Cancer.[108] Positive results on a National Cancer Institute carcinogenesis bioassay confirm

the carcinogenic potential of this substance.[233] Substantial evidence
also exists of its mutagenic and teratogenic potential.[233] It is a skin
and eye irritant and is moderately toxic by ingestion.[233]

<u>Air Emissions</u>

Chlorofluorocarbon emissions from flexible polyurethane foam slabstock
production are approximately 0.03 kg/kg of foam and emissions from molded
flexible foam production are approximately 0.02 kg/kg of foam.[282] These
emissions may result in a significant environmental or worker health problem
depending on the stream components, operating parameters for the processs,
engineering and administrative controls, and maintenance programs. According to EPA estimates, the population exposed to polyurethane foam emissions
in a 100 km^2 area around a polyurethane foam plant is: 326 persons
exposed to hydrocarbons and 2 persons exposed to particulates.[286]

<u>Wastewater Sources</u>

The only wastewater source associated with polyurethane foam production
is routine cleaning water. No wastewater data were found in the literature
consulted.

<u>Solid Wastes</u>

The solid waste generated by this process is substandard foam which
cannot be blended. This waste may be recycled and reused by adding glycol
to the foam or used to form carpet backing. This process is emission free,
not sensitive to product mixes, and the derived polyol may be completely
recycled.[276]

<u>Environmental Regulation</u>

Effluent limitations guidelines have not been set for the polyurethane
foam industry.

New source performance standards (proposed by EPA on January 5, 1981)
for volatile organic carbon (VOC) fugitive emissions include:

- Safety/release valves must not release more than 200 ppm above
  background, except in emergency pressure releases, which should not
  last more than five days; and

- Leaks (which are defined as VOC emissions greater than 10,000 ppm)
  must be repaired within 15 days.

The following compounds have been listed as hazardous wastes (46
Federal Register 27476, May 20, 1981):

di-n-butyl phthalate	- U069
dichlorodifluoromethane	- U075
dimethyl phthalate	- U102

```
dioctyl phthalate (DOP)               - U107
ethylenediamine                       - P053
hydrocyanic acid                      - P063
methyl isocyanate                     - P064
phenol                                - U188
pyridine                              - U196
resorcinol                            - U201
toluene diisocyanate (TDI)            - U223
trichlorofluoromethane                - U229
tris(2,3-dibromopropyl) phosphate     - U235
```

Disposal of these compounds or polyurethane foams containing residual amounts of these compounds must comply with the provisions set forth in the Resource Conservation and Recovery Act (RCRA).

24. Polyvinyl Acetate

INTRODUCTION

Polyvinyl acetate (PVAc) is a tasteless, odorless polymer which is
transparent, and is brittle at low temperatures. It has excellent aging
qualities since it is resistant to oxidation and inert to the effects of
ultraviolet and visible light.[234]   These characteristics make polyvinyl
acetate ideal for applications such as latex paints, adhesives, surface
coatings, and textile finishings. It is also used as an intermediate in the
manufacture of polyvinyl alcohol. The most familiar application of polyvinyl acetate is the white glue used in many homes and schools.[65] Polyvinyl acetate production in the United States for 1980 totaled 452,700
metric tons.

Polyvinyl acetate is produced by polymerizing vinyl acetate using a
radical initiator. The formula

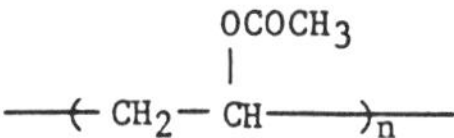

represents the repeating group which provides the backbone of the polymer
chain. With the introduction of additives, polyvinyl acetate and polyvinyl
acetate emulsions exhibit the physical properties listed in Tables A-33 and
A-34 in Appendix A.

Polyvinyl acetate has physical properties very similar to those of
aliphatic esters, which are generally soluble in organic solvents such as
ether, chloroform, and benzene. Polyvinyl acetate is insoluble in water but
soluble in most organic solvents including methanol and alcohols containing
more than five carbon atoms. The addition of water to ethanol, propanol,
and butanol enhances the solubility of polyvinyl acetate in these low
alcohol solvents. Table A-35 in Appendix A presents the solubility of
polyvinyl acetate in selected chemicals.

The properties of polyvinyl acetate can be altered by copolymerization
with another monomer. For example, vinyl acetate-ethylene and vinyl
acetate-vinyl pivalate copolymers are resistant to hydrolysis.[138] The
solubility of polyvinyl acetate polymers in a desired medium is also
influenced by copolymerization.

Polyvinyl acetate is produced by four different polymerization technologies; emulsion, solution, suspension, and mass polymerization. Emulsion polymerization accounts for as much as 90 percent of the polyvinyl acetate and vinyl acetate copolymers. The production of polyvinyl acetate in emulsions makes it very versatile in the production of paints, adhesives, and textile and paper coatings.[234]

INDUSTRY DESCRIPTION

The polyvinyl acetate industry is comprised of a diverse group of chemical and petrochemical companies, which are listed in Table 189. There are 55 polyvinyl acetate manufacturers in the United States. Of the 110 sites listed in Table 189, California, New Jersey, and Illinois contain 16, 15, and 13 sites, respectively, which total 38 percent of the total number of sites. Ohio, Texas, and Georgia have 7, 7, and 6 sites, respectively. New York and Massachusetts have five sites each, with the other fifty production sites being diversely located.

The polyvinyl acetate industry currently uses four processes: emulsion, suspension, solution, and mass (or bulk) polymerization. There is no information in the literature concerning plant capacity or a correlation between the plants listed in Table 189 and the processes discussed. However, two producers, Air Products at Calvert City, Kentucky and Monsanto at Springfield, Massachusetts, also manufacture polyvinyl alcohol at the same site and DuPont produces polyvinyl alcohol at another (LaPorte, Texas) site.[35] These three plants may use suspension or solution polymerization to manufacture PVAc since these processes are preferentially used for the production of polyvinyl acetate for polyvinyl alcohol production.

The polyvinyl acetate industry is consumer oriented. Paints, adhesives, and coatings are used in the construction, consumer goods, and textile and paper industries. PVAc production peaked in 1979 at 328,600 metric tons and dropped to 303,500 metric tons in 1980, which is still above the 286,400 metric ton level of 1976.[65] Since PVAc is used in many consumer goods, its production level reflects the general economic trends. With the extra 25,100 metric ton capacity, from 1979 to 1980, no new plants are likely in the near future.

PRODUCTION AND END USE DATA

In 1980, the sales of polyvinyl acetate resins were 303,500 metric tons [65]. An additional 134,000 metric tons of capacity were available for polyvinyl acetate hydrolysis to polyvinyl alcohol. Total 1980 polyvinyl acetate production was 452,700 metric tons.[35, 65]

Excluding polyvinyl alcohol production, polyvinyl acetate is used as:[138]

- Emulsions for adhesives;

- Dispersion resins or emulsions for paints;

TABLE 189.   MAJOR POLYVINYL ACETATE MANUFACTURERS

Manufacturer (Division)	Location
ADCO Chemical Company, Inc.	Newark, NJ
Air Products and Chemicals, Inc. Polymer Chemicals Division	Calvert City, KY City of Industry, CA Cleveland, OH Elkton, MD South Brunswick, NJ
AZS Corporation AZS Chemical Company Division	Atlanta, GA
Bennett's	Salt Lake City, UT

Borden, Inc.
  Borden Chemical Division
    Adhesives and Chemicals
    Division - East
    Thermoplastic Products
	Bainbridge, NY
	Bainbridge, NY
	Compton, CA
	Demopolis, AL
	Illiopolis, IL
	Leominster, MA

California Resin and Chemical, Co., Inc.	Vallejo, CA

Celanese Corporation
  Celanese Plastics and Specialities
    Co., subsidiary
    Celanese Resins Division
	Los Angeles, CA
	Newark, NJ

Colloids, Inc.
  North Chemical Co., Inc.,
  subsidiary
	Marietta, GA

Conchemo, Inc.
  Baltimore Operations
  Kansas City Operations
	Baltimore, MD
	Kansas City, MO

(continued)

TABLE 189 (continued)

Manufacturer (Division)	Location
Peter Cooper Corporation	Dallas, TX Gowanda, NY Pineville, NC
Dan River Inc. Chemical Products Division	Danville, VA
De Soto, Inc.	Chicago Heights, IL Garland, TX
Diamond Shamrock Corporation Process Chemicals Division	Cedartown, GA
Dutch Boy, Inc. Coatings Group	Baltimore, MD

E.I. du Pont de Nemours & Co., Inc.
  Plastics Products and Resin
    Department
	Seneca, IL

Emkay Chemical Co.	Elizabeth, NJ
Esmark Inc. Swift & Co., subsidiary	Gresham, OR Hammond, IN Houston, TX Los Angeles, CA Tampa, FL
Foy-Johnston, Inc.	Cincinnati, OH
Franklin Chemical Co.	Columbus, OH
H. B. Fuller Co. Polymer Division	Edison, NJ Atlanta, GA Blue Ash, OH
General Latex and Chemical Corp.	Ashland, OH Cambridge, MA Charlotte, NC Dalton, GA

(continued)

TABLE 189 (continued)

Manufacturer (Division)	Location
W. R. Grace and Co.	
Dewey and Almy Chemical Division	
Organic Chemicals Division	Owensboro, KY
	South Acton, MA
Gulf Oil Corporation	
Milmaster Onyx Group, subsidiary	
Lyndal Chemical Division	Lyndhurst, NJ
Hart Products Corporation	Jersey City, NJ
H & N Chemical Company	Totowa, NJ
Insilco Corporation	
Sinclair Paint Co. Division	Los Angeles, CA
International Minerals and Chemical	
Corporation	
IMC Industry Group, Inc.,	Carpentersville, IL
subsidiary	Melville, NY
IMC McWhorter Resins Division	
Jones-Blair Co.	Dallas, TX
Kelly-Moore Paint Co.	San Carlos, CA
	Hurst, TX
Kohler-McLister Paint Co.	Denver, CO
McCloskey Varnish Co.	Los Angeles, CA
	Philadelphia, PA
	Portland, OR
Monsanto Co.	
Monsanto Plastics & Resins Co.	Springfield, MA
Benjamin Moore & Co.	Los Angeles, CA
	Melrose Park, IL
	Newark, NJ
	St. Louis, MO
Napko Corporation	Houston, TX
National Casein Co.	Chicago, IL

(continued)

TABLE 189 (continued)

Manufacturer (Division)	Location
National Starch and Chemical Corp.	Meredosia, IL
Resins Division	Plainfield, NJ
Charles S. Tanner Co. Division	Enoree, SC
Norris Paints & Varnish Co.	Salem, OR
The O'Brien Corporation	
The O'Brien Corp. - Central Region	South Bend, IN
The O'Brien Corp. - Eastern Region	Baltimore, MD
The O'Brien Corp. - Western Region	South San Francisco, CA
Onyx Oils & Resins, Inc.	Brooker, FL
Philip Morris, Inc.	
Polymer Industries, Inc., subsidiary	
Adhesives and Liquid Coatings Division	Stamford, CT
Textile Chemicals Division	Greenville, SC
Raffi and Swanson, Inc.	
Polymeric Resins Division	Wilmington, MA
Reichhold Chemicals, Inc.	Azusa, CA
	Charlotte, NC
	Kansas City, KS
	Morris, IL
	South San Francisco, CA
	Tacoma, WA
Emulsion Polymers Division	Cheswold, DE
Scholler Brothers Inc.	Elwood, NJ
SCM Corporation	
Glidden Coatings & Resins Division	Chicago, IL
	Huron, OH
	Reading, PA
	San Francisco, CA
The Sherwin-Williams Co.	
Coatings Group	Chicago, IL
Stanchem, Inc.	East Berlin, CT

(continued)

TABLE 189 (continued)

Manufacturer (Division)	Location
Squibb Corporation Life Savers Inc., subsidiary	Canajoharie, NY[a]
Sybron Corporation Jersey State Chemical Co. Division	Haledon, NJ
Syncon Resins Inc. Farnow, Inc. Division	South Kearny, NJ
Union Carbide Corporation Chemicals and Plastics Division	Alsip, IL Garland, TX Somerset, NJ South Charleston, WV Torrance, CA Tucker, GA
Union Oil Co. of California Union Chemicals Division	Bridgewiew, IL Charlotte, NC Corning, NJ La Mirada, CA Lemont, IL Newark, CA

United Merchants and Manufacturers,
  Inc.
  Balchem Chemical Division
	Langly, SC

Valspar Corp. Baltimore Operations	Baltimore, MD
Yenkin-Majestic Paint Corporation Ohio Polychemicals Co. Division	Columbus, OH

[a]This plant manufactures resins for chewing gum.

Sources: *Chemical Economics Handbook*, updated annually, 1980
          data.
          Directory of Chemical Producers, 1981.

- Emulsions for paper and textile coating;

- Emulsions or dispersion resins for cement, concrete sealer, and tile cement;

- Emulsions for binding nonwoven fabrics; and

- Resins for chewing gum bases.

Table 190 delineates the sales and captive use of vinyl acetate for polyvinyl acetate production.

PROCESS DESCRIPTIONS

Polyvinyl acetate is produced by emulsion, suspension, solution, and mass (or bulk) polymerization. Of these methods, emulsion polymerization produces most of the polyvinyl acetate which is not used for hydrolysis to polyvinyl alcohol. PVAc is used as an emulsion in adhesives, paints, cement, concrete sealer, tile cement, nonwoven fabric binders, and paper and textile coatings; therefore, producing a PVAc emulsion is the most useful process for these end uses. Suspension and solution polymerization are used to produce polymer beads or a polymer solution which is in a convenient form to hydrolyze polyvinyl acetate to polyvinyl alcohol, although these resins may be used for other applications.[138, 234] Mass (or bulk) polymerization is used mainly for the production of low molecular weight resins since increasing the molecular weight of the polymerizing mass makes temperature control and pumping the mixture very difficult.[234] Tables 191 and 192 present typical input materials and operating parameters for these processes. Specialty chemicals for PVAc production are listed in Table 193.

An important economic consideration for the four PVAc processes is that because of the high monomer to polymer conversion (99.5+%), monomer stripping of the PVA is not required.

In the polymerization of vinyl acetate, the vinyl acetate monomer reacts with a free-radical initiator (R•) to start the polymerization reaction. As other vinyl acetate monomer molecules combine with the vinyl acetate radical, the polymer chain is propagated. Polymerization terminates when the radical reacts with an electron accepting group (chain transfer). The chain length is affected by the temperature of the reaction and the availability of chain transfer agents. Commercial polyvinyl acetate grades have molecular weights ranging from 11,000 to approximately 1.5 million. [138, 234] The polymerization reaction is depicted below:

<u>Initiation</u>

$$CH_2=CH-OCOCH_3 + R\bullet \rightarrow RCH_2-\overset{\bullet}{C}H-OCOCH_3$$

<u>Propagation</u>

$$RCH_2-\overset{\bullet}{C}H-OCOCH_3 + CH_2=CH-OCOCH_3 \rightarrow RCH_2-\overset{\displaystyle OCOCH_3}{\overset{|}{C}H}-CH_2-\overset{\bullet}{C}H-OCOCH$$

TABLE 190.   1980 SALES AND CAPTIVE USE OF POLYVINYL ACETATE[a]

End Product or Use	Metric Tons
Emulsion Paint	97,200
Adhesives	113,600
Paper	38,300
Textile	14,300
Non-Woven Binders	13,400
Other	18,800
TOTAL U.S. SALES	295,600
Export	7,900
TOTAL SALES	303,500

[a]These figures do not include polyvinyl acetate produced for polyvinyl alcohol.

Source: Facts and Figures of the Plastics Industry, 1981.

TABLE 191. TYPICAL INPUT MATERIALS TO POLYVINYL ACETATE
PRODUCTION PROCESSES IN ADDITION TO MONOMER

Process	Water	Initiator	Organic Solvent	Surfactants	Protective Colloids
Emulsion	X	X		X	X
Suspension	X	X			X
Solution		X	X		X
Mass		X			

TABLE 192. TYPICAL OPERATING PARAMETERS FOR POLYVINYL ACETATE
PRODUCTION PROCESSES

Process	Temperature	Reaction Time
Emulsion	80 – 90°C	4 – 5 hours
Suspension	70°C	2 hours
Solution	70 – 80°C	————a
Mass	70 – 80°C	————a

aReaction times were not available in literature sources.

Sources: Encyclopedia of Polymer Science and Technology.
Marshall Sittig, Vinyl Monomers and Polymers, 1966.

TABLE 193.   INPUT MATERIALS USED IN POLYVINYL ACETATE MANUFACTURE

Function	Compound
Comonomer	Acrylic Acid
	Alkyl Acrylates
	Butyl Acrylate
	Crotonic Acid
	Dialkyl Maleates
	Dibutyl Fumarate
	Dibutyl Maleate
	Di-2-Ethylhexyl Fumarate
	Di-2-Ethylhexyl Maleate
	Ethyl Acrylate
	Ethylene
	2-Ethylhexyl Acrylate
	N-Methylol Acrylamide
	Sodium Ethylenesulfonate
	Vinyl Caproate
	Vinyl Fumarate
	Vinyl Maleate
	Vinyl Laurate
	Vinyl Versatate
Surfactants	Anionic Sulfates
	Anionic Sulfonates
Protective Colloids	Hydroxyethyl Cellulose
	Polyvinyl Alcohol
Free Radical Initiators	Ammonium Persulfate
	Benzoyl Peroxide
	Hydrogen Peroxide
	Organometallic Compounds
	Peroxydisulfates
	Potassium Persulfate
Buffers	Acetate Salts
	Bicarbonate Salts
	Phosphate Salts

(continued)

TABLE 193 (continued)

Function	Compound
Plasticizers	Dibutyl Phthalate
	Tricresyl Phosphate
	Butyl Benzyl Phthalate
	Diethyl Phthalate
	Dimethyl Phthalate
	Diethylene Glycol Dibenzoate
	Dipropylene Glycol Dibenzoate
	Butyl Carbitol Acetate
	Cresyl Diphenyl Phosphate
	Triphenyl Phosphate
Chain Transfer Agents	Aldehydes
	Carbon Tetrachloride
	Thiols

Sources: Encyclopedia of Chemical Technology, 2nd Edition.
           Encyclopedia of Polymer Science and Technology.
           Marshall Sittig, Vinyl Monomers and Polymers, 1966.

<u>Termination via Chain Transfer</u>

$$R'R'' + R\left[CH_2-CH\underset{\underset{OCOCH_3}{|}}{}\right]_n CH_2-\overset{\bullet}{CH}-OCOCH_3 \longrightarrow$$

$$R'\bullet + R\left[CH_2-CH\underset{\underset{OCOCH_3}{|}}{}\right]_n CH_2-\underset{\underset{OCOCH_3}{|}}{CH}-R''$$

where R' and R" represent monomers, reacting polymers, or any electron accepting group.

<u>Emulsion Polymerization</u>

Emulsion polymerization produces two types of products: a fine-particle-size emulsion with 0.1 to 0.2 micron particles, used for paper coating; and a large-particle-size emulsion with a 0.5 to 3.0 micron particles used for adhesives.[138] The production of fine particle size emulsions utilizes anionic and nonionic surfactants and no protective colloid, while the large particle size emulsions are produced using protective colloids. Dispersion of the monomer in the aqueous phase without a protective colloid may reduce the number of micelles which also react to form polyvinyl acetate, since the function of the protective colloid is to keep the droplets from agglomerating.

In addition to particle size, other factors which determine the type of emulsion prepared are: the stability of the emulsion under mechanical shear, change of temperature, compounding, and passage of time; and the characteristics of the coalesced resin after application such as smoothness, opacity, and water resistance as well as characteristics of the emulsion as it is being used, such as flow properties and settling time.[234] The advantages of emulsion polymerization over mass, solution, and suspension polymerization processes include: ability to obtain a high solids content with a minimal increase in viscosity and excellent ability to maintain constant temperature during polymerization.

Figures 60 and 61 present process flow diagrams for emulsion polymerization processes. In Figure 60, the vinyl acetate monomer enters a feed tank which may have a nitrogen atmosphere. Depending on the initiator used, oxygen may inhibit the reaction. After the monomer is agitated in the feed tank, it is fed to the polymerization reactor where it is combined with a protective colloid/water mixture which may contain a surfactant. Most polyvinyl acetate emulsion processes use continuous monomer feed during the course of the reaction if small particles are desired. Batch feeding of the monomer to the reactor results in larger particles.[234] During polymerization, the reaction temperature is controlled by means of a reflux condenser, thereby removing a portion of the heat of reaction. When the polymerization is complete, the polyvinyl acetate emulsion is removed from the vented reactor and used either in this form or as spray dried resins.

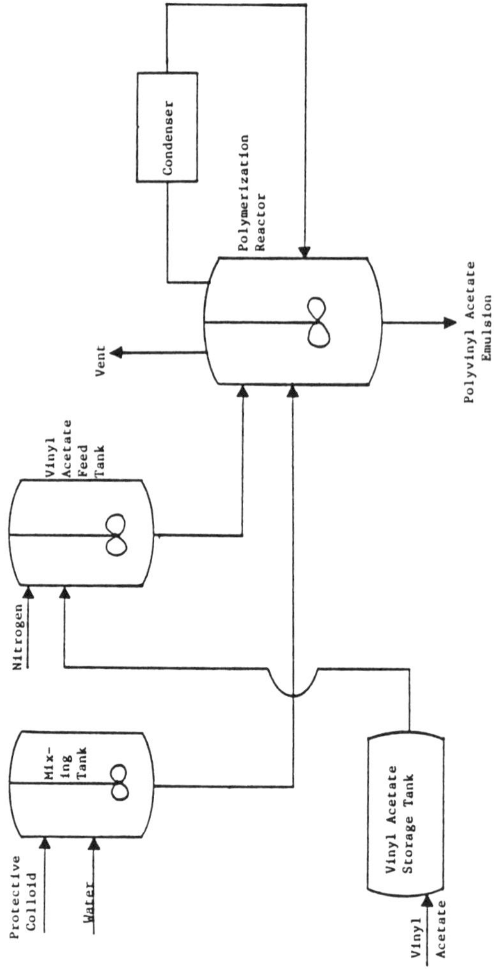

Figure 60. Polyvinyl acetate production using emulsion polymerization.

Source: Encyclopedia of Polymer Science and Technology.

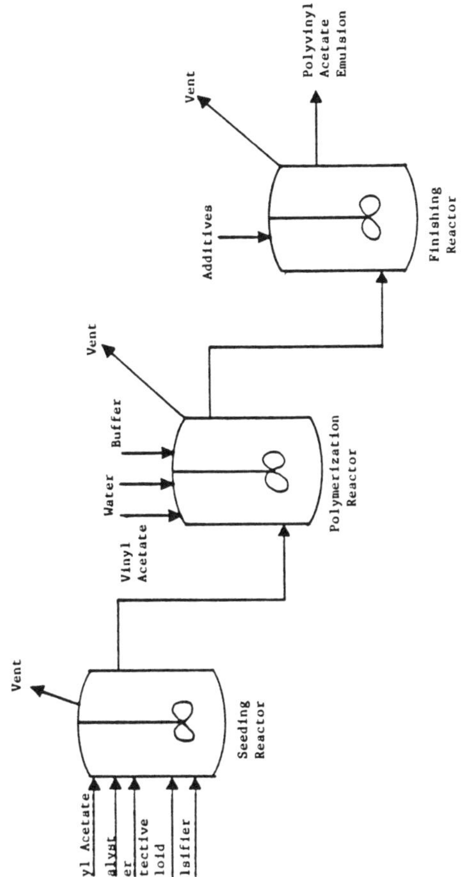

Figure 61. Polyvinyl acetate using continuous emulsion polymerization.

Source: Encyclopedia of Polymer Science and Technology.

Figure 61 presents a process flow diagram for continuous polymerization
of polyvinyl acetate. Monomer, water, catalyst, protective colloid, and
emulsifier are fed to the seeding reactor, which operates at about 70 to
75°C. The reaction mixture is then fed to a polymerization reactor where
more water, monomer, and a buffer are added. The polymerization reactor
operates at 80 to 85°C. Upon leaving the polymerization reactor, the
polymer mixture is combined with additives, which may include plasticizers,
additional stabilizers, and antifoaming agents,[138] to make the polyvinyl
acetate emulsion. Each reactor is equipped with a vent.

In emulsion polymerization processes, vinyl acetate comprises 30 to 60
percent of the charge to the reactor.[234] Up to 3 percent of the charge
may be an emulsifier.[234] Since less than 0.5 percent of the monomer is
present in the resulting polymer, monomer stripping of the polymer is
unnecessary.[138]

Comonomers may be used to alter the chemical and physical properties of
the polymer, such as hydrolysis, solubility, and water resistance. In addition to the input materials listed in Table 193, formulations for polyvinyl
acetate emulsions may include solvents, thickeners, antifoaming agents,
biocides, pigments, and fillers.

Agitated glass-lined or stainless steel reactors used range in size
from 2 to 30 m^3, the size depending on the efficient removal of heat from
the reaction mixture.[138, 234] Impellers may be turbine-type with flat,
curved vertical, or 45° blades for low-viscosity emulsions and anchor-type
agitators for high-viscosity emulsions.[138] Most emulsion polymerization
processes are performed at atmospheric pressure,[234] and Moyno® pumps are
reportedly best for pumping the resulting polymer emulsion.[138] Polyvinyl
acetate emulsions may be spray dried to produce a redispersible resin which
may be used in non-emulsion paints.[138]

A typical recipe for a large particle size emulsion follows:[125]

Material	Parts by Weight
Water	42.75
Vinyl Acetate (monomer)	55.00
Hydroxyethyl Cellulose (protective colloid) (Cellosize WP-09, Union Carbide)	2.00
Tergitol NPX (Union Carbide) (surfactant)	0.05
Potassium peroxydisulfate (initiator)	0.05
Sodium bicarbonate (buffer)	0.15

The Cellosize, which acts as a protective colloid, Tergitol (surfactant), and water are heated to 80°C for one hour, then cooled to 30°C.
After cooling, 10 percent of the vinyl acetate monomer, the potassium
peroxydisulfate (initiator), and sodium bicarbonate (buffer) are added.

This mixture is heated to 70 to 75°C. The remaining monomer is added in
increments over the next four hours while the reaction temperature is
maintained at 80°C. At the end of the monomer addition, the temperature is
raised to 90°C for 30 minutes, cooled, and filtered. The resulting solids
content of the emulsion is 56 to 57 percent, with a viscosity of 3,000 cP
and a pH of 5.0.[138]

Suspension Polymerization

The major advantage of producing polyvinyl acetate using suspension
polymerization is the manufacture of polymer beads which can be easily
removed by filtration from the reaction mixture. However, the large bead
size also renders the polymer unsuitable for use in many emulsion or dispersion type products. The equipment used for suspension polymerization,
depicted in Figure 62, is similar to that used in the emulsion polymerization process.[138] The batch process is started by mixing monomer, water,
and initiator (benzoyl peroxide or diacetyl peroxide) in a reactor and
heating the mixture to about 70°C. After approximately 10 minutes, the
protective colloid (polyvinyl alcohol) is added to the mixture and the
resultant mixture reacts for about two hours. The beads are removed from
the reactor, filtered, washed, and dried.

Wastewater generated by this process includes: the aqueous phase from
the suspension after the beads have been separated in the filter, and the
polymer wash water. Each of these streams may contain a small amount of:
vinyl acetate, initiator, protective colloid, and polyvinyl acetate. Since
the reaction is carried to high monomer conversion very little vinyl acetate
is present in the wastewater. Similarly, the small amounts of protective
colloid and initiator used preclude large amounts of this material being
released in the wastewater.

Some VOC emissions are expected from the reactor vent and the drier
outlet gas. Again, high monomer conversion and small input concentrations
keep emissions of vinyl acetate, initiator, and protective colloid very
small.

A typical suspension polymerization recipe follows:[138]

Material	Parts by Weight
Vinyl Acetate (monomer)	100
Water	200
Benzoyl peroxide (initiator)	0.3
Partially Hydrolyzed Polyvinyl Alcohol (protective colloid)	0.03

Solution Polymerization

Vinyl acetate is polymerized in solution mainly for use in polyvinyl
alcohol production.[138] Solvents include methanol, butanol, ethyl acetate,
ethanol with 5 to 10 percent water, methyl acetate, propionaldehyde, and

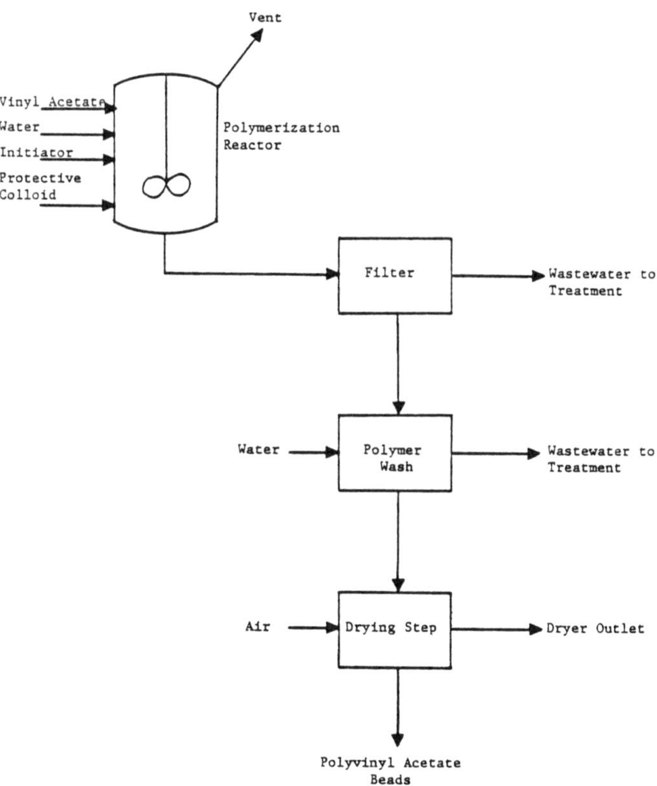

Figure 62. Polyvinyl acetate using suspension polymerization.

Source: _Encyclopedia of Polymer Science and Technology_.

ketones.[138, 234] Initiators used include benzoyl peroxide, lauroyl peroxide, t-butyl hydroperoxide, and azobisisobutyronitrile.[138] Protective colloids, such as partially hydrolyzed polyvinyl alcohol, may be used.

The major advantage of solution polymerization is the production of the polymer in a solution which can then be either reacted or formulated on site. However, the cost of the solvent is the primary disadvantage. As illustrated in Figure 63, vinyl acetate and propionaldehyde are mixed and charged to the first of two polymerization reactors in series. Ethyl acetate and initiator are combined in the catalyst solution tank and also charged to the first reactor. Both reactors are equipped with vents. Polymerization occurs at 70 to 80°C.[256] The reaction temperature is controlled by cooling a portion of the reaction solution in a reflux condenser, thereby removing a portion of the heat generated by the reaction. The solution is then fed to after-polymerization tanks where additional ethyl acetate and methanol are added, if the solution is shipped to another location for further processing. The resulting solution is filtered before use in polyvinyl alcohol production.

A typical production recipe for solution polymerization of PVAc is shown below:[54]

Material	Parts by Weight
Vinyl Acetate (monomer)	6,523
Methanol (solvent makeup)	544
Benzoyl Peroxide (initiator makeup)	60
Water (makeup)	86
Inhibitor	

Mass Polymerization

Mass (or bulk) polymerization of vinyl acetate is used to manufacture smaller amounts of the polymer than solution, suspension, or emulsion polymerization. The major advantage of this process is the production of low molecular weight polymers. The primary disadvantage of low molecular weight polymers is the difficulty in removing heat as the viscosity of the polymerizing monomer/polymer mixture increases.[234]

A continuous mass polymerization process is illustrated in Figure 64. Vinyl acetate and the initiator may be combined under a nitrogen atmosphere, depending upon the initiator used, in an initiator solution tank. The resulting mixture is passed through a screen filter before entering a mixing tank where the initiator solution is combined with additional vinyl acetate. The monomer mixture can be placed into storage for metered feed to the reactor. The stored monomer is filtered before it is charged to the reactor which is fitted with a vent. Polymerization is performed at or near atmospheric pressure and within the 70 to 80°C temperature range.[255] The reaction temperature is controlled by cooling a portion of the reaction mixture in a reflux condenser, thereby removing a portion of the heat generated by the reaction. When the polymerization is complete, the polymer is fed to a cutter which renders the polyvinyl acetate suitable for further use.

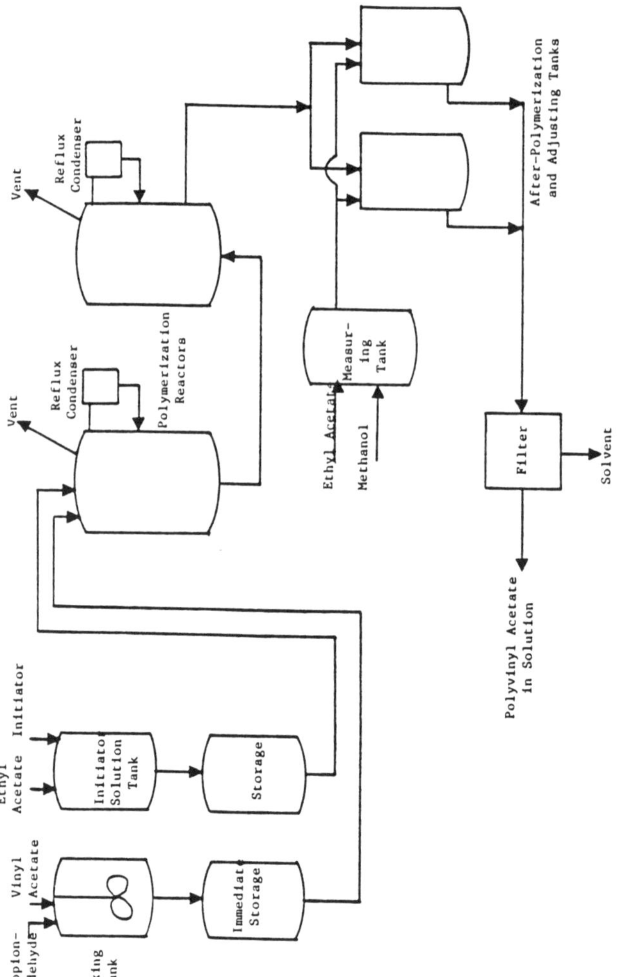

Figure 63. Polyvinyl acetate production via continuous solution polymerization.

Source: Encyclopedia of Polymer Science and Technology.

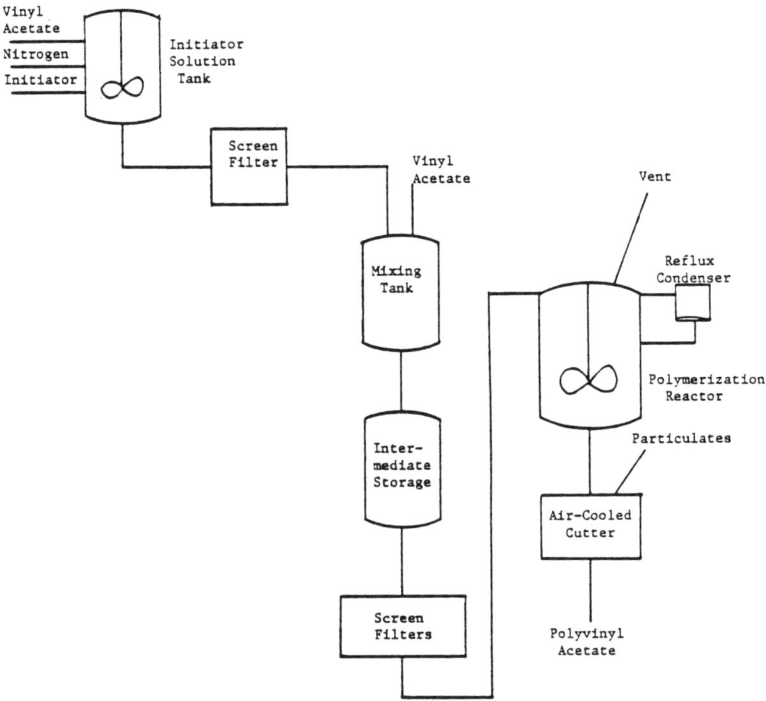

Figure 64. Polyvinyl acetate production via continuous mass polymerization.

Source: Encyclopedia of Polymer Science and Technology.

A typical production recipe for mass polymerization of PVAc was not found in the literature.

Energy Requirements

No data giving the energy required for polyvinyl acetate production were found in the literature consulted.

ENVIRONMENTAL AND INDUSTRIAL HEALTH CONSIDERATIONS

Polyvinyl acetate is nontoxic. However, the ACGIH has recommended a TLV of 10 ppm (8 hour TWA) for vinyl acetate. A significant point concerning vinyl acetate is that PVA production processes are carried to very high conversions (99.5+%). This high conversion eliminates the need for monomer stripping steps and minimizes the potential for vinyl acetate emissions or losses from downstream processing. Low monomer concentrations in the polymer also reduces vinyl acetate emissions from polymer bead drying in suspension polymerization processes.

Worker Distribution and Emissions Release Points

Worker distribution estimates for each of the five commercially viable processes for polyvinyl acetate production are shown in Table 194. We made the estimates by correlating major equipment manhour requirements with the process flow diagrams in Figures 60 through 64.

Although there are no point source air emissions from PVAc production, the fugitive emissions may pose a significant environmental and/or worker health problem based on the stream constituents, process operating parameters, engineering and administrative controls, and maintenance programs.

Process sources of fugitive emissions from PVAc manufacture are listed in Table 195. In addition to vinyl acetate, numerous additives (see Table 193) may be emitted from reactor and monomer storage tank vents. Nitrogen, used as a reactor purge, is a simple asphyxiant and may be present in reactor vent emissions. Solvents typically used in solution polymerization include methyl, butyl and ethyl alcohols, ethyl and methyl acetate propionaldehyde, and various ketone compounds. Drying and cutting operations may produce polymer particulate concentrations that pose a nuisance dust exposure hazard.

Available health effects information for input materials to PVAc production are discussed in the paragraphs below and in Table 10.

Health Effects

The manufacture of polyvinyl acetate involves the use of input materials that could adversely affect the health of workers if ingestion, inhalation, or skin contact occurs. The substances that pose the highest health risk due to carcinogenic, mutagenic, teratogenic, or toxic properties

TABLE 194. WORKER DISTRIBUTION ESTIMATES FOR POLYVINYL ACETATE PRODUCTION

Process	Unit	Workers/Unit/8-hour Shift
Emulsion Polymerization	Storage Tank	0.125
	Batch Mixer	1.0
	Batch Reactor	1.0
	Condenser	0.125
Continuous Emulsion Polymerization	Continuous Reactor	0.5
Suspension Polymerization	Batch Reactor	1.0
	Filter	0.25
	Washing	0.25
	Dryer	0.5
Solution Polymerization	Storage Tank	0.125
	Batch Reactor	1.0
	Reflux Condenser	0.125
	Batch Mixer	1.0
	Filter	0.25
Mass Polymerization	Storage Tank	0.125
	Filter	0.25
	Continuous Reactor	0.5
	Continuous Mixer	0.5
	Reflux Condenser	0.125
	Cutting	0.25

TABLE 195. SOURCES OF FUGITIVE EMISSIONS FROM
POLYVINYL ACETATE MANUFACTURE

Source	Constituent	Process			
		Emulsion	Suspension	Solution	Mass
Reactor Vents	Vinyl Acetate	X	X	X	X
	Initiators	X	X	X	X
	Emulsifiers	X			
	Protective Colloids	X	X	X	
	Solvents			X	
Monomer Storage Tanks	Vinyl Acetate and Inhibitors	X	X	X	X
Polymer Bead Drying	Polyvinyl Acetate and Vinyl Acetate Particulates and Emissions		X		
Adjusting Tanks	Solvents			X	
Polymer Cutting	Polyvinyl Acetate and Vinyl Acetate Particulates				X

include the comonomers acrylic acid and ethyl acrylate; polyvinyl alcohol,
which is used as a protective colloid; and the chain transfer agent, carbon
tetrachloride. A synopsis of the reported health effects resulting from
exposure to these subsances follows.

    Acrylic Acid is a severe skin and respiratory irritant [170] and is one
of the most serious eye injury chemicals.[157] Ingestion may produce epigastric pain, nausea, vomiting, circulatory collapse, and in severe cases,
death due to shock.[85] Although the data are not sufficient to make a
carcinogenic determination, the chemical has been associated with tumorigenic and teratogenic effects in laboratory animals.

    Carbon Tetrachloride is a suspected human carcinogen [108] which is
also extremely toxic upon inhalation or ingestion of small quantities.[241]
The toxic hazard posed by skin absorption is slight for acute exposures, but
high for chronic exposures.[241] The toxicity of carbon tetrachloride
appears to be primarily due to its fat solvent action which destroys the
selective permeability of tissue membranes and allows escape of certain
essential substances such as pyridine nucleotides.[78] The OSHA air
standard is 10 ppm (8 hour TWA) with a 25 ppm ceiling.[67]

    Ethyl Acrylate is currently being tested by the National Toxicology
Program for carcinogenesis by standard bioassay protocol. Previous test
data led to indefinite results.[107] It is very toxic by injestion,
inhalation, or skin absorption [242], producing symptoms of hypersomnia and
convulsions.[149] The OSHA air standard is 25 ppm (8 hour TWA).[67]

    Polyvinyl Alcohol has produced positive results in animal testing for
carcinogencity.[107] Its single dose toxicity, however, is presumably low.
[85] Implantation of PVA sponge as a breast prosthesis has been associated
only with foreign body type of reaction [107]; neutral solutions for PVA
have also been used in eyedrops on human eyes without difficulty.[87]

Air Emissions

    Sources of fugitive emissions from polyvinyl acetate processing are
summarized in Table 195 by process and constituent type. The majority of
fugitive emissions from polyvinyl acetate processing are generated from
reactor vents. These emissions are reported to contain 1.5 kg of vinyl
acetate per metric ton of polyvinyl acetate.[93] Reactor vent emissions may
also contain plasticizers, unreacted initiators, and, for suspension and
emulsion polymerization processes, emulsifiers and protective colloids.
These emissions may be reduced by routing the vented stream to a flare for
incineration of the hydrocarbons or to blowdown.[285]

    Polyvinyl acetate and vinyl acetate particulates result from the drying
step in suspension polymerization and the cutting step in the mass polymerization, as shown in Table 195. These particulates may be controlled by
venting these streams to a baghouse or electrostatic precipitator for particulate collection.

Wastewater Sources

Polyvinyl acetate generates several wastewater sources associated with the various production processes as shown in Table 196. The major wastewater source from polyvinyl acetate production results from the steam stripping of the vinyl acetate monomer prior to processing. Steam stripping is utilized to remove any inhibitor which may have been added to prevent polymerization during shipping and storage. The usual inhibitor is hydroquinone, and vinyl acetate will polymerize if the concentration is reduced to less than 50 ppm.[268] The steam, when condensed, typically contains 20 kg of vinyl acetate and 0.5 kg of hydroquinone per metric ton of polyvinyl acetate.[93]

Other wastewater streams during polyvinyl acetate production result from filtration operations, washing the polymer beads during suspension polymerization, and routine cleaning.

Ranges of basic wastewater parameters for wastewaters from the four production processes shown below were not separated by process by EPA for the purpose of establishing effluent limitations for the polyvinyl acetate industry:[284]

Characteristic	Unit/Metric Ton of Polyvinyl Acetate
Production	$0 - 25.03$ m^3
BOD$_5$	$0 - 2$ kg
COD	$0 - 3$ kg
TSS	$0 - 2$ kg

The control technologies common to polyvinyl acetate facilities are: equalization, chemical treatment, activated sludge, clarification, and polishing. This wastewater treatment system has been demonstrated to effectively treat the biologically degradable portions of the waste.[284]

Solid Wastes

The solid wastes generated during polyvinyl acetate production include substandard polyvinyl acetate emulsions stored in containers or beads which cannot be salvaged by blending. These wastes are not hazardous; therefore, their disposal should not pose an environmental problem.

Environmental Regulation

Effluent limitations guidelines have been set for the polyvinyl acetate industry. BPT, BAT, and NSPS call for the pH of the effluent to fall between 6.0 and 9.0 (41 Federal Register 32587, August 4, 1976).

New source performance standards (proposed by EPA on January 5, 1981) for volatile organic carbon (VOC) fugitive emissions include:

TABLE 196. SOURCES OF WASTEWATER FROM POLYVINYL ACETATE MANUFACTURE

Source	Emulsion	Suspension	Solution	Mass
Monomer Steam Stripping	X	X	X	X
Filtration		X		
Polymer Bead Washing		X		
Routine Cleaning Water	X	X	X	X

- Safety/release valves must not release more than 200 ppm above background, except in emergency pressure releases, which should not last more than five days; and

- Leaks (which are defined as VOC emissions greater than 10,000 ppm) must be repaired within 15 days.

The following compounds have been listed as hazardous (45 Federal Register 3312, May 19, 1980):

            Carbon tetrachloride - U211
            Di-n-butyl phthalate - U069
            Dimethyl phthalate   - U102
            Ethyl acrylate   - U113

Disposal of these compounds and PVAc resin containing residual amounts of these compounds must comply with the provisions set forth in the Resource Conservation and Recovery Act (RCRA).

25. Polyvinyl Alcohol

INTRODUCTION

Polyvinyl alcohol (PVA) is a water sensitive polymer that is white to
yellow in color. PVA is an unusual polymer in that it is not manufactured
from the vinyl alcohol monomer. The theoretical monomer, $CH_2=CHOH$,
rearranges to acetaldehyde, $CH_3CH=O$. Consequently, the polymer, which is
shown below, is obtained by the hydrolysis of polyvinyl acetate (PVAc).

$$-\!\!-\!(CH_2CHOH)_{\overline{n}}-\!\!-$$

The properties of PVA are primarily determined by the molecular weight
of the parent polyvinyl acetate and the extent of hydrolysis of the polyvinyl alcohol. The molecular weight of PVA ranges from a few thousand to
almost one million.[222] The degree of hydrolysis, or how many of the
acetate groups are hydrolyzed to hydroxyl groups, determines the properties
of the resulting resin.

Polyvinyl alcohol with a small percentage of the acetate groups hydrolyzed has greater flexibility, dispersing power, water sensitivity, and
adhesion to hydrophobic surfaces when compared to the highly hydrolyzed
polymer. Similarly, polyvinyl alcohol with a high percentage of hydroxyl
groups has increased tensile strength, solvent resistance, solubility in
water with resistance to additional hydrolysis, and adhesion to hydrophilic
surfaces in contrast to the properties of the polymer with less than 80
percent hydrolysis.

Physical properties of fully hydrolyzed polyvinyl alcohol are listed in
Table A-36 in Appendix A. Polyvinyl alcohol is typically classified as one
of three types: partially acetylated, which is 87 to 89 percent hydrolyzed;
fully hydrolyzed, in which 98 percent of the acetate groups have been converted to hydroxyl groups; and super hydrolyzed, which is 99.7 percent converted. Table A-37 in Appendix A gives density as a function of PVA acetate
content. Solubility of PVA in various organic solvents is presented in
Table A-38 in Appendix A.

The major uses for polyvinyl alcohol include textile warp sizing,
adhesives, suspension aids in polymerization, paper coatings, cosmetic
manufacture, optical applications, membranes, sponges, paint, and soil
erosion control. Coating, casting, and dipping applications comprise the
majority of polyvinyl alcohol uses. PVA is also used in the production of

polyvinyl butyral. In 1980, PVA production totaled 63,900 metric tons.[65]
Table 197 presents consumption of polyvinyl alcohol for 1978.

INDUSTRY DESCRIPTION

Polyvinyl alcohol is not polymerized from the vinyl alcohol monomer,
which rearranges to acetaldehyde. Rather, polyvinyl acetate is hydrolyzed
to make polyvinyl alcohol. Similarly, some of the polyvinyl alcohol
produced goes to polyvinyl butyral production. This sequence is pictured
below:

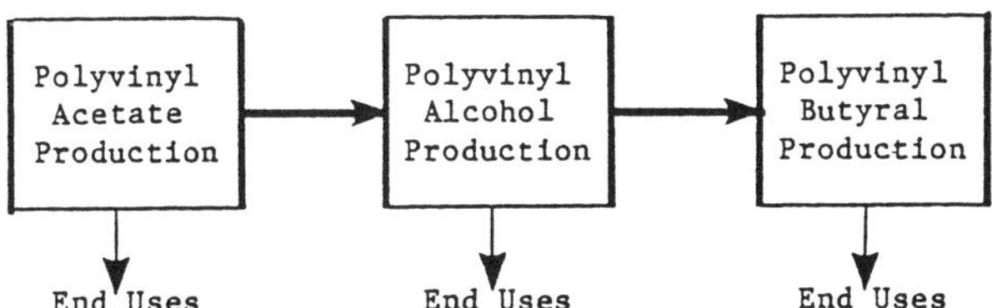

The polyvinyl alcohol industry is comprised of three companies, each of
which produces the polymer for polyvinyl butyral production as well as for
various end uses. The manufacturers for polyvinyl alcohol are listed in
Table 198. Nameplate capacity of polyvinyl alcohol in 1980 was 102,000
metric tons.[35, 56] Of this capacity, 22,000 metric tons were available
for captive polyvinyl butyral production.

The PVA industry currently uses polyvinyl acetate produced by two processes: suspension and solution polymerization. Suspension polymerization
is typically used to manufacture medium to high molecular weight (100,000 to
1,000,000) polymer while solution polymerization typically produces low
molecular weight (10,000 to 50,000) polyvinyl alcohol. The hydrolysis is
usually not a true hydrolysis reaction, where a stoichiometric quantity of
base and the presence of water are required, but rather an ester interchange
reaction. Alkaline conditions are usually used for solution polymerized
polyvinyl acetate, and acidic conditions for suspension polymerized polyvinyl acetate. The former process is sensitive to the presence of water;
therefore, it is more suited for solution polymerization products.[137]
Since process type used for polyvinyl alcohol production is not published,
there is no correlation between the plants listed in Table 198 and the
processes discussed.

PVA products are primarily consumer oriented. Adhesives, textile and
paper coatings, and paints are affected by the economy. The use of PVA as a
protective colloid in suspension and emulsion polymerization processes will
reflect the general trend of other polymers. PVA production peaked in 1979
at 67,100 metric tons, then dropped to 63,900 metric tons in 1980, but is
still above the 56,100 metric ton level of 1976. With an excess of 38,100
metric tons of capacity, no new plants are likely in the near future.

TABLE 197. 1978 CONSUMPTION OF POLYVINYL ALCOHOL[a]

Use	Thousand Metric Tons
Textile Warp Sizing	28
Adhesives	15
Polymerization Aids	8
Paper Sizings and Coatings	5
Miscellaneous	4
Exports	6
	66

[a]Excluding PVA for polyvinyl butyral.

Source: *Chemical Economics Handbook*, updated annually, 1978 data.

TABLE 198. POLYVINYL ALCOHOL PRODUCERS AND NAMEPLATE CAPACITY

Producer	Capacity (Metric Tons)
Air Products, Calvert City, KY	25,000
DuPont, LaPorte, TX	57,000[a]
Monsanto, Springfield, MA	20,000[a]
TOTAL	102,000

[a]This capacity includes 11,000 metric tons of PVA produced for polyvinyl butyral production.

Sources: *Chemical Economics Handbook*, updated annually, 1978 data.
Directory of Chemical Producers, 1981.

PRODUCTION AND END USE DATA

In 1980, production of PVA resins were 63,900 metric tons.[65] Of the 102,000 metric tons of PVA capacity in 1981, 22,000 metric tons were available for subsequent manufacture of polyvinyl butyral.[35, 56]

In addition to polyvinyl butyral production, PVA is used as:[133, 137, 222]

- Protective colloid in emulsion and suspension polymerization;

- Warp sizing in the textile industry;

- Component in aqueous adhesives;

- Binder for phosphorescent pigments and dyes in television tubes;

- Polarizing lenses;

- Film which dissolves in water for premeasured detergent packets;

- Ion exchange membranes;

- Sponge for medicinal uses;

- Paper coating to give a smooth, glossy finish which receives ink well;

- Antifogging agent for gas masks;

- Component in paint;

- Binder for sand molding;

- Photographic film base or coating;

- Sausage casings;

- Soil stabilizers; and

- Cosmetics.

PROCESS DESCRIPTIONS

Polyvinyl alcohol is produced from the hydrolysis of polyvinyl acetate since the theoretical vinyl alcohol monomer rearranges to acetaldehyde. Vinyl acetate is first polymerized to form polyvinyl acetate (see the section on polyvinyl acetate for the appropriate polymerization chemistry), which is then hydrolyzed with methanol in either an acidic or alkaline environment to produce polyvinyl alcohol and methyl acetate.

$$-(CH_2CHOCOCH_3)_{\overline{n}} + nCH_3OH \longrightarrow (CH_2CHOH)_{\overline{n}} + nCH_3OCOCH_3$$

The percentage of acetate groups replaced by hydroxyl groups gives the degree of hydrolysis, which can be controlled to produce the desired properties in the final product.

Polyvinyl acetate which is used for polyvinyl alcohol production is made by suspension or solution polymerization. Most polyvinyl alcohol processes are continuous, although there are batch processes in use.[137] If the PVA is compounded or used in polyvinyl butyral production at another site, the resin is dried and shipped in powder or granular form.[133] Comonomers may be added to produce high solvent resistant films. Tables 199 and 200 list typical input materials and operating parameters for these processes. Input materials used in addition to polyvinyl acetate for polyvinyl alcohol production and subsequent formulation are listed in Table 201.

<u>Suspension Polymerization and Hydrolysis</u>

Suspension polymerization of PVAc is typically used to produce medium to high molecular weight (100,000 to 1,000,000) polyvinyl alcohol.[133] As illustrated in Figure 65, vinyl acetate suspended in water is polymerized in the presence of a catalyst. The polyvinyl acetate beads produced are sent to a surge tank before drying. The water removed during drying is recycled to the reactor or sent to wastewater treatment. Methanol, dried beads, and the hydrolysis medium are combined to hydrolyze the polyvinyl acetate to polyvinyl alcohol. Alkaline hydrolysis results in a gel which is broken up by agitation. The polyvinyl alcohol mixture is fed to a solvent recovery column where the methanol and by-product methyl acetate are removed and either recycled, sold as paint solvent [211], or sent to an extractive distillation column where water is added to aid in the separation. Methanol and water are removed from the bottom of the column while methyl acetate and water taken from the top of the column are hydrolyzed to methanol and acetic acid. The medium and high molecular weight polyvinyl alcohol produced is removed from the solvent recovery column free of excess solvent, hydrolysis medium, and by-product. If the resin is shipped to another location for further processing, it may be dried.

A stoichiometric quantity of base and the presence of water are necessary for hydrolysis, while in solution polymerization, only a catalytic amount of base and the presence of an alcohol are required.[137]

<u>Solution Polymerization</u>

The PVAc from which low molecular weight (10,000 to 50,000) polyvinyl alcohol is made is usually produced by solution polymerization. In this process, as presented in Figure 66, methanol and vinyl acetate are fed, along with a catalyst, to a reactor. The polyvinyl acetate produced is carried along with the methanol to the hydrolyzer.

TABLE 199.   TYPICAL INPUT MATERIALS TO POLYVINYL ACETATE
             (POLYVINYL ALCOHOL) PRODUCTION PROCESSES
             IN ADDITION TO MONOMER AND INITIATOR

Polymerization Process	Water	Organic Solvent	Surfactants	Protective Colloids
Suspension	X		X	X
Solution		X		X

TABLE 200.   TYPICAL INPUT MATERIALS AND OPERATING PARAMETERS FOR
             POLYVINYL ALCOHOL PRODUCTION PROCESSES

Interchange Process	Acid	Base	Temperature	Reaction Time
Alkaline		X	Room temperature	Rapid
Acidic	X		57–59°C	3 hours

Sources: Encyclopedia of Chemical Technology, 2nd Edition.
           Encyclopedia of Polymer Science and Technology.

TABLE 201.   INPUT MATERIALS AND SPECIALTY CHEMICALS USED IN POLYVINYL
ALCOHOL MANUFACTURE IN ADDITION TO POLYVINYL ACETATE

Crosslinking Agents	dialdehyde starch
	diepoxides
	dihydroxy diphenyl sulfone
	diisocyanates
	dimethylolurea
	divinyl sulfone
	glutaraldehyde
	glyoxal
	oxalic acid
	polyacrolein
	trimethylolmelamine
Solvents	water[a]
Dyes	Congo red
	Pontamine Bordeaux B
	Pontamine Brown D3GN
	Pontamine Gast Red F
	Pontamine Green 2GB
	Pontamine Orange R
Plasticizers	dibutyl phthalate
	2,2-diethyl-1,3-propanediol
	ethoxylated phosphoric acid butyl ester
	ethyl acid phthalate
	glycols
	methylolated cyclic ethylene urea
	monophenylether of poly(oxyethylene)
	sorbitol
	tris(chloroethyl)phosphate
	tris(tetrahydrofurfuryl)phosphate
	urea
Comonomers	acrylamide
	acrylate esters
	acrylonitrile
	aldehydes
	carbamoylmethytrimethylammonium hydrochloride
	chloromethylester and trimethylamine or pyrridine
	crotonic acid
	1,2-epoxy-3-diethylaminopropane
	ethylene

(continued)

TABLE 201 (continued)

Comonomers (continued)	ethylene oxide
	ethylenimine
	formaldehyde and salicyclic acid
	methyl methacrylate
	polyfunctional acids
	styrenesulfonic acid
	thiourea
	vinyl acetate
	vinyl butyl ether
	vinyl chloride
	vinylene carbonate
	vinyl ketone
	vinyl versatate
Alkaline Hydrolysis Media	Lewis bases
	methanolic sodium methoxide
	potassium hydroxide
	sodium hydroxide
Acidic Hydrolysis Media	hydrochloric acid
	Lewis acids
	2-naphthalenesulfonic acid
	sulfuric acid

[a]See Table A-37 in Appendix A for additional solubility information. See
polyvinyl acetate, Section 23 for a list of solvents used in solution
polymerization.

Sources: Encyclopedia of Chemical Technology, 2nd Edition.
           Encyclopedia of Polymer Science and Technology.
           J. G. Prichard, Polyvinyl Alcohol: Basic Properties and Uses,
              1970.

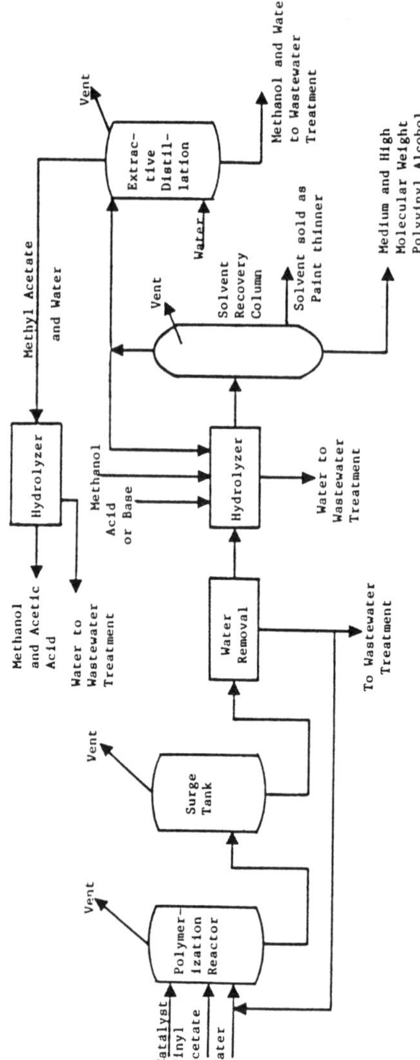

Figure 65.   Continuous polyvinyl alcohol production using the suspension polymerization process.

Sources: Hydrocarbon Processing, November 1981.
Polyvinyl Alcohol: Properties and Applications, 1973.

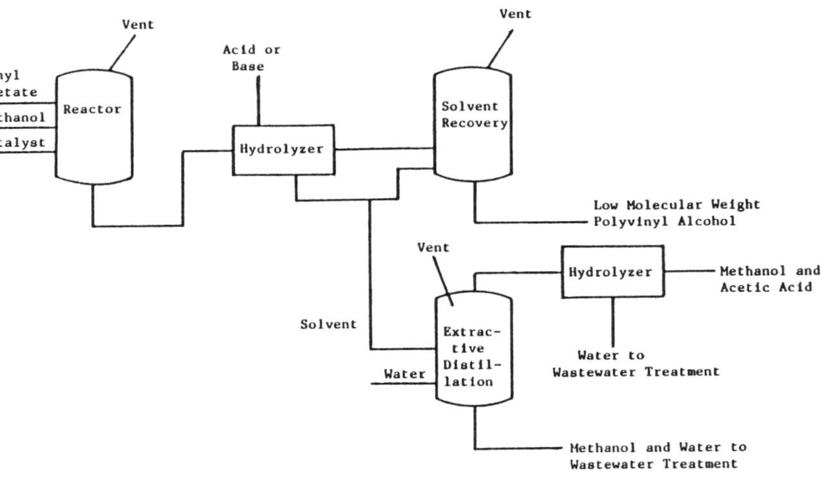

Figure 66. Continuous polyvinyl alcohol production using the solution polymerization process.

Sources: Hydrocarbon Processing, November 1981.
Polyvinyl Alcohol: Properties and Applications, 1973.

A typical recipe for the production of polyvinyl acetate by the solution process is:[54]

Material	Parts by Weight
Vinyl Acetate Monomer	6,511
2,2'-Azobisisobutyronitrile (catalyst)	24
Methanol (solvent)	4,350
Methyl Acetate (solvent)	6

The product from the polyvinyl acetate process is used along with the medium, which is either an acid or a base, in the following recipe to produce polyvinyl alcohol by the solution process:[54]

Material	Parts by Weight
Sodium Hydroxide (catalyst)	455
Methanol (solvent)	2,585
Methyl Acetate (solvent)	4
Water	21

The resulting polyvinyl alcohol mixture is sent to a solvent recovery column where the by-product methyl acetate, the solvent methanol, and the hydrolysis medium are removed and either recycled, sold as paint solvent, or sent to an extractive distillation column. The low molecular weight polyvinyl alcohol product is then free from excess solvent and by-product. If the resin is shipped to another location for further processing, it may be dried.

Energy Requirements

Data for the energy required during polyvinyl alcohol production were not found in the literature consulted.

ENVIRONMENTAL AND INDUSTRIAL HEALTH CONSIDERATIONS

Polyvinyl alcohol is an inert compound which has been approved by the FDA for use in cosmetics, bacteriostatic agents, food wrap, and other external uses. Polyvinyl alcohol film shows no ill effects when used in surgery or as a carrier for medication.[133] However, in one study performed on rats, polyvinyl alcohol exposed subcutaneously in rats caused a statistically significant increase in the incidence of metastasizing tumors in the test animals.[232] Low molecular weight PVA is absorbed by the liver and kidneys in some test animals.[137]

Since PVA is not converted from the monomer, but rather from PVAc, there are very few monomer emissions from the reactor. Polyvinyl acetate is low in vinyl acetate, thereby reducing the possibility of vinyl acetate emissions during PVA resin drying.

Worker Distribution and Emissions Release Points

Worker distribution estimates have been made using factors developed by correlating equipment manhour requirements with the process operations shown in Figures 65 and 66. Estimates for suspension and solution polymerization processes are shown in Table 202.

No major air emission point sources are associated with polyvinyl alcohol production processes. Fugitive emission sources, however, may significantly impact environmental residuals and/or worker health. The degree to which these impact the worker's environment are determined by the stream constituents, process operating parameters, engineering and administrative controls, and maintenance programs.

Table 203 lists process sources of fugitive emissions from polyvinyl alcohol production. Initiators, protective colloids and other additives potentially emitted from reactor vents are listed in Table 203.

Ethylene, a comonomer, and nitrogen, used as an inerting agent, are simple asphyxiants and their presence poses a hazard because it may limit the amount of oxygen available in the atmosphere. Typically-used solvents include methyl, butyl, and ethyl alcohols, ethyl and methyl acetate, propionaldehyde and various ketone compounds.

Health effects information regarding input materials of particular concern are discussed in the following paragraphs. Available toxicity and regulatory information for major input materials is summarized in Table 10.

Health Effects

Several of the comonomers used in polyvinyl alcohol manufacture pose a significant health risk to plant employees if exposure to even relatively low doses occurs. These comonomers include vinyl chloride, ethylene oxide, acrylonitrile, ethyleneimine, thiourea, formaldehyde, and acrylamide. The cross-linking agent divinyl sulfone is also a highly toxic compound. A synopsis of the reported health effects resulting from exposure to these substances follows.

Acrylamide is a neurotoxin, causing symptoms of ataxia, hypersomnia, and vertigo.[170] Although the polymer is nontoxic, absorption of acrylamide through skin or dusts is associated with serious neurological consequences.[88] Animal test data indicate that exposure to acrylamide might also cause mutagenic and teratogenic effects. The OSHA air standard is 0.3 mg/m^3 (8 hour TWA).[67]

Acrylonitrile, a suspected human carcinogen [107], causes irritation of the eyes and nose, weakness, labored breathing, dizziness, impaired judgment, cyanosis, nausea, and convulsions in humans. Mutagenic, teratogenic, and tumorigenic effects have been reported in the literature. The OSHA air standard is 2 ppm (8 hour TWA) with a 10 ppm ceiling.[69]

TABLE 202.   WORKER DISTRIBUTION ESTIMATES FOR POLYVINYL ALCOHOL PRODUCTION

Process	Unit	Workers/Unit/8-hour Shift
Suspension Polymerization	Continuous Reactor	0.5
	Surge Tank	0.125
	Water Removal	0.25
	Hydrolyzer	0.5
	Solvent Recovery	0.25
	Distillation Column	0.25
Solution Polymerization	Continuous Reactor	0.5
	Hydrolyzer	0.5
	Solvent Recovery	0.25
	Distillation	0.25

TABLE 203.   SOURCES OF FUGITIVE EMISSIONS FROM PVA MANUFACTURE

Source	Constituent	Process	
		Suspension	Solution
Reactor Vents	Vinyl Acetate	X	X
	Initiators	X	X
	Protective Colloids	X	X
	Solvents		X
Surge Tank Vent	Vinyl Acetate	X	
Solvent Recovery Vents	Solvents		X
	Methanol	X	X
Extractive Distillation Column Vent	Methanol	X	X
Product Drying[a]	Vinyl Acetate	X	X
	Solvents		X
	Particulates	X	X

[a]If the product is shipped.

Divinyl sulfone can cause severe skin and eye burns similar to those
from mustard gas.[87] By condensing with amino acids and other groups, it
can also cause injury and enzyme inhibition.[87] When heated to decomposition, or on contact with acid or acid fumes, it emits sulfur oxide
fumes.[242]

Ethylene oxide exposure of 500 ppm has been associated with convulsions, gastrointestinal tract effects, and pulmonary system effects in
humans.[58] Although testing for carcinogenic effects is inconclusive,
ethylene oxide causes tumorigenic, mutagenic, and teratogenic effects. The
current OSHA air standard is 50 ppm (8 hour TWA).[67]

Ethyleneimine has shown carcinogenic effects in animal tests.[101]
Exposure can cause severe eye, skin, and nose irritation; ethyleneimine is
extremely toxic if ingested with the possible lethal oral dose for humans
set between 5 and 50 mg/kg.[85] The current OSHA air standard is 1 mg/m^3
(8 hour TWA).[67]

Formaldehyde poses a high toxic hazard upon ingestion, inhalation, or
skin contact [49] with human effects reported at doses as low as 8 ppm.[127]
This suspected carcinogen is a potent mutagen and teratogen. The OSHA air
standard is 3 ppm (8 hour TWA) with a 5 ppm ceiling.[67]

Polyvinyl alcohol has produced positive results in animal testing for
carcinogenicity.[107] Its single dose toxicity, however, is presumably
low.[85] Implantation of PVA sponge as a breast prosthesis has been
associated only with foreign body type of reaction [107]; neutral solutions
of PVA have also been used in eyedrops on human eyes without difficulty.[87]

Thiourea has also shown carcinogenic effects in animal tests [100] as
well as possibly causing mutagenic effects. Death has occurred when the
human exposure level reached 147 mg/kg.[8]

Vinyl chloride is a proven human carcinogen [107] which may cause permanent injury or death after inhalation, even if the exposure is relatively
short.[242] Human reproductive effects occur at concentrations as low as 30
mg/m^3 [126]; both tumorigenic and mutagenic effects have been observed in
humans as well as test animals. The OSHA air standard for vinyl chloride is
1 ppm (8 hour TWA) with a 5 ppm ceiling.[68]

Air Emissions

Sources of fugitive VOC emissions are listed in Table 202. All of
these sources are vents which may be controlled by routing the vented stream
to a flare to incinerate the remaining hydrocarbons or to blowdown.[285]
Other fugitive VOC emissions may arise from leaks in process equipment.
Although equipment modification may be used to reduce emissions from these
sources, a routine inspection and maintenance may be the best control.

Fugitive particulates result from this process only when the product is
shipped. Venting the product dryer stream to either a baghouse or an electrostatic precipitator will control particulates.

Wastewater Sources

Polyvinyl alcohol production processes generate two major wastewater streams: one from hydrolysis and one from solvent recovery. The water removed from the beads in the suspension polymerization process may be recycled or combined with these other streams. These sources are listed in Table 204. The data found in the sources consulted listed wastewater values for a combined wastewater stream from the hydrolysis and solvent recovery sections of a polyvinyl alcohol process of an unspecified type.[93] These values are listed below.

Wastewater Component	Amount/Metric Ton of of Polyvinyl Alcohol
Methanol	0.5 kg
Acetic Acid	50 kg
Ethyl Acetate	25 kg
Sodium Sulfate	250 kg

Methyl acetate generated is converted to methanol and acetic acid; therefore, it is not expected to be a major constituent in the wastewater.

Solid Wastes

The solid wastes generated during PVA production include substandard PVA which cannot be salvaged by blending. These wastes are not hazardous; therefore, their disposal should not pose an environmental problem.

Environmental Regulation

Effluent limitations guidelines for polyvinyl alcohol have not been established.

New source performance standards (proposed by EPA on 5 January 1981) for volatile organic carbon (VOC) fugitive emissions include:

- Safety/release valves must not release more than 200 ppm above background, except in emergency pressure releases, which should not last more than five days; and

- Leaks (which are defined by VOC emissions greater than 10,000 ppm) must be repaired within 15 days.

The following compounds have been listed as hazardous (45 Federal Register 3312, May 19, 1980):

Acrylamide	– U007
Acrylonitrile	– U009
Di-n-butyl phthalate	– U069
Ethylene oxide	– U115
Formaldehyde	– U122

TABLE 204. SOURCES OF WASTEWATER FROM POLYVINYL ALCOHOL MANUFACTURE

| | Process | |
Source	Suspension	Solution
Water Removal	X	
Hydrolyzer	X	X
Solvent Recovery Train	X	X

Methanol	– U154
Methyl methacrylate	– U162
Thiourea	– U219
Vinyl chloride	– U043

Disposal of these compounds or PVA resins containing residual amounts of these compounds must comply with the provisions set forth in the Resource Conservation and Recovery Act (RCRA).

26. Polyvinyl Chloride

Polyvinyl chloride (PVC) is a very strong, moderately rigid, chemically inert, non-toxic polymer. It is also lightweight compared to metals and impact resistant when properly modified. Because of these characteristics it is used to produce a wide variety of consumer goods and industrial materials. The major use for PVC is in building and construction, which uses 50 percent of the total product. Other PVC uses include transportation, upholstery, packaging, electrical tapes, and consumer goods. PVC polymer (homopolymer) and copolymer sales in 1981 totaled 2,551,000 metric tons with the homopolymer representing 80 percent of the U.S. production. [40, 172]

PVC is manufactured from the polymerization of vinyl chloride monomer to produce the following repeating vinyl chloride structure:

$$\underrightarrow{} (CH_2 - CHCl)_{\overline{n}} \underrightarrow{}$$

With the introduction of modifiers, a wide range of properties can be achieved. Table A-39 in Appendix A presents typical properties of PVC in the rigid and plasticized form.

INDUSTRY DESCRIPTION

PVC is one of the largest segments of the plastics and resins industry. In 1980, the production of PVC was second only to low density polyethylene. [65]

There are 21 major PVC manufacturers which had a combined annual PVC manufacturing capacity of 3,589,000 metric tons ($7,896 \times 10^6$ pounds) in 1980. The projected 1981 U.S. annual capacity was 3,938,000 metric tons ($8,664 \times 10^6$ pounds) or a 10 percent increase in capacity. Major U.S. PVC manufacturers and their respective capacities are listed in Table 205.

PVC is produced by four different polymerization technologies; mass, solution, suspension, or emulsion polymerization. The literature does not specify which polymerization technology is utilized by each PVC manufacturer. However, in 1981, PVC manufactured using suspension polymerization comprised 83 percent of the United States market, PVC via emulsion polymerization added 8 percent, PVC mass polymerization contributed 7 percent, and the remaining 2 percent was manufactured by solution polymerization.[40]

TABLE 205. MAJOR POLYVINYL CHLORIDE MANUFACTURERS

Manufacturer	Location	Capacity as of October 1980 (1000 (Metric Tons)
Air Products and Chemicals, Inc. Plastics Division	Calvert City, KY	100
	Pensacola, FL	91
		100
Borden, Inc. Borden Chemical Division Thermoplastic Products	Illiopolis, IL	154
	Leominster, MA	84
	Geismar, LA	0[a]
		238
Certain-Teed Corp.	Lake Charles, LA	86
		86
Conoco Inc.[b] Conoco Chemicals Division	Aberdeen, MS	161
	Oklahoma City, OK	18[c]
		279
Diamond Shamrock[d] Industrial Chemicals and Plastics Unit Plastics Division	Deer Park, TX	213
	Delaware City, DE	55
		268
Ethyl Corp. Chemicals Group	Baton Route, LA	82
		82
Formosa Plastics, U.S.A.	Point Comfort, TX	0[e]
		0
The General Tire and Rubber Co. Chemical/Plastics Division GTR Chemical Co.	Ashtabula, OH[f]	57
	Point Pleasant, WV	27
		84

(continued)

TABLE 205 (continued)

Manufacturer	Location	Capacity as of October 1980 (1000 Metric Tons)
Georgia-Pacific Corp. Chemicals Division	Plaquemine, LA	318 318
The B. F. Goodrich Co. B. F. Goodrich Chemical Division	Avon Lake, OH	136[g]
	Convent, LA	0[g]
	Henry, IL	91
	Long Beach, CA	68
	Louisville, KY	170
	Pedricktown, NJ	68[g]
	Plaquemine, LA	86 619
The Goodyear Tire and Rubber Co. Chemical Division	Niagara Falls, NJ	32 32
Keysor Corp.	Sangus, CA	23 23
Occidental Petroleum Corp. Hooker Chemical Corp., subsidiary RUCO Division	Burlington, NJ	86
Plastics Group Plastics Division	Baton Rouge (Addis), LA	100
	Perryville, MD	118[h]
	Pottstown, PA	109[h] 413
Pantasote Inc. Film/Compound Division	Passaic, NJ	25
	Point Pleasant, WV	39 64
Rico Chemical Corp.	Guayanilla, PR	76 76
Shintech Inc.	Freeport, TX	150[i] 150

(continued)

TABLE 205 (continued)

Manufacturer	Location	Capacity as of October 1980 (1000 Metric Tons)
Stauffer Chemical Co. Plastics Division	Delaware City, DE	127
	Long Beach (Carson), CA	64
		191
Tallyrand Chemicals, Inc.[j]	New Bedford, MA	35[k]
		35
Tenneco, Inc. Tenneco Chemicals Inc.	Burlington, NJ	73
	Flemington, NJ	47
	Pasadena, TX	218[l]
		338
Union Carbide Corp. Chemicals and Plastics Division	South Charleston, WV	23
	Texas City, TX	57
		80
Whittaker Corp. Great American Chemical, subsidiary	Fitchburg, MA	34
		34
TOTAL[m]		3,589

---

[a]Company has announced plans to build a 136,000 metric tons per year plant at this location to be complete in 1983.

[b]Company will expand total PVC capacity to 325,000 metric tons per year in 1981.

[c]There is some doubt as to the actual capacity of this plant. It was reported as three different values.

[d]An expansion of 34,000 to 45,000 metric tons per year is scheduled to become effective by the end of 1981.

(continued)

TABLE 205 (continued)

[e]A 241,000 metric ton per year plant is scheduled for completion in early 1982 at this location.

[f]PVC capacity at this plant will be expanded to 68,000 metric tons per year by the fourth quarter of 1982.

[g]Plant expansions and constructions have been announced as follows:
 Avon Lake - 45,000 metric tons expansion for fourth quarter 1981,
 Pedricktown - 91,000 metric tons expansion for fourth quarter 1981,
 Convent - new 500,000 metric tons plant with plans to have first phase on
  stream by third quarter 1983, complete by 1985, has been put on
  indefinite hold.

[h]The acquisition of these plants from Firestone was announced in September 1980.

[i]Plans have been announced to double capacity by mid 1981.

[j]This plant was reported under Tallyrand Chemicals in one source and International Materials Corporation in another source.

[k]Capacity to total 45,000 metric tons per year in 1981.

[l]Total capacity at Pasadena announced to be 341,000 metric tons per year by mid-1981.

[m]Does not add to total of numbers listed due to round-off errors.

Sources: Chemical Economics Handbook, updated annually, 1980 data.
           Directory of Chemical Producers, 1981.
           Modern Plastics, January 1982.
           Organic Chemical Producers Data Base.

The major producers of PVC are major petroleum and petrochemical companies. Ten of these companies represent over 80 percent of the PVC production capacity, with other chemical and compounding companies accounting for the remainder. These companies produce PVC both for captive use and for merchant sales to PVC compounders.

Approximately 37 percent of the PVC production capacity is located in the states of Texas and Louisiana (9 sites). Another 20 percent of the capacity is located in the New Jersey, Pennsylvania, Maryland, and Delaware area (10 sites).

An important environmental aspect of PVC production at large, integrated facilities is that many waste management options exist. PVC wastes may be treated either before or after combination with the rest of the plant wastes. Treating PVC wastes before combination with the remaining plant wastes allows the removal of potentially troublesome pollutants from a relatively small stream. The treated stream is then either discharged or combined with the plant wastes sent to the plant treatment system. Waste management options are, therefore, very site and facility specific.

Since PVC is used in so many segments of the U.S. economy, the PVC market follows the general trends in the U.S. economy. With the recent downturns in the American automobile and housing markets, the PVC market has shown a similar decline. Presently about 35 percent of the available PVC manufacturing capacity is unused. These two factors suggest that new PVC facilities and process expansions will not be contemplated in the near term.

PRODUCTION AND END USE DATA

The polyvinyl chloride segment of the plastics and resins industry is the second largest after low density polyethylene.[65] PVC production was 2,551,000 metric tons (5,612 x 10^6 pounds) during 1981.

The largest markets for PVC products are construction related. Approximately 55 percent of the construction products are manufactured from rigid PVC and 45 percent are manufactured from plasticized PVC. Forty percent of the total resin used is for piping: including pressure (thickwall) pipe for water supply, irrigation, and chemical processing; and non-pressure (thinwall) pipe for drains, sewer systems and electrical conduits. Other uses for PVC are in the production of flooring, wall coverings, upholstery, appliances, housewares, automotive wire, meat wrap, blister packs, and phonograph records.

The markets, corresponding production volumes, and polymer processing operations used for PVC end product manufacture are presented in Table 206.

TABLE 206.   MAJOR MARKETS FOR POLYVINYL CHLORIDE

Market	1981 1,000 Metric Tons
CALENDERING	
Building & construction	
Flooring	68
Paneling	11
Pool-pond liners	16
Roof membranes	9
Other building	2
Transportation	
Auto upholstery/trim	32
Other upholstery/trim	8
Auto tops	5
Packaging: sheet	36
Electrical: tapes	4
Consumer & institutional	
Sporting/recreation	10
Toys	15
Baby pants	2
Footwear	14
Handbags/cases	11
Luggage	9
Bookbinding	2
Tablecloths, mats	17
Hospital & healthcare	7
Credit cards	8
Decorative film (adhesive back)	5
Stationery, novelties	2
Tapes, labels, etc.	6
Furniture/furnishings	
Upholstery	34
Shower curtains	5
Window shades/blinds/awnings	5
Waterbed sheet	4
Wallcovering	14
Other calendering	8
TOTAL CALENDERING	369

(continued)

TABLE 206 (continued)

Market	1981 1,000 Metric Tons
EXTRUSION	
Building & construction	
Pipe & conduit	
Pressure	
Water	354
Gas	9
Irrigation	111
Other	9
Drain/waste/vent	104
Conduit	168
Sewer/drain	141
Other	27
Siding & accessories	82
Window profiles	
All-vinyl windows	7
Composite windows	25
Mobile home skirt	7
Gutters/downspouts	2
Foam moldings	14
Weatherstripping	11
Lighting	8
Transportation	
Vehicle floor mats	8
Bumper strips	4
Packaging	
Film	125
Sheet	15
Electrical: wire and cable	170
Consumer & institutional	
Garden hose	18
Medical tubing	16
Blood/solution bags	20
Stationary/novelties	4
Appliances	13
TOTAL EXTRUSION	1,473

(continued)

TABLE 206 (continued)

Market	1981 1,000 Metric Tons	
INJECTION MOLDING		
Building & construction		
Pipe fittings	50	
Other building	3	
Transportation: bumper parts	5	
Electrical/electronics		
Plugs, connectors, etc.	30	
Appliances, business machines	17	
Consumer & institutional		
Footwear	20	
Hospital & healthcare	7	
Other injection	15	
TOTAL INJECTION MOLDING		147
BLOW MOLDING: bottles		45
COMPRESSION MOLDING: sound records		49
DISPERSION MOLDING		
Transportation	16	
Packaging: closures	14	
Consumer & institutional		
Toys	3	
Sporting/recreation	8	
Footwear	6	
Handles, grips	6	
Appliances	6	
Industrial: traffic cones	6	
Adhesives, etc.		
Adhesives	5	
Sealants	4	
Miscellaneous	5	
Other dispersion	5	
TOTAL DISPERSION MOLDING		84

(continued)

TABLE 206 (continued)

Market	1981 1,000 Metric Tons
DISPERSION COATING	
Building: flooring	61
Transportation	
Auto upholstery/trim	12
Other upholstery/trim	2
Anticorrosion coatings	5
Consumer & institutional	
Apparel/outerwear	6
Luggage	4
Tablecloths, mats	4
Hospital/healthcare	3
Furniture/furnishings	
Upholstery	8
Window shades/blinds/awnings	6
Wallcoverings	6
Carpet backing	7
Other	15
TOTAL DISPERSION COATING	139
SOLUTION COATING	
Packaging: cans	4
Adhesives/coatings	21
TOTAL SOLUTION COATING	25
VINYL LATEXES: adhesives/sealants	25
EXPORT	195
GRAND TOTAL	2,551

Source: _Modern Plastics_, January 1982.

PROCESS DESCRIPTIONS

PVC is produced by mass, solution, suspension, or emulsion polymerization. Of the 2,551,000 metric tons sold in 1981, PVC manufactured by suspension polymerization comprised 83 percent of the United States market.[40, 172] Emulsion, mass, and solution polymerization contributed the remaining 8, 7, and 2 percent, respectively, to the PVC market.[40] Tables 207 and 208 present typical input materials and operating parameters for these processes. Specialty chemicals and other input materials used are listed in Table 209.

Suspension polymerization is used to manufacture resins for most PVC applications, such as extrusion of pipe and siding and sheet. Although mass polymerization may also be used to produce PVC for the same applications, the polymer beads manufactured by suspension polymerization are easily separated from the aqueous medium while mass polymerization produces a resin which must be chopped, cut, or pelletized before use. Emulsion polymerization provides PVC for use in dispersion resins since the small particles may be used either in emulsion form or may be spray dried for inclusion in a dispersion. Due to the high cost of solvents, solution polymerization is used for very few PVC applications.

In PVC production, vinyl chloride monomer is polymerized to make the homopolymer, and copolymers are made from the polymerization of vinyl chloride plus a comonomer. Initially, vinyl chloride becomes reactive after combining with an initiator radical, $R\bullet$. The vinyl chloride radical then reacts with other monomers to form the polymer chain.

Initiation

$$CH_2=CHCl + R\bullet \longrightarrow RCH_2-\overset{\bullet}{C}HCl$$

Propagation

$$RCH_2-\overset{\bullet}{C}HCl + n(CH_2=CHCl) \longrightarrow R-(\!-CH_2-CHCl-\!)_{\overline{n}}-CH_2-\overset{\bullet}{C}HCl$$

The reaction can terminate in one of three ways: by radical combination, chain transfer, or disproportionation.

Radical Combination

$$RCH_2-\overset{\bullet}{C}HCl + RCH_2-\overset{\bullet}{C}HCl \longrightarrow RCH_2-CHCl-CHCl-CH_2R$$

Chain Transfer

$$R'R'' + RCH_2-\overset{\bullet}{C}HCl \longrightarrow RCH_2-CHClR'' + R'\bullet$$

where $R'R''$ represents a chain transfer agent.

Disproportionation

$$R-CH_2-\overset{\bullet}{C}HCl + R'-CH_2-\overset{\bullet}{C}HCl \longrightarrow R-CH_2-CH_2Cl + R'-CH=CHCl$$

TABLE 207.   TYPICAL INPUT MATERIALS TO PVC PRODUCTION PROCESSES
IN ADDITION TO MONOMER AND INITIATOR

Process	Input			
	Water	Organic Solvent	Emulsifier	Suspension Stabilizer
Suspension	X		X[a]	X
Emulsion	X		X	X
Mass				
Solution (Precipitation)		X		

[a]Very small concentrations, which do not provide a true emulsion, may be used.

TABLE 208.   TYPICAL OPERATING PARAMETERS FOR PVC PRODUCTION PROCESSES

Process	Temperature	Reaction Time	Conversion
Suspension	45° – 55°C	12 hours	85 – 90%
Emulsion	45° – 60°C	<1 hour	60%
Mass	40° – 70°C	8-12 hours	80 – 90%
Solution (Precipitation)	40° – 60°C	———a	90 – 95%

[a]Only one producer uses this process in the United States. Only reaction temperatures were available.

Sources: Encyclopedia of Chemical Technology, 2nd Edition.
Encyclopedia of PVC, 1976-1977.
Hydrocarbon Processing, March 1980.
Hydrocarbon Processing, November 1979.
Marshall Sittig, Vinyl Chloride and PVC Manufacture, 1978.
U.S. Environmental Protection Agency, Source Assessment:
Polyvinyl Chloride, 1978.

TABLE 209.   INPUT MATERIALS FOR PVC MANUFACTURE

Function	Compound	Process
Suspension Stabilizer	Gelatin Hydropropyl methyl cellulose Methyl cellulose Polyvinyl alcohol	Suspension Emulsion

Radical Initiator	Acetal benzoyl peroxide
	Acetyl cyclohexyl sulfonyl
	  peroxide
	t-Amyl peroxyneodecanoate
	Azobisisobutyronitrile
	t-Butyl peroxyneodecanoate
	t-Butyl peroxypivalate
	α-Cumyl peroxyneodecanoate
	α-Cumyl peroxypivalate
	Dibenzoyl peroxide
	Dibutyl peroxide
	Di-2-ethylhexyl peroxydi-
	  carbonate
	Diisononanoyl peroxide
	Diisopropyl peroxydicarbonate
	Dilauroyl peroxide
	Di-n-propyl peroxydicarbonate
	Hydrogen peroxide
	Isopropyl peroxydicarbonate
	Lauroyl peroxide
	Peroxydisulfates
		Suspension
		Emulsion
		Mass
		Solution
		(Precipitation)

Alkaline Buffer	Sodium bicarbonate Sodium carbonate Sodium phosphate	Suspension Emulsion
Comonomers	Acrylic ester Acrylonitrile Cetyl vinyl ether Ethylene Propylene Vinyl acetate Vinylidene chloride	Suspension Emulsion Mass Solution (Precipitation)

Plasticizers	Phthalates
	  Butyl benzyl
	  Butylhexyl
	  Diallyl
	  Di-n-amyl
	  Dibenzyl
	  Di-n-butyl
	  Dicapryl
		Suspension
		Emulsion
		Mass
		Solution
		(Precipitation)

(continued)

TABLE 209 (continued)

Function	Compound	Process
Plasticizers (continued)	Dicyclohexyl	
	Diethyl	
	Di-(2-ethylbutyl)	
	Di-(2-ethylhexyl)	
	Di-n-hexyl	
	Dilauryl	
	Dimethyl	
	Di-n-octyl	
	Di-n-propyl	
	Sebacates	
	Dibenzyl	
	Di-(2-butoxyethyl)	
	Dibutyl	
	Dimethyl	
	Di-(2-ethylhexyl)	
	Dihexyl	
	Phosphates	
	Tri(2-butoxyethyl)	
	Tributyl	
	Tricresyl	
	Tri(2-ethylhexyl)	
	Triphenyl	
	Dibutyl lauramide	
	Dicapryl diglycollate	
	Dioctyl maleate	
	Dioctyl thiodiglycollate	
	Ethylene glycol dipelargonate	
Emulsifiers	Alkyl sulfates	Suspension
	Alkyl sulfonates	Emulsion
	Fatty ether sulfates	
	Long chain fatty acid soaps of alkali metals	
Molecular Weight Regulators	Alkylene bis-(mercapto-alkanoates)	Emulsion
	Polybromobutenes	
Chain Transfer Agents	Carbon tetrachloride	Solution
	Dodecanethiol	

(continued)

TABLE 209 (continued)

Function	Compound	Process
Solvents	Acetone Benzene n–Butane Methanol Methanol/water n–Pentane 1,1,2,2-tetrafluoro- dichloroethane	Precipita- tion
	Chlorobenzene Cyclohexane 1,2-Dichloroethane Tetrahydrofuran	Solution

Sources: Encyclopedia of Chemical Technology, 2nd Edition.
Modern Plastics, February 1981.
W. S. Penn, PVC Technology, 1972.
Marshall Sittig, Vinyl Chloride and PVC Manufacture, 1978.

Both mechanisms are present in all of the processes discussed. Radical combination prevails in the monomer rich phase, which is within the droplets in suspension and micelles in emulsion polymerization and within the reaction medium in solution and mass polymerization. Chain transfer mechanisms dominate in a polymer system where other compounds are present to react with the radical group.

Suspension Polymerization

In suspension polymerization, the radical initiator is dissolved in the vinyl chloride monomer which is maintained as aqueous phase droplets by means of vigorous agitation and a suspension stabilizer (or protective colloid). The suspension stabilizer is added to minimize coalescence of the droplets and appreciably increases the viscosity of the water phase without changing the surface tension. The stabilizer is water soluble but insoluble in vinyl chloride. An alkaline buffer is added to neutralize hydrochloric acid as it is formed as a reaction by-product. Three recipes for suspension polymerization of PVC are listed in Table 210.

The advantages of this process over other polymerization processes include: lower cost of water as compared to organic solvents, excellent heat transfer, no solvent recovery problems, fewer safety hazards due to no volatile solvents, simpler temperature control, and less contamination of the polymer. The major disadvantage of this process is the need to separate the large amount of water from the product and the large excess reactor volume required, since 60 to 80 percent of a batch is water.

As illustrated in Figure 67, vinyl chloride monomer, an initiator, water, a suspension stabilizer, and a buffer are charged to the reactor. Ratios of water to vinyl chloride range from 1.5:1 to 4:1.[29] Low ratios facilitate larger production volumes while high ratios allow better temperature control, higher conversions, and produce a more porous resin. Suspension stabilizers are typically used in concentrations of 0.05 to 0.5 parts per hundred parts of monomer.[29] Radical initiators are employed in concentrations of 0.03 to 0.1 parts per hundred.[29]

The polymerization is performed in large agitated reactors. The reaction is carried to 85 to 90 percent conversion in either stainless steel or glass-lined carbon steel reactors ranging from 76 to 132 cubic meters (2,000 to 35,000 gallons). The polymerization rate is slower for suspension polymerization than emulsion, mass, or solution polymerization.

Since oxygen inhibits the polymerization, the reaction is normally carried out in a nitrogen atmosphere. The nitrogen atmosphere also serves to decrease the hydrochloric acid evolved and increase the polymer stability.

After the vinyl chloride droplets, which are 100 to 150 microns in diameter, have polymerized, the mixture is stripped of unreacted monomer in a flash tank and the polymer is sent to a blending tank. Air emissions may result from the blending tank vent. The blended polymer mixture is centrifuged to remove water. The resulting resin, with a water content of 18 to

TABLE 210. RECIPE FOR PVC PRODUCTION USING SUSPENSION POLYMERIZATION

Material	Parts by Weight		
	Process 1	Process 2	Process 3
Water	150 - 200	170 - 200	225 - 350
Vinyl Chloride	100	100	100
Lauroyl Peroxide (initiator)	0.1 - 0.5	0.1 - 0.3	0.2 - 0.4
Polyvinyl Alcohol (stabilizer)		0.03 - 0.06	
Gelatin (stabilizer)	0.3 - 0.6		
Methyl Cellulose (stabilizer)			0.2 - 0.9
Emulsifier	0.01 - 0.05	0.01 - 0.03	
Sulfated Ester (emulsifier)			0.05 - 0.1
Buffer	0.1 - 0.3		0.1 - 0.3

Source: W. S. Penn, _PVC Technology_, 1972.

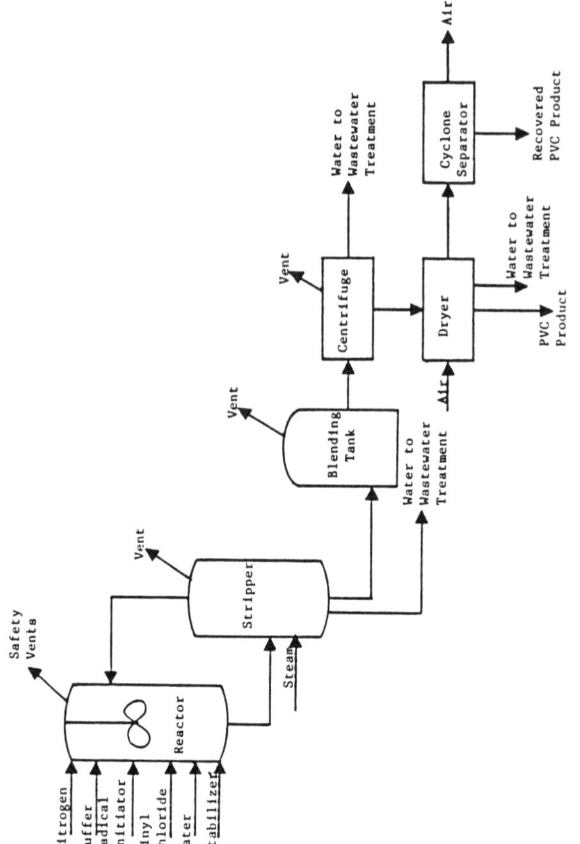

Figure 67.   PVC production using the suspension polymerization process.

Sources: Hydrocarbon Processing, March 1980.
Hydrocarbon Processing, November 1981.

25 percent [27], is dried in a dryer. Both rotary and fluidized bed dryers
are used. The dried PVC resin, which has less than 0.3 percent water[27],
is then ready for bagging. Polymer fines which remain in the dryer off gas
are removed in a cyclone separator, resulting in the recovery of additional
PVC product.

Emulsion Polymerization

Emulsion polymerization is similar to the suspension process except
that larger amounts of emulsifying agents are used. An advantage of the
emulsion polymerization process is that high-molecular-weight polymers are
produced at more rapid rates than in suspension polymerization. Vinyl
chloride monomer is emulsified by means of surface active agents in water
which distribute most of the monomer in the aqueous phase as droplets. In
contrast to suspension polymerization, the emulsion process uses a water
soluble radical initiator to start the polymerization reaction.

The emulsifier content in the reactor is typically 1.25 to 3.0 percent.
[60] Temperatures in the reactor range from 45° to 60°C.[32] The reaction
time can be as short as one-half hour. Conversion is usually held below 60
percent to ensure continued life of the emulsion.[60] Polymerization continues until the total solids content reaches 50 to 55 percent. Uniform
particles are produced which range in size from 0.05 to 10 microns.

A typical recipe for emulsion polymerization of PVC is shown below:[11]

Material	Parts by Weight
Water	158
Vinyl Chloride	100
Azobisisobutyronitrile (initiator)	1
Sodium Dodecyl Sulfate (emulsifier)	1.58
Cetyl Alcohol (emulsifier)	2

As shown in Figure 68, the emulsifier, initiator, and buffer are
dissolved in deionized water, then fed to the polymerization reactor.
Inhibitor free vinyl chloride is measured into the reactor and the mixture
agitated to form an emulsion. Polymerization is carried out to 90 percent
conversion or greater.

Following polymerization, the polymer latex is strained and the polymer
separated from the monomer and other impurities. The latex is sent to a
blend tank before it is combined with a stabilizer solution and spray dried.
The dried resin is collected and separated according to particle size before
being bagged for shipment.

The impurities removed from the polymer latex are decanted, coalesced,
and sent to a surge tank before entering the purification column. Water
mixed with some vinyl chloride monomer is recycled to the caustic decanter
while the vinyl chloride monomer is recycled to the polymerization reactor.

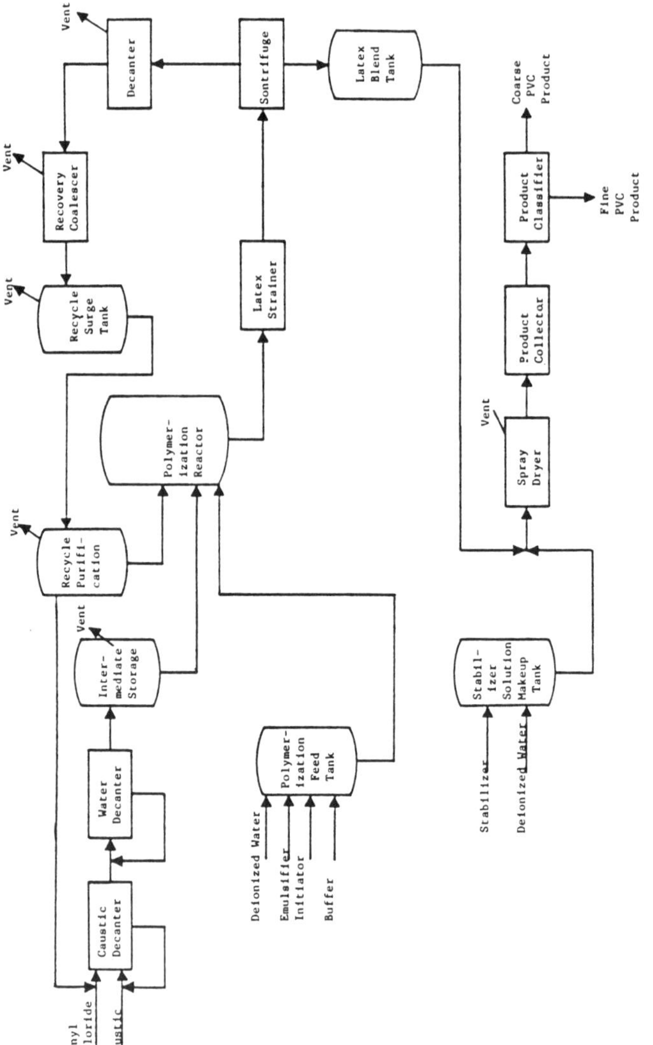

Figure 68.   PVC production using the emulsion polymerization process.

Source: U.S. Environmental Protection Agency, <u>Source Assessment: Polyvinyl Chloride</u>, 1978.

Although using an emulsion allows faster rates of reaction at lower temperatures than other processes, the residual emulsifier, about 2 to 5 percent, contaminates the polymer.[286] This contamination renders the polymer more sensitive to heat and shear stress and reduces polymer clarity, compatibility with compounding agents, electrical-insulation properties, water resistance, and weathering ability.

Mass Polymerization

Mass polymerization is, in principle, the simplest process by which PVC can be made since vinyl chloride and an initiator are the only input materials.[70] The polymer produced has high porosity and clarity, uniform shape and size, greater transparency, better impact strength, higher heat and light stability, and improved electrical-insulation properties due to the absence of contaminants. This process has lower capital and operating costs since suspension agents and organic solvents are not required. The primary disadvantage of mass polymerization is that it produces a resin which must be chopped, cut, or pelletized. The raw materials used include those listed in Table 210.

Two recipes for mass polymerization of PVC are shown below [11]:

| | Parts by Weight | |
Material	Recipe #1	Recipe #2
Vinyl Chloride	100	100
Azobisisobutyronitrile (initiator)	0.016	0.02
Diisopropyl Percarbonate (initiator)		0.00005

There are two different mass polymerization processes for PVC: the one-step process and the two-step process. The two-step process has replaced the earlier one-step process since the reactor size can be reduced using two reactors in series. This process consists of four major unit processes: prepolymerization, polymerization, devolatilization, and screening.

In the two-step process illustrated in Figure 69, the vinyl chloride monomer and low concentrations (0.03 percent) of radical initiator are fed to the prepolymerization reactor. The prepolymerization reactor is designed to provide vigorous agitation for reaction times which approach three hours. [213] When the PVC concentration reaches 7 to 12 percent, the monomer-polymer mixture is transferred to another reactor to complete the polymerization. Temperatures in the prepolymerization and polymerization stages range from 40° to 70°C and pressures range from 483 to 1,172 kPa (70 to 170 psia).[255] The reaction time ranges from five to nine hours in either a vertical or horizontal reactor which is agitated at lower speeds. The reaction temperature is controlled by cooling a portion of the unreacted vinyl chloride monomer in a reflux condenser, thereby removing a portion of the heat generated by the polymerization reaction.

When the conversion has reached 80 to 90 percent, the mixture is sent to a devolatilizer where the unreacted vinyl chloride monomer is removed

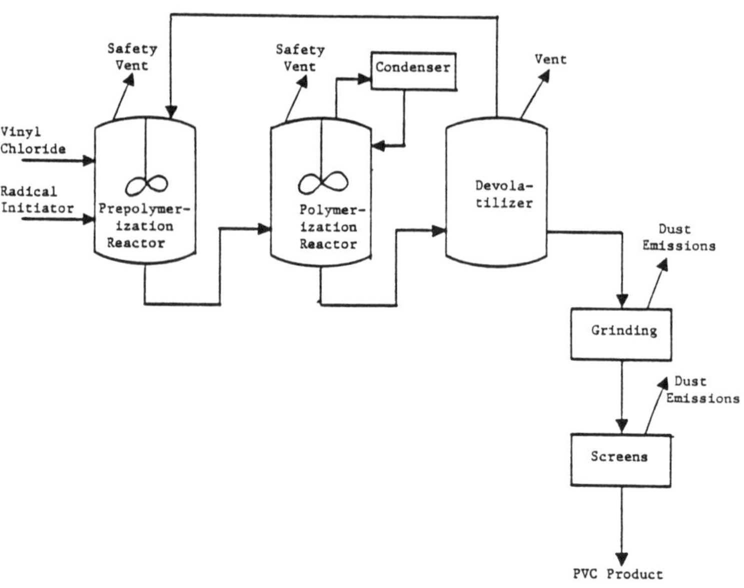

Figure 69. PVC production using the mass polymerization process.

Sources: Hydrocarbon Processing, March 1980.
          Hydrocarbon Processing, November 1981.

using low pressures and recycled. The devolatilizer uses low temperatures
(-35°C) for condensation since no water is present.[172]

The polymer is ground into particles and screened before bagging. The
average particle size of the product ranges from 50 to 200 microns.[138]

Solution and Precipitation Polymerization

Solution and precipitation polymerization processes are similar since
both use an organic solvent as the polymerization medium. The polymer
formed by solution polymerization is soluble in the solvent used, while the
polymer is insoluble in the solvent used and precipitates out of solution in
precipitation polymerization. Solvents for solution polymerization include:
chlorobenzene, chlorohexane, tetrahydrofuran, and 1,2-dichloroethane. Solvents for precipitation polymerization include: n-butane, n-pentane, methanol, benzene, acetone, methanol/water, and 1,1,2,2-tetrafluoriodichloro-
ethane.

Solution or precipitation polymerization has several advantages over
the mass polymerization process. The viscosity of the polymerization medium
is lower, the temperature control is more precise, and the molecular weight
of the polymer is lower. An advantage of solution or precipitatin polymerization over suspension and emulsion polymerization processes is the purity
of the resin since no emulsifying or suspension agents are needed. The cost
of the solvent makes the production costs higher than for mass, suspension,
or emulsion polymerization. These processes are used mostly for the production of PVC-polyvinyl acetate copolymers for applications in the phonograph
record industry which require pure resin grades as input.

As shown in Figure 70, the vinyl chloride monomer, solvent, and initiator (from 0.01 to 0.5 percent) enter a series of agitated reactors at 40°C
to 60°C.[215] After polymerization, the unreacted monomer and solvent are
removed from the polymer in a flash tank and recycled. The PVC is filtered
from the slurry and the cake is dried by flash evaporation. Solvent removed
is sent to solvent recovery. The dried resin is sent to bagging.

Figure 71 depicts precipitation polymerization. As in solution polymerization, vinyl chloride monomer, solvent, and initiator (from 0.01 to 0.5
percent) enter a series of agitated reactors at 40 to 60°C.[215] After the
polymer beads have precipitated, they are separated from the solvent in a
centrifuge and dried. The solvent is recycled to the reactors while the
dried resin is bagged.

Energy Requirements

Energy requirements for PVC are listed below according to technology:
[214, 215, 216, 217, 218]

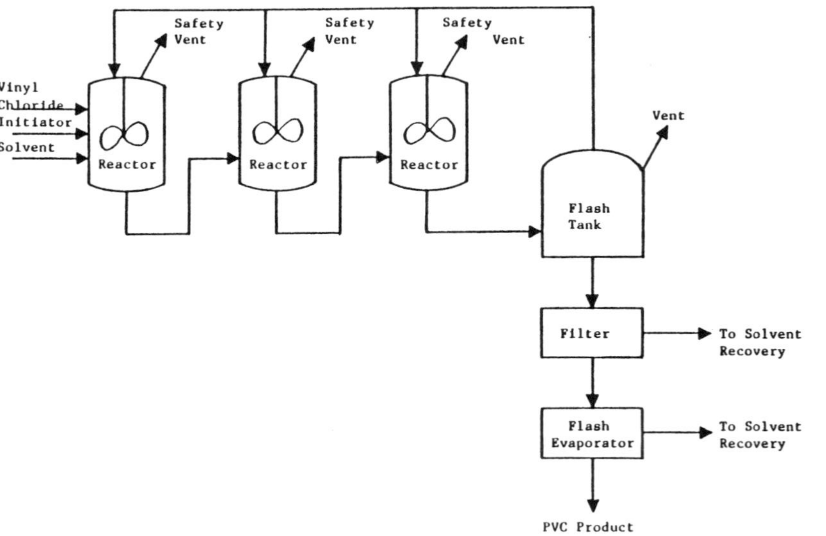

Figure 70. PVC production using the solution polymerization process.

Source: U.S. Environmental Protection Agency, Source Assessment: Polyvinyl Chloride, 1978.

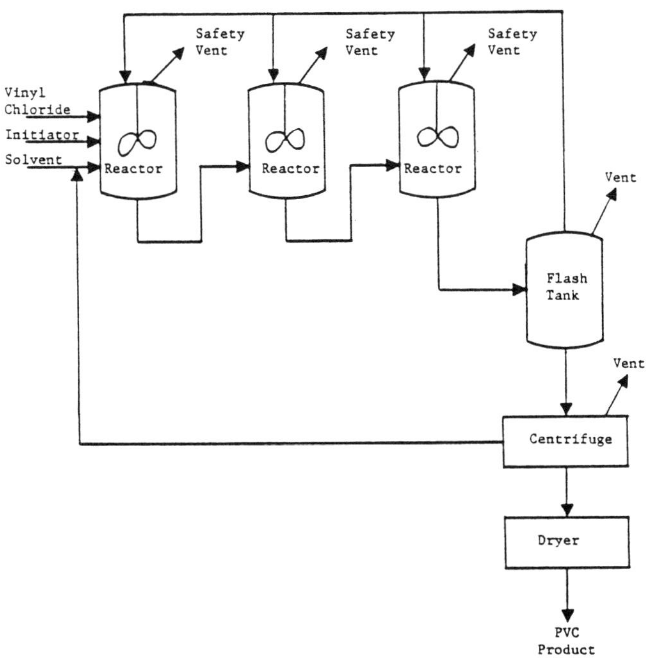

Figure 71.   PVC production using the precipitation polymerization process.

Sources: Encyclopedia of Chemical Technology, 2nd Edition.
U.S. Environmental Protection Agency, Source Assessment:
Polyvinyl Chloride, 1978.

Technology	Energy Required	Unit/Metric Ton of Product
Ato Chimie	Electricity	4.3×10^8 Joules
	Steam	1 metric ton
Hoechst AG	Electricity	7.9×10^8 Joules
	Steam	1.1 metric tons
Mitsui Toatsu Chemicals	Electricity	7.2×10^8 Joules
	Steam	1.0 metric tons
Lonza Ltd.	Electricity	10.8×10^8 Joules
	Steam	1.5 metric tons
Rhone Poulenc	Electricity	5.4×10^8 Joules
	Steam (6 atm.)	0.4 metric tons

ENVIRONMENTAL AND INDUSTRIAL HEALTH CONSIDERATIONS

A major consideration in the production and use of PVC resins and their products is the potential exposure to at least one material that is a known human carcinogen and four suspected human carcinogens. Therefore, potential human exposure to these chemicals represents the greatest environmental concern with PVC processing.

Worker Distribution and Emissions Release Points

Worker distribution estimates for polyvinyl chloride production are shown in Table 211. Estimates were made by correlating major equipment man-hour requirements with the process flow diagrams in Figures 67 through 71.

Point and fugitive emission sources associated with PVC production are shown in Tables 212 and 213, respectively. Although several potentially hazardous compounds may be emitted, the contaminant of extreme concern is vinyl chloride monomer (VCM). If control strategies and respiratory and personal protective equipment programs are directed toward minimizing exposure to VCM, exposure to other compounds will be controlled as well.

Health effects information for major contaminants from PVC production processes is discussed below and summarized in Table 10.

Control strategies in PVC manufacturing plants are among the most stringent in the industry.[280] For this reason, the controls discussion in Section 1 uses PVC production as an example.

Health Effects

Four of the input materials used in the manufacture of PVC are offici-ally classified by the International Agency for Research on Cancer (IARC) as suspected human carcinogens. They are acrylonitrile, which is a comonomer

TABLE 211. WORKER DISTRIBUTION ESTIMATES FOR PVC PRODUCTION

Process	Unit	Workers/Unit/8-hour Shift
Suspension Polymerization	Batch Reactor	1.0
	Steam Stripper	0.25
	Blending Tank	1.0
	Centrifuge	0.25
	Dryer	0.5
	Cyclone	0.125
Emulsion Polymerization	Storage Tank	0.125
	Decanter	0.25
	Batch Reactor	1.0
	Strainer	0.25
	Centrifuge	0.25
	Batch Mixer	1.0
	Spray Dryer	1.0
	Recycle Purification	0.25
	Recovery Coalescer	0.25
	Screens	0.25
	Collector	0.5
Mass Polymerization	Batch Reactor	1.0
	Devolatilizer	0.25
	Condenser	0.125
	Grinding	0.25
	Screens	0.25
Solution Polymerization	Batch Reactor	1.0
	Flash Tank	0.125
	Filter	0.25
	Evaporator	0.25
Precipitation Polymeriza- tion	Batch Reactor	1.0
	Flash Tank	0.125
	Centrifuge	0.25
	Dryer	0.5

TABLE 212.　SOURCES OF AIR EMISSIONS FROM PVC MANUFACTURE

| | VCM Emissions in kg/100 kg PVC Produced | | | |
Source	Suspension	Emulsion	Mass	Solution
Fugitive Emissions	1.50	1.13	0.48	0.03
Opening Loss – Reactor	0.14	0.15	0.08	0.50
Opening Loss – Monomer Recovery Vessel[a]	0.32	1.23	—	0.05
Monomer Recovery Vent[a]	0.48	0.50	1.50	0.31
Slurry Blend Tank Vent	0.42	0.34	—	—
Centrifuge Vent	0.13	—	—	—
Silo Storage Vent	0.70	2.41	—	0.83
Reactor Safety Valve Vents	0.20	0.22	0.10	0.06
Bagger Area Emissions	—	—	0.23	—
Emissions From Process Water and Reactor Washings[b]	0.03	0.03	<0.01	<0.01
Total	3.92	6.01	2.40	1.78

[a]The monomer recovery vessel is either a stripper, devolatilizer or a flash tank.
[b]The only wastewater associated with the mass and solution polymerization processes results from reactor washing.

Source:　Marshall Sittig, _Vinyl Chloride and PVC Manufacture_, 1978.

TABLE 213.   SOURCES OF FUGITIVE EMISSIONS FROM PVC MANUFACTURE

Source	Constituents	Suspension	Mass	Emulsion	Solution
Agitator Seals	Vinyl Chloride Solvent	X	X	X	X X
Pump Seals	Vinyl Chloride Solvent	X	X	X	X X
Valves	Vinyl Chloride Solvent	X	X	X	X X
Leaks in Polymeri-zation Equipment	Vinyl Chloride Solvent	X	X	X	X X
Dryer Vent	Vinyl Chloride Particulates	X X		X X	
Bagging Process (Product Prepara-tion)	Particulates	X	X	X	X
Cutting/ Grinding/ Screening			X	X	
Emulsion Blend Tank	Vinyl Chloride			X	
Filtration	Vinyl Chloride Solvent				X X
Evaporation	Vinyl Chloride Solvent				X X

Source: Marshall Sittig, _Vinyl Chloride and PVC Manufacture_, 1978.

in the process; carbon tetrachloride, which is used as a chain transfer
agent; and the two solvents benzene and 1,2-dichloroethane. In addition,
polyvinyl alcohol, a suspension stabilizer, and vinylidene chloride, a
comonomer, are known animal carcinogens. Although health effects data are
relatively limited for the solvent tetrahydrofuran, the substance is considered highly toxic. The paragraphs below summarize the reported health
effects of exposure to these substances.

Acrylonitrile, a suspected human carcinogen [107], causes irritation
of the eyes and nose, weakness, labored breathing, dizziness, impaired
judgment, cyanosis, nausea, and convulsions in humans. Mutagentic,
teratogenic, and tumorigenic effects have been reported in the literature.
The OSHA air standard is 2 ppm (8 hour TWA) with a 10 ppm ceiling.[69]

Benzene is a suspected human carcinogen [101] which also exhibits
mutagenic and teratogenic properties. It poses a moderate toxic hazard for
acute exposures and a high hazard for chronic exposures through ingestion,
inhalation, and skin adsorption.[242] Effects on the central nervous system
have been observed in humans upon exposure to concentrations of 100 ppm.
[98] Intermittent exposure to 100 ppm over 10 years has been linked with
the development of cancers in humans.[271] The OSHA standard in air is 10
ppm (8 hour TWA) with a ceiling of 25 ppm.[67]

Carbon tetrachloride is a suspected human carcinogen [108] which is
also extremely toxic upon inhalation or injection of small quantities.[241]
The toxic hazard posed by skin adsorption is slight for acute exposures, but
high for chronic exposures.[241] The toxicity of carbon tetrachloride
appears to be primarily due to its fat solvent action which destroys the
selective permeability of tissue membranes and allows escape of certain
essential substances such as pyridine nucleotides.[78] The OSHA air standard
is 10 ppm (8 hour TWA) with a 25 ppm ceiling[67].

1,2-Dichloroethane is a suspected human carcinogen [108], in addition
to exhibiting mutagenic properties. It is several times more toxic than
carbon tetrachloride when taken in a single oral dose.[170] Toxic effects
include gastrointestinal tract upset, central nervous system depression,
mental confusion, dizziness, nausea, and vomiting.[170] Liver, kidney, and
adrenal injuries may occur at subacute levels.[170] Deaths from accidental
ingestion and from inhalation have been reported.[4] The OSHA air standard
is 50 ppm (8 hour TWA) with a ceiling of 100 ppm.[67].

Polyvinyl alcohol has produced positive results in animal testing for
carcinogenicity.[107] Its single dose toxicity, however, is presumably
low.[85] Implantation of PVA sponge as a breast prosthesis has been
associated only with foreign body type of reaction [107]; neutral solutions
of PVA have also been used in eyedrops on human eyes without difficulty.[87]

Polyvinyl chloride is a suspected human carcinogen.[107] Although the
finished foam resin may cause allergic dermatitis, it is generally free from
health hazard unless comminuted or strongly heated.[278] It burns readily
and gives off objectionable smoke.[278] Several cases of angiosarcoma of
the liver have been detected among long-time workers in plants that make
polyvinyl chloride from monomeric vinyl chloride.[85]

Tetrahydrofuran is very toxic by ingestion and inhalation.[242] When
heated to decomposition, it emits toxic fumes and it can react with oxidiz-
ing materials.[242] The OSHA air standard is 200 ppm (8 hour TWA).[67]

Vinyl Chloride is a proven human carcinogen [107] which may cause permanent injury or death after inhalation, even if the exposure is relatively
short.[242] Human reproductive effects occur at concentrations as low as 30
mg/m^3 [126]; both tumorigenic and mutagenic effects have been observed in
humans as well as test animals. The OSHA air standard for vinyl chloride is
1 ppm (8 hour TWA) with a 5 ppm ceiling.[68]

Vinylidene chloride has produced mutagenic, teratogenic, and carcinogenic effects in laboratory animals.[107] In addition, concentrations as
low as 25 ppm have been associated with systemic effects in humans.[37] If
ignited, the highly toxic hydrogen chloride will likely evolve.[157]

Air Emissions

Air emissions for suspension, emulsion, mass, and solution polymerization are listed in Table 212. The values listed are reported to be the
average value of the data collected from surveys of PVC processing plants
and producer reported data.[255]

Of the four processes used, emulsion polymerization emits the greatest
amount of vinyl chloride monomer from point sources followed by suspension,
mass, and solution polymerization, respectively. However, solution polymerization has added solvent emissions, not only from point sources, but also
from fugitive emission sources. Although these two polymerization processes
produce more emissions than suspension or mass polymerization, only 10
percent of the PVC produced is manufactured by these processes.

Considering the data cited in Table 212 and the fact that suspension
polymerization accounts for 83 percent of the PVC production, fugitive
emissions associated with the suspension polymerization process would appear
to be the significant air emission issue for PVC processes. Fugitive emissions account for approximately 38 percent of the VCM emissions associated
with the suspension polymerization process. In addition, fugitive emissions
are the source of greatest concern from a worker exposure perspective.

Sources of fugitive emissions from PVC processing are presented in
Table 213. Major fugitive emission sources are the valves, pump seals, and
dryer vent.

Wastewater Sources

In addition to air emissions, PVC production generates several wastewater sources associated with the different processes listed in Table 214.
Suspension and emulsion polymerization processes produce large quantities of
wastewater since both of these processes use water in the reaction to suspend the monomer droplets. This water is separated from the polymer beads
in suspension polymerization and contains trace amounts of suspension
stabilizers. PVC emulsions may be shipped as such, or the resin may be

TABLE 214.   SOURCES OF WASTEWATER FROM PVC MANUFACTURE

Source	Sus-pension	Emul-sion	Mass	Solu-tion	Precipi-tation
Centrifuge	X	N/A	N/A	N/A	N/A
Monomer Stripping	X	N/A	N/A	N/A	N/A
Polymer Screening	N/A	X	N/A	N/A	N/A
Drier	X	X	N/A	N/A	N/A
Reactor Washing	X	X	X	X	X

N/A — Not applicable

removed from the aqueous phase in a spray drier. If the resin is dried, the resulting drier off gases, when condensed, may contain trace amounts of emulsifiers as well as suspension stabilizers. The only wastewater associated with mass and solution polymerization processes is from reactor washing.

PVC reactor wash waters and monomer stripping wastewater will contain a certain small amount of vinyl chloride. Plasticizers, buffers, and comonomers will also be present in small amounts, if used in the polymerization. In solution polymerization, a small amount of solvent will be present in the reactor wash water.

Ranges of basic wastewater parameters for waters from the four processes, which were not separated by process for the purpose of establishing effluent limitations for the industry, are as follows:[284]

Characteristic	Unit/Mg PVC Produced
Wastewater Production	2.5 – 41.72 m^3
BOD_5	0.1 – 48 kg
COD	0.2 – 100 kg
TSS	1 – 30 kg

Due to the volume of wastewater generated by suspension and emulsion polymerization, these processes, which comprise over 90 percent of the industry, have the potential for providing the greatest impact to receiving waters.

Solid Wastes

The solid wastes generated by this process are mostly PVC. Solid waste streams arise from reactor cleaning, product blending, particulate removal, and spillage. The quantity of these wastes range from 0.001 to 0.03 kg of PVC per kg of PVC product. The related vinyl chloride losses range from 0.0001 to 0.003 kg of vinyl chloride per kg of PVC product.[255] Solid wastes resulting from off-grade product are less than 2 percent of PVC produced.[61]

Environmental Regulation

New source performance standards proposed by EPA on January 5, 1981 for volatile organic carbon (VOC) fugitive emissions include:

- Safety/release valves must not release more than 200 ppm above background, except in emergency pressure releases, which should not last more than five days; and

- Leaks (which are defined as VOC emissions greater than 10,000 ppm) must be repaired within 15 days.

Proposed regulations for vinyl chloride provide a limitation of 0.2 g of vinyl chloride per kg PVC produced for reactor emissions. Included in these regulations are the following limits for emissions occurring after the reaction process:

- In polyvinyl chloride plants using stripping technology (incinerators): 2,000 ppm for polyvinyl chloride dispersion resins (excluding latex resins); 400 ppm for all other polyvinyl chloride resins, including latex resins, averaged separately for each type of resin.

- In polyvinyl chloride plants without strippers or with technology in addition to strippers: 2 grams/kilogram (0.002 lb/lb) of product from the strippers (or reactors if strippers are not used) for dispersion polyvinyl chloride resins, excluding latex, with the product determined on a dry solids basis; 0.4 g/kg (0.0004 lb/lb) product for all other polyvinyl resins, including latex.

The emissions standards for vinyl chloride sources also include specific operational and control requirements designed to reduce fugitive emissions to no more than 10 ppm of vinyl-chloride/exhaust gas.

Additional standards are required for EPA or state-approved leak detection and elimination programs. Plant personnel must perform continuous monitoring and recordkeeping procedures.

Effluent limitations guidelines have been set for the polyvinyl chloride industry. BPT, BAT, and NSPS call for the effluent to fall within a pH range of 6.0 to 9.0 (41 FR 32587, August 4, 1976).

The following compounds are listed as hazardous wastes (46 FR 27476, May 20, 1981):

Acetone	- U002
Acrylonitrile	- U009
Benzene	- U019
Carbon tetrachloride	- U211
Chlorobenzene	- U037
Di-n-butyl phthalate	- U069
1,2-Dichloroethane	- U077
Dimethyl phthalate	- U102
Di-n-octyl phthalate	- U107
Methanol	- U154
Tetrahydrofuran	- U213
Vinyl chloride	- U043
Vinlyidene chloride	- U078

All disposal of these materials or PVC which contains residual amounts of these materials must comply with the provisions set forth in the Resource Conservation and Recovery Act (RCRA).

27. Polyvinylidene Chloride (PVDC)

Polyvinylidene chloride (PVDC) is a thermally stable, chemical resistant polymer which exhibits high impermeability to gases and vapors. Although the homopolymer has these valuable properties as well as combustion resistance, the difficulties encountered during homopolymer processing led to the development of three major copolymers: vinylidene chloride-vinyl chloride, vinylidene chloride-alkyl acrylate, and vinylidene chloride-acrylonitrile. The incorporation of the comonomers permits polymer processing at lower melt temperatures due to decreased polymeric crystallinity. These products were originally marketed under the Saran trademark. In the United States, Saran is now a generic term for PVDC and its copolymers; however, Saran is still a registered trademark outside the United States.

The homopolymer, Saran A, is not commercially produced due to the difficulties encountered during processing. Saran B, Saran C, and Saran F are designations for vinylidene chloride-vinyl chloride, vinylidene chloride-alkyl acrylates, and vinylidene chloride-acrylonitrile copolymers, respectively. Other commercially available grades of PVDC contain more than one comonomer; the third monomer is introduced to improve the processability, improve the solubility, or modify the specified end-use properties of the polymer.[292] PVDC copolymer compositions range from 73 to 95 percent vinylidene chloride, with a typical content being 85 percent.[247] Typical properties of PVDC are listed in Table A-40 in Appendix A.

Polyvinylidene chloride is produced on a commercial scale using emulsion and suspension polymerization processes. Emulsion polymerization products are used either in the latex as coating vehicles or as a powder when separated from the latex and dried. Suspension polymerization polymers are typically used in molding and extrusion applications. Each of these processes offers different advantages. For example, emulsion polymerization reaction times are shorter; however, suspension polymerization requires fewer additives that detract from the polymer properties.[292, 293]

INDUSTRY DESCRIPTION

Three chemical manufacturers comprise the polyvinylidene chloride industry: Borden, Dow Chemical, and W. R. Grace. Borden and Dow Chemical produce vinylidene chloride-vinyl chloride copolymers while W. R. Grace is classified as a polyvinylidene chloride manufacturer.[56] Production facilities are located in California, Illinois, Kentucky, and Michigan. Table 215 lists these producers and their locations.

Production of polyvinylidene chloride polymers and copolymers totaled approximately 78,000 metric tons in 1980.[65] Miscellaneous vinyl resin production, which includes PVDC as well as polyvinyl butyral and polyvinyl formal, has been generally declining since 1976.

PRODUCTION AND END USE DATA

In 1980, total polyvinylidene chloride production was 78,000 metric tons.[65] Polyvinylidene chloride is used primarily as a film or coating on flexible surfaces for food packaging since it exhibits high oxygen impermeability and oil resistance. Other uses include:[24, 65, 292, 293]

- Unit dose packaging;

- Drum and pack liners;

- Laminating;

- Chemical resistant molding, pipe, and tubing as well as chemical resistant tubing and pipe liners;

- Fire resistant fiber for draperies;

- Combustion modifiers for foams;

- Wire coating, cable jacketing, and other low temperature uses;

- Film coating to give resistance to fats, oils, oxygen, and water vapor while remaining inert, tasteless, and nontoxic;

- Binders for iron-oxide pigmented coatings of magnetic tapes, such as audio, video, and computer tapes;

- Binders for paints and non-woven fabrics; and

- Fiber for screen cloth, furniture and automotive upholstery, furniture webbing, venetian blind tape, filter cloth, agricultural shade cloth, window awning fabrics, brush bristles, doll hair, dust mops, and industrial fabrics.

PROCESS DESCRIPTIONS

Polyvinylidene chloride copolymers are manufactured on a commercial scale using emulsion or suspension polymerization. The end use intended for the polymer determines, in part, which process is used.

Polyvinylidene chloride is produced by free radical polymerization. The reaction steps include initiation, propagation, and termination of the

TABLE 215. U.S. POLYVINYLIDENE CHLORIDE PRODUCERS

Producer	Location
Borden, Inc. Borden Chemical Division Thermoplastic Products	Compton, CA Illiopolis, IL
Dow Chemical, USA	Midland, MI
W. R. Grace & Co. Industrial Chemicals Group Organic Chemicals Division	Owensboro, KY

Source: _Directory of Chemical Producers_, 1982.

polymer chain. Saran B is used in the following discussion to illustrate the reaction chemistry.

The reaction begins with the decomposition of a free radical initiator. During initiation, the comonomers vinylidene chloride and vinyl chloride combine with the free radical ($R\bullet$) to begin the formation of the polymer chain:

$$CH_2 = CCl_2 + R\bullet \longrightarrow R - CH_2CCl_2\bullet$$

vinylidene chloride

$$CH_2 = CHCl + R\bullet \longrightarrow R - CH_2CHCl\bullet$$

vinyl chloride

During propagation, reactive comonomers combine with additional comonomer molecules as the radical is transferred along the polymer chain. The units add in a head to tail fashion:

$$R - CH_2CCl_2\bullet + CH_2 = CCl_2 + CH_2 = CHCl \longrightarrow$$

$$R - CH_2CCl_2 - CH_2CCl_2 - CH_2CHCl\bullet$$

Polymerization is terminated when the free radical is consumed by either radical combination or chain transfer. In combination, two reactive copolymers combine as shown below:

$$R-CH_2CCl_2-CH_2CCl_2-CH_2CHCl\bullet + R'-CH_2CCl_2-CH_2CCl_2-CH_2CCl_2\bullet \longrightarrow$$

$$R-CH_2CCl_2-CH_2CCl_2-CH_2CHCl-CCl_2CH_2-CCl_2CH_2-CCl_2CH_2-R'$$

polyvinylidene chloride

A chain transfer agent ($R'-R"$) may accept the radical, terminating the growing polymer chain as illustrated below:

$$R'-R" + R-CH_2CCl_2-CH_2CCl_2-CH_2CHCl\bullet \longrightarrow$$

$$R"\bullet + R-CH_2CCl_2-CH_2CCl_2-CH_2CHCl-R'$$

polyvinylidene chloride

Oxygen reacts with the vinylidene chloride monomer used, making a peroxide radical shown below:

$$CH_2=CCl_2 + O_2 \longrightarrow CH_2CCl_2O_2\bullet$$

This slows the reaction rate. Monomer storage and polymerization are performed in an inert atmosphere to minimize this reaction.

Tables 216 and 217 list typical input materials and operating parameters for polyvinylidene chloride processing. Other input materials used in these processes such as free radical initiators, redox initiation systems, emulsifiers, and specialty chemicals are listed in Table 218.

Emulsion Polymerization

Emulsion polymerization is used to produce either a polyvinylidene chloride latex or, more commonly, a powder which may be used in other applications. The latex is used in lacquers, bindings, and coatings while the recovered polymer is used in other products. Emulsion polymerization typically takes only one-fourth to one-eighth of the reaction time required for suspension polymerization, and reaches a higher percentage conversion.[292] However, the additives used in emulsion polymerization (for example, the emulsifier or surfactant) decrease the stability and increase the water sensitivity of the polymer.

This process has two major manufacturing advantages: high molecular weight polymers may be produced since the initiation and propagation reaction steps may be controlled more independently; and the monomer may be added during the polymerization to control the comonomer distribution along the chain.[292] These advantages are offset by the reduction in polymer properties, as mentioned above. Recovery of the polymer from the emulsion improves the heat stability, light stability, water resistance, and electrical properties.

As shown in Figure 72, the monomer, comonomer, and specialty chemicals needed for this reaction are fed to a reactor which contains water and an inert atmosphere. The monomer and comonomer are distributed as an emulsion through the aqueous phase by the protective colloid and emulsifier. A typical recipe for the production of PVDC by the batch emulsion process is given below:[59]

Material	Parts by Weight
Vinylidene Chloride (monomer)	78
Vinyl Chloride (comonomer)	22
Water	180
Potassium Persulfate (initiator)	0.22
Sodium Bisulfite (activator)	0.11
Dihexyl Sodium Sulfosuccinate, 80% (emulsifier)	3.58
Nitric Acid, 69% (pH control)	0.07

The reaction is carried to a high conversion before the emulsion is discharged. Normally, the polymer is recovered by coagulation, then washed and dried. Reaction temperatures are kept below 80°C since polymer degradation occurs at this temperature. Relatively short reaction times at these low temperatures are attained by using redox initiator systems.[292, 293]

TABLE 216.   TYPICAL INPUT MATERIALS TO POLYVINYLIDENE CHLORIDE PROCESSING

Process	Water	Monomer	Surfactant	Protective Colloid	Initiator
Emulsion	X	X	X	X	X
Suspension	X	X		X	X

TABLE 217.   TYPICAL OPERATING PARAMETERS FOR POLYVINYLIDENE CHLORIDE PROCESSING

Process	Temperature	Conversion	Reaction Time
Emulsion	30°C	95 - 98%	7-8 hours
Suspension	60°C	85 - 90%	30-60 hours

Sources: Encyclopedia of Chemical Technology, 2nd Edition.
Encyclopedia of Polymer Science and Technology.

TABLE 218.   INPUT MATERIALS AND SPECIALTY CHEMICALS USED IN
POLYVINYLIDENE CHLORIDE MANUFACTURE

Function	Compound
Monomer	vinylidene chloride
Comonomer	acrylates
	acrylate esters
	acrylic acid
	acrylonitrile
	alkyl maleates
	allyl chloride
	α-alkylacrylates
	α-methyl styrene
	butadiene
	1-chloro-1-bromoethylene
	chloroprene
	diallyl fumarate
	ethyl acrylate
	glycidyl methacrylate
	isobutylene
	isopropenyl acetate
	methacrylic acid
	methyl acrylate
	methyl methacrylate
	(2-methacryloyloxyethyl)-diethyl- ammonium methyl sulfate
	N-(2-formamidoethyl)acrylamide
	styrene
	trichloroethylene
	vinyl acetate
	vinyl chloride and other vinyl halides
	vinyl esters
	vinylidene bromide
	5-vinyl-2-picoline
Plasticizer	dibutyl sebacate
	diisobutyl adipate
Heat Stabilizer	epoxides
	tetrasodium pyrophosphate
Reinforcing Agent	calcium carbonate
	glass
	wollastonite

(continued)

TABLE 218 (continued)

Function	Compound

Initiator	ammonium persulfate
	azo compounds
	organic peroxides
	  lauroyl peroxide
	  percarbonates
	peroxydisulfate
	potassium persulfate

Redox System	ammonium persulfate - sodium metabisulfite
Organometallic	butyllithium
Emulsifier	dihexyl sodium sulfosuccinate, 80% sodium lauryl sulfate
Suspension Stabilizer (Protective Colloid)	carboxymethylcellulose gelatin methyl(hydroxypropyl)cellulose polyvinyl alcohol Tergitol NP35® (Union Carbide Corp.)
Initiator Activator	sodium bisulfite sodium formaldehyde sulfoxylate
pH Control	nitric acid

Stabilizer	acid acceptors
	  alkaline - earth salts
	  heavy metal salts
	    barium fatty acid salts
	    cadmium fatty acid salts
	    lead fatty acid salts
	epoxy compounds
	  epoxidized soybean oil
	  glycidyl esters
	  glycidyl ethers
	organotin compounds
	  organotin mercaptides
	  salts of carboxylic acids

UV Absorber	benzophenone derivatives benzotriazole derivatives resorcylic acid derivatives salicylic acid derivatives

(continued)

TABLE 218 (continued)

Function	Compound
Antioxidant	phenols 2,6-di-t-butyl-4-methylphenol substituted bisphenols organic phosphites organic sulfur compounds
Solvent for Lacquer Coatings	acetone dimethylformamide ethyl acetate methyl ethyl ketone methyl isobutyl ketone tetrahydrofuran toluene

Sources: Encyclopedia of Chemical Technology, 2nd Edition.
Encyclopedia of Polymer Science and Technology.
Modern Plastics Encyclopedia, 1981-1982.

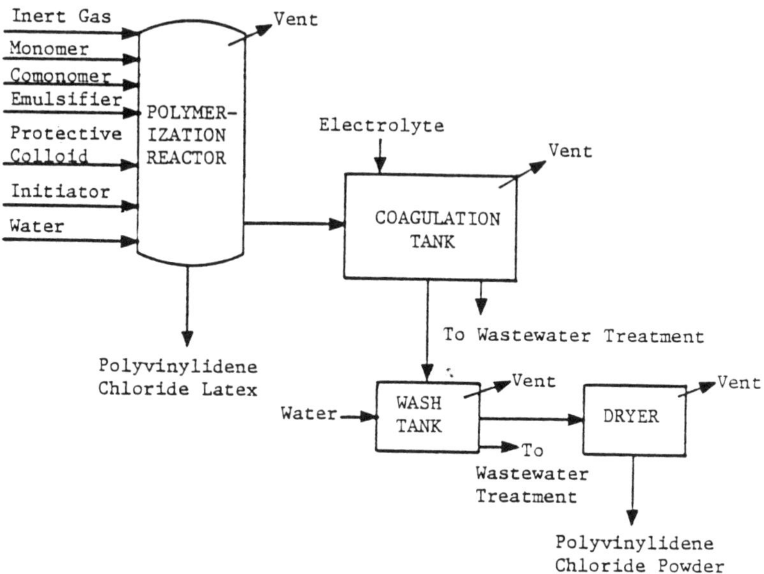

Figure 72.   Polyvinylidene chloride production using emulsion
             polymerization.

Sources: <u>Encyclopedia of Chemical Technology</u>, 2nd Edition.
           <u>Encyclopedia of Polymer Science and Technology</u>.

Suspension Polymerization

Suspension polymerization is typically used to produce extrusion and molding resins of polyvinylidene chloride. Fewer inputs to the reactor reduce the adverse effects on polymer properties as well as the cost. Generally, water sensitivity is less than that of emulsion polymerized resins, but the stability is improved. An added advantage of the suspension polymerization process is the manufacture of polymer beads which are easily separated, dried, and used in the production of polyvinylidene chloride end products. The major disadvantages of this process are the increased reaction times and lower molecular weight polymers.[292]

The suspension polymerization process is shown in Figure 73. The monomer and comonomer are fed to a polymerization reactor where, with the aid of a suspension stabilizer, they are suspended as droplets in an aqueous medium. The inert gas prevents polymerization of vinylidene chloride before the desired initiator is added. Due to the inability to control the initiation and propagation steps as easily as in the emulsion polymerization process, the conversion only reaches 85 to 90 percent.

The monomer is removed by vacuum distillation, condensed, and recycled while the polymer beads are centrifuged and dried in a flash or fluidized bed dryer. The suspension polymerization process is typically a batch process as compared to the emulsion polymerization process that may be either batch or continuous.

A typical recipe for the production of PVDC by the suspension process is given below:[59]

Material	Parts by Weight
Vinylidene Chloride (monomer)	85
Vinyl Chloride (comonomer)	15
Water	200
(Methyl)(Hydroxypropyl) Cellulose, 400 cp (protective colloid)	0.05
Lauroyl Peroxide (initiator)	0.3

Energy Requirements

No data for the energy required for polyvinylidene chloride production were found in the literature consulted.

ENVIRONMENTAL AND INDUSTRIAL HEALTH CONSIDERATIONS

Although PVDC and its copolymers are considered to be nontoxic, several of the input materials used have been listed as hazardous under RCRA: acetone, acrylic acid, acrylonitrile, ethyl acetate, ethyl acrylate, methyl ethyl ketone, methyl isobutyl ketone, tetrahydrofuran, toluene, trichloroethylene, vinyl chloride, and vinylidene chloride. Vinylidene chloride is highly volatile; inhalation of the vapor is consideed to be hazardous, but

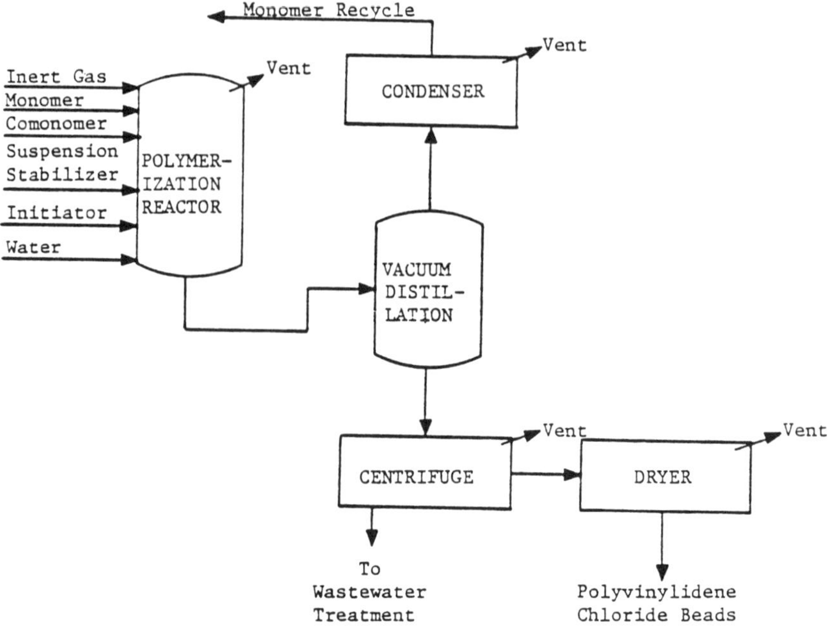

Figure 73. Polyvinylidene chloride production using suspension polymerization.

Sources: Encyclopedia of Chemical Technology, 2nd Edition.
Encyclopedia of Polymer Science and Technology.

is readily controlled by precautions commonly taken in the chemical industry. Chloroprene and acrylonitrile (comonomers) are suspected human carcinogens. Other potential carcinogens include vinylidene chloride, trichloroethylene, and polyvinyl alcohol. Acrylic acid, ethyl acrylate, methyl
acrylate, styrene, and toluene all pose significant health hazards.

<u>Worker Distribution and Emissions Release Points</u>

We have estimated worker distribution for PVDC production by correlating major equipment manhour requirements with the process flow diagrams
in Figures 72 and 73. Estimates are shown in Table 219.

No major point source air emissions are associated with PVDC processing. However, fugitive and process air emissions from such sources as
reactor vents, wash tank vents, and dryer vents may pose a significant
environmental and/or worker health problem based on the components present
in the stream, operating parameters for the process, engineering and
administrative controls, and maintenance programs.

Process sources of fugitive emissions are listed in Table 220.
Numerous compounds may be used as comonomers and additives (see Table 218).
Emissions of nitrogen gas, used as an inerting agent in the reactor, may
pose a simple asphyxiation hazard if elevated levels reduce the oxygen
concentration of workplace air. Particulate emissions from centrifuging and
drying operations may result in elevated nuisance dust concentrations.

Health effects information for contaminants of concern is provided in
the paragraphs below and in Table 10.

<u>Health Effects</u>

The manufacture of polyvinylidene chloride involves the use of two
suspected human carcinogens. They are chloroprene and acrylonitrile, both
comonomers in the process. Three other input materials or specialty chemicals have produced positive results in animal testing for carcinogenicity--
vinylidene chloride, a monomer; trichloroethylene, a comonomer; and the
suspension stabilizer, polyvinyl alcohol. Several other process chemicals
are currently being tested for carcinogenic potential. Highly toxic substances used by the industry include four comonomers--acrylic acid, ethyl
acrylate, methyl acrylate, and styrene--and the solvent toluene. The
reported health effects of exposure to these substances are summarized
below.

<u>Acrylic Acid</u> is a severe skin and respiratory irritant [170] and is one
of the most serious eye injury chemicals.[157] Ingestion may produce epigastric pain, nausea, vomiting, circulatory collapse, and in severe cases,
death due to shock.[85] Although the data are not sufficient to make a
carcinogenic determination, the chemical has been associated with tumorigenic and teratogenic effects in laboratory animals.

TABLE 219.   WORKER DISTRIBUTION ESTIMATES FOR PVDC PRODUCTION

Process	Unit	Workers/Unit/8-hour Shift
Emulsion Polymerization	Batch or	
	Continuous Reactor	1.0 or 0.5
	Coagulation Tank	1.0
	Washing	0.25
	Dryer	0.5
Suspension Polymerization	Batch Reactor	1.0
	Condenser	0.125
	Distillation Column	0.25
	Centrifuge	0.25
	Dryer	0.5

TABLE 220. SOURCES OF FUGITIVE EMISSIONS FROM PVDC MANUFACTURE

Source	Constituent	Emulsion	Suspension
Reactor Vent	Vinylidene Chloride	X	X
	Comonomer	X	X
Coagulation Tank Vent	Vinylidene Chloride	*	
	Comonomer	*	
	Electrolyte	*	
Wash Tank Vent	Vinylidene Chloride	*	
	Comonomer	*	
Monomer Condenser Vent	Vinylidene Chloride		X
	Comonomer		X
Centrifuge Vent	Particulates		X
	Vinylidene Chloride		*
	Comonomer		*
Dryer Vent	Particulates	X	X
	Vinylidene Chloride	*	*
	Comonomer	*	*

*Trace amounts expected due to high conversion in the reactor.

TABLE 221. SOURCES OF WASTEWATER FROM PVDC MANUFACTURE

Source	Emulsion	Suspension
Coagulation Tank	X	
Wash Tank	X	
Centrifuge		X
Routine Cleaning Water	X	X

Acrylonitrile, a suspected human carcinogen [107], causes irritation of the eyes and nose, weakness, labored breathing, dizziness, impaired judgment, cyanosis, nausea, and convulsions in humans. Mutagenic, teratogenic, and tumorigenic effects have been reported in the literature. The OSHA air standard is 2 ppm (8 hour TWA) with a 10 ppm ceiling.[69]

Chloroprene has been classified by the International Agency for Research on Cancer as a suspected human carcinogen.[107] It has also produced mutations in rats at an exposure by inhalation of less than 2 mg/m^3 over a 16-week period.[233] Pregnant rats given a single oral dose of 1 mg/kg or an inhaled dose of 4 mg/m^3 for 24 hours have experienced abnormal development of the fetus.[233] The OSHA air standard is 25 ppm (8 hour TWA).[67]

Ethyl Acrylate is currently being tested by the National Toxicology Program for carcinogenesis by standard bioassay protocol. Previous test data led to indefinite results.[107] It is very toxic by injection, inhalation, or skin absorption [242], producing symptoms of hypersomnia and convulsions.[149] The OSHA air standard is 25 ppm (8 hour TWA).[67]

Methyl Acrylate is highly toxic by ingestion, inhalation, and skin absorption upon acute exposure to relatively low doses.[242] The monomer is very irritating to eyes, skin, and mucous membranes; lethargy and convulsions may occur if the vapor is inhaled in high concentrations.[149] Toxic effects have been observed in humans at a concentration of 75 ppm.[76] Methyl acrylate also produces tumors in laboratory animals, but sufficient data do not exist to make a carcinogenic determination.[107] The OSHA air standard is 10 ppm (8 hour TWA).[67]

Polyvinyl Alcohol has produced positive results in animal testing for carcinogenicity.[107] Its single dose toxicity, however, is presumably low. [85] Implantation of PVA sponge as a breast prosthesis has been associated only with foreign body type of reaction [107]; neutral solutions of PVA have also been used in eyedrops on human eyes without difficulty.[87]

Styrene has been linked with increased rates of chromosomal aberrations in persons exposed in an occupational setting.[107] Animal test data strongly support epidemiological evidence of its mutagenic potential.[233] Styrene also produces tumors and affects reproductive fertility in laboratory animals.[233] Toxic effects of exposure to styrene usually involve the central nervous system.[233] The OSHA air standard is 100 ppm (8 hour TWA) with a ceiling of 200 ppm.[67]

Toluene exhibits tumorigenic, mutagenic, and teratogenic potential in laboratory tests.[233] Human exposure data indicate that toluene is also very toxic. Exposures at 100 ppm have produced psychotropic effects and central nervous system effects have been observed at 200 ppm.[233] Symptoms of exposure include headache, nausea, vomiting, fatigue, vertigo, paresthesia, anorexia, mental confusion, drowsiness, and loss of consciousness. The OSHA air standard is 200 ppm (8 hour TWA) with a 300 ppm ceiling.[67]

Trichloroethylene has produced positive results in animal testing for
carcinogenicity.[108] It has also produced mutagenic and teratogenic
effects in laboratory animals [233] and is classified as an acute narcotic
which causes death from respiratory failure if exposure is severe and prolonged.[4] Since trichloroethylene has a rather long biologic half-life,
major consideration must be given to cumulative effects of this compound.
[31] Sublethal exposure may cause liver and kidney lesions, reversible
nerve damage, and psychic disturbances.[85] OSHA has set an air standard of
100 ppm (8 hour TWA) with a 200 ppm ceiling.[67]

Vinylidene Chloride has produced mutagenic, teratogenic, and carcinogenic effects in laboratory animals.[107] In addition, concentrations as
low as 25 ppm have been associated with systemic effects in humans.[37] If
ignited, the highly toxic hydrogen chloride will likely evolve.[157]

In addition to its health hazards, vinylidene chloride is flammable
between 6.5 percent and 15.5 percent concentration (by volume) in air.
Vinylidene chloride will spontaneously polymerize, forming peroxides and
creating a potentially explosive situation. Vinylidene chloride also reacts
with copper and aluminum; the impurities in the monomer form copper acetyl-
ides while aluminum reacts with the monomer to produce aluminum chloralkyls.
Both of these compounds are extremely reactive and potentially hazardous.
[293]

Air Emissions

Sources of fugitive emissions from PVDC manufacture are summarized in
Table 220 by process and constituent type. Suspension polymerization does
not carry the reaction to the high level of conversion achieved in the emulsion polymerization process. Therefore, fugitive VOC emissions are expected
downstream of the reactor for the suspension polymerization process while
sources upstream of the reactor in emulsion polymerization are the major
contributors to the fugitive VOC emissions. The remaining hydrocarbons in
these streams may be removed by venting to either a flare or a blowdown.
[285]

Fugitive particulate emission sources are also listed in Table 220.
These particulates may be collected by venting to a baghouse or electrostatic precipitator.

Wastewater Sources

The wastewater sources associated with PVDC processing are associated
with removal of the polymer from the aqueous medium. These sources are
listed in Table 221. No data are available in the literature consulted on
the wastewater produced or the associated parameters.

Solid Waste

The solid wastes generated by this process include polymer lost due to
spillage and reactor cleaning, substandard polymer which cannot be blended,
and collected particulates.

Environmental Regulation

Effluent limitations guidelines have not been set for the polyvinylidene chloride industry.

New source performance standards (proposed by EPA on January 5, 1981) for volatile organic carbon (VOC) fugitive emissions include:

- Safety/release valves must not release more than 200 ppm above background, except in emergency pressure releases, which should not last more than five days; and

- Leaks (which are defined as VOC emissions greater than 10,000 ppm) must be repaired within 15 days.

The following compounds have been listed as hazardous wastes (46 Federal Register 27476, May 20, 1981):

Acetone	- U002
Acrylic acid	- U008
Acrylonitrile	- U009
Ethyl acetate	- U112
Ethyl acrylate	- U113
Methyl ethyl ketone	- U159
Methyl isobutyl ketone	- U161
Methyl methacrylate	- U162
Phenol	- U188
Tetrahydrofuran	- U213
Toluene	- U220
Trichloroethylene	- U228
Vinyl chloride	- U043
Vinylidene chloride	- U078

All disposal of these compounds or polyvinylidene chloride resins containing residual amounts of these compounds must comply with the provisions set forth in the Resource Conservation and Recovery Act (RCRA).

28. Styrene-Acrylonitrile (SAN)

Styrene-acrylonitrile (SAN) resins combine the high clarity and gloss of styrene with the added chemical resistance, heat resistance, and toughness of acrylonitrile. These rigid, hard, transparent resins are easily processed and possess good dimensional stability. This combination of properties is unique for a transparent resin.

SAN is a random copolymer of styrene and acrylonitrile which has an amorphous structure. The molecular weight and the acrylonitrile content of the resin play the major role in determining polymer properties. Table A-41 in Appendix A lists some typical properties of SAN resins. Table A-42 in Appendix A presents tensile strength and elongation as a function of acrylonitrile content.

SAN is resistant to aliphatic hydrocarbons, alkalies, battery acids, vegetable oils, foods, and detergents.[64] This insolubility makes SAN suitable for use in applications requiring contact with petroleum products.[18] SAN is soluble in acetone, chloroform, dioxane, methyl ethyl ketone, and pyridine. Swelling of the resin occurs on contact with benzene, ether, or toluene; SAN is insoluble in carbon tetrachloride, ethyl alcohol, gasoline, kerosene, and lubricating oil.

SAN may be produced using emulsion, suspension, or continuous mass (bulk) polymerization. The latex manufactured via emulsion polymerization is used primarily as an input to ABS manufacture. This section presents the polymerization processes used to produce SAN. Acrylonitrile-butadiene-styrene (ABS) processes are discussed separately.

INDUSTRY DESCRIPTION

The SAN industry is comprised of two producers that sell the resin on the merchant market, Dow Chemical and Monsanto. All ABS producers have SAN production capacity which is normally used in the production of ABS resins. These companies sell small quantities of SAN resin on the merchant market. Table 222 lists the location and production capacity of the two SAN producers.

SAN is used in appliances, automobiles, housewares, packaging, and construction. These industries reflect the general trend of the U.S.

TABLE 222. U.S. PRODUCERS OF SAN

Producer	January 1, 1982 Capacity Thousand Metric Tons
Dow Chemical USA	
Midland, MI	31.8
Pevely, MO	29.5
COMPANY TOTAL	61.3
Monsanto Co.	
Monsanto Plastics and Resins Co.	
Addyston, OH	22.7
COMPANY TOTAL	22.7
TOTAL	84.0

Sources: <u>Chemical Economics Handbook</u>, updated annually, 1981 data.
<u>Directory of Chemical Producers</u>, 1982.

economy. In 1981, SAN production fell from the 1979 level of 56,400 metric
tons to 51,000 metric tons, which is 400 metric tons below the 1976 production level of 51,400 metric tons.[65, 143] The present SAN capacity of
84,000 metric tons is sufficient to support growth in the near future.

One environmentally significant portion of SAN processing is the
devolatilization step used in continuous mass polymerizaton. Residual
monomer in the polymer is reported to be reduced to 0.7 percent, which
reduces the population exposure risk.[153] This reduced risk is offset by
the additional processing step which increases the sources of fugitive VOC
emissions.

PRODUCTION AND END USE DATA

In 1981, SAN production totaled 51,000 metric tons.[143] Other SAN end
uses are:[64, 153]

- Knobs, refrigerator meat and vegetable compartments, blender and
  mixer bowls, and other appliance uses;

- Instrument lenses, dash components, and glass-filled support panels
  for automotive use;

- Battery cases and meter lenses for electronics applications;

- Tumblers, mugs, and other housewares;

- Syringes, blood aspirators, artificial kidney devices, and other
  medical products;

- Cosmetic containers, closures, bottles, jars, and other packaging
  materials;

- Safety glazing and water filter bowls for construction use; and

- Specialty products, such as brush block and bristles, typewriter
  keys, and pen and pencil barrels.

Table 223 lists the major markets for SAN and the amount contributed to each
for 1981.

PROCESS DESCRIPTIONS

SAN is produced by emulsion, suspension, or mass polymerization. Emulsion polymerized resins are especially suitable for ABS production; however,
the residual emulsifying agent makes the polymer less suitable for high
transparency applications. Both mass and suspension polymerization produce
resins which are typically used for molding applications. As with the
emulsion resins, suspension polymerized resins contain residual suspending
agents which make the resin less desirable for transparent uses. Emulsion
and suspension polymerization may be either batch or continuous processes;
mass polymerization processes used are continuous.

TABLE 223. SAN MAJOR MARKETS AND THE 1981 MARKET SHARE

Market	Thousand Metric Tons
Appliances	8
Automotive	4
Batteries (non-automotive)	2
Compounding	8
Housewares	9
Packaging, Molded	6
Other	6
Export	8
	51

Source: _Modern Plastics_, January 1982.

Mass polymerization has fewer waste treatment and environmental impacts due to the simplified process and the absence of water in the polymerization. This process does not require polymer drying, which in turn reduces the energy input to the process. The disadvantages of this process include: longer time required to reach steady state and difficult mixing and heat removal as the viscosity of the polymerizing mass increases with increasing conversion.[153]

SAN is produced by the polymerizaton of styrene and acrylonitrile. A free radical initiator (R•) is used to initiate the reaction by reacting with a styrene molecule. The styrene radical then combines with either styrene or acrylonitrile monomers to create the random copolymer. The reaction is terminated by either two radical groups combining or the introduction of a chain transfer agent (R'R"). The polymerization reaction is shown below:

Initiation

$$CH=CH_2 + \overset{\bullet}{R} \longrightarrow \overset{\bullet}{C}HCH_2R$$

Propagation

$$\overset{\bullet}{C}HCH_2R + CH=CH_2 \longrightarrow \overset{\bullet}{C}HCH_2-CHCH_2R$$

$$\overset{\bullet}{C}HCH_2-CHCH_2R + CH_2=CHCN \longrightarrow \overset{\bullet}{C}H_2-CH-CHCH_2-CHCH_2R$$
(CN)

Termination

Radical Combination

$$\overset{\bullet}{C}HCH_2R + \overset{\bullet}{C}H_2CH-CHCH_2-CHCH_2R \longrightarrow RCH_2CH-CH_2CH-CHCH_2-CHCH_2R$$
(CN)

Chain Transfer

$$R'R" + \overset{\bullet}{C}H_2CH-CHCH_2CHCH_2 \longrightarrow R"CH_2CH-CHCH_2-CHCH_2 + R'\bullet$$
(CN)

Oxygen reacts with the free radical initiators used, thus making a peroxide radical shown below, which slows the reaction rate:

$$R\bullet + O_2 \longrightarrow RO_2\bullet$$

Polymerization is performed in a nitrogen atmosphere to minimize this reaction.

Tables 224 and 225 list typical input materials and operating conditions for the three SAN production processes. Other input materials used in SAN production such as free radical initiators, emulsifiers, chain transfer agents, and suspension stabilizers are listed in Table 226.

Emulsion Polymerization

The SAN resin produced by emulsion polymerization may be incorporated into ABS production as an emulsion or it may be recovered for use in molding or extrusion processes. As illustrated in Figure 74, emulsion polymerization performed batchwise is started by charging the reactor with part of the necessary monomers, chain transfer agent, initiator, emulsifier, and water. The reactor is purged with nitrogen and heated to reflux while the remaining portion of the input materials are fed to the reactor. After the reactor is completely charged, the reaction is carried to the desired conversion at reflux conditions.

Continuous emulsion polymerization uses two stirred-tank reactors in series followed by a large hold tank.[153] Either the batch or continuous process may provide SAN emulsions for use in ABS production. For other uses, the particles may be recovered by coagulation using an electrolyte, washed, and dried.

A typical batch emulsion polymerization recipe follows:[153]

Material	Parts by Weight
Initiator-Emulsifier Solution	
Dresinate (rosin soap)	3
$K_2S_2O_8$ (initiator)	0.06
Water	146
Monomer Solution	
Styrene	68.5
Acrylonitrile	31.5
Tert-Dodecyl Mercaptan (chain transfer agent)	0.4

Suspension Polymerization

Suspension polymerization differs from emulsion polymerization since very small amounts (typically 0.01 to 0.05 percent) of a suspending agent

TABLE 224.   TYPICAL INPUT MATERIALS TO SAN PRODUCTION PROCESSES
IN ADDITION TO STYRENE, ACRYLONITRILE AND INITIATOR

Process	Water	Chain Transfer Agent	Suspension Stablilizer	Emulsifier
Emulsion	X	X		X
Suspension	X	X	X	
Mass		X		

TABLE 225.   TYPICAL OPERATING PARAMETERS FOR SAN PRODUCTION PROCESSES

Process	Temperature	Reaction Time	Conversion
Emulsion	70 - 100°C[a]	1-3 hours	97% or higher
Suspension	60 - 150°C	5 hours	---[b]
Mass	100 - 200°C	1 hour	40 - 70%

[a]With redox systems, the temperature may be lowered to 38°C.

[b]No conversion is given in the literature.

Source: Encyclopedia of Chemical Technology, 3rd Edition.

TABLE 226.   INPUT MATERIALS AND SPECIALTY CHEMICALS USED IN SAN MANUFACTURE
             (IN ADDITION TO SYTRENE AND ACRYLONITRILE)

Free Radical Initiators	Di-tert-butyl peroxide Potassium peroxydisulfate (aqueous potassium persul- fate solution)
Emulsifiers	Rosin soap
Chain Transfer Agents	Dipentene Tert-dodecyl mercaptan
Suspending Agents	Acryclic acid-2-ethylhexyl acrylate (90:10)

Source: Encyclopedia of Chemical Technology, 3rd Edition.

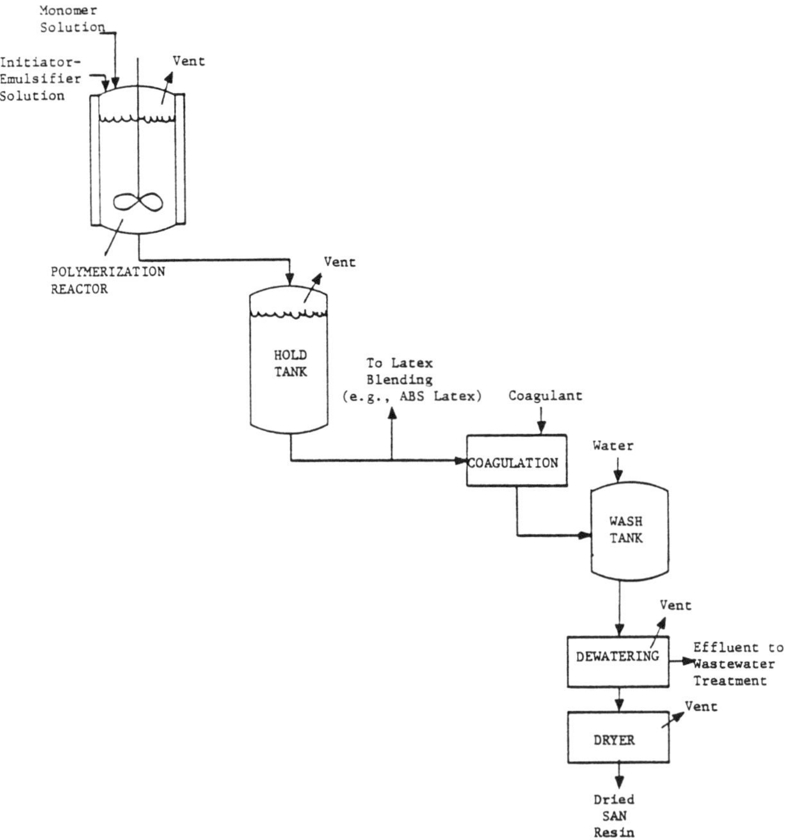

Figure 74. SAN emulsion process.

Source: Encyclopedia of Chemical Technology, 3rd Edition.

are used to maintain the monomer droplets in suspension polymerization.
Emulsifier levels of 1 to 5 percent are used for emulsion polymerization.[153]

In the suspension polymerization process shown in Figure 75, the reactor is charged with input materials and purged with nitrogen. The reaction
mixture is heated and the reaction proceeds to the desired conversion. The
resulting polymer beads are stripped of unreacted monomer by distillation.
The styrene and acrylonitrile monomers are recycled to the reactor while the
polymer beads are centrifuged and dried.

A typical recipe for suspension polymerization of SAN follows:[153]

Material	Parts by Weight
Styrene	70.00
Acrylonitrile	30.00
Dipentene (chain transfer)	1.20
Di-Tert-Butyl Peroxide (initiator)	0.03
Acrylic Acid-2-Ethylhexyl Acrylate (90:10) (suspending agent)	0.03
Water	100.00

Mass Polymerization

Mass polymerization differs from emulsion and suspension polymerization
in that water is not used in the process.   Mass polymerization may also be
initiated with a free radical initiator.

As illustrated in Figure 76, styrene, acrylonitrile, and an initiator
are fed to a polymerization reactor. When the reaction reaches the desired
conversion, the polymer melt is pumped to a devolatilizer. Unreacted monomers are removed and recycled to the reactor. The stripped polymer is then
pelletized in an extruder before it is bagged for shipment.

A typical production recipe for mass polymerization of SAN was not
found in the literature.

Energy Requirements

Data for the energy required during SAN production were not found in
the literature consulted.

ENVIRONMENTAL AND INDUSTRIAL HEALTH CONSIDERATIONS

SAN is a nontoxic compound which is used in medical applications. Styrene is considered to be a relatively safe organic chemical.[17] However,
prolonged or repeated contact with the skin may cause skin irritation and
over exposure to vapors may cause eye and nasal irritation. Styrene odors
are detectable at 60 ppm and are quite strong at 100 ppm.[17] Acrylonitrile
can be handled in industrial processes with relative safety although it is
toxic and a suspected carcinogen.[153]

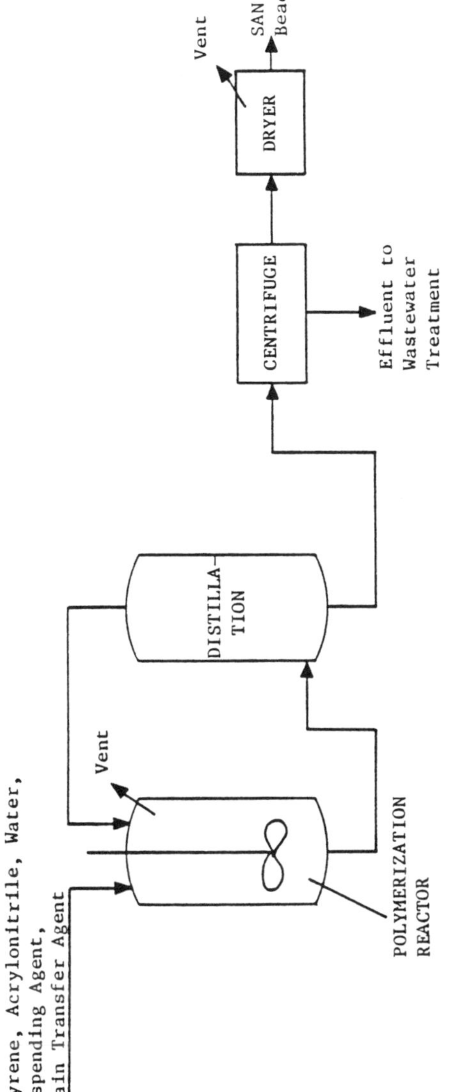

Figure 75. SAN suspension process.

Source: Encyclopedia of Chemical Technology, 3rd Edition.

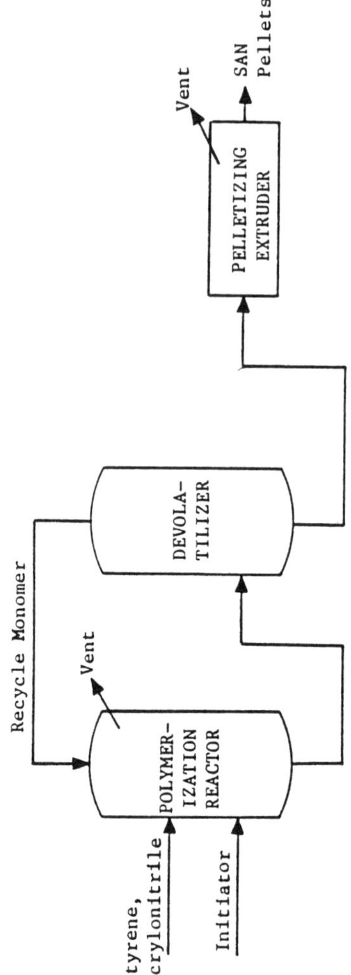

Figure 76. SAN mass process.

Source: Encyclopedia of Chemical Technology, 3rd Edition.

Emulsion polymerization SAN processes carry the polymerization reaction to high conversion, 97 percent. Due to this high percentage of monomers converted to polymer, the upstream sources of air emissions contribute significantly more to the total VOC emissions than the sources downstream from the reactor.

Unreacted monomers are recycled to the reactor in suspension and mass polymerization. Low conversion (40 to 70 percent) in the mass polymerization process makes the reactor vent more suspect as a source of acrylonitrile and styrene emissions. Similarly, suspension polymerization reactor vents are also a probable source of significant emissions since the distillation step recycles unreacted monomers to the reactor.

The lower levels of conversion which are achieved by the mass polymerization process require a devolatilization step. Although the mass polymerization process generates significantly less wastewater, the added devolatilization step adds additional fugitive VOC emissions.

<u>Worker Distribution and Emissions Release Points</u>

Estimates of worker distribution for SAN production are shown in Table 227. Estimates were derived by correlating major equipment manhour requirements with the process flow diagrams in Figures 74 through 76.

The closed process conditions associated with SAN production minimize employee exposure potential to hazardous chemical substances and physical agents. Fugitive emissions from the sources listed in Table 228 and from leaks in valves, pump and compressor seals, and drains are the major sources of workplace contamination.

Styrene and acrylonitile are the major contaminants of concern associated with SAN production. Health effects information about these chemicals is summarized in the paragraphs below. Additionally, nitrogen gas used as an inerting agent in the reactor may pose a simple asphyxiation hazard if fugitive emissions result in reduced atmospheric oxygen.

<u>Health Effects</u>

Sufficient data do not exist with which to evaluate the potential health impacts of several input materials and specialty chemicals used in SAN manufacture. Of the chemicals which have a significant history of investigation, acrylonitrile and acrylic acid pose the most serious risk to the health of plant employees. A summary of the reported health effects resulting from exposure to these substances follows.

<u>Acrylic Acid</u> is a severe skin and respiratory irritant [170] and is one of the most serious eye injury chemicals.[157] Ingestion may produce epigastric pain, nausea, vomiting, circulatory collapse, and in severe cases, death due to shock.[85] Although the data are not sufficient to make a carcinogenic determination, the chemical has been associated with tumorigenic and teratogenic effects in laboratory animals.

TABLE 227.   WORKER DISTRIBUTION ESTIMATES FOR SAN PRODUCTION

Process	Unit	Workers/Unit/8-hour Shift
Emulsion Polymerization	Batch Reactor or Continuous Reactor	1.0 or 0.5
	Hold Tank	0.125
	Coagulation	1.0
	Washing	0.25
	Dewatering	0.25
	Dryer	0.5
Solution Polymerization	Batch Reactor	1.0
	Distillation	0.25
	Centrifuge	0.25
	Dryer	0.5
Mass Polymerization	Batch Reactor	1.0
	Devolatilizer	0.25
	Extruder	1.0

TABLE 228.   SOURCES OF FUGITIVE EMISSIONS FROM SAN MANUFACTURE

Source	Constituent	Process Emulsion	Suspension	Mass
Reactor Vents	Styrene	X	X	X
	Acrylonitrile	X	X	X
	Initiators	X	X	X
	Emulsifiers	X		
	Suspension Stabilizers		X	
Hold Tank	Styrene	X		
	Acrylonitrile	X		
Dewatering	SAN and Monomer Emissions and Particulates	X[a]	X	
Dryer Vent	SAN and Monomer Emissions and Particulates	X[a]	X	
Extruder	SAN and Monomer Emissions and Particulates			X

[a]Present if the SAN emulsion resin is recovered.

Acrylonitrile, a suspected human carcinogen [107], causes irritation of
the eyes and nose, weakness, labored breathing, dizziness, impaired judgment, cyanosis, nausea, and convulsions in humans. Mutagenic, teratogenic,
and tumorigenic effects have been reported in the literature. The OSHA air
standard is 2 ppm (8 hour TWA) with a 10 ppm ceiling.[69]

Styrene has been linked with increased rates of chromosomal aberrations
in persons exposed in an occupational setting.[107] Animal test data
strongly support epidemiological evidence of its mutagenic potential.[233]
Styrene also produces tumors and affects reproductive fertility in laboratory animals.[233] Toxic effects of exposure to styrene usually involve the
central nervous system.[233] The OSHA air standard is 100 ppm (8 hour TWA)
with a ceiling of 200 ppm.[67]

Air Emissions

Sources of fugitive VOC emissions are listed in Table 228. These
sources are vents and may be controlled by venting these streams to a flare
to incinerate the remaining hydrocarbons or a blowdown.[285] Other equipment sources, including valves, flanges, and pumps, may also contribute to
total VOC emissions. The best control for these sources may be a routine
inspection and maintenance program.

Sources of fugitive particulates are also listed in Table 228. Particulates may be collected by routing these streams to a baghouse or electrostatic precipitator.

Wastewater Sources

There are several wastewater sources associated with the various SAN
polymerization processes shown in Table 229. The major wastewater sources
result from emulsion and suspension polymerization which use water as a
process medium for the production of SAN resin. Ranges of several wastewater parameters for wastewaters from SAN production are shown below.
Values for the wastewater from the processes presented were not distinguished by process type by EPA for the purpose of establishing effluent
limitations for the SAN industry:[284]

SAN Wastewater Characteristics	Unit/Metric Ton of ABS
Production	$1.67 - 24.03 \ m^3$
BOD_5	2 - 20.7 kg
COD	5 - 33.5 kg
TSS	0 - 30 kg

Solid Wastes

The solid wastes generated by this process are mostly SAN. Solid waste
streams originate from reactor cleaning, product blending, particulate
removal, and spillage.

TABLE 229.   SOURCES OF WASTEWATER FROM SAN MANUFACTURE

	Process		
Source	Emulsion	Suspension	Mass
Polymer Dewatering	X[a]	X	
Spent Water Bath			
Routine Cleaning Water	X	X	X

[a]Present if the SAN emulsion resin is recovered.

Environmental Regulation

Effluent limitations guidelines have been set for the SAN industry. BPT, BAT, and NSPS call for the pH of the effluent to fall between 6.0 and 9.0 (41 Federal Register 32587, August 4, 1976).

New source performance standards (proposed by EPA on January 5, 1981) for volatile organic carbon (VOC) fugitive emissions include:

- Safety/release valves must not release more than 200 ppm above background, except in emergency pressure releases, which should not last more than five days; and

- Leaks (which are defined as VOC emissions greater than 10,000 ppm) must be repaired within 15 days.

Acrylonitrile is listed as a hazardous waste (46 Federal Register 27476, May 20, 1981); it is designated as U009. All disposal of acrylonitrile or ABS which contains any residual acrylonitrile must comply with the provisions set forth in the Resource Conservation and Recovery Act (RCRA).

References

(1)   Agranoff, Joan, ed. <u>Modern Plastics Encyclopedia</u>, 1982-1983.
      Volume 59, Number 10A. pp. 74, 187.

(2)   Albright, Lyle F. <u>Processes for Major Addition-Type Plastics and
      Their Monomers</u>. McGraw-Hill, New York, New York, 1974.

(3)   Altman, J. W. "Acrylic." <u>Modern Plastics Encyclopedia</u>. McGraw-
      Hill, New York, New York, 1981-1982. pp. 8-10.

(4)   American Conference of Governmental Industrial Hygienists. <u>Documen-
      tation of Threshold Limit Values for Substances in Workroom Air</u>, 3rd
      Edition with Supplements. 1977. pp. 79, 194, 263, 298, 407, 413.

(5)   American Conference of Governmental Industrial Hygienists. <u>Documen-
      tation of Threshold Limit Values for Substances in Workroom Air</u>, 4th
      Edition. 1980. pp. 21, 22, 27, 86, 121, 212, 263, 413, 431.

(6)   American Conference of Governmental Industrial Hygienists. <u>Documen-
      tation of Threshold Limit Values for Substances in Workroom Air</u>,
      Supplements for Those Substances Added or Changed Since 1981, Volume
      4. 1981. p. 185.

(7)   "...And Carbide Closes an LDPE Plant." <u>Plastics Technology</u>, May
      1981, p. 111.

(8)   Arena, J. M. <u>Poisoning; Toxicology, Symptoms, Treatments</u>, 2nd
      Edition. C. C. Thomas, Springfield, Illinois, 1970.

(9)   Avery, J. A. "Thermoplastic Polyester: PBT." <u>Modern Plastics
      Encyclopedia</u>. McGraw-Hill, New York, New York, 1981-1982. pp.
      49-50.

(10) Backus, J. K. "Flame Resistant Rigid Foams." <u>Advances in Urethane
      Science and Technology</u>, Volume 1, K. C. Frisch and S. L. Reegen
      (eds). TECHNOMIC Publishing Co., Inc., Westport, Connecticut, 1971.
      pp. 145-155.

(11) Benton, J. L. and C. A. Brighton. "Vinyl Chloride Polymers (Polym-
      erization)." Encyclopedia of Polymer Science and Technology, Vol.
      14. John Wiley and Sons, New York, New York, 1971. pp. 320-345.

(12) Bevington, J. C. and H. May. "Aldehyde Polymers." <u>Encyclopedia of
      Polymer Science and Technology</u>, Volume 1, Norman G. Gaylord (ed).
      Interscience Publishers, New York, New York. pp. 609-628.

(13)   Blais, John F.   _Amino Resins_.   Reinhold Publishing Corporation, New York, New York, 1959.

(14)   Boeke, P. J.   "Polyphenylene Sulfide."   _Modern Plastics Encyclopedia_. McGraw-Hill, New York, New York, 1981-1982.   p. 78.

(15)   Boenig, H. V.   "Unsaturated Polyesters."   _Encyclopedia of Polymer Science and Technology_, Volume 11, Norbert M. Bikales (ed).   Interscience Publishers, New York, New York.   pp. 129-168.

(16)   Bottenbruch, L.   "Polycarbonates."   _Encyclopedia of Polymer Science and Technology_, Volume 10, Norbert M. Bikales (ed).   Interscience Publishers, New York, New York.   pp. 710-761.

(17)   Boyer, Raymond F., Henno Keskkula and Alan Platt.   "Styrene Polymers."   _Encyclopedia of Polymer Science and Technology_, Volume 13, Norbert M. Bikales (ed).   Interscience Publishers, New York, New York.   pp. 128-440.

(18)   Brighton, C. A., G. Pritchard and G. A. Skinner.   _Styrene Polymers: Technology and Environmental Aspects_.   1979.

(19)   Bringer, Robert P.   "Fluorine-Containing Polymers: Chlorotrifluoro-ethylene Polymers."   _Encyclopedia of Polymer Science and Technology_, Volume 7, Norbert M. Bikales (ed).   Interscience Publishers, New York, New York.   pp. 204-219.

(20)   _British Journal of Industrial Medicine_, Volume 23, 1966, p. 318.

(21)   _British Medical Journal_, Volume 2, 1970, p. 406.

(22)   Brode, George L.   "Phenolic Resins."   _Kirk-Othmer Encyclopedia of Chemical Technology_, 2nd Edition, Volume 7, Martin Grayson (ed). John Wiley and Sons, New York, New York, 1982.   pp. 384-416.

(23)   Brown, Henry C.   "Fluorine-Containing Polymers: Survey."   _Encyclopedia of Polymer Science and Technology_, Volume 7, Norbert H. Bikales (ed).   Interscience Publishers, New York, New York.   pp. 179-204.

(24)   Brown, W. E.   "Vinylidene Chloride Polymers and Copolymers."   _Modern Plastics Encyclopedia_.   McGraw-Hill, New York, New York, 1981-1982. pp. 119-120.

(25)   Buckley, D. J.   "Butylene Polymers."   _Encyclopedia of Polymer Science and Technology_, Volume 2, Norbert M. Bikales (ed).   Interscience Publishers, New York, New York.   pp. 754-795.

(26)   Buist, J. M. and H. Gudgeon.   _Advances in Polyurethane Technology_. John Wiley and Sons, New York, New York, 1968.

(27) Cameron, J. B., A. J. Lundeen, and J. H. McCulley, Jr. "Trends in Suspension PVC Manufacture." _Hydrocarbon Processing_, March 1980, p. 39.

(28) Campbell, P. E. and R. V. Jones. "High-Molecular-Weight Ethylene Polymers: Prepared by the Phillips Process." _Encyclopedia of Polymer Science and Technology_, Volume 6, Norbert M. Bikales (ed). Interscience Publishers, New York, New York. pp. 332-336.

(29) Cantow, Manfred J. R. "Vinyl Polymers (Chloride)." _Kirk-Othmer Encyclopedia of Chemical Technology_, 2nd Edition, Volume 21, Anthony Standen (ed). John Wiley and Sons, New York, New York, 1968. pp. 369-412.

(30) Carhart, R.O., G. P. Diehl, and J. E. Kochanowski. "Polycarbonate." _Modern Plastics Encyclopedia_. McGraw-Hill, New York, New York, 1981-1982. pp. 46.

(31) Cassarett, Louis and Doull. _Toxicology_, 2nd Edition. McMillan, New York, New York, 1980. pp. 475, 534.

(32) Chatterjee, A. M. "Polybutylene." _Modern Plastics Encyclopedia_. McGraw-Hill, New York, New York, 1981-1982. p. 44.

(33) _Chemcyclopedia 1985_. J. H. Kuney, ed. American Chemical Society © 1984. p. 239.

(34) "Chementator." _Chemical Engineering_, November 21, 1977, p. 111.

(35) _Chemical Economics Handbook_, SRI International, updated annually.

(36) _Chemico-Biological Interactions_, Volume 6, 1973, p. 375.

(37) _Chemistry and Industry_, Volume 11, 1976, p. 463.

(38) Cipriani, Cipriano and Charles A. Trischman, Jr. "New Catalyst Controls LLDPE's Particle Geometry." _Chemical Engineering_, May 17, 1982, p. 66.

(39) Cipriani, Cipriano and Charles A. Trischman, Jr. "New Catalyst Cuts Polypropylene Costs and Energy Requirements." _Chemical Engineering_, April 20, 1981, p. 80.

(40) Cohan, G. F. "Polyvinyl and Vinyl Copolymers." _Modern Plastics Encyclopedia_. Mc-Graw-Hill, New York, New York, 1981-1982. pp. 100-108.

(41) Concha, Mario. "Linear Low Density Polyethylene." _Modern Plastics Encyclopedia_. McGraw-Hill, New York, New York, 1981-1982. p. 66.

(42) *Condensed Chemical Dictionary*, 9th Edition, Gessner G. Hawley (ed). Van Nostrand Reinhold, New York, New York, 1977. pp. 546, 889, 917.

(43) Conference Board. *Energy Consumption in Manufacturing*. Ballinger Publishing Co., Cambridge, Massachusetts, 1974.

(44) Connolly, E. M. and E. Helmes. "Polyimide Resins." *Chemical Economics Handbook*. Stanford Research Institute, Menlo Park, California, April 1984.

(45) Cordts, H. P. and J. A. Bauer. "Unsaturated Polyester." *Modern Plastics Encyclopedia*. McGraw-Hill, New York, New York, 1981-1982. p. 54.

(46) Crespi, Giovanni and Luciano Luciani. "Olefin Polymers: Polypropylene." *Kirk-Othmer Encyclopedia of Chemical Technology*, 3rd Edition, Volume 16, Martin Grayson (ed). John Wiley and Sons, New York, New York, 1981. pp. 453-469.

(47) Crull, Anna W. *Polyurethane and Other Foams*. Business Communications, Co., Inc., Stamford, Connecticut, 1979.

(48) Crull, Anna W. *Polyurethanes: Developments, Processes*. Business Communications Co., Inc., Stamford, Connecticut, 1978.

(49) "Cut Polyester Costs – Use PO." *Hydrocarbon Processing*, December 1977, p. 115.

(50) Cystic Research Center. *Cystic Monograph No. 2, Polyester Handbook*. Wellingborough, England, 1977.

(51) Davis, John C. "LDPE Goes Low Pressure." *Chemical Engineering*, January 2, 1978, p. 25.

(52) Deichmann, W. B. *Toxicology of Drugs and Chemicals*. Academic Press, New York, New York, 1969. pp. 75, 519.

(53) deLeeuw, P. W., C. R. Lindegren, and R. F. Schimber. "Polybutylene: Market Status and Outlook." *Chemical Engineering Progress*, January 1980, p. 57.

(54) Denny, Richard G. "Polyvinyl Acetate and Polyvinyl Alcohol." *Process Economics Program*, Report No. 57. Stanford Research Institute, Menlo Park, California, 1970.

(55) Di Drusco, G. and R. Rinaldi. "High Yield Polypropylene." *Hydrocarbon Processing*, May 1981, p. 153.

(56) *Directory of Chemical Producers*. SRI International, 1982.

(57) Doyle, E. N. _The Development and Use of Polyurethane Products_.
      McGraw-Hill, New York, New York, 1971.

(58) _Drug Intelligence and Chemical Pharmacy_, Volume 15, 1981, p. 384.

(59) Edwards, F. G. and R. A. Wessling. "Vinylidene Chloride Polymers."
      Encyclopedia of Polymer Science and Technology. John Wiley and Sons,
      New York, New York, 1971. pp. 540-577.

(60) Emmerich, Anthony. "Processing the 'Linear' LDPE's: There Are Some
      Differences." _Plastics Technology_, February 1980, p. 33.

(61) _Encyclopedia of PVC_, Leonard Nass (ed). M. Dekker, New York, New
      York, 1976-1977.

(62) _Epoxy Resin Technology_, Paul F. Bruins (ed). John Wiley and Sons,
      New York, New York, 1968.

(63) _Epoxy Resin Technology: Developments Since 1979_, J. I. DiStasio
      (ed). Noyes Data Corporation, Park Ridge, New Jersey, 1982.

(64) Evans, T. E. "Styrene-Acrylonitrile." _Modern Plastics Encyclopedia_.
      McGraw-Hill, New York, New York, 1981-1982. pp. 110-112.

(65) _Facts & Figures of the Plastics Industry_, 1981 Edition. The Society
      of the Plastics Industry, Inc., New York.

(66) Farber, Elliott. "Suspension Polymerization." _Encyclopedia of
      Polymer Science and Technology_, Volume 18, Norbert M. Bikales (ed).
      Interscience Publishers, New York, New York. pp. 552-571.

(67) _Federal Register_, Volume 39, 1974, p. 23540.

(68) _Federal Register_, Volume 40, 1975, p. 23073.

(69) _Federal Register_, Volume 43, 1979, p. 45809, 57601.

(70) Fisher, Nicolas and Louis Goiran. "Trends in Mass PVC." _Hydrocarbon
      Processing_, May 1981, p. 143.

(71) Floryan, D. G. and I. W. Serfaty. "Polyetherimide: More Information
      on a New High-Performance Resin." _Modern Plastics_, June 1982.
      pp. 146-151.

(72) _Fluoropolymers_, Leo A. Wall (ed). John Wiley and Sons, New York, New
      York, 1972.

(73) Fong, Wing Sien. "Polyacrylates." _Process Economics Program_.
      Report No. 65, Standford Research Institute, Menlow Park, California,
      1970.

(74)   "For ABS Resins, Some Good News and Some Bad." _Chemical Week_, September 9, 1981, p. 77.

(75)   Fox, D. W. "Polycarbonates." _Kirk-Othmer Encyclopedia of Chemical Technology_, 3rd Edition, Volume 18, Martin Grayson (ed). John Wiley and Sons, New York, New York, 1982. pp. 479-494.

(76)   Gaines, L. L. and S. Y. Shen. _Energy and Materials Flows in the Production of Olefins and Their Derivatives_. Argonne National Laboratory, August 1980.

(77)   Gangal, S. V. "Fluorine Compounds, Organic: Polytetrafluoroethylene." _Kirk-Othmer Encyclopedia of Chemical Technology_, 3rd Edition, Volume 11, Martin Grayson (ed). John Wiley and Sons, New York, New York. pp. 1-24.

(78)   Garner, Rubin J. _Veterinary Toxicology_. Williams and Wilkins, Baltimore, Maryland, 1967. p. 139.

(79)   Gmitter, George T. and E. M. Maxey. "One-Shot Slab Polyether Urethane." _Polyurethane Technology_, Paul F. Bruins (ed). John Wiley and Sons, New York, New York, 1969. pp. 39-65.

(80)   Goodman, I. "Polyesters." _Encyclopedia of Polymer Science and Technology_, Volume 11, Norbert M. Bikales (ed). Interscience Publishers, New York, New York. pp. 62-128.

(81)   Goodman, I. "Polyesters, Thermoplastics." _Encyclopedia of Polymer Science and Technology Supplement_, Vol. 2. Interscience Publishers, New York, New York, 1976. pp. 447-448.

(82)   Goodman, I. "Polyesters." _Kirk-Othmer Encyclopedia of Chemical Technology_, Volume 16, Anthony Standen (ed). John Wiley and Sons, New York, New York. pp. 159-189.

(83)   Goodman, Louis S. _Pharmacological Basis of Therapeutics_, 5th Edition. MacMillian Publishers, New York, New York 1975. p. 904.

(84)   Gordian Associates. _Industrial International Data Base: The Plastics Industry_. Technical Information Center, Oak Ridge, Tennessee, September 1977.

(85)   Gosselin, R., et. al. _Chemical Toxicology of Commercial Products_, 4th Edition. Williams and Wilkins, Baltimore, Maryland, 1976. pp. 139, II-75, 77, 79, 92, 97, 100, 112, 126-128, 133, 245, III-4, 223.

(86)   Graff, Gordon M. "Engineering Plastics: Primed for Key CPI Role." _Chemical Engineering_, August 23, 1982, p. 42.

(87)   Grant, W. Norton. _Toxicology of the Eye_. C. C. Thomas, 1974. pp. 425, 849.

(88) Hamilton, Alice and Harriet L. Hardy. Industrial Toxicology, 3rd
      Edition. PSG Publishers, 1974. pp. 87, 216, 341.

(89) Harker, L. A., et. al. New Approaches to Reducing Cardiovascular
      Death. International Symposium 59, 1979.

(90) Hattori, Kiyo. "Polypropylene." Modern Plastics Encyclopedia.
      McGraw-Hill, New York, New York, 1981-1982. pp. 80-89.

(91) Hawley, G. G., ed. The Condensed Chemical Dictionary, 10th Edition.
      Van Nostrand Reinhold Company, New York, New York, © 1981.
      pp. 813-814, 1051.

(92) Hay, A. S., P. Shenian, A. C. Gowan, P. F. Erhardt, W. R. Haaf, and
      J. E. Theberge. "Phenols, Oxidative Polymerization." Encyclopedia
      of Polymer Science and Technology, Volume 10, Norbert M. Bikales
      (ed). Interscience Publishers, New York, New York. pp. 92-111.

(93) Hedley, W. H., et. al. Potential Pollutants from Petro chemical
      Processes, Final Report, Contract 68-02-0226, Task 9, MRC-DA-406.
      Monsanto Research Corporation, Dayton, Lab., Dayton, Ohio, December
      1973.

(94) Hill, H. Wayne Jr. and D. G. Brady. "Polymers Containing Sulfur:
      Poly(phenylene sulfide)." Kirk-Othmer Encyclopedia of Chemical
      Technology, 3rd Edition, Volume 18, Martin Grayson (ed). John Wiley
      and Sons, New York, New York. pp. 793-814.

(95) Hoff, Melvern C., Un K. Im, William F. Hauschildt, and Imre Puskas.
      "Butylenes." Kirk-Othmer Encyclopedia of Chemical Technology, 3rd
      Edition, Volume 4, Martin Grayson (ed). John Wiley and Sons, New
      York, New York. pp. 346-375.

(96) Hygiene and Sanitation (English Translation of Gigiene Sanitariya),
      Volume 36, 1971, p. 178.

(97) Imhausen, Karl-Heinz, Fritz Schoffel, Jurgen Zink, Wolfgang Payer,
      and Wilhelm Zoller. "Build Larger LDPE Plants." Hydrocarbon
      Processing, November 1976, p. 155.

(98) Industrial Medicine, Volume 17, 1948, p. 199.

(99) International Agency for Research on Cancer. IRAC Monographs on the
      Evaluation of Carcinogenic Risk of Chemicals to Man, Volume 2, 1973.
      p. 74.

(100) International Agency for Research on Cancer. IRAC Monographs on the
      Evaluation of Carcinogenic Risk of Chemicals to Man, Volume 4, 1974.
      p. 137.

(101)   International Agency for Research on Cancer. IARC Monographs on the
        Evaluation of Carcinogenic Risk of Chemicals to Man, Volume 7, 1977.
        pp. 95, 203, 223, 227.

(102)   International Agency for Research on Cancer. IARC Monographs on the
        Evaluation of Carcinogenic Risk of Chemicals to Man, Volume 9, 1975.
        p. 37.

(103)   International Agency for Research on Cancer. IRAC Monographs on the
        Evaluation of Carcinogenic Risk of Chemicals to Man, Volume 11, 1976.
        pp. 131, 141, 147.

(104)   International Agency for Research on Cancer. IRAC Monographs on the
        Evaluation of Carcinogenic Risk of Chemicals to Man, Volume 14, 1977.
        p. 1.

(105)   International Agency for Research on Cancer. IRAC Monographs on the
        Evaluation of Carcinogenic Risk of Chemicals to Man, Volume 15, 1977.
        pp. 155, 255.

(106)   International Agency for Research on Cancer. IRAC Monographs on the
        Evaluation of Carcinogenic Risk of Chemicals to Man, Volume 16, 1978.
        pp. 221, 233.

(107)   International Agency for Research on Cancer. IARC Monographs on the
        Evaluation of Carcinogenic Risk of Chemicals to Man, Volume 19, 1979.
        pp. 52, 57, 73, 131, 213, 231, 244, 285, 296, 297, 303, 341, 377,
        439.

(108)   International Agency for Research on Cancer. IARC Monographs on the
        Evaluation of Carcinogenic Risk of Chemicals to Man, Volume 20, 1979.
        pp. 371, 401, 415, 428, 545, 429.

(109)   International Agency for Research on Cancer. IRAC Monographs on the
        Evaluation of Carcinogenic Risk of Chemicals to Man, Volume 23, 1980.
        pp. 205, 325.

(110)   Jones, Robert W. and K. T. Chandy. "Synthetic Plastics." Riegel's
        Handbook of Industrial Chemicals, 7th Edition, James A. Kent (ed).
        Van Nostrand Reinhold, New York, New York, 1974. p. 264.

(111)   Journal of Pediatrics, Volume 94, 1979. p. 147.

(112)   Kamath, Vasanth R. "New Vinyl Chloride Initiators Improve PVC Heat
        Stability." Modern Plastics, February 1981, p. 54.

(113)   Kanner, B. and B. Prokai. "Silicone Surfactants for Urethane Foam."
        Advances in Urethane Science and Technology, Volume 2, K. C. Frisch
        and S. L. Reegen (eds). TECHNOMIC Publishing Co., Inc., Westport,
        Connecticut, 1973. pp. 221-240.

(114)   Kaplan, M. "Chemical and Mechanical Factors Affecting One-Shot Rigid
        Urethane Foam." Polyurethane Technology, Paul F. Bruins (ed). John
        Wiley and Sons, New York, New York, 1969. pp. 77-93.

(115)   Kenson, R. E. and R. O. Hoffland. "Control of Toxic Air Emissions in
        Chemical Manufacture." Chemical Engineering Progress, February 1980,
        p. 80.

(116)   Kent, J. A., ed. Riegel's Handbook of Industrial Chemistry.
        Van Nostrand Reinhold, New York, New York, 1983.

(117)   Keskkula, Henno, Alan E. Platt and Raymond F. Boyer. "Styrene
        Plastics." Kirk-Othmer Encyclopedia of Chemical Technology, 2nd
        Edition, Volume 19, Anthony Standen (ed). John Wiley and Sons, New
        York, New York. pp. 85-134.

(118)   Keutgen, W. A. "Phenolic Resins." Encyclopedia of Polymer Science
        and Technology, Volume 10, Norbert M. Bikales (ed). Interscience
        Publishers, New York, New York. pp. 1-73.

(119)   "Key Polymers." Chemical and Engineering News, August 30, 1982.
        p. 11.

(120)   Kine, Benjamin B. and Ronald W. Novak. "Acrylic Ester Polymers:
        Survey." Kirk-Othmer Encyclopedia of Chemical Technology, 3rd
        Edition, Volume 1, Martin Grayson (ed). John Wiley and Sons, New
        York, New York. pp. 386-408.

(121)   Kine, Benjamin B. and R. W. Novak. "Methacrylic Polymers." Kirk-
        Othmer Encyclopedia of Chemical Technology, 3rd Edition, Volume 15,
        Martin Grayson (ed). John Wiley and Sons, New York, New York.
        pp. 377-398.

(122)   Kirshenbaum, G. S. and J. M. Rhodes. "Thermoplastic Polyester: PET."
        Modern Plastics Encyclopedia. McGraw-Hill, New York, New York,
        1981-1982. pp. 50-51.

(123)   Knop, Andre and W. Scheib. Chemistry and Application of Phenolic
        Resins. Springer-Vorlag, 1979. pp. 60-66.

(124)   Kochhar, R. K., Yuri V. Kissin, and David L. Beach. "Olefin Poly-
        mers: Polymers of Higher Olefins." Kirk-Othmer Encyclopedia of
        Chemical Technology, 3rd Edition, Volume 16, Martin Grayson (ed).
        John Wiley and Sons, New York, New York. pp. 470-479.

(125)   Kohan, Melvin I. Nylon Plastics. John Wiley and Sons,
        New York, New York, 1973.

(126)   Labor Hygiene and Occupational Diseases, Volume 24, 1980, p. 28.

(127) _Lancet_, Volume 1, 1981, p. 1099.

(128) Lanson, H. J. "Alkyd Resins." _Kirk-Othmer Encyclopedia of Chemical Technology_, 3rd Edition, Volume 2, Martin Grayson (ed). John Wiley and Sons, New York, New York. pp. 18-24.

(129) "A Leap Ahead in Polyethylene Technology." _Chemical & Engineering News_, December 5, 1977, p. 21.

(130) Lebovits, A. "Acrylonitrile Polymers: Acrylonitrile-Butadiene-Styrene Copolymers." _Encyclopedia of Polymer Science and Technology_, Volume 1, Norman G. Gaylord (ed). Interscience Publishers, New York, New York. pp. 436-444.

(131) Lee, H. and K. Neville. "Epoxy Resins." _Encyclopedia of Polymer Science and Technology_, Volume 6, Norbert M. Bikales (ed). Interscience Publishers, New York, New York. pp. 209-269.

(132) Lee, Henry and Kris Neville. _Handbook of Epoxy Resins_. McGraw-Hill, New York, New York, 1967.

(133) Leeds, Morton. "Poly(vinyl alcohol)." _Kirk-Othmer Encyclopedia of Chemical Technology_, 2nd Edition, Volume 21, Anthony Standen (ed). John Wiley and Sons, New York, New York. pp. 353-366.

(134) Lefaux. _Practical Toxicology of Plastics_. 1968. p. 108.

(135) Lenz, Robert. "Polymerization Mechanisms and Processes." _Kirk-Othmer Encyclopedia of Chemical Technology_, 2nd Edition, Volume 16, Anthony Standen (ed). John Wiley and Sons, New York, New York. pp. 219-242.

(136) Lichtenberg, D. W. "Amino." _Modern Plastics Encyclopedia_. McGraw-Hill, New York, New York, 1981-1982. pp. 12-14.

(137) Lindemann, Martin K. "Vinyl Alcohol Polymers: Polyvinyl Alcohol." _Encyclopedia of Polymer Science and Technology_, Volume 14, Norbert M. Bikales (ed). Interscience Publishers, New York, New York. pp 149-208.

(138) Lindemann, Martin K. "Vinyl Ester Polymers: Vinyl Acetate Polymers." _Encyclopedia of Polymer Science and Technology_, Volume 15, Norbert M. Bikales (ed). Interscience Publishers, New York, New York. pp. 577-677.

(139) Loncrini, D. F. "Aromatic Polyesterimides." Journal of Polymer Science: Part A-1, Vol. 4, 1966. pp. 1531-1541.

(140)   Luskin, Leo S. and Robert J. Myers. "Acrylic Ester Polymers: Sur-
        vey." Encyclopedia of Polymer Science and Technology, Volume 1,
        Norman G. Gaylord (ed). Inter-Science Publishers, New York, New
        York. pp. 246-327.

(141)   Martino, Robert. "How Crisis in Resin Business Affects Users of PP
        Materials." Modern Plastics, April 1981, p. 48.

(142)   Martino, Robert. "Polypropylene: Enough Vigor for Resin Shortage?"
        Modern Plastics, May 1982, p. 58.

(143)   "Materials 1982." Modern Plastics. January 1982, p. 55-87.

(144)   Matsuyama, Kiyoshi, Akinobu Shiga, Masahiro Kakugo, and Hitoshi
        Hashimoto. "Use BPP for Polypropylene." Hydrocarbon Processing,
        November 1980, p. 131.

(145)   McCane, Donald I. "Tetrafluoroethylene Polymers." Encyclopedia of
        Polymer Science and Technology, Volume 13, Norbert M. Bikales (ed).
        Interscience Publishers, New York, New York. pp. 523-670.

(146)   McCrann, L. J. "Polystyrene." Modern Plastics Encyclopedia, Joan
        Agranoff (ed). McGraw-Hill, New York, New York. pp. 90-94.

(147)   McCurdy, J. L. "Mass Polymerization of Styrene: Route to Economy
        and Efficiency." Modern Plastics, September 1979, p. 97.

(148)   Merck Index, 8th Edition. Merck and Co., Inc., 1968. p. 506.

(149)   Merck Index, 9th Edition. Merck and Co., Inc., 1976. pp. 471, 495,
        741, 786, 941, 986.

(150)   Modern Plastics. "GE Adds a High-Heat Polyetherimide to Its Engi-
        neering Thermoplastics Lineup." March 1982. pp. 16-18.

(151)   Moran, Bob. "Modified Phenylene Oxide." Modern Plastics Encyclo-
        pedia. McGraw-Hill, New York, New York, 1981-1982. pp. 30-32.

(152)   "More Details on a Polypropylene Process." Chemical Week, February
        25, 1981, p. 49.

(153)   Morneau, G. A., W. A. Pavelich, L. G. Roettger. "Acrylonitrile
        Polymers." Kirk-Othmer Encyclopedia of Chemical Technology, 3rd
        Edition, Volume 1, Martin Grayson (ed). John Wiley and Sons, New
        York, New York. pp. 427-456.

(154)   Mraz, Richard G. and Raymond P. Silver. "Alkyd Resins." Encyclo-
        pedia of Polymer Science and Technology, Volume 1, Norman G. Gaylord
        (ed). Interscience Publications, New York, New York. pp. 663-734.

(155) <u>Mutation Research</u>, Volume 17, 1980, p. 109.

(156) National Clearinghouse for Poison Control Centers. <u>Bulletin</u>. U.S.
       Department of Health, Education and Welfare, January/February 1969.

(157) National Fire Protection Association. <u>Fire Protection Guide to
       Hazardous Materials</u>. 1978. pp. 35, 55, 94, 122.

(158) National Institute for Occupational Safety and Health. <u>Criteria
       Document for Occupational Exposure to Inorganic Mercury</u>. U.S.
       Department of Health, Education, and Welfare, NIOSH 73-11-024, 1973.

(159) Nelson, W. E. <u>Nylon Plastics Technology</u>. Newnes-Butterworths,
       Published for the Plastics and Rubber Institute, London, 1976.

(160) "New Materials: New Asbestos-Free Phenolics for Metal Replacement."
       <u>Modern Plastics</u>, April 1979, p. 98.

(161) "New Materials: Resins & Compounds: Improved Properties Are Offered
       in New PE and PP Materials." <u>Modern Plastics</u>, May 1981, p. 80.

(162) "New PP Process Offers Energy Savings as Lure." <u>Chemical Week</u>, July
       9, 1980, p. 30.

(163) "Now - A High-Heat, Impact-Modified Phenolic for Injection Molding."
       <u>Plastics Technology</u>, March 1980, p. 29.

(164) Nowak, R. M. and L. C. Rubens. "Unsaturated Polyesters." <u>Kirk-
       Othmer Encyclopedia of Chemical Technology</u>, 2nd Edition, Volume 20,
       Anthony Standen (ed). John Wiley and Sons. pp. 791-839.

(165) Nurse, R. H. "HMW - High Density Polyethylene." <u>Modern Plastics
       Encyclopedia</u>. McGraw-Hill, New York, New York, 1981-1982. p. 70.

(166) O'Sullivan, Dermont A. "Polypropylene Catalyst: A Peek From
       Montedison." <u>Chemical and Engineering News</u>, May 2, 1977, p. 28.

(167) Parkyn, Brian, F. Lamb, and B. V. Clifton. <u>Polyesters</u>, Volume 2,
       Unsaturated Polyesters and Polyester Plasticizers. The Plastics
       Institute, London, England, 1967. pp. 22-28 and 60-67.

(168) Paschke, Eberhard. "Olefin Polymers." <u>Kirk-Othmer Encyclopedia of
       Chemical Technology</u>, 3rd Edition, Volume 16, Martin Grayson (ed).
       John Wiley and Sons, New York, New York. pp. 385-452.

(169) Paschke, Ed. "The Outlook for High Density Polyethylene."
       <u>Chemical Engineering Progress</u>, January 1980, p. 74.

(170) Patty, F. A. <u>Industrial Hygiene and Toxicology</u>, 2nd Edition, Volume
       2. Wiley-Interscience, 1958. pp. 847, 848, 1281, 1283, 1386, 1610,
       1794, 1832.

(171) Peerman, D. E. "Polyamides from Fatty Acids." Encyclopedia of
       Polymer Science and Technology, Volume 10, Norbert M. Bikales (ed).
       Interscience Publishers, New York, New York. pp. 597-614.

(172) Penn, W. S. PVC Technology, 3rd Edition. John Wiley and Sons, New
       York, New York, 1972.

(173) Persak, K. J. and L. M. Blair. "Acetal Resins." Kirk-Othmer
       Encyclopedia of Chemical Technology, 3rd Edition, Volume 1, Martin
       Grayson (ed). John Wiley and Sons, New York, New York. pp. 112-123.

(174) Persak, K. J. and S. I. Wilson. "Acetal Homopolymer." Modern
       Plastics Encyclopedia, McGraw-Hill, New York, New York, 1981-1982.
       pp. 5-7.

(175) "PET Bottle Capacity Will Expand Again." Chemical Week, August 13,
       1980, p. 18.

(176) "PET Bottle Market Healthy Here, Abroad." Chemical Week, June 11,
       1980, p. 18.

(177) "Phthalates are alive but being watched." Chemical Week, June 24,
       1981, p. 18.

(178) Pigott, K. A. "Polyurethans." Encyclopedia of Polymer Science and
       Technology, Volume 11, Norbert M. Bikales (ed). Interscience
       Publishers, New York, New York. pp. 506-561.

(179) Pigott, K. A. "Urethan Polymers." Kirk-Othmer Encyclopedia of
       Chemical Technology, 2nd Edition, Volume 21, Anthony Standen (ed).
       John Wiley and Sons, New York, New York. pp. 56-106.

(180) "Polyacetal Resins - Euteco Impianti S.p.A." Hydrocarbon Processing,
       November 1981, p. 202.

(181) Polycarbonate - Montedison S.p.A." Hydrocarbon Processing, November
       1979, p. 216.

(182) "Polyester - Hoechst, AG." Hydrocarbon Processing, November 1977,
       p. 202.

(183) "Polyesters - Inventa AG." Hydrocarbon Processing, November 1981,
       p. 206.

(184) "Polyethylene--Phillips Petroleum Co." Hydrocarbon Processing,
       November 1981, p. 213.

(185) "Polyethylene (HD)--Chemische Werke Huels AG." Hydrocarbon Proces-
       sing, November 1981, p. 209.

(186) "Polyethylene (HD)--Hoechst AG." Hydrocarbon Processing, November
       1981, p. 211.

(187) "Polyethylene (HD)--Montedison S.p.A." Hydrocarbon Processing,
       November 1979, p. 224.

(188) "Polyethylene (HD)--Napthachimie." Hydrocarbon Processing, November
       1979, p. 225.

(189) "Polyethylene (HD)." Hydrocarbon Processing, November 1983.

(190) "Polyethylene (LD, HD - Unipol Process)--Union Carbide Corp." Hydro-
       carbon Processing, November 1979, p. 227.

(191) "Polyethylene (LD)--ANIC." Hydrocarbon Processing, November 1981,
       p. 207.

(192) "Polyethylene (LD)--ARCO Technology, Inc." Hydrocarbon Processing,
       November 1981, p. 208.

(193) "Polyethylene (LD)--Ato Chemie." Hydrocarbon Processing, November
       1979, p. 220.

(194) "Polyethylene (LD)--El Paso Polyolefins Co." Hydrocarbon Processing,
       November 1981, p. 210.

(195) "Polyethylene (LD)--Gulf Chemicals Co." Hydrocarbon Processing,
       November 1979, p. 222.

(196) "Polyethylene (LD)--Imhausen International Co." Hydrocarbon Proces-
       sing, November 1981, p. 212.

(197) "Polyethylene (HD, MD, LLDPE)." Hydrocarbon Processing, November
       1983.

(198) "Polyethylene (LLD)." Hydrocarbon Processing, November 1983.

(199) "Polypropylene - El Paso Polyolefins Co." Hydrocarbon Processing,
       November 1981, p. 214.

(200) "Polypropylene--Hoechst AG." Hydrocarbon Processing, November 1977,
       p. 213.

(201) "Polypropylene." Hydrocarbon Processing, November 1983.

(202) "Polypropylene--Mitsui Petrochemical Industries, Ltd. and Montedison
       S.p.A." Hydrocarbon Processing, November 1979, p. 228.

(203) "Polypropylene--Mitsui Toatsu Chemicals, Inc." Hydrocarbon Proces-
       sing, November 1979, p. 229.

(204) "A Polypropylene Route Revamped by a New High-Yield Catalyst Was
       Detailed..." Chemical Engineering, March 9, 1981, p. 18.

(205) "Polypropylene--Sumitomo Chemical Co., Ltd." Hydrocarbon Processing,
       November 1979, p. 231.

(206) "Polystyrene--Arco Technology, Inc." Hydrocarbon Processing,
       November 1979, p. 232.

(207) "Polystyrene--Ato Chemie." Hydrocarbon Processing, November 1977,
       p. 216.

(208) "Polystyrene--Cosden Technology, Inc." Hydrocarbon Processing,
       November 1981, p. 219.

(209) "Polystyrene--Gulf Oil Chemicals Co." Hydrocarbon Processing,
       November 1979, p. 234.

(210) "Polystyrene--Mitsui Toatsu Chem., Inc." Hydrocarbon Processing,
       November 1971, p. 202.

(211) "Polystyrene - Societe Chimique des Charbonnages (CdF) - The Badger
       Co., Inc." Hydrocarbon Processing, November 1981, p. 218.

(212) Polyvinyl Alcohol: Properties and Applications, C. A. Finch (ed).
       John Wiley and Sons, London, 1973.

(213) "Polyvinyl Alcohol (PVA Thiokol)--Halcon SD Group, Inc." Hydrocarbon
       Processing, November 1981, p. 220.

(214) "Polyvinylchloride--Ato Chemie." Hydrocarbon Processing, November
       1979, p. 236.

(215) "Polyvinylchloride--Hoechst AG." Hydrocarbon Processing, November
       1981, p. 221.

(216) "Polyvinylchloride--Lonza, Ltd." Hydrocarbon Processing, November
       1979, p. 237.

(217) "Polyvinylchloride--Mitsui Toatsu Chemicals, Inc." Hydrocarbon
       Processing, November 1981, p. 222.

(218) "Polyvinylchloride--Rhone-Poulenc." Hydrocarbon Processing, November
       1979, p. 238.

(219) Potter, W. G. Epoxide Resins. Iliffe Books, (Published for the
       Plastics Institute, London, 1970.

(220) Potter, W. G. Uses of Epoxy Resins. Newnes-Butterworths, London,
       1975.

(221) Preston, F. J. "Polyurethane." Modern Plastics Encyclopedia.
       McGraw-Hill, New York, New York, 1981-1982. pp. 94-100.

(222) Pritchard, J. G. Polyvinyl Alcohol: Basic Properties and Uses.
       Gordon and Breach Science Publishers, London, 1970.

(223) "Process Engineering News: Four New PP Resins Arrive for Injection
       and Blow Molding." Plastics Technology, October 1980, p. 22.

(224) "Process Engineering News: New Line of PE, PP Film Resins Intro-
       duced." Plastics Technology, October 1979, p. 15.

(225) "Propylene Catalyst: A Peek from Montedison." Chemical and Engi-
       neering News, May 2, 1977.

(226) Prout, E. O., "UHMW Polyethylene." Modern Plastics Encyclopedia.
       McGraw-Hill, New York, New York, 1981-1982. pp. 70-72.

(227) Putscher, Richard E. "Polyamides: General." Kirk-Othmer Encyclo-
       pedia of Chemical Technology, 3rd Edition, Volume 18, Martin Grayson
       (ed). John Wiley and Sons, New York, New York. pp. 328-371.

(228) Raff, R. A. V. "Polyethylene." Encyclopedia of Polymer Science and
       Technology, Volume 6, Norbert M. Bikales (ed). Interscience
       Publishers, New York, New York. pp. 275-332.

(229) Ranney, M. William. Epoxy Resins and Products: Recent Advances.
       Noyes Data Corporation, Park Ridge, New Jersey, 1977.

(230) Ranney, M. W., Dr. Polyimide Manufacture 1971. Noyes Data
       Corporation, Park Ridge, New Jersey. pp. 57-77.

(231) Reed, George V. "Improving PP with Phosphite Stabilizers." Modern
       Plastics, November 1979, p. 76.

(232) Registry of Toxic Effects of Chemical Substances, 1979 Edition. U.S.
       Department of Health and Human Services, Public Health Service,
       Center for Disease Control, National Institute for Occupational
       Safety and Health, Cincinnati, Ohio, September 1980.

(233) Registry of Toxic Effects of Chemical Substances, On-Line File. U.S.
       Department of Health and Human Services, Public Health Service,
       Center for Disease Control, National Instituent for Occupational
       Safety and Health, Cincinnati, Ohio, updated monthly. Data
       retrieved September 1982.

(234) Rhum, David. "Vinyl Polymers: Poly(vinyl) Acetate." Kirk-Othmer
       Encyclopedia of Chemical Technology, 2nd Edition, Volume 21, Anthony
       Standen (ed). John Wiley and Sons, New York, New York. pp. 317-353.

(235) Riley, J. B. "Polyethylene and Ethylene Copolymers." _Modern Plastics Encyclopedia_. McGraw-Hill, New York, New York, 1981-1982. pp. 58-66.

(236) Rodriguez, Ferdinand. _Principles of Polymer Science_, 2nd ed. McGraw-Hill, New York, New York, 1982. pp. 462-463.

(237) Rodriguez, F. _Principles of Polymer Systems_. McGraw-Hill, New York, New York, 1970.

(238) Rumack, Barry H. _Poisindex_. Micromedics, Englewood, Colorado.

(239) Sager, Fredrik. "Cut Polymer Costs with Cyclones." _Hydrocarbon Processing_, February 1978, p. 119.

(240) Sailors, H. R. and J. P. Hogan. "History of Polyolefins." _Journal of Macromolecular Science - Chemistry_, A15(7), 1981, p. 1377.

(241) Sax, N. Irving. _Dangerous Properties of Industrial Materials_, 3rd Edition. Van Nostrand Reinhold, New York, New York. p. 535.

(242) Sax, N. Irving. _Dangerous Properties of Industrial Materials_, 4th Edition. Van Nostrand Reinhold, New York, New York. pp. 370, 413, 440, 506, 539, 582, 610, 671, 680, 717, 725, 776, 780, 816, 883, 908, 926, 1020, 1071, 1151, 1235.

(243) Sax, N. Irving. _Dangerous Properties of Industrial Materials_, 5th Edition. Van Nostrand Reinhold, New York, New York. pp. 447, 584, 792, 882, 1050.

(244) Sayad, Richard S. and Ken W. Williams. "Methylene Chloride as a Auxiliary Blowing Agent for Flexible Slab stock Polyurethane Foam." _Advances in Urethane Science and Technology_, Volume 7, K. C. Frisch and S. L. Reegen (eds). TECHNOMIC Publishing Co., Inc., Westport, Connecticut, 1978. pp. 92.

(244A) Schildknecht, Calvin E. and Irving Skeist (eds.). _Polymerization Processes_, John Wiley and Sons, Inc., New York, New York. 1977.

(245) Schlanger, L. M., E. R. Baumgaertner, and W. A. Miller. "Fluoro-plastics." _Modern Plastics Encyclopedia_. McGraw-Hill, New York, New York, 1981-1982. pp. 24-26.

(246) Schule, E. D. "Polyamide Plastics." _Encyclopedia of Polymer Science and Technology_, Volume 10, Norbert M. Bikales (ed). Interscience Publishers, New York, New York. pp. 460-482.

(247) Schwartz, Seymour S. and Sidney H. Goodman. _Plastics Materials and Processes_. Van Nostrand Reinhold, New York, New York, 1982.

(248) Scroog, C. E. "Polyimides." Encyclopedia of Polymer Science and Technology, Vol. 11. John Wiley and Sons, New York, New York, © 1969. pp. 247-272.

(249) Sherman, Stanley, John Gannon, Gordon Buchi, and W. R. Howell. "Epoxy Resins." Kirk-Othmer Encyclopedia of Chemical Technology, 3rd Edition, Volume 9, Martin Grayson (ed). John Wiley and Sons, New York, New York. pp. 267-290.

(250) Sherwood, Dr. L. F., Jr. "Polyamide: Nylon, Crystalline." Modern Plastics Encyclopedia. McGraw-Hill, New York, New York, 1981-1982. pp. 32-40.

(251) Shreve, R. N. and J. A. Brink. Chemical Process Industries. McGraw-Hill, New York, New York, 1977. pp. 57, 577, 581-584.

(252) Sittig, Marshall. Pollution Control in the Plastics and Rubber Industry. Noyes Data Corporation, Park Ridge, New Jersey, 1975.

(253) Sittig, Marshall. Polyacetal Resins. Princeton University, Gulf Publishing Company, Houston, Texas, 1963. pp. 41-81.

(254) Sittig, Marshall. Practical Techniques for Saving Energy in the Chemical, Petroleum and Metals Industries. Noyes Data Corporation, Park Ridge, New Jersey, 1977. pp. 270-273.

(255) Sittig, Marshall. Vinyl Chloride and PVC Manufacture. 1978. pp. 94-343.

(256) Sittig, Marshall. Vinyl Monomers and Polymers. Noyes Data Corporation, Park Ridge, New Jersey, 1966.

(257) Skeist, Irving, ed. Handbook of Adhesives, 2nd ed. Van Nostrand Reinhold Company, New York, New York, 1977.

(258) Smith, Harry A. "Poly(arylene sulfides)." Encyclopedia of Polymer Science and Technology, Volume 10, Norbert M. Bikales (ed). Interscience Publishers, New York, New York. pp. 653-695.

(259) Smoluk, George. "New Candidate for Insulation: Phenolic Foam." Modern Plastics, July 1980, p. 48.

(260) Soder, Sara L. "Polycarbonate Resins." Chemical Economics Handbook, April 1976.

(261) Speirs, R. G., III. "Polyamide: Nylon, Amorphous (Transparent)." Modern Plastics Encyclopedia. McGraw-Hill, New York, New York, 1981-1982. pp. 40-42.

(262) Speit, G., et al. Mutation Research, Volume 78, 1980. p. 267.

(263) Steiner, T. E., "Alkyd Polyester." Modern Plastics Encyclopedia.
       McGraw-Hill, New York, New York, 1981-1982. p. 55.

(264) Steiner, T. E. "Phenolic." Modern Plastics Encyclopedia.
       McGraw-Hill, New York, New York, 1981-1982. pp. 29-30.

(265) Storck, William J. "ABS Market a Mix of Optimism and Problems."
       Chemical and Engineering News, July 30, 1979, p. 8.

(266) Stover, Robert R. "Polypropylene." Chemical Engineering Progress,
       July 1977, p. 29.

(267) Sweeny, W. and J. Zimmerman. "Polyamides." Encyclopedia of Polymer
       Science and Technology, Volume 10, Norbert M. Bikales (ed).
       Interscience Publishers, New York, New York, pp. 483-597.

(268) Synthetic Organic Chemicals: U.S. Production and Sales. United
       States International Trade Committee, Washington, D.C., USITC
       Publication 1099, 1979.

(269) "The Technology and Economics of Polyester Intermediates." Chemical
       Engineering Progress, August 1977, p. 74.

(270) Teer, G. E. "ABS and Related Multipolymers." Modern Plastics
       Encyclopedia. McGraw-Hill, New York, New York, 1981-1982. pp. 4-5.

(271) Texas Reports on Biology and Medicine, Volume 37, 1978, p. 153.

(272) "There's Still Growth in Nylon: Fibers are Slow, But Resins Gain."
       Chemical Week, September 1, 1982, p. 36.

(273) Thienes, Clinton H. and Haley Thomas. Clinical Toxicology, 5th
       Edition. Lee-Frebiger, Philadelphia, Pennsylvania, 1972. p. 192.

(274) Tomura, Y., H. Nagashima, K. Shirayama, and I. Suzuki. "New
       Processes for LDPE." Hydrocarbon Processing, November 1978, p. 151.

(275) "Tough, Glass-Reinfored Thermoplastic Resins for the Auto Indus-
       try..." Chemical Engineering, October 23, 1978, p. 94.

(276) Ulrick, Henri. "Recycling of Polyurethane and Isocyanurate Foam."
       Advances in Urethane Science and Technology, Volume 5, K. C. Frisch
       and S. L. Reegen (eds). TECHNOMIC Publishing Co., Inc., Westport,
       Connecticut, 1978. pp. 49-57.

(277) Ulrick, Henri. "Urethane Polymers." Kirk-Othmer Encyclopedia of
       Chemical Technology, 3rd ed., Vol. 23. John Wiley and Sons,
       New York, New York.

(278) United Nations, International Labor Office. Encyclopedia of
       Occupational Health and Safety. McGraw-Hill, New York, New York,
       1971. pp. 562, 861, 1013, 1043, 1068.

(279) Updegraff, Ivor H., Sewell T. Moore, William F. Herbes, and Philip B.
       Roth. "Amino Resins and Plastics." Kirk-Othmer Encyclopedia of
       Chemical Technology, 3rd Edition, Volume 2, Martin Grayson (ed).
       John Wiley and Sons, New York, New York. pp. 440-469.

(280) U.S. Department of Health, Education, and Welfare (Public Health
       Service, Center for Disease Control, National Institute for Occupa-
       tional Safety and Health, Division of Physical Sciences and Engineer-
       ing). Engineering Control Technology Assessment for the Plastics
       and Resins Industry, DHEW (NIOSH) Publication No. 78-159. March
       1978.

(281) U.S. Environmental Protection Agency. "Amino-Resins - Urea and
       Melamine." Development Document for Effluent Limitations Guidelines
       and New Source Performance Standards for the Synthetic Resins Segment
       of the Plastics and Synthetic Materials Manufacturing Point Source
       Category, EPA-440/1-74-010-a. March 1974. pp. 63-70.

(282) U.S. Environmental Protection Agency (Industrial Environmental
       Research Laboratory). Control of Chlorofluorocarbon Emissions From
       Flexible Polyurethane Slabstock Manufacture. May 1982.

(283) U.S. Environmental Protection Agency (Office of Air Quality and
       Planning Standards, Industrial Environmental Research Laboratory).
       VOC Fugitive Emissions Data--High Density Polyethylene Process Unit,
       EPA-600/2-81-109. June 1981.

(284) U.S. Environmental Protection Agency (Office of Air and Water Pro-
       grams, Effluent Guidelines Division). Development Document for
       Effluent Limitations Guidelines and New Source Performance Standards
       for the Synthetic Resins Segment of the Plastics and Synthetic
       Materials Manufacturing Point Source Category, EPA 440/1-74-010-a.
       1974.

(285) U.S. Environmental Protection Agency (Office of Research and
       Development). The Assessment of Atmospheric Emissions from
       Petroleum Refining. Contract No. 68-02-2147. July 1980.

(286) U.S. Environmental Protection Agency (Office of Research and Develop-
       ment, Industrial Environmental Research Laboratory). Source Assess
       ment: Plastics Processing, State of the Art, EPA-600/2-78-004c.
       March 1978.

(287) U.S. Environmental Protection Agency (Office of Research and Develop-
       ment, Industrial Environmental Research Laboratory). Source Assess-
       ment: Polyvinyl Chloride, EPA-600/2-78-0041. May 1978.

(288) Vale, C. P. and W. G. K. Taylor. Aminoplastics. The Plastics
       Institute by J. Iliffe Books, Ltd., London, 1964.

(289)   Villani, Tom. "Epoxy." Modern Plastics Encyclopedia. McGraw-Hill, New York, New York, 1981-1982. pp. 18-23.

(290)   Weismantel, Guy E. "Plastics Producers Search for Paths to Linear LDPE." Chemical Engineering, August 24, 1981, p. 47.

(291)   Welgos, R. J. "Polyamides: Plastics." Kirk-Othmer Encyclopedia of Chemical Technology, 3rd Edition, Volume 18, Martin Grayson (ed). John Wiley and Sons, New York, New York. pp. 406-425.

(292)   Wessling, R. A. and F. G. Edwards. "Vinylidene Chloride Polymers." Encyclopedia of Polymer Science and Technology, Volume 14, Norbert M. Bikales (ed). Interscience Publishers, New York, New York. pp. 540-577.

(293)   Wessling, R. and F. G. Edwards. "Vinylidene Polymers: Polyvinylidene Chloride." Kirk-Othmer Encyclopedia of Chemical Technology, 2nd Edition, Volume 21, Anthony Standen (ed.) John Wiley and Sons, New York, New York. pp. 275-303.

(294)   West, A. C. "Fluorine Compounds, Organic: Polychlorotrifluoroethylene." Kirk-Othmer Encyclopedia of Chemical Technology, 3rd Edition, Volume 11, Martin Grayson (ed). John Wiley and Sons, New York, New York. pp. 49-54.

(295)   White, Dwain M. and Glenn D. Cooper. "Polyethers: Aromatic Polyethers." Kirk-Othmer Encyclopedia of Chemical Technology, 3rd Edition, Volume 18, Martin Grayson (ed). John Wiley and Sons, New York, New York. pp. 594-605.

(296)   Widmer, Gustave. "Amino Plastics." Encyclopedia of Polymer Science and Technology, Volume 2, Norman G. Gaylord (ed). Interscience Publishers, New York, New York. pp. 1-94.

(297)   Wilson, M. M. "Acetal Copolymer." Modern Plastics Encyclopedia. McGraw-Hill, New York, New York, 1981-1982. pp. 7-8.

(298)   Wirth, Joseph G. and D. R. Heath. "Process for Making Polyetherimides and Products Derived Therefrom." U.S. Patent No. 3,838,097, Patented September 24, 1974.

(299)   Wood, A. Stuart. "More Muscle, Higher Heat: They Power a Phenolic Molding 'Revival'." Modern Plastics, July 1979, p. 50.

(300)   Wood, A. S. "Why All the Action in Extrusion-Grade Nylons?" Modern Plastics, April 1982, p. 78.

(301)   Wood, A. S. "Why All the Excitement About Phenolic Foam?" Modern Plastics, October 1981, p. 60.

(302) Yaws, Carl L. <u>Physical Properties</u>. McGraw-Hill, New York, New York, 1977.

(303) Zumba, Gerald P. "Acrylonitrile Polymers: Acrylonitrile-Styrene Copolymers." <u>Encyclopedia of Polymer Science and Technology</u>, Volume 1, Norman G. Gaylord (ed). Interscience Publishers, New York, New York. pp. 425-435.

Appendix A: Typical Physical and Chemical Properties of Polymers

TABLE A-1. TYPICAL PROPERTIES OF POLYMETHYL METHACRYLATE

Property	Cast Resin	Molding Resin
Density, g/cm^3	1.17 – 1.20	1.17 – 1.20
Tensile Strength, psi	8,000 – 11,000	7,000 – 11,000
Elongation, %	2.0 – 7.0	2.0 – 10.0
Tensile Modulus, psi x 10^5	3.5 – 4.5	3.8 – 4.5
Flexural Strength, psi	12,000 – 17,000	13,000 – 19,000
Izod Impact Strength, ft-lb/in of notch	0.3 – 0.4	0.3 – 0.5
Rockwell M Hardness	80 – 100	85 – 105
Flexural Modulus, psi x 10^5	3.90 – 4.75	4.2 – 4.6
Thermal Conductivity, °C/cm x 10^{-4}	4.0 – 6.0	4.0 – 6.0
Coefficient of Linear Thermal Expansion, $(°C)^{-1}$ x 10^{-5}	5.0 – 9.0	5.0 – 9.0

Source: Seymour S. Schwartz and Sidney H. Goodman, Plastics Materials and Processes, 1982.

TABLE A-1.   TYPICAL PROPERTIES OF POLYMETHYL METHACRYLATE

Property	Cast Resin	Molding Resin
Density, g/cm^3	1.17 - 1.20	1.17 - 1.20
Tensile Strength, psi	8,000 - 11,000	7,000 - 11,000
Elongation, %	2.0 - 7.0	2.0 - 10.0
Tensile Modulus, psi x 10^5	3.5 - 4.5	3.8 - 4.5
Flexural Strength, psi	12,000 - 17,000	13,000 - 19,000
Izod Impact Strength, ft-lb/in of notch	0.3 - 0.4	0.3 - 0.5
Rockwell M Hardness	80 - 100	85 - 105
Flexural Modulus, psi x 10^5	3.90 - 4.75	4.2 - 4.6
Thermal Conductivity, °C/cm x 10^{-4}	4.0 - 6.0	4.0 - 6.0
Coefficient of Linear Thermal Expansion, $(°C)^{-1}$ x 10^{-5}	5.0 - 9.0	5.0 - 9.0

Source: Seymour S. Schwartz and Sidney H. Goodman, Plastics Materials and Processes, 1982.

TABLE A-2. PROPERTIES OF IMPACT MODIFIED ACRYLIC MOLDING COMPOUNDS

Property	Value				
	MMA-Styrene	Impact Acrylic	MMA-Methyl Styrene	Acrylic Multi-Polymer	Acrylic PVC
Density, g/cm^3	1.09	1.11–1.18	1.09–1.16	1.10–1.12	1.35
Tensile Strength, psi	10,000	5,000–9,000	10,000	6,000–8,000	6,500
Elongation, %	3.0	20.0–70.0	3.0	3.0–40.0	100
Tensile Modulus, psi x 10^5	4.3	2.0–4.0	4.3–4.6	3.1–4.3	3.31–3.35
Flexural Strength, psi	16,000–19,000	7,000–13,000	12,000–19,000	9,500–13,000	10,700
Izod Impact Strength, ft-lb/in of notch	0.3	0.8–2.5	0.3	1.0–2.0	15.0
Rockwell M Hardness	M75	R105–R120	M75–M104	R108–R119	R99–R105
Flexural Modulus, psi x 10^5	2.6–3.8	2.0–3.8	2.6–3.8	3.0–4.0	3.3–4.0
Thermal Conductivity, °C/cm x 10^{-4}	4.0–5.0	4.0–5.0	4.0–5.0	5.3	--
Thermal Expansion, $(°C)^{-1}$ x 10^{-5}	6.0–8.0	5.0–8.0	6.0–8.0	6.0–9.0	—
Clarity	Transparent	Transparent, can be translucent opaque	Transparent	Transparent	Opaque

Source: Seymour S. Schwartz and Sidney H. Goodman, _Plastics Materials and Processes_, 1982.

TABLE A-3.   TYPICAL PROPERTIES OF ABS RESINS ACCORDING TO RESIN GRADE

Property	Value			
	High Impact	Medium Impact	Low Impact	Heat Resistant
Izod Impact Strength, ft-lb/in	7-12	4-7	2-4	2-6
Tensile Strength, psi x 10^3	4.8-6.0	6.0-7.0	6.0-7.5	6.0-7.5
Elongation, %	15-70	10-50	5-30	5-20
Tensile Modulus, psi x 10^5	2.5-3.0	3.0-3.6	3.0-3.8	3.0-3.8
Rockwell Hardness	88-90	95-105	105-110	105-110
Density, g/cm^3 at 23°C	1.02-1.04	1.04-1.05	1.05-1.07	1.04-1.06
Linear Coefficient of Thermal Expansion, (°C)$^{-1}$ x 10^{-5}	9.5-11.0	7.8-8.8	7.0-8.2	6.5-9.3

Source: _Encyclopedia of Chemical Technology_, 3rd Edition.

TABLE A-4. TYPICAL PROPERTIES OF ALKYD MOLDING RESINS

Property	Value
Impact Strength, ft-lb	0.3 - 2.0
Flexural Strength, psi	10,000 - 20,000
Tensile Strength, psi	8,000 - 15,000
Density, g/cm^3	1.70 - 2.20
Molecular Weight	24,000 - 35,000

Sources: Encyclopedia of Polymer Science and Technology.
Modern Plastics Encyclopedia, 1981-1982.

TABLE A-5.   TYPICAL PROPERTIES OF FILLED AMINO RESIN MOLDED PRODUCTS

| | Value | | | | |
| Property | Urea Formaldehyde Resins | Melamine-Formaldehyde Resins | | | |
	α-Cellulose	α-Cellulose	Macerated Fabric	Asbestos	Glass Fiber
Density, g/cm^3	1.47–1.52	1.47–1.52	1.5	1.7–2.0	1.8–2.0
Tensile Strength, psi x 10^3	5.5–7.0	7–13	8–10	5.5–6.5	5–10
Elongation, %	0.5–1.0	0.6–0.9	0.6–0.8	0.3–0.45	N/A
Tensile Modulus, psi x 10^5	13–14	13.5	14–16	19.5	24
Rockwell M Hardness	110–120	120	120	110	115
Flexural Strength, psi x 10^3	11–18	12–15	12–15	7.4–10	13–24
Flexural Modulus, psi x 10^5	14–15	11	14	18	24
Impact Strength, ft-lb/in of notch	0.27–0.34	0.24–0.35	0.6–1.0	0.3–0.4	0.6–18
Thermal Conductivity, 10^{-4} x cal/(sec)(cm)(°C)	10.1	7.0–10.1	10.6	13.0–17.0	11.5
Linear Coefficient of Thermal Expansion, 10^{-5} x $(°C)^{-1}$	2.2–3.6	2.0–5.7	2.5–2.8	2.0–4.5	1.5–1.7

N/A – Not available.

Source: _Encyclopedia of Chemical Technology_, 3rd Edition.

TABLE A-6.   TYPICAL PROPERTIES OF MODIFIED POLYPHENYLENE OXIDE AND
POLYPHENYLENE SULFIDE

| | Value | |
| | Modified Polyphenylene Oxide | Polyphenylene Sulfide |
Property		
Density, g/cm^3	1.06 - 1.36	1.34 - 2.67*
Tensile Strength, psi x 10^3	7.8 - 17	10.8
Elongation, %	4.0 - 60.0	2 - 3
Flexural Modulus, psi x 10^5	3.6 - 11.0	6 - 11.6
Flexural Strength, psi x 10^3	12.8 - 20.0	14 - 20
Tensile Modulus, psi x 10^5	3.55 - 12.0	4.8 - 7.0
Rockwell R Hardness	106 - 119	123 - 124
Compressive Strength, psi x 10^3	16.0 - 17.9	16 - 24

*Filled.

Source: Seymour S. Schwartz and Sidney H. Goodman, _Plastics Materials and
Processes_, 1982.

TABLE A-7. TYPICAL PROPERTIES OF EPOXY RESINS

	Value				
Property	Encapsulating Grades	Phenol Novolak Epoxy Resins	Cyclo-Aliphatic Epoxy Resins	Casting Resins	Molding Compounds
Density, g/cm^3	1.7-2.1	1.16-1.21	1.16-1.21	1.05-2.0	0.75-2.0
Tensile Strength, psi	4,000-15,000	6,000-12,000	8,000-12,000	2,000-13,000	2,500-20,000
Elongation, %	---	2.0-6.0	2.0-10.0	13.9-24.9	13.4-15.4
Izond Impact Strength, ft/lb per in. of notch	0.3-2.0	0.5-2.0	---	0.2-1.0	0.15-30.0
Rockwell M Hardness	100-112	---	---	55-120	100-110
Thermal Conductivity, °C/cm	4.0-10.0	---	---	4.0-25.0	4.0-30.0
Specific Heat, cal/°C/g	---	---	---	0.20-0.27	0.19
Thermal Expansion, $(°C)^{-1}$ x 10^5	3.0-6.0	---	---	2.0-10.0	1.1-5.0

Source: Seymour S. Schwartz and Sidney H. Goodman, Plastics Materials and Processes, 1982.

TABLE A-8.   TYPICAL PROPERTIES OF PTFE AND PCTFE

Property	Value	
	PTFE	PCTFE
Tensile Yield Strength, psi x 10^3	2.8 - 3.0	3.5 - 3.7
Compressive Strength (at 1% offset), psi x 10^3	0.7 - 1.8	2.0
Tensile Modulus, psi x 10^5	0.4 - 0.9	1.9 - 3.0
Compressive Modulus, psi x 10^5	0.7 - 0.9	1.8
Flexural Modulus, psi x 10^5	0.9	2.2
Izod Impact Strength, ft-lb/in of notch	2.5 - 4.0	3.5 - 3.6
Coefficient of Thermal Expansion, $(°C)^{-1}$ x 10^{-5}	2.8 - 5.6	2.2

Source: Seymour S. Schwartz and Sidney H. Goodman, _Plastics Materials and Processes_, 1982.

TABLE A-9.   COMPARISON OF DENSITY AND ELONGATION OF FLUOROPOLYMERS

Property	Value							
	PTFE	PCTFE	FEP	PVDF	PETFE	PECTFE	PVA	PVF
Density, g/cm^3	2.1– 2.2	2.10– 2.15	2.14– 2.17	1.75– 1.78	1.70	1.68	2.16	1.38– 1.57
Elongation, % Ultimate	250– 400	125– 175	160	40– 100	100– 400	200	300	110– 260

Source: Seymour S. Schwartz and Sidney H. Goodman, _Plastics Materials and Processes_, 1982.

TABLE A-10.   TYPICAL PROPERTIES OF PHENOLIC MOLDING COMPOUNDS
ACCORDING TO GRADE

| | Resin Grade | | | |
Property	General-Purpose	Impact	Heat-Resistant	Special Electrical
Flexural Strength, psi	7,000–9,000	7,000–9,000	6,000–9,000	8,000
Tensile Strength, psi	3,500–6,500	4,000–6,000	4,000–4,500	4,000–4,500
Izod Impact Strength, ft-lb/in of notch	0.24–0.46	0.8–4.0	0.23–0.64	0.3

Source: *Encyclopedia of Chemical Technology*, 3rd Edition.

TABLE A-11.   TYPICAL PROPERTIES OF POLYACETALS

	Value	
	Homopolymer	Copolymer with Ethylene Oxide
Density, g/cm^3	1.42	1.41
Tensile Strength (at break, 23°C), psi x 10^3	10	8.8
Elongation (at break), %	40	60
Modulus of Elasticity (23°C), psi x 10^3	450	410
Flexural Modulus, psi x 10^3		
23°C	410	375
70°C	260·	
71°C		180
Coefficient of Linear Thermal Expansion, $(°C)^{-1}$ x 10^{-5}		
-30° to +30°C	10.4	8.5

Source: _Encyclopedia of Chemical Technology_, 3rd Edition.

TABLE A-12.   TYPICAL PROPERTIES OF POLYAMIDE RESINS

	Value	
	Nylon 6	Nylon 66
Density, g/cm^3	1.13	1.14
Tensile Strength, psi x 10^3	11.8	12.5
Elongation at Break, %	50 – 200	40 – 80
Young's Modulus, over 1% extension at 20°C, psi x 10^5	3.5	4.3
Compressive Strength, psi x 10^3	14.0	16.0
Shear Strength, psi x 10^3	8.5	9.5
Izod Impact Strength, ft-lb/0.5 in of notch	1.1	1.1

Source: Encyclopedia of Chemical Technology, 2nd Edition.

TABLE A-13.   TYPICAL PROPERTIES OF POLYBUTYLENE

Property	Value
Density, g/cm^3	0.910 - 0.915
Tensile Strength, psi	
Yield	1,990
Break	4,550
Elongation at Break, %	350
Modulus of Elasticity, psi	3,550
Izod Impact Strength, ft-lb/in	no break
Shore D Hardness	55
Environmental Stress Crack Resistance, hours	no failure

Source: *Encyclopedia of Chemical Technology*, 3rd Edition.

TABLE A-14.   TYPICAL PROPERTIES OF POLYCARBONATE

Property	Value
Density, g/cm^3	1.200
Linear Coefficient of Thermal Expansion, $(°C)^{-1} \times 10^{-5}$	6.75
Tensile Strength, psi x 10^3	
Yield	9.00
Ultimate	9.50
Elongation, %	
Yield	6 - 8
Break	110
Tensile Modulus, psi x 10^3	345
Flexural Strength, psi	13.5
Flexural Modulus, psi	339
Compressive Strength, psi	12.5
Izod Impact Strength, ft-lb/in	
Notched	16.1
Unnotched	No Failure
Tensile Impact, $ft-lb/in^2$	226 - 300

Source: Encyclopedia of Chemical Technology, 3rd Edition.

TABLE A-15. TYPICAL PROPERTIES OF POLY(ETHER-IMIDE) RESINS

	Unfilled	30% Glass Fiber-Reinforced
Melting temperature, °C (amphorous)	215	215
Specific gravity	1.27	1.51
Processing temperature range, °F	640-800	640-800
Molding pressure range, 10^3 psi	10-18	10-18
Tensile yield strength, psi	15,200	24,500
Compressive strength, psi	20,300	23,500
Izod impact (ft-lb/in)	1.0	2.0
Hardness, Rockwell	M109	M125
Coefficient of linear thermal expansion, 10^{-6} in/in/°C	56	20
Thermal conductivity, 10^{-4} cal/sec,cm,°C	1.6	

Source: Modern Plastics Encyclopedia, 1982-83, McGraw-Hill, Inc.,
New York, October 1982, Vol. 59, No. 10A, p. 483.

TABLE A-16.   TYPICAL PROPERTIES OF PET RESINS

Property	Value
Density, g/cm^3	1.2 - 1.4
Specific Heat, cal/(g)(°C)	85
Coefficient of Linear Thermal Expansion, $(°C)^{-1} \times 10^{-5}$	7
Elongation (break), %	12 - 130
Rockwell M Hardness	106
Tensile Strength, psi	8,500 - 11,000
Injection Molded Resin	
Tensile Strength (Yield), psi	2,350
Tensile Strength, (Break) psi	1,320
Elongation (Break), %	100 - 200

Flexural Modulus, psi
  At 20°C
  At 75°C
	
	83,580
	23,460

Flexural Yield Strength, psi	3,430
Izod Impact Strength, ft-lb/in of notch	1.8
Shear Strength, psi	1,850

Sources: Encyclopedia of Polymer Science and Technology.
Modern Plastics Encyclopedia, 1981-1982.

TABLE A-17.   TYPICAL PROPERTIES OF POLYBUTYLENE TEREPHTHALATE RESINS

Property	Value
Tensile Strength, psi	7,500 – 19,000
Flexural Modulus, psi	340,000 – 1,200,000
Izod Impact Strength, ft-lb/in of notch	
Unreinforced resin	1.0
Reinforced resin grades	1.8 – 2.0
Impact modified unreinforced	16
Impact modified reinforced resin	3.5
Coefficient of Linear Thermal Expansion, $(°C)^{-1} \times 10^{-5}$ over the range from -30 to 30°C	5.94

Source: <u>Modern Plastics Encyclopedia</u>, 1981–1982.

TABLE A-18.   CROSSLINKING AGENTS USED FOR UNSATURATED POLYESTER
RESIN PRODUCTION

Monomer	Important Characteristics Relative to Styrene
Styrene	
Vinyltoluene, 60/40 m/p	Lower vapor pressure, slightly higher reactivity
Chlorostyrene, 67/33 o/p	Fast cure, improved hot strength, better surface
t-butylstyrene, 95/5 p/m	Low vapor pressure, high flash point, high heat-distortion temperature
Divinylbenzene (m-isomer)	Increased crosslinking
α-methyl styrene	Increased storage time
Methyl methacrylate	Improved weatherability when used in a 50:50 ratio with styrene
Methyl acrylate	Improved strength and weatherability
Diallyl phthalate	Very low volatility
Triallyl cyanurate	High temperature properties

Source: Encyclopedia of Polymer Science and Technology.

TABLE A-19.   TYPICAL PROPERTIES OF REINFORCED UNSATURATED
              POLYESTER RESINS

Property	Value
Flexural Strength, psi x 10^{-3}	0.6 – 20.7
Flexural Modulus, psi x 10^{-3}	0.3 – 660
Impact Strength, ft-lb/in	1.3 – 11.2
Rockwell M Hardness	0 – 111
Tensile Strength, psi x 10^{-3}	2.55 – 9.86
Tensile Modulus, psi x 10^{-3}	0.4 – 870
Elongation, %	1.20 – 1.85

Source: _Encyclopedia of Polymer Science and Technology._

TABLE A-20. TYPICAL PROPERTIES OF HIGH DENSITY POLYETHYLENE

Property	Value
Density, g/cm^3	0.941 - 0.965
Elongation, %	15 - 100
Heat Capacity, cal/(°C)(g)	0.55
Impact Strength, ft-lb/in	1.5 - 20
Melt Index, g/10 min	0.1 - 4.0
Molecular Weight, weight average	125,000
Tensile Impact Strength, $ft-lb/in^2$	60
Tensile Modulus, psi x 10^5	0.6 - 1.5
Tensile Strength, psi	3,100 - 5,500
Thermal Conductivity	
20°C	0.44
50°C	0.40
100°C	0.33
Coefficient of Linear Thermal Expansion, 20-90°C, $(°C)^{-1}$ x 10^4	1.7

Sources: *Encyclopedia of Chemical Technology*, 3rd Edition.
Encyclopedia of Polymer Science and Technology.

TABLE A-21. DENSITY DEPENDENT PROPERTIES OF HIGH DENSITY POLYETHYLENE

Property	Density, g/cm^3		
	0.96	0.95	0.94
		Value	
Stiffness, psi	145,000	113,100	79,800
Tensile Strength, psi[a]	4,400	3,700	2,900
Softening Temperature, Vicat, °C	127	124	121
Shore D Hardness	68	67	63
Elongation, %	20	50	120
Environmental Stress Crack Resistance, Time for 50% of the Samples to Fail, hours	20	100	700

[a]At 2 in/min.

Source: Encyclopedia of Chemical Technology, 3rd Edition.

TABLE A-22.   MOLECULAR WEIGHT DEPENDENT PROPERTIES OF HIGH DENSITY POLYETHYLENE[a]

| | | | Molecular Weight | | |
Property	175,000	142,000	135,000	105,000	90,000
			Value		
Tensile Impact, ft-lb/in^2	100	64	59	41	30
Elongation at Break, %[b]	14	4.0	2.0	1.5	1.2
Environmental Stress Crack Resistance, Time for 50% of the Samples to Fail, hours	60	14	10	2	1
Brittleness Temperature, °C	<-118	<-118	<-118	-101	-73

[a]For compression molded resin.

[b]At 2 in/min.

Source: Encyclopedia of Chemical Technology, 3rd Edition.

TABLE A-23. TYPICAL HIGH DENSITY POLYETHYLENE PROPERTIES ACCORDING
TO RESIN GRADE

Property	Injection Molding	Resin Grade Blow Molding	Extrusion
		Value	
Density, g/cm^{3}[a]	0.957 - 0.958	0.946 - 0.959	0.944 - 0.955
Yield Stress, psi	4205 - 4350	3480 - 4350	3190 - 3625
Elongation, (yield), %	12	12 - 14	14 - 16
Ultimate Tensile Strength, psi	3915 - 4350	4785 - 5800	4640 - 5800
Elongation (rupture), %	>500 - >800	>800	>800
Shore D Hardness	68 - 69	64 - 68	60 - 66

[a]At 23°C

Source: <u>Encyclopedia of Chemical Technology</u>, 3rd Edition.

TABLE A-24. TYPICAL PROPERTIES OF ULTRA HIGH MOLECULAR WEIGHT-
HIGH DENSITY POLYETHYLENE

Property	Ziegler Catalyst	Phillips Catalyst	
		Copolymer	Homopolymer
		Value	
Density, g/cm3	0.94	0.941	0.955
Tensile Strength, psi	5,400	3,000	4,000
Elongation, %	525	>600	>600
Izod Impact Strength, ft-lb/in	>20[a]	7	16

[a]Specimen deflects but does not break.

Source: Encyclopedia of Polymer Science and Technology.

TABLE A-25.   COMPARISON OF LINEAR LOW DENSITY POLYETHYLENE
AND LOW DENSITY POLYETHYLENE TYPICAL PROPERTIES

	Density, g/cm^3			
	Linear Low Density Polyethylene		Low Density Polyethylene	
	0.922	0.926	0.918	0.927
Property		Value		
Tensile Strength, psi	1,885	2,900	1,740	2,320
Elongation, %	800	600	550	600
Flexural Modulus, psi	33,930	73,950	20,010	60,030
Shore D Hardness	--	58	--	58
Environmental Stress- Crack Resistance	--	1,000	--	65

Source: Encyclopedia of Chemical Technology, 3rd Edition.

TABLE A-26. TYPICAL PROPERTIES OF LOW DENSITY POLYETHYLENE

| | Density, g/cm^3 | |
Property	0.910 - 0.925	0.926 - 0.940
	Value	
Tensile Yield Stress, psi	11.6 - 17.4	14.5 - 26.1
Elongation, % (yield)	20 - 40	10 - 20
Tensile Ultimate Stress, psi	14.5 - 24.7	16.0 - 25.4
Elongation, % (ultimate)	400 - 700	100 - 500
Shear Strenght, psi	10.2 - 37.7	30.5 - 43.5
Linear Coefficient of Thermal Expansion, 20-60°C, $(°C)^{-1}$ x 10^5	15 - 30	15 - 30
Heat Capacity, cal/(°C)(g)	0.55	
Impact Strength, ft-lb/in	>16	0.5 - >16
Molecular Weight, weight average	14,000 - 1,400,000	14,000 - 1,400,000
Tensile Impact Strength, $ft-lb/in^2$	180	60
Tensile Modulus, psi x 10^5	0.17 - 0.35	0.25 - 0.55
Tensile Strength, psi	1,000 - 2,300	1,200 - 3,500

Sources: Encyclopedia of Chemical Technology, 3rd Edition.
Encyclopedia of Polymer Science and Technology.

TABLE A-27.   TYPICAL PROPERTIES OF POLYPROPYLENE

Property	Homopolymer	Random Copolymer	Block Copolymer
Tensile Strength, psi	4,900	4,000	3,400 - 4,300
Flexural Modulus, psi	200,000	160,000	120,000 - 180,000
Ethylene Content, %	0	1 - 4	5 - 20
Melt Flow Range, g/10 min	3 - 12	2 - 10	2 - 12

Source: Modern Plastics Encyclopedia, 1981-1982.

TABLE A-28.   TYPICAL PROPERTIES OF POLYPROPYLENE ACCORDING TO RESIN GRADE

| | Grade | | | |
Property	Pipe and Sheet	Film	Injection Molding	Coat and Tubular Film
Melt Flow Rate, g/10 minutes	0.25–0.35	1.5–2.0	5–7	7–10
Density, g/cm^3	0.900–0.905	0.900–0.905	0.900–0.905	0.900–0.905
Tensile Yield Strength (at 2 inches), psi	4,205–4,495	4,640–4,930	4,785–5,075	4,785–5,075
Elongation (yield), %	10–12	10–12	8–10	8–10
Flexural Modulus, psi	159,500–188,500	188,500–203,000	217,500–232,000	232,000–246,500
Izod Impact Strength, ft-lb/in at 23°C	0.028–0.037	0.0094–0.013	0.0056–0.0094	0.0037–0.0056
Rockwell C Hardness	55–60	55–60	60–65	60–65
Coefficient of Linear Thermal Expansion, x 10^{-5}				
from −30 to 0°C	6.5	6.5	6.5	6.5
from 0 to 30°C	10.5	10.5	10.5	10.5
from 30 to 60°C	14.0	14.0	14.0	14.0

Source: Encyclopedia of Chemical Technology, 3rd Edition.

TABLE A-29.   TYPICAL PHYSICAL PROPERTIES OF GENERAL PURPOSE POLYSTYRENE

Property	Value
Specific Gravity	1.05
Tensile Stress, psi Yield and Rupture	4,710 − 8,000
Elongation, %	0.9 − 2.0
Tensile Modulus, psi	400,000 − 570,000
Izod Impact Strenght, ft-lb/in	0.2 − 0.5
Thermal Coefficient of Linear Expansion, $(°C)^{-1}$ x 10^5	6 − 8
Specific heat, cal/g	
0°C	0.283
50°C	0.300
100°C	0.439
Thermal Conductivity, cal/(sec)(cm)(°C)	
0°C	2.51×10^{-4}
50°C	2.78×10^{-4}
100°C	3.06×10^{-4}

Sources: Modern Plastics Encyclopedia, 1981-1982.
         Encyclopedia of Polymer Science and Technology.

TABLE A-30.   TYPICAL PROPERTIES OF RUBBER MODIFIED (IMPACT) POLYSTYRENE

Property	Value
Density, g/cm^3	1.04 – 1.065
Tensile Modulus, psi	250,000 – 350,000
Izod impact strength, ft-lb/in	0.6 – 4
Elongation, %	10 – 50
Tensile stress, psi	
Rupture	2,820
Yield	2,920
Linear coefficient of thermal expansion $(°C)^{-1} \times 10^5$	6 – 8
Specific heat, cal/g	
0°C	0.283
50°C	0.300
100°C	0.439
Thermal conductivity, cal/(sec)(cm)(°C)	
0°C	2.51×10^{-4}
50°C	2.78×10^{-4}
100°C	3.06×10^{-4}

Sources: Encyclopedia of Polymer Science and Technology.
Modern Plastics Encyclopedia, 1981-1982.

TABLE A-31.   TYPICAL PROPERTIES OF IMPACT (RUBBER MODIFIED)
POLYSTYRENE GRADES

Property	Medium Impact PS	High Impact PS	Super Impact PS
Specific gravity	1.05	1.05	1.02
Rubber, %	3.4	5.1	14.5
Tensile yield, psi	3,700	3,000	2,000
Elongation, % (rupture)	1.4	35	17
Tensile modulus, psi x 10^5	4.5	3.4	2.5
Izod impact strength, ft-lb/in	0.6	1.3	4.5

Source: Encyclopedia of Polymer Science and Technology.

TABLE A-32. TYPICAL PROPERTIES FOR POLYURETHANE FOAM FORMULATIONS

	Value			
Property	High-Density	Medium Density	High Load Bearing	Self Extinguishing
Density, lb/ft^3	2.40 – 2.44	1.80	2.3	1.5
Tensile Strength, psi	14.9 – 17.5	19.9	18.7	12
Elongation, %	107 – 248	258	65	200

Sources: Encyclopedia of Chemical Technology, 3rd Edition.
Encyclopedia of Polymer Science and Technology.

TABLE A-33.   TYPICAL PROPERTIES OF POLYVINYL ACETATE

	Value
Density, g/cm^3	1.05 to 1.19
Tensile Strength[a], psi x 10^3	4.26 to 7.10
Elongation[a,b], %	10 to 20
Young's Modulus[a], psi x 10^5	1.85 to 3.26
Charpy Impact Strength[a], ft-lb/in^2	>47
Thermal Coefficient of linear expansion,$(°C)^{-1}$ x 10^5	7 to 22
Specific Heat, cal/g°C	0.10
Thermal conductivity, cal/(sec)(cm)(°C)	38 x 10^{-5}
Dielectric constant, 10^3 Hz	3.15

[a]20°C

[b]0% relative humidity

Sources: Encyclopedia of Chemical Technology, 3rd Edition.
           Encyclopedia of Polymer Science and Technology.

TABLE A-34.   PHYSICAL PROPERTIES OF POLYVINYL ACETATE EMULSIONS

Property	Units	Polyvinyl Acetate for Adhesive Value	Polyvinyl Acetate for Paper Coating Value
Viscosity	Brookfield LVT,[a] 60 rpm, cP	800 - 1,200	50 - 80
Specific Gravity		1.11	1.09
Average Particle Size	Microns	1 - 3	0.15
Average Density	g/cc at 20°C	5.35	5.23

[a]This test measures the torque produced by the emulsion, resulting from the rotation of a spindle inside a sample chamber through which the sample flows.

Source: Encyclopedia of Polymer Science and Technology.

TABLE A-35. SOLUBILITY OF POLYVINYL ACETATE IN SELECTED ORGANIC SOLVENTS

Solvent	Soluble or Insoluble
Aromatics	Soluble
Butanol	Soluble When Heated
Butanol with 5 to 10 percent water	Soluble
Carboxylic Acids	Soluble
Esters	Soluble
Ethanol with 5 to 10 percent water	Soluble
Halogenated Hydrocarbons	Soluble
Ketones	Soluble
Low Carbon Alcohols (Except Methanol)	Insoluble
Methanol	Soluble
Nonpolar Liquids	
Ether	Insoluble
Carbon Disulfide	Insoluble
Aliphatic Hydrocarbons	Insoluble
Oils	Insoluble
Fats	Insoluble
Propanol with 5 to 10 percent water	Soluble
Water	Insoluble
Xylene	Soluble When Heated

Source: Encyclopedia of Chemical Technology, 2nd Edition.

TABLE A-36.   PHYSICAL PROPERTIES OF FULLY HYDROLYZED POLYVINYL ALCOHOL

Property	Value
Specific gravity	1.19–1.31
Tensile strength, psi	up to 22,000[a]

Elongation, %
  unplasticized film
  plasticized film
	
	up to 300[a]
	up to 600[a]

Thermal coefficient of linear expansion, 0–50°C, plasticized, $(°C)^{-1} \times 10^5$	7–12
Specific heat, cal/g°C	0.4
Compression molding temperature, °C	120–150

[a]At 50 percent relative humidity.

Sources: Encyclopedia of Chemical Technology, 2nd Edition.
           Encyclopedia of Polymer Science and Technology.

TABLE A-37.   POLYVINYL ALCOHOL DENSITY AS A FUNCTION OF
             REMAINING ACETATE GROUPS

% Acetate in Resin	Density, g/cm^3 at 20°C
0	1.329
5	1.322
10	1.316
20	1.301
30	1.288
40	1.274
50	1.260
60	1.246
70	1.232

Source: Encyclopedia of Chemical Technology, 2nd Edition.

TABLE A-38. SOLUBILITY OF POLYVINYL ALCOHOL IN SELECTED ORGANIC SOLVENTS
ACCORDING TO ACETATE CONTENT

Chemical	Soluble or Insoluble	Polyvinyl Alcohol Acetate Content
acetone	Insoluble	Low
acetylene	Insoluble	Low
anhydrous ethanol	Insoluble	Low
anhydrous methanol	Insoluble	Low
benzene	Insoluble	Low
butane	Insoluble	Low
1-butanol	Insoluble	Low
carbon tetrachloride	Insoluble	Low
chlorofluorocarbons	Insoluble	Low
cyclohexanol	Insoluble	Low
diacetone alcohol	Insoluble	Low
diethylene glycol	Insoluble	Low
diethylene glycol monobutyl ether (butyl Carbitol)	Insoluble	Low
dioxane	Insoluble	Low
ethyl acetoacetate	Insoluble	Low
ethylene glycol	Insoluble	Low
fenchyl alcohol	Insoluble	Low
formamide	Insoluble	Low
furfural	Insoluble	Low
gasoline	Insoluble	Low

(continued)

TABLE A-38 (continued)

Chemical	Soluble or Insoluble	Polyvinyl Alcohol Acetate Content
kerosene	Insoluble	Low
methyl acetate	Insoluble	Low
methylene dichloride	Insoluble	Low
monochlorobenzene	Insoluble	Low
propane	Insoluble	Low
sulfur dioxide	Insoluble	Low
trichloroethylene	Insoluble	Low
xylene	Insoluble	Low
acetamide	Soluble when heated[a]	Low
acetic acid	Soluble when heated[a]	Low
amino hydroxy compounds	Soluble when heated[a,b]	Low
ethanolamine salts and amides	Soluble	Low
formamide	Soluble[a]	Low
N-(2-hydroxyethyl)acetamide	Soluble	Low
N-(2-hydroxyethyl)formamide	Soluble	Low
ethylene glycol	Soluble	Low
glycerol	Soluble[a]	Low
glycol	Soluble when heated[a]	Low

(continued)

TABLE A-38 (continued)

Chemical	Soluble or Insoluble	Polyvinyl Alcohol Acetate Content
lower polyethylene glycols	Soluble when heated[a]	Low
phenol	Soluble when heated[a]	Low
sorbitol	Slightly soluble	Low
2,2'-thiodiethanol	Soluble when heated[a]	Low
urea	Slight soluble[a]	Low
cresylic acid	Soluble	Above 45%
liquid sulfur dioxide	Soluble	Above 80%
piperazine	Soluble	–[c]
N-methyl pyrrolidone		–[c]
tris(dimethylamido)phosphoric acid	Soluble	–[c]
dimethyl sulfoxide	Soluble	–[c]

[a]Forms a gel upon cooling.

[b]If the phenol contains water, it will stay liquid when cooled.

[c]No percentage of acetate designation was given.

Sources: Encyclopedia of Chemical Technology, 2nd Edition.
Encyclopedia of Polymer Science and Technology.
J. G. Prichard, Polyvinyl Alcohol: Basic Properties and Uses, 1970.

TABLE A-39. TYPICAL PROPERTIES OF POLYVINYL CHLORIDE

	Rigid	Plasticized
Specific gravity	1.35 to 1.45	1.16 to 1.35
Tensile strength,[a] psi x 10^3	5.0 to 9.0	1.5 to 3.5
Elongation %	2.0 to 40	200 to 450
Tensile modulus,[a] psi x 10^5	3.5 to 6.0	----
Compressive strength, psi x 10^3	8.0 to 13	0.9 to 1.7
Flexural yield strength,[a] psi x 10^3	10 to 16	----
Izod impact strength,[a] ft-lb/in	0.4 to 1.2	----
Thermal conductivity cal/(sec)(cm)(°C) x 10^4	3.0 to 7.0	3.0 to 4.0
Specific heat, cal/(g)(°C)	0.2 to 0.28	0.3 to 0.5
Linear thermal coefficient of expansion (°C)$^{-1}$ x 10^5	5.0 to 18.5	7.0 to 25
Continuous use temp, °C	66 to 80	66 to 80
Dielectric constant, 10^3 Hz	3.0 to 3.3	4.0 to 8.0

[a]Structural properties are reported in English units, according to convention.

Source: F. Rodriguez, _Principles of Polymer Systems_, 1970.

TABLE A-40.   TYPICAL PROPERTIES OF POLYVINYLIDENE CHLORIDE

	Value	
Property	PVDC	PVDC Molding Compounds
Tensile Strength, psi x 10^3		
Unoriented	5 – 10	
Oriented	30 – 60	
Molded		3.5 – 4.5
Elongation, %		
Unoriented	10 – 20	
Oriented	15 – 40	
Molded		15 – 25
Izod Impact Strength, ft-lb/in	0.5 – 1.0	2 – 8

Sources: _Encyclopedia of Chemical Technology_, 2nd Edition.
            Encyclopedia of Polymer Science and Technology.

TABLE A-41. TYPICAL PROPERTIES OF STYRENE-ACRYLONITRILE RESINS

Property	Value
Tensile Strength, psi	8,300 - 11,000
Elongation, %	2.1 - 3.7
Tensile Modulus, psi	520,000 - 540,000
Izod Impact Strength, ft-lb/in.	0.30 - 0.45
Rockwell Hardness	80 - 83
Coefficient of Linear Thermal Expansion $(°C)^{-1}$ x 10^5	6.85 - 6.67
Specific Heat, cal/(g)(°C)	0.31
Density, g/cm^3	1.06 - 1.08

Sources: *Encyclopedia of Chemical Technology*, 3rd Edition.
Encyclopedia of Polymer Science and Technology.

TABLE A-42. TENSILE STRENGTH AND ELONGATION VALUES FOR STYRENE-
ACRYLONITRILE RESINS WITH VARYING ACRYLONITRILE CONTENT

SAN Acrylonitrile Content, %	Tensile Strength, psi	Elongation, %
5.5	6,130	1.6
9.8	7,922	2.1
14.0	8,321	2.2
21.0	9,259	2.5
27.0	10,511	3.2

Source: Encyclopedia of Chemical Technology, 3rd Edition.

Appendix B: Input Monomers

<u>Function</u>	<u>Compound</u>
Monomers and Comonomers	Acetaldehyde

Acrylamide
Acrylates
Acrylate esters
Acrylic acid
Acrylic esters
Acrylonitrile
Adipic acid
Aldehydes
Alkyl acrylates
Alkyl maleates
Allyl chloride
Allyl esters of dicarboxylic acids
Allyl ethers of dihydric alcohols
Allyl methacrylate
α-alkylacrylates
α-methyl styrene
Amino resins
Benzoquanamine
2,2-bis(4-hydroxyphenyl)-1,1-dichloroethylene
Bisphenol-A
Butadiene
1,4-butanediol
1-butene
Butyl acrylate
t-butylaminoethyl methacrylate
1,4-butylene dimethacrylate
Carbamoylmethytrimethylammonium hydrochloride
Cetyl vinyl ether
1-chloro-1-bromoethylene
Chloromethylester and trimethylamine or
  pyrridine
Chloroprene
Chlorotriflurorethylene (CTFE)
Cresols
Crotonic acid
Dialkyl fumarates
Dialkyl maleates
Diallyl fumarate
Diallyl phthalate (DAP)
Dibutyl fumarate
Dibutyl maleate
Diethylenetriamine
Di-2-ethylhexyl fumarate
Di-2-ethylhexyl maleate
ε-caprolactam

(continued)

703

APPENDIX B (continued)

Function	Compound
Monomers and Comonomers (continued)	Dihydroxyethyleneurea

Dimethylaminoethyl methacrylate
Dimethyl terephthalate
1,3-dioxolane
2,6-Diphenylphenol
Divinyl benzene
Epichlorohydrin
1,2-epoxy-3-diethylaminopropane
Ethyl acrylate
Ethylene
Ethylene glycol
Ethylene oxide
2-ethylhexyl acrylate
Ethylenimine
Formaldehyde
Formaldehyde and salicyclic acid
Furfuraldehyde
Glycidyl methacrylate
Glycol dimethacrylates
Hexafluoropropylene
Hexamethylenediamine
1-hexene
High molecular weight fatty acid
2-hydroxyethyl acrylate
N-hydroxymethyl acrylamide
N-hydroxymethyl methacrylamide
Isobutylene
Isopropenyl acetate
Itaconic acid
Methacrylate esters
Methacrylic acid
(2-methacryloyloxyethyl)-diethylammonium
  methyl sulfate
Methyl acrylate
Methyl methacrylate
2-methyl-6-phenylphenol
N-(2-formamidoethyl)acrylamide
N-methylol acrylamide
Nonylphenol
1-octene
Octylphenol
1-pentene
Phenol
Phosgene
p-dichlorobenzene

(continued)

APPENDIX B (continued)

Function   Compound

Monomers and Comonomers   Phenolic resins
  (continued)   Polyurethane
                                      Polyvinyl acetate
                                      p-phenylphenol
                                      Propylene
                                      p-tert-butylphenol
                                      Resorcinol
                                      Sodium ethylenesulfonate
                                      Sodium sulfide
                                      Styrene
                                      Styrenesulfonic acid
                                      t-butylphenol
                                      Terephthalic acid
                                      Tetrabromobisphenol-A
                                      Tetrafluoroethylene (TFE)
                                      Tetramethylbisphenol-A
                                      Thiourea
                                      Triallyl cyanurate
                                      Trichloroethylene
                                      Triethylenetetramine
                                      Trioxane
                                      Unsaturated polyesters
                                      Vinyl acetate
                                      Vinyl butyl ether
                                      Vinyl caproate
                                      Vinyl chloride
                                      Vinyl esters
                                      Vinyl fumarate
                                      Vinyl ketone
                                      Vinyl laurate
                                      Vinyl maleate
                                      Vinyl toluene
                                      Vinyl versatate
                                      Vinylene carbonate
                                      Vinylidene bromide
                                      Vinylidene chloride
                                      Vinylidene fluoride
                                      5-vinyl-2-picoline
                                      Xylenols
                                      2,6-xylenol (2,6-dimethylphenol)

(continued)

APPENDIX B (continued)

Function	Compound
Polybasic Acids	Adipic acid
	Azelaic acid
	Camphoric acid
	Chlorendic anhydride
	Citric acid
	Cyclopentadiene
	Diallyl phthalic acids and anhydrides
	Dimerized fatty acid
	Fumaric acid
	Glutaric acid
	Hexahydrophthalic anhydride
	Isophthalic acid
	Isosebacic acid
	Maleic acid
	Maleic anhydride
	α-methyl adipic acid
	Nitrophthalic anhydride
	Phthalic anhydride
	Pimelic acid
	Sebacic acid
	Succinic acid
	Tartaric acid
	Terephthalic acid
	Tetrachlorophthalic anhydride
	Tetrahydrophthalic anhydride
	Trimellitic anhydride
Oils	China wood
	Coconut
	Cottonseed
	Dehydrated castor
	Fish
	Linseed
	Oiticica
	Safflower
	Soya
	Soybean
	Sunflower
	Tung
	Walnut

(continued)

APPENDIX B (continued)

Function	Compound

Monobasic Acids	Benzoic acid
	Fatty acis and fractionated fatty
	acids obtained from oils
	  eleostearic
	  lauric
	  licanic
	  linoleic
	  linoleic (conjugated)
	  linolenic
	  oleic
	  ricinoleic
	  palmitic
	  stearic
	p-tert-butylbenzoic acid
	Synthetic saturated fatty acids
	  2-ethyl hexanoic
	  isodecanoic
	  isononanoic
	  isooctanoic
	  pelargonic
	Tall oil fatty acids

Polyols (Polyhydric
  Alcohols)
	1,4-butylene glycol
	2,3-butylene glycol
	Cyclohexanediol
	Diethylene glycol
	Dipentaerythritol
	Dipropylene glycol
	Ethylene glycol
	Glycerol
	Mannitol
	Neopentylene glycol (2,2-dimethyl-1,3-
	  propanediol)
	Pentaerythritol
	Polyethylene glycol
	Poly(oxypropylene) adducts of glycerol
	Poly(oxypropylene) adducts of
	  1,2,6-hexanetriol
	Poly(oxypropylene) adducts of pentaerythritol
	Poly(oxypropylene) adducts of sorbitol
	Poly(oxypropylene) adducts of trimethylol
	  propane
	Poly(oxypropylene) glycols

(continued)

APPENDIX B (continued)

Function	Compound
Polyols (Polyhydric Alcohols) (continued)	Poly(oxypropylene-b-oxyethylene) adducts of ethylenediamine
	Poly(oxypropylene-b-oxyethylene) adduct of trimethylolpropane
	Poly(oxypropylene-b-oxyethylene) glycols
	Propylene glycol
	Sorbitol
	Triethylene glycol
	Trimethylolethane (2-(hydroxymethyl)-2-methyl-1,3-propanediol)
	Trimethylolpropane (2-ethyl-2-(hydroxymethyl)-1,3-propanediol)
Isocyanates	dianisidine diisocyanate
	2,5-dichlorophenyl isocyanate
	3,4-dichlorophenyl isocyanate
	4,4'-diphenylmethane diisocyanate
	ethyl isocyanate
	hexamethylene diisocyanate
	hydrogenated methylene diphenyl isocyanate
	isophorone diisocyanate
	m-chlorophenyl isocyanate
	m-xylene diisocyanate
	methyl isocyanate
	methylene diphenyl isocyanate (MDI)
	n-butyl isocyanate
	n-propyl isocyanate
	o-chlorophenyl isocyanate
	octadecyl isocyanate
	p-chlorophenyl isocyanate
	phenyl isocyanate
	tolidine diisocyanate
	toluene diisocyanate (TDI)
Polyesters	Polyesters made from:
	adipic acid
	1,3-butylene glycol
	1,4-butylene glycol
	caprolactone
	diethylene glycol
	ethylene glycol
	phthalic glycol
	propylene glycol

(continued)

APPENDIX B (continued)

Function	Compound
Polyethers	Propylene oxide adducts of
	α-methylglucoside
	ethylenediamine
	glycerol
	pentaerythritol
	sorbitol
	sucrose
	trimethylol propane

Appendix C: Companies that Produce Plastics

(continued)

Companies 1–13:

	ADCO Chemical Co., Inc.	Adall Plastics, Inc.	A&E Plastic Pak Co.	Air Products & Chemicals, Inc.	Akzona Inc.	Allied Corp.	The Alpha Corp.	American Cyanamid Co.	American Grilon, Inc.	American Hoechst Corp.	American Petrofina, Inc.	AMETEK Inc.	Apex Chemical Corp.
Styrene-Acrylonitrile													
Polyvinylidene Chloride													
Polyvinyl Chloride				●									
Polyvinyl Alcohol				●									
Polyvinyl Acetate	●			●									
Polyurethane											●		
Polystyrene		●						●	●				
Polypropylene													
Low Density Polyethylene													
Linear Low Density Polyethylene													
High Density Polyethylene						●				●			
Unsaturated Polyester Resins								●	●				
Polybutylene Terephthalate, Polyethylene Terephthalate							●	●		●			
Poly(Ester-Imide) and Poly(Ether-Imide) Resins													
Polycarbonate													
Polybutylene													
Polyamide Resins		●					●			●	●		
Polyacetal													
Phenolic Resins									●		●		●
Fluoropolymers							●						
Epoxy Resins													
Modified Polyphenylene Oxide and Polyphenylene Sulfide													
Amino Resins									●		●		●
Alkyd Resins	●												
Acrylonitrile-Butadiene-Styrene													
Acrylic Resins	●								●				

Companies 14–26:

	Applied Plastics Co., Inc.	Ashland Oil, Inc.	Atlantic Richfield Co.	Auralux Chemical Corp., Inc.	AZS Corp.	Badische Corp.	Baker Int'l. Corp.	Ball Chemical Co.	Barton Chemical Corp.	BASF Wyandotte Corp.	Beatrice Foods Co.	Belding Heminway Co, Inc.	Beals Co., Inc.
Styrene-Acrylonitrile													
Polyvinylidene Chloride													
Polyvinyl Chloride													
Polyvinyl Alcohol													
Polyvinyl Acetate					●								
Polyurethane													
Polystyrene			●								●		
Polypropylene			●										
Low Density Polyethylene			●										
Linear Low Density Polyethylene													
High Density Polyethylene			●										
Unsaturated Polyester Resins		●				●				●		●	
Polybutylene Terephthalate, Polyethylene Terephthalate					●								
Poly(Ester-Imide) and Poly(Ether-Imide) Resins													
Polycarbonate													
Polybutylene													
Polyamide Resins				●	●							●	●
Polyacetal													
Phenolic Resins			●										
Fluoropolymers													
Epoxy Resins													
Modified Polyphenylene Oxide and Polyphenylene Sulfide													
Amino Resins			●										
Alkyd Resins						●			●		●		
Acrylonitrile-Butadiene-Styrene													
Acrylic Resins						●					●		

APPENDIX C (continued)

Columns 1–13:

	Acrylic Resins	Acrylonitrile-Butadiene-Styrene	Alkyd Resins	Amino Resins	Modified Polyphenylene Oxide and Polyphenylene Sulfide	Epoxy Resins	Fluoropolymers	Phenolic Resins	Polyacetal	Polyamide Resins	Polybutylene	Polycarbonate	Poly(Ester-Imide) and Poly(Ether-Imide) Resins
The Bendix Corp.				●				●					
Bennett's			●										
Bisonite Co. Inc.			●										
Borden Inc.	●			●				●		●			
Borg Warner Corp.		●											
Brand-S Corp.						●							
M.A. Bruder & Sons, Inc.			●										
California Resin and Chemical Co., Inc.			●										
Cargill Inc.			●	●									
E. R. Carpenter Co., Inc.													
Celanese Corp.	●		●	●		●			●	●			
Certain-Teed Corp.													
Chagrin Valley Co. Ltd.			●					●					
Chemical Products Corp.	●		●										
Chemplex Co.													
Ciba-Geigy Corp.						●							
Cities Service Co.													
Clark Oil and Refinery Corp.			●					●					
C.N.C. Chemical Corp.			●										
Colloids, Inc.													
Commercial Products Co., Inc.			●										
Conchemo, Inc.													
Conoco, Inc.													
Consolidated Papers, Inc.				●									
Cook Industrial Coatings Co.	●		●	●									
Cook Paint and Varnish Co.	●		●	●									

Columns 14–26:

	Polyethylene Terephthalate, Polybutylene Terephthalate	Unsaturated Polyester Resins	High Density Polyethylene	Linear Low Density Polyethylene	Low Density Polyethylene	Polypropylene	Polystyrene	Polyurethane	Polyvinyl Acetate	Polyvinyl Alcohol	Polyvinyl Chloride	Polyvinylidene Chloride	Styrene-Acrylonitrile
The Bendix Corp.													
Bennett's									●				
Bisonite Co. Inc.													
Borden Inc.									●		●	●	
Borg Warner Corp.													
Brand-S Corp.													
M.A. Bruder & Sons, Inc.													
California Resin and Chemical Co., Inc.													
Cargill Inc.		●											
E. R. Carpenter Co., Inc.								●					
Celanese Corp.	●								●				
Certain-Teed Corp.											●		
Chagrin Valley Co. Ltd.													
Chemical Products Corp.													
Chemplex Co.			●		●								
Ciba-Geigy Corp.													
Cities Service Co.			●		●								
Clark Oil and Refinery Corp.													
C.N.C. Chemical Corp.													
Colloids, Inc.									●				
Commercial Products Co., Inc.													
Conchemo, Inc.										●			
Conoco, Inc.											●		
Consolidated Papers, Inc.													
Cook Industrial Coatings Co.		●											
Cook Paint and Varnish Co.		●						●					

(continued)

APPENDIX C (continued)

Company	Acrylic Resins	Acrylonitrile-Butadiene-Styrene	Alkyd Resins	Amino Resins	Modified Polyphenylene Oxide and Polyphenylene Sulfide	Epoxy Resins	Fluoropolymers	Phenolic Resins	Polyacetal	Polyamide Resins	Polybutylene	Polycarbonate	Poly(Ester-Imide) and Poly(Ether-Imide) Resins
Peter Cooper Corp.													
Cooper Polymers, Inc.										●			
Core-Lube Inc.								●					
CPC Int'l. Inc.				●									
Crosby Chemicals, Inc.										●			
Crown-Metro Inc.	●			●									
Dan River, Inc.				●									
Degen Oil and Chemical Co.		●											
The Derby Co., Inc.	●												
DeSoto, Inc.	●		●	●									
The Dexter Corp.	●		●					●					
Diamond Shamrock										●			
Dock Resins Corp.	●		●	●									
Dow Chemical		●				●							
E.I. du Pont de Nemours & Co., Inc.	●		●				●		●	●			
Dutch Boy, Inc.	●		●										
Eastern Color and Chemical Co.				●									
Eastman Kodak Co.													
El Paso Natural Gas Co.													
Emhart Corp.										●			
Emkay Chemical Co.		●											
Esmark, Inc.													
Ethyl Corp.													
Exxon Corp.													
Fiber Industry, Inc.													
Firestone Tire & Rubber Co.										●			

Company	Polyethylene Terephthalate, Polybutylene Terephthalate	Unsaturated Polyester Resins	High Density Polyethylene	Linear Low Density Polyethylene	Low Density Polyethylene	Polypropylene	Polystyrene	Polyurethane	Polyvinyl Acetate	Polyvinyl Alcohol	Polyvinyl Chloride	Polyvinylidene Chloride	Styrene-Acrylonitrile
Peter Cooper Corp.									●				
Cooper Polymers, Inc.													
Core-Lube Inc.													
CPC Int'l. Inc.													
Crosby Chemicals, Inc.													
Crown-Metro Inc.													
Dan River, Inc.									●				
Degen Oil and Chemical Co.													
The Derby Co., Inc.													
DeSoto, Inc.									●				
The Dexter Corp.													
Diamond Shamrock									●		●		
Dock Resins Corp.													
Dow Chemical			●	●	●	●	●					●	●
E.I. du Pont de Nemours & Co., Inc.	●		●		●				●	●			
Dutch Boy, Inc.									●				
Eastern Color and Chemical Co.													
Eastman Kodak Co.					●	●							
El Paso Natural Gas Co.					●	●							
Emhart Corp.		●											
Emkay Chemical Co.									●				
Esmark, Inc.									●				
Ethyl Corp.											●		
Exxon Corp.						●							
Fiber Industry, Inc.	●												
Firestone Tire & Rubber Co.	●												

(continued)

APPENDIX C (continued)

Company	Acrylic Resins	Acrylonitrile-Butadiene-Styrene	Alkyd Resins	Amino Resins	Modified Polyphenylene Oxide and Polyphenylene Sulfide	Epoxy Resins	Fluoropolymers	Phenolic Resins	Polyacetal	Polyamide Resins	Polybutylene	Polycarbonate	Poly(Ester-imide) and Poly(Ether-imide) Resins	Polyethylene Terephthalate, Polybutylene Terephthalate	Unsaturated Polyester Resins	High Density Polyethylene	Linear Low Density Polyethylene	Low Density Polyethylene	Polypropylene	Polystyrene	Polyurethane	Polyvinyl Acetate	Polyvinyl Alcohol	Polyvinyl Chloride	Polyvinylidene Chloride	Styrene-Acrylonitrile
Formosa Plastics, U.S.A.																								●		
Foy-Johnston, Inc.			●																			●				
Franklin Chemical Co.																						●				
H. B. Fuller Co.	●																					●				
GAF Corp.														●												
General Electric Co.			●	●	●			●				●	●	●												
General Latex and Chemical Corp.	●																				●	●				
General Motors Corp.																					●					
General Tire and Rubber Co.																								●		
The P.D. George Co.			●					●					●		●											
Georgia Pacific Corp.				●				●		●														●		
Getty Oil Co.				●				●																		
The B. F. Goodrich Co.	●																				●			●		
The Goodyear Tire and Rubber Co.														●							●			●		
Carl Gordon Industries Inc.																			●							
W. R. Grace and Co.	●																					●			●	
Great Western Carpet Cushion Co., Inc.																					●					
Guardsman Chemicals, Inc.	●		●	●																						
Gulf Oil Corp.	●		●					●								●			●	●	●	●				
Handschy Chemical Co.			●																							
Hanna Chemical Coatings Corp.	●		●	●																						
Hart Products Corp.	●																					●				
Henkel of America, Inc.	●									●																
Hercules Inc.	●		●							●						●		●								
Heresite-Saekaphen Inc.								●																		
H&N Chemical Co.				●																						

(continued)

APPENDIX C (continued)

	Acrylic Resins	Acrylonitrile-Butadiene-Styrene	Alkyd Resins	Amino Resins	Modified Polyphenylene Oxide and Polyphenylene Sulfide	Epoxy Resins	Fluoropolymers	Phenolic Resins	Polyacetal	Polyamide Resins	Polybutylene	Polycarbonate	Poly(Ester-Imide) and Poly(Ether-Imide) Resins	Polyethylene Terephthalate, Polybutylene Terephthalate	Unsaturated Polyester Resins	High Density Polyethylene	Linear Low Density Polyethylene	Low Density Polyethylene	Polypropylene	Polystyrene	Polyurethane	Polyvinyl Acetate	Polyvinyl Alcohol	Polyvinyl Chloride	Polyvinylidene Chloride	Styrene-Acrylonitrile
Hugh J.-Resins Co.	•		•					•							•											
Huntsmann Goodson Chemical Corp.																				•						
ICI Americas Inc.							•							•	•											
Inland Steel Co.								•																		
Insilco Corp.			•																			•				
Int'l. Minerals and Chemical Corp.																						•				
InterNorth Inc.																			•							
Iovite Chemicals Inc.			•																							
The Ironsides Co.								•																		
Jones-Blair Co.			•																			•				
S.C. Johnson & Son, Inc.	•																									
Kama Corp.																				•						
Kelly-Moore Paint Co.			•																			•				
Keysor Corp.																								•		
Kohler-McLister Paint Co.																						•				
Komac Paint, Inc.	•		•																							
Koppers Co., Inc.								•							•											
Lawter Int'l. Inc.			•					•																		
Libby-Owens-Ford Co.		•						•	•																	
Lilly Industrial Coatings, Inc.			•																							
LNP Corp.												•														
Masonite Corp.								•																		
McCloskey Varnish Co.			•																			•				
Minnesota Mining and Manufacturing Co.	•						•	•							•											
Mobay Chemical Corp.	•	•									•												•			
Mobil Corp.	•		•	•				•		•					•					•						

(continued)

APPENDIX C (continued)

	Acrylic Resins	Acrylonitrile-Butadiene-Styrene	Alkyd Resins	Amino Resins	Modified Polyphenylene Oxide and Polyphenylene Sulfide	Epoxy Resins	Fluoropolymers	Phenolic Resins	Polyacetal	Polyamide Resins	Polybutylene	Polycarbonate	Poly(Ester-imide) and Poly(Ether-imide) Resins	Polyethylene Terephthalate, Polybutylene Terephthalate	Unsaturated Polyester Resins	High Density Polyethylene	Linear Low Density Polyethylene	Low Density Polyethylene	Polypropylene	Polystyrene	Polyurethane	Polyvinyl Acetate	Polyvinyl Alcohol	Polyvinyl Chloride	Polyvinylidene Chloride	Styrene-Acrylonitrile
Monogram Industries Inc.								●																		
Monsanto Co.		●		●				●		●				●						●		●	●			●
Benjamin Moore & Co.			●																			●				
Morton-Norwich Products, Inc.	●																					●				
Napko Corp.																						●				
National Casein Co.				●																		●				
National Distillors & Chem. Corp.										●																
National Petrochemicals Corp.																●										
National Starch and Chemical Corp.	●			●																		●				
Niles Chemical Paint Co.			●					●																		
Norris Paints & Varnish Co.	●																					●				
The O'Brien Corp.	●		●					●		●					●							●				
Occidental Petroleum Corp.							●	●																●		
Olin Corp.																					●					
Onyx Oils & Resins Inc.																						●				
C. J. Osborn Chemicals Inc.			●																							
Owens-Corning Fiberglas Corp.										●					●											
Pacific Anchor Chemical Corp.										●																
Pantasote Inc.																								●		
Pennwalt Corp.							●																			
Perry & Derrick Co.			●																							
Perstorp Inc.				●																						
Pervo Paint Co.			●																							
Phillip Morris Inc.	●																					●				
Phillips Petroleum Co.		●								●						●	●		●							
Plaskon Products Inc.				●																						

(continued)

APPENDIX C (continued)

Company	Acrylic Resins	Acrylonitrile-Butadiene-Styrene	Alkyd Resins	Amino Resins	Modified Polyphenylene Oxide and Polyphenylene Sulfide	Epoxy Resins	Fluoropolymers	Phenolic Resins	Polyacetal	Polyamide Resins	Polybutylene	Polycarbonate	Poly(Ester-Imide) and Poly(Ether-Imide) Resins
Plastics Engineering Co.								●					
Plastic Management Corp.				●									
Polychrome Corp.			●										
Polymer Applications Inc.								●					
Polyrez Co., Inc.								●					
Polysar Group													
PPG Industries Inc.	●		●	●									
Pratt & Lambert Inc.			●										
Purex Corp.	●												
K. J. Quinn & Co., Inc.	●												
Raffi and Swanson Inc.	●												
Raybestos-Manhattan, Inc.								●					
Reeves Bros. Inc.													
Reichhold Chemicals, Inc.	●		●	●		●		●		●			
Rico Chemical Corp.													
Rilsan Corp.										●			
H. H. Robertson Co.	●		●										
Rogers Corp.								●					
Rohm & Haas Co.	●											●	
Schenectady Chemicals Inc.			●					●					●
Scholler Bros. Inc.													
SCM Corp.	●							●					
Scott Paper Co.				●									
Shakespeare Co.										●			
Shell Chemical Co.					●						●		
Sheller-Globe Corp.													

Company	Polyethylene Terephthalate, Polybutylene Terephthalate	Unsaturated Polyester Resins	High Density Polyethylene	Linear Low Density Polyethylene	Low Density Polyethylene	Polypropylene	Polystyrene	Polyurethane	Polyvinyl Acetate	Polyvinyl Alcohol	Polyvinyl Chloride	Polyvinylidene Chloride	Styrene-Acrylonitrile
Plastics Engineering Co.													
Plastic Management Corp.								●					
Polychrome Corp.													
Polymer Applications Inc.													
Polyrez Co., Inc.													
Polysar Group							●						
PPG Industries Inc.		●											
Pratt & Lambert Inc.													
Purex Corp.													
K. J. Quinn & Co., Inc.													
Raffi and Swanson Inc.								●					
Raybestos-Manhattan, Inc.													
Reeves Bros. Inc.								●					
Reichhold Chemicals, Inc.		●						●					
Rico Chemical Corp.													●
Rilsan Corp.													
H. H. Robertson Co.		●						●					
Rogers Corp.													
Rohm & Haas Co.													
Schenectady Chemicals Inc.		●											
Scholler Bros. Inc.											●		
SCM Corp.		●									●		
Scott Paper Co.								●					
Shakespeare Co.													
Shell Chemical Co.									●	●			
Sheller-Globe Corp.								●					

(continued)

APPENDIX C (continued)

Company	Acrylic Resins	Acrylonitrile-Butadiene-Styrene	Alkyd Resins	Amino Resins	Modified Polyphenylene Oxide and Polyphenylene Sulfide	Epoxy Resins	Fluoropolymers	Phenolic Resins	Polyacetal	Polyamide Resins	Polybutylene	Polycarbonate	Poly(Ester-Imide) and Poly(Ether-Imide) Resins
Sherwin-Williams Co.	•		•					•					
Shintech Inc.													
Simpson Timber Co.								•					
D. Sobek Co.													
Soltek Polymer Corp.													
Southeastern Adhesives Co.				•									
Squibb Corp.													
Stanchem Inc.													
Standard Oil Co. of IN													
The Standard Oil Co. (OH)			•					•					
Stauffer Chemical Co.													
Sullivan Chemical Coatings			•										
Sun Chemical Corp.	•									•			
Sybron Corp.	•		•	•									
Syncon Resins Inc.													
Synray Corp.			•										
Synthron, Inc.				•									
Tallyrand Chemicals, Inc.													
Tenneco, Inc.													
Texapol Corp.										•			
Textron Inc.	•	•											
Texstyrene Plastics, Inc.													
Tyler Corp.			•	•									
Union Camp Corp.										•			
Union Carbide Corp.						•		•					
Union Oil Co. of CA	•												

Company	Polyethylene Terephthalate, Polybutylene Terephthalate	Unsaturated Polyester Resins	High Density Polyethylene	Linear Low Density Polyethylene	Low Density Polyethylene	Polypropylene	Polystyrene	Polyurethane	Polyvinyl Acetate	Polyvinyl Alcohol	Polyvinyl Chloride	Polyvinylidene Chloride	Styrene-Acrylonitrile
Sherwin-Williams Co.		•							•				
Shintech Inc.											•		
Simpson Timber Co.													
D. Sobek Co.								•					
Soltek Polymer Corp.			•		•								
Southeastern Adhesives Co.													
Squibb Corp.									•				
Stanchem Inc.									•				
Standard Oil Co. of IN			•			•	•						
The Standard Oil Co. (OH)		•											
Stauffer Chemical Co.											•		
Sullivan Chemical Coatings													
Sun Chemical Corp.													
Sybron Corp.									•				
Syncon Resins Inc.									•				
Synray Corp.													
Synthron, Inc.													
Tallyrand Chemicals, Inc.											•		
Tenneco, Inc.											•		
Texapol Corp.													
Textron Inc.							•						
Texstyrene Plastics, Inc.								•					
Tyler Corp.													
Union Camp Corp.													
Union Carbide Corp.			•	•	•				•		•		
Union Oil Co. of CA									•				

(continued)

APPENDIX C (continued)

Company	Acrylic Resins	Acrylonitrile-Butadiene-Styrene	Alkyd Resins	Amino Resins	Modified Polyphenylene Oxide and Polyphenylene Sulfide	Epoxy Resins	Fluoropolymers	Phenolic Resins	Polyacetal	Polyamide Resins	Polybutylene	Polycarbonate	Poly(Ester-Imide) and Poly(Ether-Imide) Resins
United Foam Corp.													
United Merchants & Manufacturers Inc.	•			•									
United Technologies Corp.	•		•					•					
The Upjohn Co.													
U.S. Oil Co.				•									
U.S. Polymers Inc.		•											
U.S. Steel Corp.		•	•										
Valentine Sugars, Inc.								•					
Valspar Corp.	•		•	•									
Vititek, Inc.													
Wellman, Inc.										•			
West Coast Adhesives Co.								•					
West Point-Pepperell, Inc.				•									
Westinghouse Electric Corp.			•	•				•					
Weyerhauser Co.				•				•					
Whittaker Corp.													
Witco Chemical Corp.													
Yenkin-Majestic Paint Corp.	•		•										
TOTAL COMPANIES	15	5	63	53	2	6	5	50	2	32	1	2	5

Company	Polyethylene Terephthalate, Polybutylene Terephthalate	Unsaturated Polyester Resins	High Density Polyethylene	Linear Low Density Polyethylene	Low Density Polyethylene	Polypropylene	Polystyrene	Polyurethane	Polyvinyl Acetate	Polyvinyl Alcohol	Polyvinyl Chloride	Polyvinylidene Chloride	Styrene-Acrylonitrile
United Foam Corp.								•					
United Merchants & Manufacturers Inc.								•	•				
United Technologies Corp.													
The Upjohn Co.								•					
U.S. Oil Co.													
U.S. Polymers Inc.													
U.S. Steel Corp.						•	•						
Valentine Sugars, Inc.													
Valspar Corp.							•						
Vititek, Inc.						•							
Wellman, Inc.													
West Coast Adhesives Co.													
West Point-Pepperell, Inc.													
Westinghouse Electric Corp.													
Weyerhauser Co.													
Whittaker Corp.											•		
Witco Chemical Corp.								•					
Yenkin-Majestic Paint Corp.								•					
TOTAL COMPANIES	15	26	14	2	9	12	18	23	54	3	21	3	2

Polyamide Resins: 15 Nylon, 17 Other

Polyethylene Terephthalate / Polybutylene Terephthalate: 12 PET, 3 PBT